T0370443

Recent Advancements in Graph Theory

Mathematical Engineering, Manufacturing, and Management Sciences

Series Editor: Mangey Ram, Professor, Assistant Dean (International Affairs), Department of Mathematics, Graphic Era University, Dehradun, India

The aim of this new book series is to publish the research studies and articles that bring up the latest development and research applied to mathematics and its applications in the manufacturing and management sciences areas. Mathematical tool and techniques are the strength of engineering sciences. They form the common foundation of all novel disciplines as engineering evolves and develops. The series will include a comprehensive range of applied mathematics and its application in engineering areas such as optimization techniques, mathematical modelling and simulation, stochastic processes and systems engineering, safety-critical system performance, system safety, system security, high assurance software architecture and design, mathematical modelling in environmental safety sciences, finite element methods, differential equations, reliability engineering, etc.

Sustainable Procurement in Supply Chain Operations
Edited by Sachin Mangla, Sunil Luthra, Suresh Jakar, Anil Kumar, and Nirpendra Rana

Mathematics Applied to Engineering and Management
Edited by Mangey Ram and S.B. Singh

Mathematics in Engineering Sciences
Novel Theories, Technologies, and Applications
Edited by Mangey Ram

Advances in Management Research
Innovation and Technology
Edited by Avinash K. Shrivastava, Sudhir Rana, Amiya Kumar Mohapatra, Mangey Ram

Market Assessment with OR Applications
Adarsh Anand, Deepti Aggrawal, Mohini Agarwal

Recent Advances in Mathematics for Engineering
Edited by Mangey Ram

Probability, Statistics, and Stochastic Processes for Engineers and Scientists
Aliakbar Montazer Haghighi and Indika Rathnathungalage Wickramasinghe

Total Quality Management (TQM)
Principles, Methods, and Applications
Sunil Luthra, Dixit Garg, Ashish Aggarwal, and Sachin K. Mangla

Recent Advancements in Graph Theory
Edited by N. P. Shrimali and Nita H. Shah

Mathematical Modeling and Computation of Real-Time Problems: An Interdisciplinary Approach
Edited by Rakhee Kulshrestha, Chandra Shekhar, Madhu Jain, & Srinivas R. Chakravarthy

For more information about this series, please visit: https://www.crcpress.com/Mathematical-Engineering-Manufacturing-and-Management-Sciences/book-series/CRCMEMMS

Recent Advancements in Graph Theory

Edited by

N. P. Shrimali

Nita H. Shah

CRC Press is an imprint of the
Taylor & Francis Group, an **informa** business

First edition published 2020
by CRC Press
6000 Broken Sound Parkway NW, Suite 300, Boca Raton, FL 33487-2742

and by CRC Press
2 Park Square, Milton Park, Abingdon, Oxon, OX14 4RN

Library of Congress Cataloging-in-Publication Data
Names: Shrimali, N. P., editor. | Shah, Nita H., editor.
Title: Recent advancements in graph theory / [edited by] N. P. Shrimali, Nita H. Shah.
Description: Boca Raton : CRC Press, 2020. | Series: Mathematical engineering, manufacturing, and management sciences | Includes bibliographical references and index.
Identifiers: LCCN 2020023816 (print) | LCCN 2020023817 (ebook) | ISBN 9780367458867 (hardback) | ISBN 9781003038436 (ebook)
Subjects: LCSH: Graph theory. | Discrete mathematics.
Classification: LCC QA166 .R42 2020 (print) | LCC QA166 (ebook) | DDC 511/.5--dc23
LC record available at https://lccn.loc.gov/2020023816
LC ebook record available at https://lccn.loc.gov/2020023817

ISBN: 978-0-367-45886-7 (hbk)
ISBN: 978-1-003-03843-6 (ebk)

Typeset in LMRoman
by Nova Techset Private Limited, Bengaluru & Chennai, India

Visit the Taylor & Francis Web site at
http://www.taylorandfrancis.com

and the CRC Press Web site at
http://www.crcpress.com

Contents

Preface

There are numerous reasons for the spurt of interest in graph theory. It has many applications to the areas of physics, chemistry, communication science, computer technology, electrical and civil engineering, architecture, operation research, genetics, psychology, sociology, economics, anthropology and linguistics. It is also closely connected to many branches of mathematics namely matrix theory, group theory, topology, numerical analysis, probability, and combinatorics.

The theory has many branches such as decomposition of graphs, domination in graphs, chromatic graph theory, theory of hypergraphs, graph labeling, algebraic graph theory, graph coloring, graph energy, and many more. The paper written by Leonhard Euler on the Seven Bridges of Königsberg and published in 1736 is regarded as the first paper in the history of graph theory. Since then plenty of research papers on graph theory have been published by many researchers.

The editors thank DST-FIST file # MSI-097 for technical support to the department of Mathematics, Gujarat University, Ahmedabad.

A national conference on “Recent Advancements in Graph Theory (RAGT-2019)” was held on November 9-10, 2019 at the Department of Mathematics, Gujarat University, Ahmedabad. More than 80 participants attended the conference. 48 research papers were presented by research scholars and 12 invited speakers delivered their talks on various topics of graph theory. This book is an outcome of some papers of participants presented in RAGT-2019. The book comprises four main topics: graph labeling, operations on graphs, graph coloring and energy of graphs. Chapters 1 to 16 address some results on graph labeling techniques such as graceful labeling, $L(2,1)$-labeling, V_4-cordial labeling, SD-prime cordial labeling, total edge product cordial labeling, product cordial labeling, sum divisor cordial labeling, Fibonacci cordial labeling, parity combination cordial labeling, total neighborhood prime labeling, Gaussian vertex prime labeling, vertex magic total labeling, antimagic labeling, distance magic and distance antimagic labeling. Chapter 17 deals with graphs from subgraphs. Chapter 18 discusses unit graphs having their domination number half their order. Chapters 19 to 21 present some results on pendant number and Wiener index. Chapter 22 studies algebraic signed graphs. Chapter 23 presents the nullity and energy of complete tripartite graphs. Chapter 24 examines some new results on restrained edge domination number of graphs. Chapter 25 discusses some new graph coloring problems. Chapter 26 deals with total global dominator coloring of graphs. Chapters 27 and 28 deal with the rainbow vertex connection number of a class of triangular snake graph and Hamiltonian chromatic number of trees. Chapters 29 and 30 discuss graph energy. Chapter 31 presents L-spectra of graphs obtained by duplicating graphs elements.

It is hoped that this book will provide a useful resource to the younger generation of researchers in graph theory. The book comprises results derived by the contributors. Editors are not responsible for output.

We are thankful to all contributors for their chapters in this book. We are thankful to the Department of Mathematics, Gujarat University, for providing generous support towards the successful conduct of the event and to all the participants and invited speakers for their active involvement and efforts; we are also grateful to reviewers of the chapters for reviewing the process in the given time. The editors thank research scholar Siddhant Trivedi for compiling this book.

N. P. Shrimali

Nita H. Shah

Biography of Editors

Dr. N. P. Shrimali joined the Department of Mathematics, Gujarat University as a lecturer on 16th January, 1998. Presently, he is working as an associate professor in the same department. He did his Ph.D. in point-set topology in the year 2013 from Saurashtra University, Rajkot. Four research papers were published during his Ph.D. Presently, he is working in the area of graph theory especially in graph labeling. Three students completed their M.Phil. under his supervision. Five Ph.D. students are registered under his supervision. Four students are doing research in graph labeling and one student is researching point-set topology. Dr. Shrimali has published 12 research papers in graph labeling in national as well as international joumals. He has presented 7 research papers in national as well as international conferences. He is involved in preparing various syllabi in his department. He is a life member of Gujarat Ganit Mandal, Indian Mathematical Society, Academy of Discrete Mathematics & its Applications (ADMA) and Gwalior Academy of Mathematical Sciences (GAMS). He also serves as the convener of the P C Vaidya Sanman Nidhi Trust. This trust publishes a journal titled *Mathematics Today.*

Prof. Nita H. Shah is head of the Department of Mathematics in Gujarat University, India. She received her Ph.D. in Statistics from Gujarat University in 1994. She is a post-doctoral visiting research fellow of the University of New Brunswick, Canada. Her research interests include inventory modeling in supply chain, robotic modeling, mathematical modeling of infectious diseases, image processing, dynamical systems and their applications etc. She has completed 3 - UGC sponsored projects. She has published 13 monographs, 5 textbooks, and 475+ peer-reviewed research papers. Five edited books are prepared for IGI-Global and Springer with co-editor Dr. Mandeep Mittal. Her papers are published in high impact Elsevier, Inderscience and Taylor and Francis journals. According to Google scholar, her total number of citations is over 2919 and the maximum number of citations for a single paper is over 174. The H-index is 24 as of March 2020 and i-10 index is 73. She has guided 28 Ph.D. students and 15 M.Phil students thus far. Eight students are pursuing research for their Ph.D. degree along with one post-doctoral fellow of D. S. Kothari-UGC. She has travelled in the USA, Singapore, Canada, South Africa, Malaysia, and Indonesia for speaking engagements. She is Vice-President of the Operational Research Society of India. She is a council member of the Indian Mathematical Society.

Contributors

V. J. Kaneria
Saurashtra University
Gujarat, India

J. C. Kanani
L. E. College (Diploma)
Gujarat, India

Hiren P. Chudasama
Government Polytechnic
Gujarat, India

Divya K. Jadeja
V. V. P. Engineering College
Gujarat, India

U. M. Prajapati
St. Xavier's College
Gujarat, India

N. B. Patel
Gujarat University
Gujarat, India

Neha B. Rathod
Government Engineering College
Gujarat, India

Kailas K. Kanani
Government Engineering College
Gujarat, India

A. V. Vantiya
Gujarat University
Gujarat, India

Chirag M. Barasara
Hemchandracharya North Gujarat University
Gujarat, India

M. I. Bosmia
Government Engineering College
Gujarat, India

D. G. Adalja
Marwadi Education Foundation
Gujarat, India

G. V. Ghodasara
H. & H. B. Kotak Institute of Science
Gujarat, India

K. K. Raval
Gujarat University
Gujarat, India

K. P. Shah
Gujarat University
Gujarat, India

N. P. Shrimali
Gujarat University
Gujarat, India

A. K. Rathod
Gujarat University
Gujarat, India

S. K. Singh
Gujarat University
Gujarat, India

S. T. Trivedi
Gujarat University
Gujarat, India

Tarunkumar Chhaya
L. E. Polytechnic
Gujarat, India

Y. M. Parmar
Gujarat University
Gujarat, India

Joseph Varghese Kureethara
CHRIST (Deemed to be University)
Bengaluru, India

Johan Kok
City of Tshwane
South Africa

Pranjali
University of Rajasthan
Rajasthan, India

Amit Kumar
Banasthali Vidyapith
Rajasthan, India

Mukti Acharya
Christ University
Bengaluru, India

Pooja Sharma
Banasthali Vidyapith
Rajasthan, India

Jomon K Sebastian
Savio HSS
Kerala, India

Sudev Naduvath
CHRIST (Deemed to be University)
Bengaluru, India

H. S. Mehta
Sardar Patel University
Gujarat, India

J. George
Shri Alpesh N. Patel P. G. Institute of Science & Research
Gujarat, India

S. K. Vaidya
Saurashtra University
Gujarat, India

M. R. Jadeja
Shree Manibhai Virani and Smt. Navalben Virani Science College(Autonomous)
Gujarat, India

Renu Naresh
Banasthali Vidyapith
Rajasthan, India

P. D. Ajani
Atmiya University
Gujarat, India

K.P. Chithra
CHRIST (Deemed to be University)
Bengaluru, India

Mayamma Joseph
CHRIST (Deemed to be University)
Bengaluru, India

Dharamvirsinh Parmar
C. U. Shah University
Gujarat, India

Bharat Suthar
C. U. Shah University
Gujarat, India

Devsi Bantva
Lukhdhirji Engineering College
Gujarat, India

Mitesh J. Patel
Tolani College of Arts and Science
Gujarat, India

G. K. Rathod
V.V.P. Engineering College
Gujarat, India

K. M. Popat
Atmiya University
Gujarat, India

1 Graceful Labeling for Eight Sprocket Graph

V. J. Kaneria
Department of Mathematics,
Saurashtra University,
Rajkot, Gujarat (INDIA)
E-mail: kaneriavinodraya@gmail.com

J. C. Kanani
General Department (Mathematics),
L. E. College (Diploma),
Morbi, Gujarat (INDIA)
E-mail: kananijagrutic@gmail.com

We have obtained a new graph which is called an eight sprocket graph. We prove that the eight sprocket graph is graceful. We investigate some eight sprocket graph related families of connected graceful graphs. We prove that the path union of the eight sprocket graph, the cycle of the eight sprocket graph and the star of the eight sprocket graph are graceful.

1.1 INTRODUCTION

The concept of graceful labeling was introduced by Rosa [6] in 1967 and numbering in a graph was defined by S. W. Golomb [3]. Many researchers have studied gracefulness of graphs; refer to J. A. Gallian survey (2018) [2]. A good number of papers are found with a variety of applications in coding theory, radar communication, cryptography etc. In-depth details about applications of graph labeling is found in Bloom and Golomb [1]. We accept all notations and terminology from Harary [4]. We recall some definitions from [5], which are used in this chapter.

A function f is called graceful labeling of a graph $G = (V, E)$ if $f : V \to \{0, 1, \ldots, q\}$ is injective and the induce function $f^* : E \to \{1, \ldots, q\}$ defined as $f^*(e) = |f(u) - f(v)|$ is bijective for every edge $e = (u, v) \in E(G)$. A graph G is called a graceful graph if it admits a graceful labeling.

Let G be a graph and G_1, G_2, $\ldots$, G_n, $n \geq 2$ be n copies of graph G. Then the graph obtained by adding an edge from G_i to G_{i+1} $(1 \leq i \leq n-1)$ is called the path union of G. For a cycle C_n, each vertex of C_n is replaced by connected graphs G_1, G_2, $\ldots$, G_n and is known as a cycle of graphs. We shall denote it by C (G_1, G_2, ..., G_n). If we replace each vertex by a graph G, *i.e.* $G_1 = G, G_2 = G, \ldots, G_n = G$. Such a cycle of a graph G is denoted by C (n . G).

Let G be a graph on n vertices. The graph obtained by replacing each vertex of the star $K_{1,n}$ by a copy of G is called a star of G and is denoted by G^*.

An eight sprocket graph is a union of eight copies of C_{4n}. If $V_{i,j}$($\forall$ $i = 1,2,\ldots,8, \forall\ j = 1,2,\ldots,4n$) are vertices of ith copy of C_{4n} then we shall combine $V_{1,4n}$ and $V_{2,1}$, $V_{2,4n}$ and $V_{3,1}$, $V_{3,4n}$ and $V_{4,1}$, $V_{4,4n}$ and $V_{5,1}$, $V_{5,4n}$ and $V_{6,1}$, $V_{6,4n}$ and $V_{7,1}$, $V_{7,4n}$ and $V_{8,1}$ and $V_{8,4n}$ and $V_{1,1}$, by a single vertex, where $n \in N-1$. So the graph seems like a sprocket shape, and here the number of sprockets is eight, therefore we gave it the name eight sprocket. It is denoted by SC_n of $16n$ size, where $n \in N-1$, $\mid V(SC_n) \mid = 16n-8$, $\mid E(SC_n) \mid = 16n$. In this chapter we introduced the gracefulness of the eight sprocket graph, path union of the eight sprocket graph, cycle of the eight sprocket graph and star of the eight sprocket graph. For a detailed survey of graph labeling we refer to Gallian [2].

1.2 MAIN RESULTS

Theorem 1.1

An eight sprocket graph SC_n is a graceful graph, where $n \in N-\{1\}$. ■

Proof. Let G $=$ SC_n be any eight sprocket graph of size $16n$, where $n \in N-\{1\}$. We mention all vertices of SC_n like $V_{i,j}$ ($\forall\ i = 1,2,\ldots,8, \forall\ j = 1,2,\ldots,4n$). We see the numbers of vertices in G are $\mid V(SC_n) \mid = p = 16n-8$ and $\mid E(SC_n) \mid = q = 16n$.
We define labeling function $f : V(G) \to \{0,1,\ldots,q\}$ as follows

$$f(v_{1,j}) = \begin{cases} q-(\frac{j-1}{2}), & \text{if } j = 1,\ 3,\ \ldots,\ 2n-1 \\ \dfrac{j-2}{2}, & \text{if } j = 2,\ 4,\ 6,\ 8 \\ \dfrac{j}{2}, & \text{if } j = n+2,\ n+4,\ \ldots,\ 2n, \end{cases}$$

$$f(v_{2,j}) = \begin{cases} n+(\frac{j-1}{2}), & \text{if } j = 1,\ 3,\ \ldots,\ 2n-1 \\ q-n-(\frac{j-2}{2}), & \text{if } j = 2,\ 4,\ 6,\ 8 \\ q-n-(\frac{j}{2}), & \text{if } j = n+2,\ n+4,\ \ldots.,\ 2n, \end{cases}$$

$$f(v_{3,j}) = \begin{cases} q-2n-(\frac{j-1}{2}), & \text{if } j=1,\ 3,\ \ldots,\ 2n-1 \\ 2n+(\frac{j-2}{2}), & \text{if } j=2,\ 4,\ 6,\ 8 \\ 2n+(\frac{j}{2}), & \text{if } j=n+2,\ n+4,\ \ldots.,\ 2n, \end{cases}$$

$$f(v_{4,j}) = \begin{cases} 3n+(\frac{j-1}{2}), & \text{if } j=1,\ 3,\ \ldots,\ n+1 \\ 3n+(\frac{j+1}{2}), & \text{if } j=n+3,\ n+5,\ \ldots,\ 2n-1 \\ q-3n-(\frac{j}{2}), & \text{if } j=n+2,\ n+4,\ \ldots,\ 2n \\ q-3n-(\frac{j-2}{2}), & \text{if } j=2,\ 4,\ 6,\ 8, \end{cases}$$

$$f(v_{5,j}) = \begin{cases} q-4n-(\frac{j-1}{2}), & \text{if } j=1,\ 3,\ \ldots,\ 2n-1 \\ 4n+2+(\frac{j-2}{2}), & \text{if } j=2,\ 4,\ 6,\ 8 \\ 4n+2+(\frac{j}{2}), & \text{if } j=n+2,\ n+4,\ \ldots.,\ 2n, \end{cases}$$

$$f(v_{6,j}) = \begin{cases} 5n+2+(\frac{j-1}{2}), & \text{if } j=1,\ 3,\ \ldots,\ 2n-1 \\ q-5n-(\frac{j-2}{2}), & \text{if } j=2,\ 4,\ 6,\ 8 \\ q-5n-(\frac{j}{2}), & \text{if } j=n+2,\ n+4,\ \ldots.,\ 2n, \end{cases}$$

$$f(v_{7,j}) = \begin{cases} q-6n-(\frac{j-1}{2}), & \text{if } j=1,\ 3,\ \ldots,\ 2n-1 \\ 6n+2+(\frac{j-2}{2}), & \text{if } j=2,\ 4,\ 6,\ 8 \\ 6n+2+(\frac{j}{2}), & \text{if } j=n+2,\ n+4,\ \ldots.,\ 2n, \end{cases}$$

$$f(v_{8,j}) = \begin{cases} 7n+2+(\frac{j-1}{2}), & \text{if } j=1,\ 3,\ \ldots,\ 2n-1 \\ q-7n-(\frac{j-2}{2}). & \text{if } j=2,\ 4,\ 6,\ \ldots,\ (2n-2). \end{cases}$$

The above labeling pattern gives rise to a graceful labeling to the graph G. So G is a graceful graph. □

Illustration 1.2.1. The graceful labeling of eight sprocket graph SC_8 is shown in Figure 1.1.

Theorem 1.2

The path union of finite copies of the eight sprocket graph SC_n is a graceful graph, where, $n \in N-\{1\}$. ■

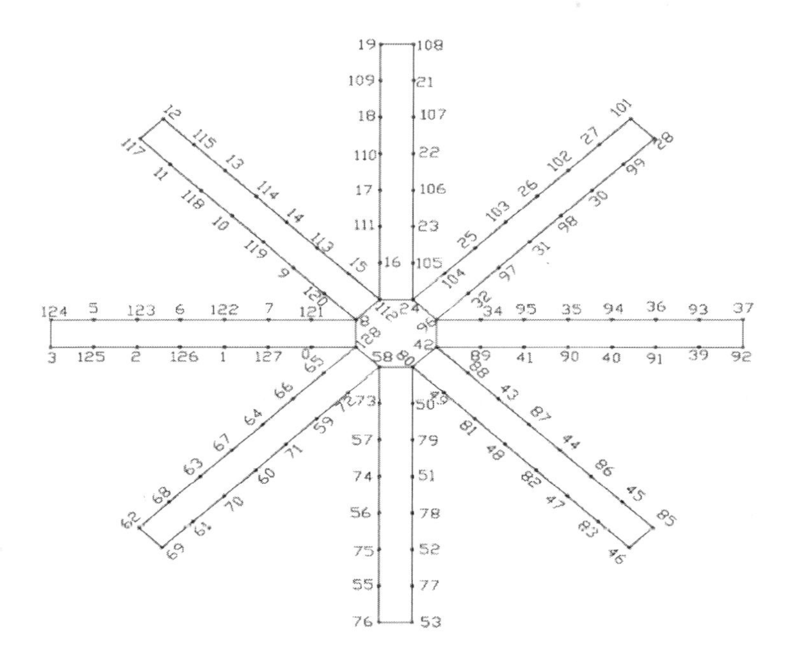

Figure 1.1: Graceful labeling of eight sprocket graph with $p = 120$ and $q = 128$

Proof. Let G = P (r . SC_n) be a path union of r copies for the eight sprocket graph SC_n, where $n \in N - \{1\}$. Let f be the graceful labeling of SC_n as we mentioned in Theorem 1.1 In graph G, we see that the vertices $\mid V(G) \mid = P = r(16(n) - 8)$ and the edges $\mid E(G) \mid = Q = (r-1) + r16(n)$. Let $u_{k,i,j}$ ($\forall$ $i = 1,2,\ldots,8, \forall$ $j = 1,2,\ldots,4n$) be the vertices of kth copy of SC_n, ($\forall$ $k = 1,2,\ldots,r$) where the vertices of kth copy of SC_n are $p = 16(n) - 8$ and edges of kth copy of SC_n are $q = 16n$. Join vertices $u_{k,1,2n+1}$ with $u_{k+1,1,2n+1}$ for $k = 1,2,\ldots,r-1$ by an edge to form the path union of r copies of the eight sprocket graph. Define labeling function $g : V(G) \rightarrow \{0,1,\ldots,Q\}$ as follows

$$g(u_{1,i,j}) = \begin{cases} f(u_{i,j}), & if \quad f(u_{i,j}) \leq \frac{q}{2}+1, \\ f(u_{i,j}) + (Q-q), & if \quad f(u_{i,j}) > \frac{q}{2}+1, \\ (\forall\ i = 1,2\ldots,8, \forall\ j = 1,2,\ldots,4n;) & \end{cases}$$

$$g(u_{2,i,j}) = \begin{cases} g(u_{1,i,j}) + (Q-q), & if \quad g(u_{1,i,j}) \leq \frac{q}{2}+1, \\ g(u_{1,i,j}) - (Q-q), & if \quad g(u_{1,i,j}) > \frac{q}{2}+1, \\ (\forall\ i = 1,2\ldots,8, \forall\ j = 1,2,\ldots,4n;) & \end{cases}$$

$$g(u_{k,i,j}) = \begin{cases} g(u_{k-2,i,j}) + (q+1), & if \quad g(u_{k-2,i,j}) \leq \frac{q}{2}+1, \\ g(u_{k-2,i,j}) - (q+1), & if \quad g(u_{k-2,i,j}) > \frac{q}{2}+1, \\ (\forall\ i = 1,2\ldots,8, \forall\ j = 1,2,\ldots,4n, \forall\ k = 3,4,\ldots,r.) & \end{cases}$$

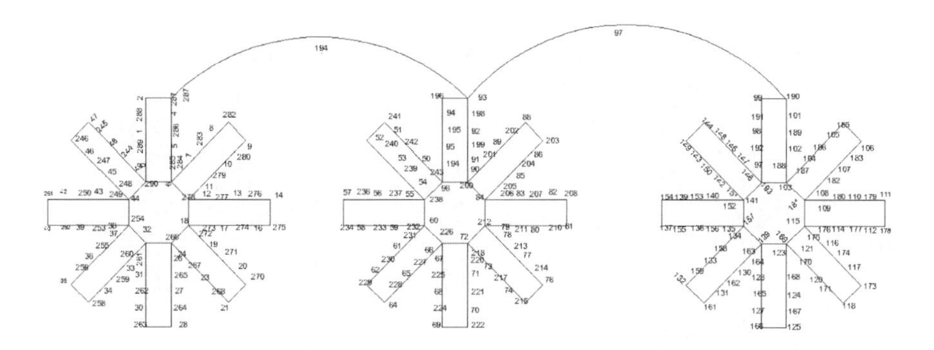

Figure 1.2: A path union of 3 copies of SC_6 and its graceful labeling with $p = 264$ and $q = 290$

The above labeling pattern gives rise to a graceful labeling of the given graph G. So the path union of finite copies of the eight sprocket graph is a graceful graph. □

Illustration 1.2.2. The path union of 3 copies of SC_6 and its graceful labeling are shown in Figure 1.2.

Theorem 1.3

The cycle of r copies of the eight sprocket graph C(r . SC_n) is a graceful graph, where $n \in N - \{1\}$ and $r \equiv 0, 3 \pmod 4$. ■

Proof. Let G = C(r . SC_n) be a cycle of eight sprocket graph SC_n, where $n \in N - \{1\}$. Let f be the graceful labeling for SC_n as we mentioned in Theorem 1.1. In graph G, we see that the vertices $| V(G) | = P = r(16(n) - 8)$ and the edges $| E(G) | = Q = r(16(n) + 1)$. Let $u_{k,i,j}$ $(\forall\, i = 1, 2, \ldots, 8, \forall\, j = 1, 2, \ldots, 4n)$ be the vertices of kth copy of SC_n, $(\forall\, k = 1, 2, \ldots, r)$ where the vertices of kth copy of SC_n are $p = 16(n) - 8$ and edges of k^{th}copy of SC_n are q = 16n. Join vertices $u_{k,1,2n+1}$ with $u_{k+1,1,2n+1}$ for $k = 1, 2, \ldots, r-1$ and $u_{r,1,2n+1}$ with $u_{1,1,2n+1}$ by an edge to form C(r . SC_n). We define labeling function $g : V(G) \to \{0, 1, \ldots, Q\}$ as follows

$$g(u_{1,i,j}) = \begin{cases} f(u_{i,j}), & if \;\; f(u_{i,j}) \leq \frac{q}{2} + 1, \\ f(u_{i,j}) + (Q - q), & if \;\; f(u_{i,j}) > \frac{q}{2} + 1, \\ (\forall\, i = 1, 2 \ldots, 8, \forall\, j = 1, 2, \ldots, 4n;) \end{cases}$$

$$g(u_{2,i,j}) = \begin{cases} g(u_{1,i,j}) + (Q - q), & if \;\; g(u_{1,i,j}) \leq \frac{q}{2} + 1, \\ g(u_{1,i,j}) - (Q - q), & if \;\; g(u_{1,i,j}) > \frac{q}{2} + 1, \\ (\forall\, i = 1, 2 \ldots, 8, \forall\, j = 1, 2, \ldots, 4n;) \end{cases}$$

$$g(u_{\lceil\frac{k}{2}\rceil+1,i,j})=\begin{cases} g(u_{\lceil\frac{k}{2}\rceil-1,i,j})+(q+2), & if \quad g(u_{\lceil\frac{k}{2}\rceil-1,i,j})\leq\frac{q}{2}+1,\\ g(u_{\lceil\frac{k}{2}\rceil-1,i,j})-(q+1), & if \quad g(u_{\lceil\frac{k}{2}\rceil-1,i,j})>\frac{q}{2}+1,\\ (\forall\ i=1,2\ldots,8,\forall\ j=1,2,\ldots,4n,\forall\ k=3,4,\ldots,\lceil\frac{r}{2}\rceil;) \end{cases}$$

$$g(u_{\lceil\frac{k}{2}\rceil+2,i,j})=\begin{cases} g(u_{\lceil\frac{k}{2}\rceil,i,j})+(q+2), & if \quad g(u_{\lceil\frac{k}{2}\rceil,i,j})\leq q,\\ g(u_{\lceil\frac{k}{2}\rceil,i,j})-(q+1), & if \quad g(u_{\lceil\frac{k}{2}\rceil,i,j})>q,\\ (\forall\ i=1,2\ldots,8,\forall\ j=1,2,\ldots,4n,\forall\ k=3,4,\ldots,\lceil\frac{r}{2}\rceil;) \end{cases}$$

$$g(u_{k,i,j})=\begin{cases} g(u_{k-2,i,j})+(q+2), & if \quad g(u_{k-2,i,j})\leq\frac{q}{2}+1,\\ g(u_{k-2,i,j})-(q+1). & if \quad g(u_{k-2,i,j})>\frac{q}{2}+1,\\ (\forall\ i=1,2\ldots,8,\forall\ j=1,2,\ldots,4n,\forall\ k=\lceil\frac{r}{2}\rceil+3,\lceil\frac{r}{2}\rceil+5,..,r) \end{cases}$$

$$g(u_{k,i,j})=\begin{cases} g(u_{k-2,i,j})+(q+2), & if \quad g(u_{k-2,i,j})\leq q,\\ g(u_{k-2,i,j})-(q+1), & if \quad g(u_{k-2,i,j})>q,\\ (\forall\ i=1,2\ldots,8,\forall\ j=1,2,\ldots,4n,\forall k=\lceil\frac{r}{2}\rceil+4,\lceil\frac{r}{2}\rceil+6,\ldots,r) \end{cases}$$

The above labeling pattern gives rise to a graceful labeling to the cycle of r copies for the eight sprocket graph. □

Illustration 1.2.3. The graceful labeling of the cycle of 4 copies for the eight sprocket graph is shown in Figure 1.3.

Theorem 1.4

The star of the eight sprocket graph $(SC_n)^*$ is graceful, where $n \in N-\{1\}$. ■

Proof. Let G $= (SC_n)^*$ be a star of eight sprocket graph SC_n, where $n \in N-\{1\}$, let f be the graceful labeling for SC_n as we mention in Theorem 1.1. In graph G, we see that the vertices $\mid V(G) \mid = P = p(p+1)$ and the edges $\mid E(G) \mid = Q = (p+1)q+p$, where $p = 16(n)-8$ and $q = 16(n)$. Let $u_{k,i,j}$ $(\forall\ i = 1,2,\ldots,8, \forall\ j = 1,2,\ldots,4n)$ be the vertices of kth copy of SC_n, $(\forall\ k = 1,2,\ldots p)$ where the vertices of kth copy of SC_n are $p = 16(n)-8$ and edges of kth copy of SC_n are $q = 16(n)$. We mention that the central copy of $(SC_n)^*$ is $(SC_n)^0$ and other copies of $(SC_n)^*$ are $(SC_n)^{(k)}$ $(\forall\ k = 1,2,\ldots,p)$. We define labeling function $g: V(G) \to \{0,1,\ldots.Q\}$ as follows

$$g(u_{0,i,j})=\begin{cases} f(u_{i,j}), & if \quad f(u_{i,j})\leq\frac{q}{2}+1,\\ f(u_{i,j})+(Q-q), & if \quad f(u_{i,j})>\frac{q}{2}+1,\\ (\forall\ i=1,2\ldots,8,\forall\ j=1,2,\ldots,4n;) \end{cases}$$

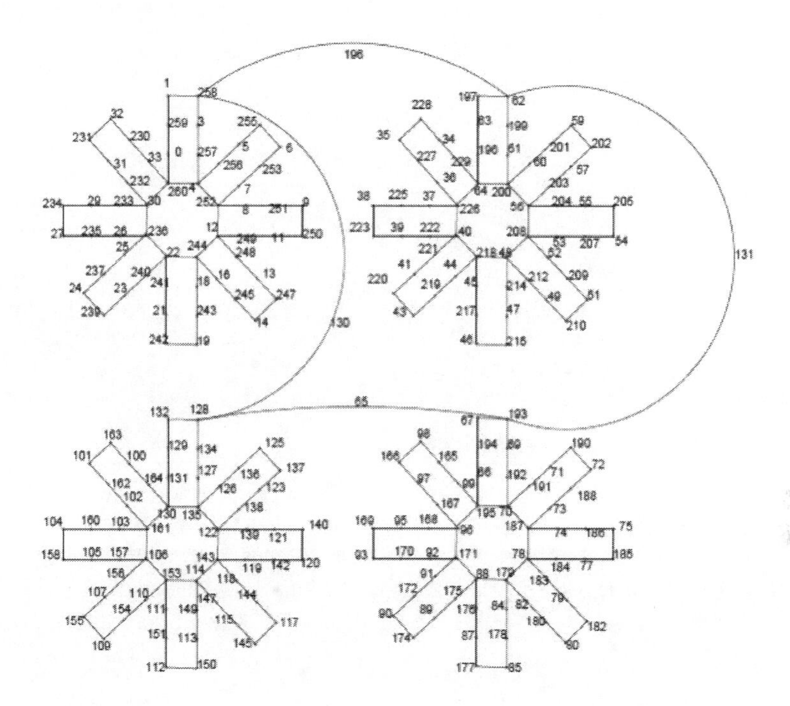

Figure 1.3: Cycle of 4 copies of eight sprocket graph SC_4 with $p = 224$ and $q = 260$ is graceful labeling

$$g(u_{1,i,j}) = \begin{cases} g(u_{0,i,j}) + p(q+1), & if \quad g(u_{0,i,j}) \le \frac{p}{2}, \\ g(u_{0,i,j}) - p(q+1), & if \quad g(u_{0,i,j}) > \frac{p}{2}, \\ (\forall\ i = 1,2\ldots,8, \forall\ j = 1,2,\ldots,4n;) \end{cases}$$

$$g(u_{k,i,j}) = \begin{cases} g(u_{k-2,i,j}) + (q+1), & if \quad g(u_{k-2,i,j}) < np, \\ g(u_{k-2,i,j}) - (q+1), & if \quad g(u_{k-2,i,j}) > np. \\ (\forall\ i = 1,2,\ldots,8, \forall\ j = 1,2,\ldots,4n, \forall\ k = 2,3,..p, \\ \forall\ n = 1,2,\ldots,(p-1)) \end{cases}$$

We see that the difference of vertices for the central copy $(SC_n)^0$ of G and its other copies $(SC_n)^k, (1 \le k \le p)$ is precisely following the sequence

$$p(q+1)$$

$$(q+1)$$

$$(p-1)(q+1)$$

$$\vdots$$

$$\lfloor \tfrac{p}{2} \rfloor\ (q+1).$$

Using this sequence we can produce the required edge label by joining corresponding vertices of $(SC_n)^0$ with its other copy $(SC_n)^k$ $(1 \leq k \leq p)$ in G.Thus G admits graceful labeling. □

1.3 CONCLUDING REMARK

Here we identified a new graph is called the eight sprocket graph. Present work contributes some new results. We discussed gracefulness of eight sprocket graphs, path union of the eight sprocket graph, cycle of the eight sprocket graph and star of the eight sprocket graph. The labeling pattern is demonstrated by means of illustrations which provide better understanding of the derived results.

REFERENCES

1. G. S. Bloom and S. W. Golomb. Application of numbered undirected graphs. *IEEE*, 65(4): 562-570, 1977.
2. J. A. Gallian, A dynamic survey of graph labeling. *The Electronics Journal of Combinatorics*, 25(4), 2018.
3. S. W. Golomb. How to number a graph. *Graph Theory and Computing (R. C. Read. Ed.)* Academic Press, New York, 23-37, 1972.
4. F. Harary. *Graph Theory and its Application.* Narosa Publishing House, New Delhi, 2001.
5. V. J. Kaneria and H. M. Makadia. Graceful labeling for double step grid graph. *International Journal of Mathematics and its Applications*, 3(1): 33-38, 2015.
6. A. Rosa. On certain valuations of the vertices of a graph. *Theory of Graphs (Internat. Symposium, Rome, July 1966)* Gordon and Breach, NY and Dunod Paris, 349-355, 1967.

2 Universal Absolute Mean Graceful Graphs

Hiren P. Chudasama
Government Polytechnic,
Rajkot, Gujarat (INDIA)
E-mail: hirensrchudasama@gmail.com

Divya K. Jadeja
V. V. P. Engineering college,
Rajkot, Gujarat (INDIA)
E-mail: divyajadeja89@gmail.com

The present chapter is an advancement of research in absolute mean graceful labeling. We introduce the new concept of extreme points in an absolute m graceful graph. We prove that any absolute mean graceful graph has at least two extreme points. We also prove every path P_n, cycle graph, complete bipartite graph $K_{m,n}$, star graph $K_{1,n}$ and coconut tree $CT_{m,n}$ are universal absolute mean graceful graphs. We also discuss an illustration T shaped graph $S(2,0,1)$ that is a universal absolute mean graceful graph while not a universal graceful graph.

2.1 ESSENTIAL PREFACE

The beginning of the 1960's is supposed to have been the birth of graph labeling. Most of the research study in graph labeling has initiatives to one introduced by Rosa [5]. Essentially, there are two major kinds of labelings of graphs studied: quantitative labelings and qualitative labelings. In the present chapter, the author focuses on quantitative labeling. The quantitative labelings have been inspired by ample diversity of applications in missile guidance coding technology, radar location coding, designing of circuits, X-ray crystallography, communication networking, astronomy, etc. One of the most beloved quantitative labelings is graceful labeling.

Kaneria and Chudasama [3] applied more liberty to vertex labeling in graceful labeling and introduced absolute mean graceful labeling. Kaneria and Chudasama [3] proved that all path P_n, cycle C_n, complete bipartite graph $K_{m,n}$, grid graph $P_m \times P_n$, step grid graph St_n and double step grid graph DSt_n are absolute mean graceful graphs. Makadia et al. [4] introduced the concept of the graceful centre and universal graceful graph. The present chapter shows the way ahead to new researches for the concept of universal graceful graphs. For conceptual parts and notations, we refer to J. A. Gallian [1] and Harary [2].

All the graphs which are discussed in the present chapter are finite, simple and undirected. Let $G = (V,E)$ be a (p,q) graph having p number of vertices and q number of edges. A function f is called an ***absolute mean graceful labeling*** of a graph $G = (V,E)$, if $f : V(G) \rightarrow \{0,\pm 1,\pm 2,\ldots,\pm q\}$ is injective and the induced function $f^* : E(G) \rightarrow \{1,2,\ldots,q\}$ defined as $f^*(e) = \left\lceil \frac{|f(u) - f(v)|}{2} \right\rceil$ is bijective for every edge $e = uv \in E(G)$. A graph G is called absolute mean graceful, if it admits absolute mean graceful labeling. This labeling f is said to be α-labeling, if there exists an integer k such that for each edge uv either $f(u) \leq k <| f(v) |$ or $f(v) \leq k <| f(u) |$, $\forall u,v \in V(G)$. The graph which holds absolute mean graceful labeling as well as α-labeling is called an α-absolute mean graceful graph.

Let G be an absolute mean graceful graph with the vertex labeling $f : V(G) \rightarrow \{0,\pm 1,\pm 2,\ldots,\pm q\}$. A vertex $v \in V(G)$ is called an extreme vertex of graph G if $f(v) = q$. Any absolute mean graceful graph G has at least two extreme vertices as it is apparent to check that f is also absolute mean graceful labeling if f holds. If a graph G has precisely two extreme vertices, then they are adjacent in G, as they produce the edge label q under f.

A graph G is said to be a universal absolute mean graceful graph if for any $v \in V(G)$, v is an extreme vertex for G with respect to some absolute mean graceful labeling of G. Similarly, the graph G is called a universal α-absolute mean graceful graph if for any $v \in V(G)$, v is an extreme vertex for G with respect to some α-absolute mean graceful labeling of G.

2.2 α-ABSOLUTE MEAN GRACEFUL GRAPHS CONSISTING OF THE PROPERTY OF UNIVERSALITY

An absolute mean graceful labeling serves a wide scope of family of graphs satisfying the labeling. Kaneria and Chudasama [3] already proved that every cycle $C_n, n \geq 3$ and complete bipartite graph $K_{m,n}$ are absolute mean graceful graphs. These families of graphs are universal absolute mean graceful graphs as well as they are universal α-absolute mean graceful graphs because of their symmetric structures. In the present chapter, we have proved the universality property in path P_n, star $K_{1,n}$, coconut tree $CT_{m,n}$ and one specific T-shaped graph.

2.2.1 EXTREME VERTICES IN α-ABSOLUTE MEAN GRACEFUL GRAPH

Theorem 2.1

Any α-absolute mean graceful graph G has at least two extreme vertices. ■

Proof. Let G be an α-absolute mean graceful graph and $f : V(G) \rightarrow \{0, \pm 1, \pm 2, ..., \pm q\}$ be an α-labeling for G.

Define $g : V(G) \rightarrow \{0, \pm 1, \pm 2, ..., \pm q\}$ such that

$$g(v) = \begin{cases} f(v) - 1, & if \quad f(v) \neq -q \\ -q, & if \quad f(v) = -q. \end{cases}$$

Note that g is injective, as f is an injective map. Further for any $uv \in E(G)$

$$g^*(uv) = \left\lceil \frac{| g(u) - g(v) |}{2} \right\rceil$$
$$= \left\lceil \frac{| (f(u) - 1) - (f(v) - 1) |}{2} \right\rceil$$
$$= \left\lceil \frac{| f(u) - f(v) |}{2} \right\rceil$$
$$= f^*(uv).$$

Therefore, $g^* : E(G) \rightarrow \{1, 2, ..., q\}$ is also a bijection, as f^* is a bijective map. Thus, g is also an absolute mean graceful labeling for G. In fact g is an α-absolute mean graceful labeling for G, as $min\ (g(u), g(v)) \leq k \leq max(g(u),\ g(v)),\ \forall\ uv \in E(G)$. It is clear to see both labelings f and g have two different extreme vertices. So, G admits at least two extreme vertices. □

2.2.2 UNIVERSALITY IN THE PATH GRAPH

Theorem 2.2

Every path graph P_n is a universal α-absolute mean graceful graph. ■

Proof. Let P_n be a path graph with $V(P_n) = \{v_1,\ v_2, \ldots,\ v_n\}$ such that $| V(P_n) | = n$ and $| E(P_n) | = n - 1$. Let v_k, $(1 \leq k \leq n)$ be any vertex with label q. We will show that for any arbitrary k it holds absolute mean graceful labeling.

Define $f : V(P_n) \rightarrow \{0, \pm 1, \pm 2, ..., \pm q\}$ such that

Case I: If $k = 1$

$$f(v_i) = \begin{cases} q, & i = 1 \\ (-1)^{i+1}[\ |f(v_{i-1})| - 1], & \forall \quad i = 2,\ 3,\ \ldots,\ n. \end{cases}$$

Case II: If $k = 2$

$$f(v_i) = \begin{cases} -q, & i = 1 \\ q, & i = 2 \\ (-1)^i[\ |f(v_{i-1})| - 2], & i = 3 \\ (-1)^i[\ |f(v_{i-1})| - 1], & \forall\ \ i = 4,\ 5,\ \dots,\ n. \end{cases}$$

Case III: If $3 \leq k \leq \left\lceil \frac{n}{2} \right\rceil$

$$f(v_i) = \begin{cases} (-1)^{n-k+1}(k-2), & i = 1 \\ (-1)^{k-i+1}[|f(v_{i-1})| - 1], & i = 2,\ 3,\ \dots,\ k-1,\ if\ n \equiv 1\ (mod\ \ 2) \\ (-1)^{k-i}[|f(v_{i-1})| - 1], & i = 2,\ 3,\ \dots,\ k-1,\ if\ n \equiv 0\ (mod\ \ 2) \\ q,\ \ i = k & \\ (-1)^{i-k}[\ |f(v_{i-1})| - 1], & \forall\ \ i = k+1,\ k+2,\ \dots,\ k + \left\lceil \frac{q-1}{2} \right\rceil \\ (-1)^{i-k}[\ |f(v_{i-1})| - 2], & \forall\ \ i = k + \left\lceil \frac{q+1}{2} \right\rceil \\ (-1)^{i-k}[\ |f(v_{i-1})| - 1], & \forall\ \ i =\ k + \left\lceil \frac{q+3}{2} \right\rceil,\ k + \left\lceil \frac{q+5}{2} \right\rceil,\ \dots,\ n. \end{cases}$$

Thus, by choosing any vertex with label k, we get vertex labeling function f as an injective function. Define the induced edge labeling function by f^* : $E(P_n) \to \{1,\ 2,\ \dots,\ q\}$ as $f^*(e) = \left\lceil \frac{|\ f(u) - f(v)\ |}{2} \right\rceil$ which is a bijective for every edge $e = (u, v) \in E(P_n)$. Therefore, P_n admits absolute mean graceful labeling by selecting any vertex as an extreme vertex. Hence, every vertex is an extreme vertex in P_n. Therefore, P_n is a universal absolute mean graceful graph. □

Illustration 2.2.1. Absolute mean graceful labeling in path P_{10}, by selecting extreme vertex as v_4 with vertex label q is shown in Figure 2.1.

2.2.3 UNIVERSALITY IN THE STAR GRAPH

Theorem 2.3

Every star graph $K_{1,n}$ is a universal absolute mean graceful graph. ■

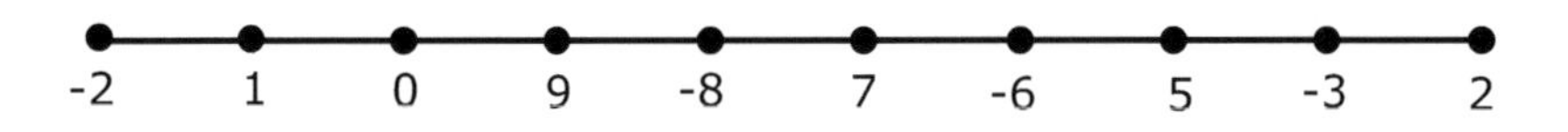

Figure 2.1: Absolute mean graceful labeling in path P_{10} with $p = 10$ and $q = 9$

Proof. Let $K_{1,n}$ be a star graph with $V(K_{1,n}) = \{u, v_1, v_2, \dots, v_n\}$ such that $|V(K_{1,n})| = n+1$ and $|E(K_{1,n})| = n$. We will show that by selecting either u or any arbitrary vertex v_k, $(1 \le k \le n)$ with label q, it holds absolute mean graceful labeling.
Define $f : V(K_{1,n}) \to \{0, \pm 1, \pm 2, ..., \pm q\}$ such that

Case I: If u is extreme vertex.

$$f(u) = q$$

$$f(v_i) = \begin{cases} -q, & i = 1 \\ f(v_{i+2}), & \forall \ i = 2, 3, \dots, n. \end{cases}$$

Case II: If v_k, $(1 \le k \le n)$ is extreme vertex.

$$f(u) = -q$$

$$f(v_i) = \begin{cases} -q+2i, & i < k \\ q, & i = k \\ -q+2(i-1), & i > k. \end{cases}$$

By choosing any vertex with label u or v_k, we get the vertex labeling function f as injective function. Define induced edge labeling function by f^* : $E(K_{1,n}) \to \{1, 2, \dots, q\}$ as $f^*(e) = \left\lceil \frac{|f(u) - f(v)|}{2} \right\rceil$ which gives a bijective map for every edge $e = (u, v) \in E(K_{1,n})$. Therefore, $K_{1,n}$ admits absolute mean graceful labeling by selecting any vertex as an extreme vertex. Hence, every vertex is an extreme vertex in $K_{1,n}$. Therefore, $K_{1,n}$ is a universal absolute mean graceful graph. □

Illustration 2.2.2. Absolute mean graceful labeling in $K_{1,7}$, by selecting extreme vertex as u with vertex label 7 is shown in Figure 2.2.

2.2.4 UNIVERSALITY IN THE COCONUT TREE GRAPH

Theorem 2.4

Every coconut tree graph $CT_{m,n}$ is a universal absolute mean graceful graph. ■

Proof. Let $CT_{m,n}$ be a coconut tree graph with $V(CT_{m,n}) = \{u_1, u_2, \dots, u_n, v_1, v_2, \dots, v_m\}$ where $u_1, u_2, \dots, u_n$ are vertices of path P_n and $v_1, v_2, \dots, v_m$ are pendent vertices being adjacent with u_1. Clearly, $|V(CT_{m,n})| = m+n$ and $|E(CT_{m,n})| = m+n-1$. We will show that by selecting any arbitrary vertex u_k $(1 \le k \le n)$ or v_k $(1 \le k \le m)$ with label q, it holds absolute mean graceful labeling.

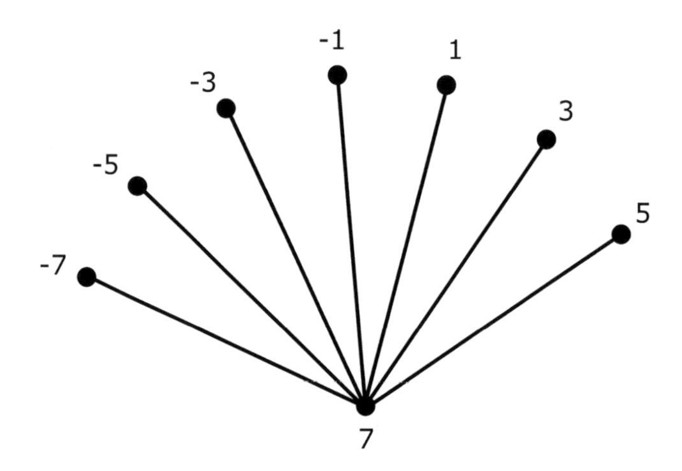

Figure 2.2: Absolute mean graceful labeling in $K_{1,7}$ with $p=8$ and $q=7$

Define $f: V(CT_{m,n}) \rightarrow \{0, \pm 1, \pm 2, ..., \pm q\}$ such that

Case I : If u_1 is an extreme vertex.
Case II : If u_k, $(k \neq 1,\ n)$ is an extreme vertex.
Case III : If u_n is an extreme vertex.
Case IV : If v_k, $(1 \leq k \leq m)$ is an extreme vertex.

Case I: If u_1 is an extreme vertex.
We will prove that that u_1 is an extreme vertex for $m \geq n$ and $m < n$.

Subcase I': If $m \geq n$

$$f(u_i) = \begin{cases} q, & i=1 \\ 0, & i=2 \\ (-1)^q, & i=3 \\ (-1)^{q+i+1}[\,|\,f(u_{i-1})\,|+1], & \forall\ i=4,\ 5,\ \ldots,\ n. \end{cases}$$

$$f(v_i) = \begin{cases} 1-q, & i=1 \\ f(v_{i-1})+2, & i=2,\ 3,\ \ldots,\ \left\lceil \frac{q-1}{2} \right\rceil \\ f(v_{i-1})+3, & i=\left\lceil \frac{q+1}{2} \right\rceil \\ f(v_{i-1})+2, & i=\left\lceil \frac{q+3}{2} \right\rceil,\ \left\lceil \frac{q+5}{2} \right\rceil,\ \ldots,\ m. \end{cases}$$

Subcase I": If $m < n$

$$f(u_i) = \begin{cases} q, & i = 1 \\ (-1)^{i+1}[\,|\,f(u_{i-1})\,|-1], & i = 2,\ 3,\ ,\ldots,\ n. \end{cases}$$

$$f(v_i) = \begin{cases} q-1, & i = 1 \\ f(u_{i-1})-2, & i = 2,\ 3,\ \ldots,\ m. \end{cases}$$

Case II: If u_k, $(k \neq 1,\ n)$ is an extreme vertex.

$$f(u_i) = \begin{cases} (-1)^k(q-n+2), & i = 1 \\ (-1)^{i-k-1}[|f(u_{i-1})|+1], & i = 2,\ 3,\ \ldots,\ k-1 \\ q, & i = k \\ (-1)^{i-k}[|f(u_{i-1})|-1], & i = k+1,\ k+2,\ \ldots,\ n-1 \\ -f(u_{n-1}), & i = n,\ if\ f(u_{n-1}) < 0 \\ 1-f(u_{n-1}), & i = n,\ if\ f(u_{n-1}) > 0 \end{cases}$$

$$f(v_i) = \begin{cases} (-1)^{k+1}(q-n+1), & i = 1 \\ f(v_{i-1})+(-1)^k \cdot 2, & i = 2,\ 3,\ \ldots,\ m-\left\lceil \dfrac{n-k-2}{2} \right\rceil \\ f(v_{i-1})+(-1)^k \cdot 4, & i = m-\left\lceil \dfrac{n-k-4}{2} \right\rceil \\ f(v_{i-1})+(-1)^k \cdot 2, & i = m-\left\lceil \dfrac{n-k-6}{2} \right\rceil,\ m-\left\lceil \dfrac{n-k-8}{2} \right\rceil,\ \ldots,\ m \end{cases}$$

Case III: If u_n is an extreme vertex.

$$f(u_i) = \begin{cases} (-1)^{n+1}(q-n+1), & i = 1 \\ (-1)^{n+i}[\ |f(u_{i-1})|+1], & i = 2,\ 3,\ \ldots,\ n. \end{cases}$$

$$f(v_i) = \begin{cases} (-1)^n[\,|\,f(u_i)\,|-1\,], & i = 1 \\ f(v_{i-1})+2\cdot(-1)^{n+1}, & i = 2,\ 3,\ \ldots,m. \end{cases}$$

Case IV: If v_k, $(1 \leq k \leq m)$ is an extreme vertex.

Let v_k be an extreme vertex for any arbitrary k. Assign other pendent vertices of $K_{1,m}$ which are adjacent with u_1 as $v_1,\ v_2,\ \ldots,\ v_{m-1}$ in any arbitrary order.

$$f(v_i) = \begin{cases} q-2i, & 1 \leq i \leq \left\lceil \dfrac{q-3}{2} \right\rceil \\ f(v_{i-1})-5, & i = \left\lceil \dfrac{q-1}{2} \right\rceil \\ f(v_{i-1})-2, & i = \left\lceil \dfrac{q+1}{2} \right\rceil,\ \left\lceil \dfrac{q+3}{2} \right\rceil,\ \ldots,\ m-1. \end{cases}$$

$$f(u_i)=\begin{cases} -q, & i=1 \\ 0, & i=2 \\ (-1)^{q+i+1}[\,|f(u_{i-1})|+1], & i=3,\,4,\,\ldots,\,\left\lceil\frac{q+2}{2}\right\rceil \\ (-1)^{q+i+1}[\,|f(u_{i-1})|+3], & i=\left\lceil\frac{q+4}{2}\right\rceil \\ (-1)^{q+i+1}[\,|f(u_{i-1})|+1], & i=\left\lceil\frac{q+6}{2}\right\rceil,\,\left\lceil\frac{q+8}{2}\right\rceil,\,\ldots,\,n-1 \\ (-1)^{q+i+1}\left\lceil\frac{q}{2}\right\rceil, & i=n. \end{cases}$$

Thus, by choosing any vertex with label u_k, $(1 \leq k \leq n)$ or v_k, $(1 \leq k \leq m)$, we get vertex labeling function f as injective function. Define induced edge labeling function by $f^* : E(CT_{m,n}) \to \{1,\,2,\,\ldots,\,q\}$ as $f^*(e) = \left\lceil\frac{|f(u)-f(v)|}{2}\right\rceil$ which gives a bijective map for every edge $e=(u,v) \in E(CT_{m,n})$. Therefore, $CT_{m,n}$ admits absolute mean graceful labeling by selecting any vertex as an extreme vertex. Hence, every vertex is an extreme vertex in $CT_{m,n}$. Therefore, $CT_{m,n}$ is a universal absolute mean graceful graph. □

Illustration 2.2.3. Absolute mean graceful labeling in $CT_{9,4}$, by selecting extreme vertex as u_1 with vertex label 12 is shown in Figure 2.3.

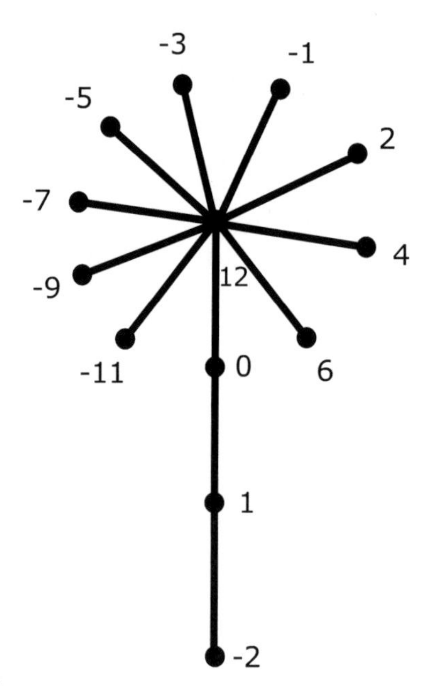

Figure 2.3: Absolute mean graceful labeling in $CT_{9,4}$ with $p=13$ and $q=12$

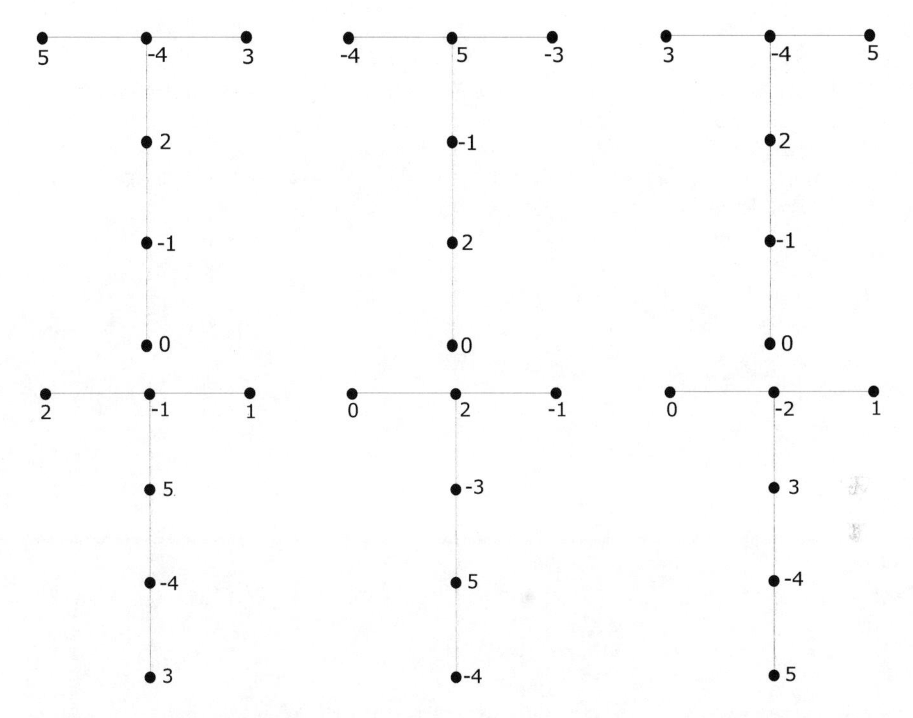

Figure 2.4: α-absolute mean graceful labeling in T shaped graph $S(2,0,1)$ for each of six vertices as an extreme vertex

2.2.5 SPECIAL GRAPH CONSISTING OF UNIVERSALITY IN ABSOLUTE MEAN GRACEFUL LABELING

Let T be a caterpillar $S(2,0,1)$ graph. Then T is an α-graceful graph, but it is not a universal graceful graph, as one of the vertex v cannot be a graceful center for T with respect to any graceful labeling for T which was proved by Makadia et al.[5].

We will show that the same graph T is a universal α-absolute mean graceful graph. The graphs in figure 2.4 show α-absolute mean graceful labeling for each vertex of graph T. Here $\mid V(T) \mid = 6$ and $\mid E(T) \mid = 5$.

REFERENCES

1. J. A. Gallian. A dynamic survey of graph labeling. *The Electronics Journal of Combinatorics*, #DS6, 1-535, 2019.
2. F. Harary. *Graph Theory.* Addison-Wesley Publishing Company, Massachusetts, 1-281, 1969.

3. V. J. Kaneria and H. P. Chudasama. Absolute mean graceful labeling in various graphs. *Int. J. of Mathematics and Its Applications*, 5 (4-E): 723-726, 2017.
4. H. M. Makadia et al. Graceful centres of graceful graphs and universal graceful graphs. *Proyecciones Journal of Mathematics*, 38(2): 305-314, 2019.
5. A. Rosa, "On certain valuations of the vertices of a graph," Theory of Graphs (Internat. Symposium, Rome, July 1966), Gordon and Breach, N. Y. and Dunod Paris (1967), 349–355.

3 Universal α-graceful Gear related Graphs

Divya K. Jadeja
Department of Mathematics,
Saurashtra University,
Rajkot, Gujarat (INDIA)
E-mail:divyajadeja89@gmail.com

V. J. Kaneria
Department of Mathematics,
Saurashtra University,
Rajkot, Gujarat (INDIA)
E-mail: kaneriavinodray@gmail.com

In 1967 Rosa defined graceful labeling and α-labeling. A graph G, a graceful labeling f is called α-labeling if there is a non-negative integer $k(0 \leq k \leq |E(G)|)$ such that $\min\{f(u), f(v)\} \leq k < \max\{f(u), f(v)\}, \forall uv \epsilon E(G)$. Here we call a graph G which admits a α -labeling as a α-graceful graph. A vertex $v \in V(G)$ is called an extreme vertex for G, if there is an α -graceful labeling f on G such that $f(v) = 0$. A graph G is called a universal α-graceful graph if all of its vertices are extreme. In this chapter we have proved G_n^* is a universal α-graceful graph when n is even and G_n^* is the graph obtained by taking duplication of a vertex of gear graph G_n, whose degree is two. Also the graph obtained by 2-super subdivision of all the rim edges of W_n is a universal α-graceful graph, when n is even. We have shown that the graph obtained by t-super subdivision of all the rim edges of W_n is a universal α-graceful graph, when n is even.

3.1 INTRODUCTION

In this chapter, we shall consider a simple, finite and undirected (p,q) graph $G = (V,E)$ with $p = |V|$ vertices and $q = |E|$ edges. For all terminologies, notations and basic definitions we follow Harary [2]. For a comprehensive bibliography and references of papers, we have referred to Gallian [1]. Rosa [3] defined α-labeling which here we call α-graceful labeling. Kaneria, Teraiya and Meera [5] obtained α-graceful labeling for a double path union of some α-graceful graphs. Kaneria and Makadia [6] proved that t-super subdivision of the grid graph is graceful. A function f is called graceful labeling of graph G if $f : V(G) \rightarrow \{0, 1, 2, \ldots, q\}$ is injective and its induced function

$f^* : E(G) \to \{1, 2, \ldots, q\}$ defined as $f^*(uv) = |f(u) - f(v)|$, for all $uv \in E(G)$, is bijective. A graph G is called a graceful graph if it admits a graceful labeling f of G. A graceful graph G with a graceful labeling f of G is called an α-graceful graph if there exists a non-negative integer k $(0 \leq k < q)$ such that $min\{f(u), f(v)\} \leq k < max\{f(u), f(v)\}$, $\forall$ $uv \in E(G)$. Makadia, Karavadiya and Kaneria [7] defined an extreme vertex for a graceful graph G and they proved that any α-graceful graph has at least four extreme vertices.

Let G be a graceful graph with a graceful labeling f. A vertex $v \in V(G)$ is called graceful extreme for G if $f(v) = 0$. If $f(w) = q$, for some $w \in V(G)$ in a graceful graph G with a graceful labeling f, then w is an extreme vertex for G with respect to the graceful labeling $q - f$ of G. If f is an α-graceful labeling for G, then $q - f$, h and $q - h$ are all α-graceful labelings for G, where k is a non-negative integer such that $h = k - f$, when $f \leq k$ and $h = q + k + 1 - f$, when $k < f \leq q$. Also these α-graceful labelings for G create extreme vertices for G whose labels are q, k and $k + 1$ under f. The wheel graph W_n $(n \geq 3)$ is $K_1 + C_n$ and the vertex corresponding to K_1 is known as the apex for W_n; as well the vertices and edges corresponding to C_n are known as rim vertices and rim edges for W_n respectively. By duplication of vertex $v \in V(G)$ of a graph G we mean a super graph H of G with $V(H) = V(G) \cup \{v'\}$ and $E(H) = E(G) \cup \{uv' / \ \forall \ u \in N(v)\}$. Here v' is the duplicate vertex of v in H and $N_H(v') = N_G(v)$, where $N_G(v)$ is called neighbourhood of v in G which is the set of all vertices $u \in V(G)$ with $uv \in E(G)$. A graph H is said to be a t-super subdivision of G if H is obtained from G by replacing every edge $e \in E(G)$ by a complete bipartite graph $K_{2,t}$, for some $t \in N$. If $t = 1$, then H is called a Bary centric subdivision of G.

A graph G is called a universal graceful graph if each vertex of G is an extreme vertex for G with respect to some graceful labeling f of G and it is called a universal α-graceful graph if for each vertex v of G, there is an α-graceful labeling f of G with $f(v) = 0$ or $f(v) = q$ or $f(v) = k$ or $f(v) = k+1$, where k is a non-negative integer as discussed earlier. Cycle C_n is a universal graceful graph when $n \equiv 3 \ (mod \ 4)$ and it is a universal α-graceful graph when $n \equiv 0 \ (mod \ 4)$, as C_n has symmetric structure.

Ma and Feng [4] discussed the gracefulness of gear graph G_n. Let n be a positive even integer. Take $V(G_n) = \{u_i / \ 0 \leq i \leq n\} \cup \{v_i / \ 1 \leq i \leq n\}$ and $E(G_n) = \{u_0u_i / \ 1 \leq i \leq n\} \cup \{u_iv_i / \ 1 \leq i \leq n\} \cup \{u_{i+1}v_i / \ 1 \leq i < n\} \cup \{u_1v_n\}$. If we define $f(u_0) = 0$, $f(u_i) = q + 1 - i$ $(1 \leq i \leq n)$, $f(v_i) = n - 1 + i$ $(1 \leq i \leq \frac{n}{2})$ and $f(v_i) = n + i$ $(\frac{n}{2} < i \leq n)$, where $f : V(G_n) \to \{0, 1, 2, \ldots, 3n\}$, then f is an α-graceful labeling for G_n and G_n is a universal graceful graph, as $f(u_0) = 0$, $f(u_1) = 3n$, $f(u_n) = 2n + 1 = k + 1$, $f(v_n) = k = 2n$ as G_n has symmetric structure.

3.2 UNIVERSAL α-GRACEFUL GEAR RELATED GRAPHS

Theorem 3.1

Let n be a positive even integer and G_n^* be the graph obtained by taking duplication of a vertex (whose degree is two) of gear graph G_n, then G_n^* is a universal α-graceful graph. ■

Proof. Let n be a positive even integer and $G = G_n^*$. It is obvious that $|V(G)| = 2n+2$ and $|E(G)| = 3n+2$. Let $V(G_n) = \{u_i/\ 0 \leq i \leq n\} \cup \{v_i/\ 1 \leq i \leq n\}$ and $E(G_n) = \{u_0u_i/\ 1 \leq i \leq n\} \cup \{u_iv_i/\ 1 \leq i \leq n\} \cup \{u_{i+1}v_i/\ 1 \leq i < n\} \cup \{u_1v_n\}$. To obtain all the vertices of G as extreme vertices for some α-graceful labeling of G, the following two cases are essential.
Case-1: We obtain G from G_n by taking duplication of vertex v_1 which is a new vertex w, i.e. $V(G) = V(G_n) \cup \{w\}$ and $E(G) = E(G_n) \cup \{u_1w,\ wu_2\}$. In this case we define $f : V(G) \to \{0,\ 1,\ 2,\ \ldots,\ 3n+2\}$ as follows

$$f(u_0) = 2;\ f(w) = 0;$$

$$f(u_i) = 3n+3-i, \text{ where } 1 \leq i \leq n;$$

$$f(v_i) = \begin{cases} n+1+i, & \text{where } 1 \leq i \leq \dfrac{n}{2} \\ n+2+i, & \text{where } \dfrac{n}{2} < i \leq n \end{cases}$$

For the above defined labeling pattern f is injective, as $f(V(G)) \subseteq \{0,\ 1,\ 2,\ \ldots,\ 3n+2\}$ and $|V(G)|=$ number of elements of range of the function f. Moreover $\{f^*(u_0u_i)/\ 1 \leq i \leq n\} = \{2n+1,\ 2n+2,\ \ldots,\ 3n\}$, $\{f^*(u_1w),\ f^*(wu_2)\} = \{3n+1,\ 3n+2\}$, $f^*(u_1v_n) = n$, $\{f^*(u_iv_i)/\ 1 \leq i \leq n\} = \{1,\ 3,\ \ldots,\ n-3,\ n-1,\ n+2,\ n+4,\ \ldots,\ 2n-2,\ 2n\}$ and $\{f^*(u_{i+1}v_i)/\ 1 \leq i < n\} = \{2,\ 4,\ \ldots,\ n-4,\ n-2,\ n+1,\ n+3, \ldots,\ 2n-3,\ 2n-1\}$. Thus, $\{f^*(e)/\ e \in E(G)\} = \{1,\ 2,\ \ldots,\ 3n+2\}$ so, f^* is a bijection as the number of elements of domain of f^* are the number of elements of range of f^* and the co-domain of f^* is the range of f^*. Therefore, f is a graceful labeling for G. Since, G is a bipartite graph and by taking $k = 2n+2$, it can be observed that for any $uv \in E(G)$, $min\{f(u), f(v)\} \leq k < max\{f(u), f(v)\}$, f is an α-labeling for G. Since $f(w) = 0$, $f(u_1) = q$, $f(u_n) = k+1$ and $f(v_n) = k$, G is an α-graceful graph with extreme vertices u_1, v_1, u_n and v_n. By replacing labels of v_1 and w, it is easy to get that w is also an extreme vertex with respect to new α-graceful labeling of G.
Case-2: We obtain G from G_n by taking duplication of $v_k \in V(G_n)$ which is a new vertex w i.e. $V(G) = V(G_n) \cup \{w\}$ and $E(G) = E(G_n) \cup \{u_kw,\ wu_{k+1}\}$ without loss of generality we assume here $k \leq \dfrac{n}{2}$. In this case we define $g : V(G) \to \{0,\ 1,\ 2,\ \ldots,\ q\}$ as follows,

$$g(u_0) = 0;\ g(w) = g(v_k)+2 = n+k+1;$$
$$g(u_i) = 3n+3-i, \text{ where } 1 \leq i \leq n;$$

$$g(v_i) = \begin{cases} n-1+i, & \text{where } 1 \leq i \leq k \\ n+1+i, & \text{where } k < i \leq \dfrac{n}{2} \\ n+2+i, & \text{where } \dfrac{n}{2} < i \leq n \end{cases}$$

For the above defined labeling pattern g is injective, as $g(V(G)) \subseteq$ co-domain of g and $|V(G)|$ = number of elements of range of the function g. Moreover $\{g^*(u_0u_i)/\ 1 \leq i \leq n\} = \{2n+3,$ 2n+4, $\ldots,\ 3n+2\}$, $g^*(u_1v_n) = n$, $g^*(u_kw) = 2n+2-2k$, $g^*(wu_{k+1}) = 2n+1-2k$ and $\{g^*(u_iv_i)/\ 1 \leq i \leq n\} \cup \{g^*(v_iu_{i+1})/\ 1 \leq i < n\} = \{1,\ 2,\ \ldots,$ n-1, n+1, n+2, $\ldots,$ 2n-2k, 2n-2k+3, $\ldots,\ 2n+2\}$. Thus $\{g^*(e)/\ e \in E(G)\}$ = co-domain of g^* and so g is bijective mapping as $|E(G)| = q = 3n+2$ = number of elements of the range of g^*. Since G is a bipartite graph, it can be observed that for any $uv \in E(G)$, $min\ \{g(u), g(v)\} \leq k = 2n+2 < max\ \{g(u), g(v)\}$, which gives g is an α-graceful labeling for G. Since $g(u_1) = q$, $g(u_0) = 0$, $g(u_n) = 2n+3 = k+1$ and $g(v_n) = k = 2n+2$, G is an α-graceful graph with extreme vertices u_0, u_1, u_n and v_n. Also, we have taken duplication of arbitrary vertex v_k of G_n in G, all the vertices of G except v_k and w are extreme vertices of G with respect to the α-graceful labeling g of G. Therefore,G is a universal α-graceful graph.

□

Theorem 3.2

Let n be a positive even integer. Then the graph obtained by 2-super subdivision of all the rim edges of W_n is a universal α-graceful graph. ■

Proof. Let n be a positive even integer and G be the graph obtained by 2-super subdivision of all the rim edges of W_n. It is obvious that $V(G) = 3n+1$ and $q = |E(G)| = 5n$. Let $V(G) = \{u_i/\ 0 \leq i \leq n\} \cup \{v_i, w_i/\ 1 \leq i \leq n\}$ and $E(G) = \{u_0u_i, u_iv_i, u_iw_i/\ 1 \leq i \leq n\} \cup \{v_iu_{i+1}, w_iu_{i+1}/\ 1 \leq i < n\} \cup \{u_1v_n, u_1w_n\}$. We define $f : V(G) \rightarrow \{0,\ 1,\ 2,\ \ldots,\ 5n\}$ as follows,

$$f(u_0) = 0;$$
$$f(u_i) = 5n+1-i, \text{ where } 1 \leq i \leq n;$$

$$f(v_i) = \begin{cases} n-1+i, & \text{where } 1 \leq i \leq \dfrac{n}{2} \\ n+i, & \text{where } \dfrac{n}{2} < i \leq n \end{cases}$$

$$f(w_i) = \begin{cases} 3n-1+i, & \text{where } 1 \leq i \leq \dfrac{n}{2} \\ 3n+i, & \text{where } \dfrac{n}{2} < i \leq n \end{cases}$$

The above defined labeling pattern gives that f is injective, as no vertex labeling of G is repeated under f. Moreover we see that $\{f^*(u_0u_i)/\ 1 \leq i \leq n\} = \{4n+1,$

4n+2, ..., $5n\}$, $\{f^*(u_iv_i), f^*(u_iw_i)/\ 1 \le i \le \frac{n}{2}\} \cup \{f^*(u_{i+1}v_i\}, f^*(u_{i+1}w_i\}/\ 1 \le i \le \frac{n}{2}\}$ = $\{n+1$, n+2, ..., 2n, 3n+1, 3n+2, ..., $4n\}$ and $\{f^*(u_iv_i\}$, $f^*(u_iw_i)/\ \frac{n}{2} < i \le n\} \cup \{f^*(u_{i+1}v_i\}, f^*(u_{i+1}w_i\}/\ \frac{n}{2} < i < n\} \cup \{f^*(u_1v_n), f^*(u_1w_n)\}$ = {1, 2, ..., n, 2n+1, 2n+2, ..., $3n\}$. Thus, $\{f^*(e)/\ e \in E(G)\}$ = $\{1, 2, \ldots, 5n\}$ = co-domain of f^* and so f^* is bijection. Since G is a bipartite graph, by taking $k = 4n$ it can be observed that for any $uv \in E(G)$, $min\ \{f(u), f(v)\} \le k < max\ \{f(u), f(v)\}$, which gives that f is an α-graceful labeling for G. Now $f(u_0) = 0$, $f(u_1) = q$, $f(u_n) = k+1$ and $f(w_n) = k$, which gives that G is an α-graceful graph with extreme vertices u_0, u_1, u_n and w_n. By replacing labels of v_i and $w_i (1 \le i \le n)$, it is easy to get that v_n is also an extreme vertex with respect to new α-graceful labeling of G and due to the symmetric structure of G, it is a universal α-graceful graph. □

Theorem 3.3

Let n be a positive even integer. Then the graph obtained by t-super subdivision of all the rim edges of W_n is a universal α-graceful graph. ■

Proof. Let n be a positive even integer and G be the graph obtained by t-super subdivision of all the rim edges of W_n. It is obvious that $|V(G)| = (t+1)n+1$ and $q = |E(G)| = (2t+1)n$. Let $V(G) = \{u_i/\ 0 \le i \le n\} \cup \{v_{ji}/\ 1 \le i \le n, 1 \le j \le t\}$ and $E(G) = \{u_0u_i, u_iv_{ji}\ (1 \le j \le t)/\ 1 \le i \le n\} \cup \{u_{i+1}v_{ji}\ (1 \le j \le t)/\ 1 \le i < n\} \cup \{u_1v_{jn}/\ 1 \le j \le t\}$. We define $f : V(G) \to \{0, 1, 2, \ldots, (2t+1)n\}$ as follows,

$$f(u_0) = 0;$$
$$f(u_i) = q+1-i, \text{ where } 1 \le i \le n;$$
$$f(v_{ji}) = \begin{cases} (2j-1)n-1+i, & \text{where } 1 \le i \le \frac{n}{2} \\ (2j-1)n+i, & \text{where } \frac{n}{2} < i \le n \end{cases}$$

For the above defined labeling pattern f is a graceful labeling for G. Since G is a bipartite graph, it can be observed that for $uv \in E(G)$, $min\{f(u), f(v)\} \le k = 2tn < max\{f(u), f(v)\}$, where f is an α-graceful labeling for G with u_0, u_1, v_{tn} and u_{tn} are extreme vertices for G. Due to symmetric structure of G, it is a universal α-graceful graph. □

REFERENCES

1. J. A. Gallian. A dynamic survey of graph labeling. *The Electronics Journal of Combinatorics*, #DS6, 21, 2018.
2. F. Harary. *Graph Theory*. Addition Wesley, Massachusetts, 1972.

3. A. Rosa. On certain valuation of the vertices of a graph. *Theory of Graphs (International Symposium, Rome, July 1966)*, Goden and Breach, N. Y. and Dunod Paris, 349-355, 1967.
4. K.J. Ma and C. J. Feng. On the gracefulness of gear graphs. *Math. Practice Theory*, 4: 72-73, 1984.
5. V. J. Kaneria, O. Teraiya and M. Meghpara. Double path union of α-graceful graph with α-labeling. *Journal of Graph Labeling*, 2016.
6. H. M. Makadia. Some results on graceful labeling for step grid related graphs. *International Journal of Mathematics Trends and Technology*, 65(11): 29-38, 2019.
7. H. M. Makadia, Karavadia and V. J. Kaneria. Graceful centres of graceful graphs and universal graceful graph. *Proyecciones(Antofagasta) Journal of Mathematics*, 38(22): 305-314, 2019.

4 $L(2,1)$ - Labeling on Jahangir Graph

U. M. Prajapati
Department of Mathematics,
St. Xavier's College,
Ahmedabad, Gujarat (INDIA)
E-mail: udayan.prajapati@sxca.edu.in

N. B. Patel
Department of Mathematics,
Gujarat University,
Ahmedabad, Gujarat (INDIA)
E-mail: nittalbpatel000@gmail.com

$L(2,1)$ - labeling problems consist of an assignment of non-negative integers to the nodes of a graph such that the adjacent nodes have labels which differ by at least two, and the nodes at distance two must have different labels. The span of $L(2,1)$ - labeling is the difference between the minimum and maximum labels which are assigned to the nodes. The minimum span is called $L(2,1)$-labeling number or λ - number. In this chapter, $L(2,1)$-labeling number of Jahangir graph ($J_{n,m}$) for $n \geq 3$ and $m \geq 5$ is discussed.

4.1 INTRODUCTION

A frequency assignment problem was given by Hale [1], in order to assign a frequency (non-negative integer) to each TV or radio transmitter, located at various places such that communication does not interfere. Hale [1] introduced the notion of T-coloring of a graph in 1980, to formulate the frequency assignment problem as a graph coloring problem. In 1988, Roberts proposed a variation of the frequency assignment problem in which close transmitters must receive frequencies that are at least two apart. In a graph model of this problem, the transmitters are represented by the nodes of a graph. Two nodes are very close if they are adjacent in the graph and close if they are at a distance two apart in the graph. Motivated by the problem of assigning frequency to radio or TV transmitters, Griggs and Yeh [2] introduced $L(2,1)$-labeling in 1992. For standard terminology Bondy and Murty [3] is used. For a detailed survey on graph labeling we refer to [4] is refered. Griggs and Yeh [2] showed that the $L(2,1)$ - labeling problem is NP-complete for general graphs and determined the exact values of $\lambda(P_n)$, $\lambda(C_n)$ and $\lambda(W_n)$. He also showed that for n-cube Q_n, $\lambda(Q_n) \leq 2n+1$ for $n \geq 5$; for a tree T with maximum degree

$\Delta \geq 1$, $\lambda(T)$ is either $\Delta+1$ or $\Delta+2$; for a general graph G with maximum degree Δ, $\lambda(G) \leq \Delta^2+2\Delta$; for 3 - connected graph G, $\lambda(G) \leq \Delta^2+2\Delta-3$; for a graph G of diameter - 2 graph, $\lambda(G) \leq \Delta^2$. In 1996, Chang and Kuo [5] improved the upper bound for general graph G with maximum degree Δ, i.e $\lambda(G) \leq \Delta^2+\Delta$. He also provided polynomial-time algorithm for the $L(2,1)$ - labeling problem on trees. In 2004, Kuo and Yan [6] have studied the $L(2,1)$ - labeling problem for the cartesian products of paths and cycles. In 2011, Vaidya and Bantva [7] have proved the λ - number of an n - ary, Δ - regular cactus is $\Delta+1$ or $\Delta+2$, $\lambda(C_n^k) = 2k+1$ where C_n^k is the one point union of k cycles C_n. In 2016, Baby Smitha [8] and K. Thirusangu have determined the $L(2,1)$ - labeling number for the line graph of triangular snake graph and spiked snake graphs. In 2019, Prajapati and Patel [10] have determined the $L(2,1)$ - labeling number $\lambda(G)$ for the line graph of a crown graph and the line graph of an armed crown graph.

4.2 DEFINITIONS

Definition. An $L(2,1)$ - labeling of graph G with $V(G)$ and $E(G)$ as vertex set and edge set respectively, is a function $f : V(G) \to \{0\} \cup N$ such that for every pair of vertices x and y the following condition must be satisfied:

1. If $d(x,y) = 1$ then $\|f(x) - f(y)\| \geq 2$ and
2. If $d(x,y) = 2$ then $\|f(x) - f(y)\| \geq 1$,

where $d(x,y)$ denotes the distance between x and y. The $L(2,1)$ - labeling number of G, denoted by $\lambda(G)$, is the smallest number k such that G admits an $L(2,1)$ - labeling with maximum label k.

As bandwidth is a limited resource, the main target in FAP is to come up with a frequency assignment using a minimum number of frequencies, i.e. one needs to minimize the span of the labeling proposed. For convenience, without loss of generality, the smallest label is considered to be zero, so that the span is the highest label assigned.

In this chapter, $L(2,1)$ - labeling of a generalized Jahangir graph which has been introduced by Ali et al. [9] is discussed.

Definition. Jahangir graph $J_{n,m}$ for $m \geq 3$ and $n \geq 2$ is a graph on $nm+1$ vertices consisting of a cycle C_{nm} with one additional vertex which is adjacent to m vertices of C_{nm} at distance n to each other on C_{nm}.

Jahangir graph $J_{2,8}$ appears on Jahangir's tomb in his mausoleum. It is situated 5 km northwest of Lahore, Pakistan, across the River Ravi.

Remark. Wheel graph W_n and gear graph G_n are special cases of $J_{n,m}$ for $n = 1$ and $n = 2$ respectively.

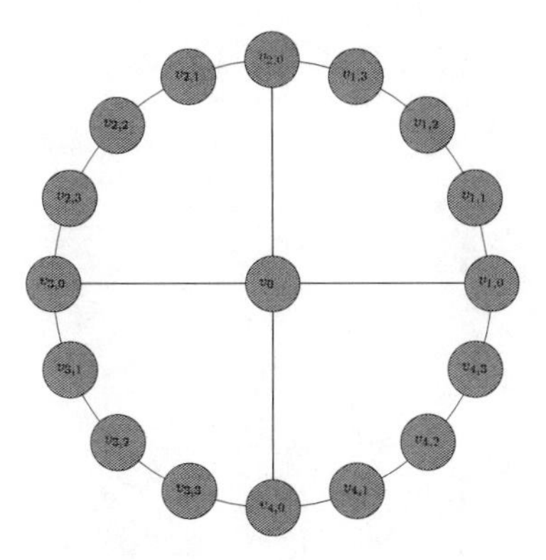

Figure 4.1: Graph $J_{4,4}$

Definition. The region between the edge formed by joining central vertex to vertices on C_{nm} at distance n from each other is known as petals of $J_{n,m}$. Let P_m be the petal formed by $v_0 v_{i,0} v_{i+1,0} \pmod{m}$ for $1 \leq i \leq m$ i.e $P_i = v_0 v_{i,0} v_{i+1,0} \pmod{m}$ for $1 \leq i \leq m$. Each petal has a path of even or odd length.

Notation. In $J_{n,m}$, v_0 is adjacent to $v_{i,0}$ for $1 \leq i \leq m$ which are at distance n from each other and $P_i = v_0 v_{i,0} v_{i+1,0} \pmod{m}$. The vertex set of $J_{n,m}$ is $V(J_{n,m}) = \{v_{i,0} | 1 \leq i \leq m\} \cup \{v_{i,j} | 1 \leq i \leq m, 1 \leq j \leq n-1\} \cup \{v_0\}$ and the edge set of $J_{n,m}$ is $E(G) = \{v_0 v_{i,0} | 1 \leq i \leq m\} \cup \{v_{i,0} v_{i,1} | 1 \leq i \leq m\} \cup \{v_{i,n-1} v_{i+1,0} | 1 \leq i \leq m\} \cup \{v_{i,j} v_{i,(j+1)} | 1 \leq i \leq m, 1 \leq j \leq n-2\}$. To form $J_{n,m}$ join $v_{i,j}$ to $v_{i,j+1}$ for $1 \leq i \leq m$ and $o \leq j \leq n-2$ and join $v_{i,n-1}$ to $v_{i+1,0} \pmod{m}$ for $1 \leq i \leq m$. So the distance between the vertices is $d(v_0, v_{i,0}) = 1$ for $1 \leq i \leq m$; $d(v_{i,0}, v_{j,0}) = 2$ for $1 \leq i$ and $j \leq m$ with $i \neq j$; $d(v_{i,0}, v_{i,1}) = 1$ for $1 \leq i \leq m$; $d(v_{i,n-1}, v_{i+1,0} \pmod{m}) = 1$ for $1 \leq i \leq m$; $d(v_{i,j}, v_{i,j+1}) = 1$ for $1 \leq i \leq m$ and $1 \leq j \leq n-2$. $J_{n,m}$ for $n = 4$ and $m = 4$ can be shown in the Figure 4.1

4.3 LABELING NUMBER OF GENERALIZED JAHANGIR GRAPH

Theorem 4.1

For $m \geq 5$ and $n \equiv 0 \pmod{3}$, $\lambda(J_{n,m}) \leq m+1$. ∎

Proof. In $J_{n,m}$, v_0 is adjacent to $v_{i,0}$ for $1 \leq i \leq m$ which are at distance n from each other and $P_i = v_0 v_{i,0} v_{i+1,0} \pmod m$. Define $f : V(G) \to N \cup \{0\}$ as follows:

$$
\begin{aligned}
f(v_0) &= 0; \\
f(v_{i,0}) &= i+1 \quad \text{for } 1 \leq i \leq m; \\
f(v_{i,j}) &= f(v_{i,j-1}) + 1 \pmod{m+1}) + 1 \quad \text{for } 1 \leq j \leq 2,\ 1 \leq i \leq m-1; \\
f(v_{m,1}) &= f(v_{m,0}) \pmod{m+1}) + 1; \\
f(v_{m,2}) &= f(v_{m,1}) + 3 \pmod{m+1}) + 1; \\
f(v_{i,j}) &= f(v_{i,0}) \quad \text{for } j \equiv 0 \pmod 3,\ 3 \leq j \leq n-1,\ 1 \leq i \leq m; \\
f(v_{i,j}) &= f(v_{i,1}) \quad \text{for } j \equiv 1 \pmod 3,\ 3 \leq j \leq n-1,\ 1 \leq i \leq m; \\
f(v_{i,j}) &= f(v_{i,2}) \quad \text{for } j \equiv 2 \pmod 3,\ 3 \leq j \leq n-1,\ 1 \leq i \leq m.
\end{aligned}
$$

Hence $\lambda(J_{n,m}) \leq m+1$ for $m \geq 5$ and $n \equiv 0 \pmod 3$. □

Illustration 4.3.1. The $L(2,1)$ - labeling of $J_{6,5}$ is shown in Figure 4.2.

Theorem 4.2

For $m \geq 5$ and $n \equiv 1 \pmod 3$, $\lambda(J_{n,m}) \leq m+1$. ■

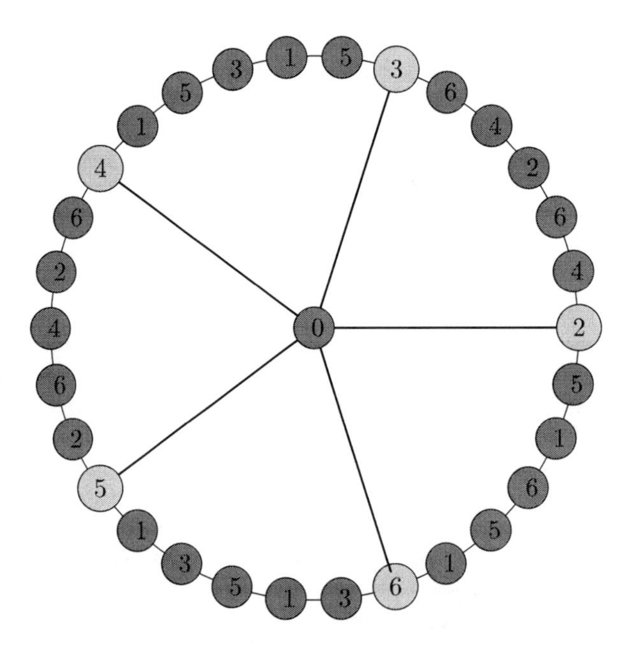

Figure 4.2: $L(2,1)$ - labeling of $J_{6,5}$

Proof. Define $f : V(G) \to N \cup \{0\}$ as follows:

$$\begin{aligned} f(v_0) &= 0; \\ f(v_{i,0}) &= i+1 \quad \text{for } 1 \le i \le m; \end{aligned}$$

Now, we take different cases:

Case-1: If $m = 5$ then define

$$\begin{aligned} f(v_{1,1}) &= f(v_{1,0}) + 1 \pmod{m+1}) + 1; \\ f(v_{1,2}) &= f(v_{1,1}) + 1 \pmod{m+1}) + 1; \\ f(v_{1,n-1}) &= f(v_{1,n-2}) + 1 \pmod{m+1}); \\ f(v_{i,1}) &= f(v_{i,0}) + 1 \pmod{m+1}) + 1,\ 2 \le i \le 3; \\ f(v_{i,2}) &= f(v_{i,1}) + 1 \pmod{m+1}),\ 2 \le i \le 3; \\ f(v_{i,n-1}) &= f(v_{i,n-2}) + 1 \pmod{m+1}) + 1,\ 2 \le i \le 3; \\ f(v_{4,1}) &= f(v_{4,0}) + 1 \pmod{m+1}) + 1; \\ f(v_{4,2}) &= f(v_{4,1}) + 2 \pmod{m+1}) + 1; \\ f(v_{4,n-1}) &= f(v_{4,n-2}) + 3 \pmod{m+1}) + 1; \\ f(v_{5,1}) &= f(v_{5,0}) + 1 \pmod{m+1}); \\ f(v_{5,2}) &= f(v_{5,1}) + 1 \pmod{m+1}) + 1; \\ f(v_{5,n-1}) &= f(v_{5,n-2}) + 1 \pmod{m+1}) + 1; \\ f(v_{i,j}) &= f(v_{i,0}) \quad \text{for } j \equiv 0 \pmod 3,\ 3 \le j \le n-2,\ 1 \le i \le m; \\ f(v_{i,j}) &= f(v_{i,1}) \quad \text{for } j \equiv 1 \pmod 3,\ 3 \le j \le n-2,\ 1 \le i \le m; \\ f(v_{i,j}) &= f(v_{i,2}) \quad \text{for } j \equiv 2 \pmod 3,\ 3 \le j \le n-2,\ 1 \le i \le m. \end{aligned}$$

Case-2: If $m \ge 6$ then define

$$\begin{aligned} f(v_{i,j}) &= f(v_{i,j-1}) + 1 \pmod{m+1}) + 1 \quad \text{for } 1 \le j \le 2,\ 1 \le i \le m; \\ f(v_{i,j}) &= f(v_{i,0}) \quad \text{for } j \equiv 0 \pmod 3,\ 3 \le j \le n-2,\ 1 \le i \le m; \\ f(v_{i,j}) &= f(v_{i,1}) \quad \text{for } j \equiv 1 \pmod 3,\ 3 \le j \le n-2,\ 1 \le i \le m; \\ f(v_{i,j}) &= f(v_{i,2}) \quad \text{for } j \equiv 2 \pmod 3,\ 3 \le j \le n-2,\ 1 \le i \le m; \\ f(v_{i,n-1}) &= f(v_{i,n-2}) + 1 \pmod{m+1}) + 1,\ 1 \le i \le m. \end{aligned}$$

Hence $\lambda(J_{n,m}) \le m+1$ for $m \ge 5$ and $n \equiv 1 \pmod 3$. □

Illustration 4.3.2. The $L(2,1)$ - labeling of $J_{7,6}$ is shown in the Figure 4.3.

Theorem 4.3

For $m \ge 5$ and $n \equiv 2 \pmod 3$, $\lambda(J_{n,m}) \le m+1$. ■

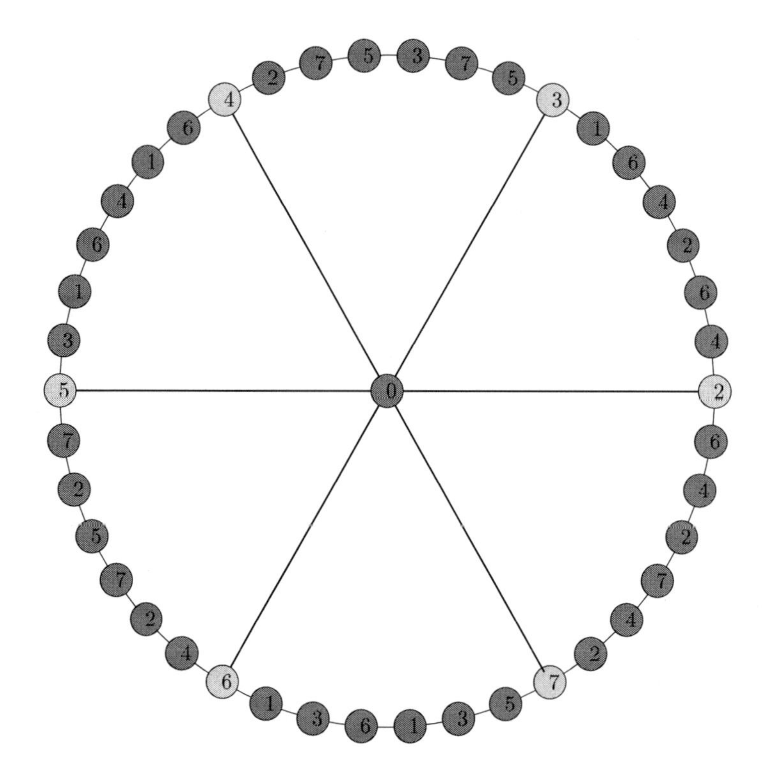

Figure 4.3: $L(2,1)$ - labeling of $J_{7,6}$

Proof. Define $f : V(G) \to N \cup \{0\}$ as follows:

$$f(v_0) = 0;$$
$$f(v_{i,0}) = i+1 \quad \text{for } 1 \leq i \leq m;$$

Now take different cases:

Case-1: If $m = 5$ then define

$$f(v_{i,j}) = f(v_{i,j-1}) + 3 \,(\bmod\, m+1)) + 1 \quad \text{for } 1 \leq j \leq 2,\ 1 \leq i \leq m;$$
$$f(v_{i,j}) = f(v_i, 0) \quad \text{for } j \equiv 0 \,(\bmod\, 3),\ 3 \leq j \leq n-3,\ 1 \leq i \leq m;$$
$$f(v_{i,j}) = f(v_i, 1) \quad \text{for } j \equiv 1 \,(\bmod\, 3),\ 3 \leq j \leq n-3,\ 1 \leq i \leq m;$$
$$f(v_{i,j}) = f(v_i, 2) \quad \text{for } j \equiv 2 \,(\bmod\, 3),\ 3 \leq j \leq n-3,\ 1 \leq i \leq m;$$
$$f(v_{i,n-1}) = f(v_{i,n-2}) + 3 \,(\bmod\, m+1)) + 1,\ 1 \leq i \leq m;$$
$$f(v_{i,n-2}) = f(v_{i,n-3}) + 3 \,(\bmod\, m+1)) + 1,\ 1 \leq i \leq m.$$

Case-2: If $m = 6$ then define

$$f(v_{i,j}) = f(v_{i,j-1}) + 1 \,(\bmod\, m+1)) + 1 \quad \text{for } 1 \leq j \leq 2,\ 1 \leq i \leq m;$$

$$\begin{aligned}
f(v_{i,j}) &= f(v_i, 0) \quad \text{for } j \equiv 0 \pmod 3,\ 3 \le j \le n-3,\ 1 \le i \le m;\\
f(v_{i,j}) &= f(v_i, 1) \quad \text{for } j \equiv 1 \pmod 3,\ 3 \le j \le n-3,\ 1 \le i \le m;\\
f(v_{i,j}) &= f(v_i, 2) \quad \text{for } j \equiv 2 \pmod 3,\ 3 \le j \le n-3,\ 1 \le i \le m;\\
f(v_{i,n-2}) &= f(v_{i,n-1}) + 2 \ (\text{mod}\, m+1)) + 1,\ 1 \le i \le m;\\
f(v_{i,n-1}) &= f(v_{i,n-2}) + 4 \ (\text{mod}\, m+1)) + 1,\ 1 \le i \le m.
\end{aligned}$$

Case-3: If $m = 7$ then define

$$\begin{aligned}
f(v_{i,j}) &= f(v_{i,j-1}) + 1 \ (\text{mod}\, m+1)) + 1 \quad \text{for } 1 \le j \le 2,\ 1 \le i \le m;\\
f(v_{i,j}) &= f(v_i, 0) \quad \text{for } j \equiv 0 \pmod 3,\ 3 \le j \le n-3,\ 1 \le i \le m;\\
f(v_{i,j}) &= f(v_i, 1) \quad \text{for } j \equiv 1 \pmod 3,\ 3 \le j \le n-3,\ 1 \le i \le m;\\
f(v_{i,j}) &= f(v_i, 2) \quad \text{for } j \equiv 2 \pmod 3,\ 3 \le j \le n-3,\ 1 \le i \le m;\\
f(v_{i,n-2}) &= f(v_{i,n-1}) + 3 \ (\text{mod}\, m+1)) + 1,\ 1 \le i \le m;\\
f(v_{i,n-1}) &= f(v_{i,n-2}) + 4 \ (\text{mod}\, m+1)) + 1,\ 1 \le i \le m.
\end{aligned}$$

Case-4: If $m \ge 8$ then define

$$\begin{aligned}
f(v_{i,j}) &= f(v_{i,j-1}) + 1 \ (\text{mod}\, m+1)) + 1 \quad \text{for } 1 \le j \le 2,\ 1 \le i \le m;\\
f(v_{i,j}) &= f(v_i, 0) \quad \text{for } j \equiv 0 \pmod 3,\ 3 \le j \le n-3,\ 1 \le i \le m;\\
f(v_{i,j}) &= f(v_i, 1) \quad \text{for } j \equiv 1 \pmod 3,\ 3 \le j \le n-3,\ 1 \le i \le m;\\
f(v_{i,j}) &= f(v_i, 2) \quad \text{for } j \equiv 2 \pmod 3,\ 3 \le j \le n-3,\ 1 \le i \le m;\\
f(v_{i,n-2}) &= f(v_{i,n-1}) + 1 \ (\text{mod}\, m+1)) + 1,\ 1 \le i \le m;\\
f(v_{i,n-1}) &= f(v_{i,n-2}) + 1 \ (\text{mod}\, m+1)) + 1,\ 1 \le i \le m.
\end{aligned}$$

Hence $\lambda(J_{n,m}) \le m+1$ for $m \ge 5$ and $n \equiv 2 \pmod 3$. □

Illustration 4.3.3. The $L(2,1)$ - labeling of $J_{8,6}$ is shown in Figure 4.4.

Theorem 4.4

The $L(2,1)$ - labeling number of $J_{n,m}$ for $m \ge 5$ and $n \ge 3$ is $m+1$. ■

Proof. Define $f : V(G) \to N \cup \{0\}$ such that $f(v_0) = 0$. Since $d(v_0, v_{i,0}) = 1$ for $1 \le i \le m$, in order to preserve $L(2,1)$ - labeling, $f(v_{1,0}) \ge 2$ is assumed. Now according to the definition of $L(2,1)$ - labeling, if the distance between any two vertices is 2 then the label must differ by at least 1. Since $f(v_{1,0}) \ge 2$ there are $m-1$ remaining vertices $v_{2,0}, v_{3,0}, \ldots, v_{m,0}$ which have to be labeled. By labeling the remiaing $m-1$ vertices consecutively, one of the vertices will get a label greater than or equal to $(m-1)+2$. Thus $\lambda(J_{n,m}) \ge m+1$ for $m \ge 5$ and $n \ge 3$. Now, from the Theorems 4.1 to 4.3, it can be concluded that $\lambda(J_{n,m}) \le m+1$ for $m \ge 5$ and $n \ge 3$. Hence $\lambda(J_{n,m}) = m+1$ for $m \ge 5$ and $n \ge 3$. □

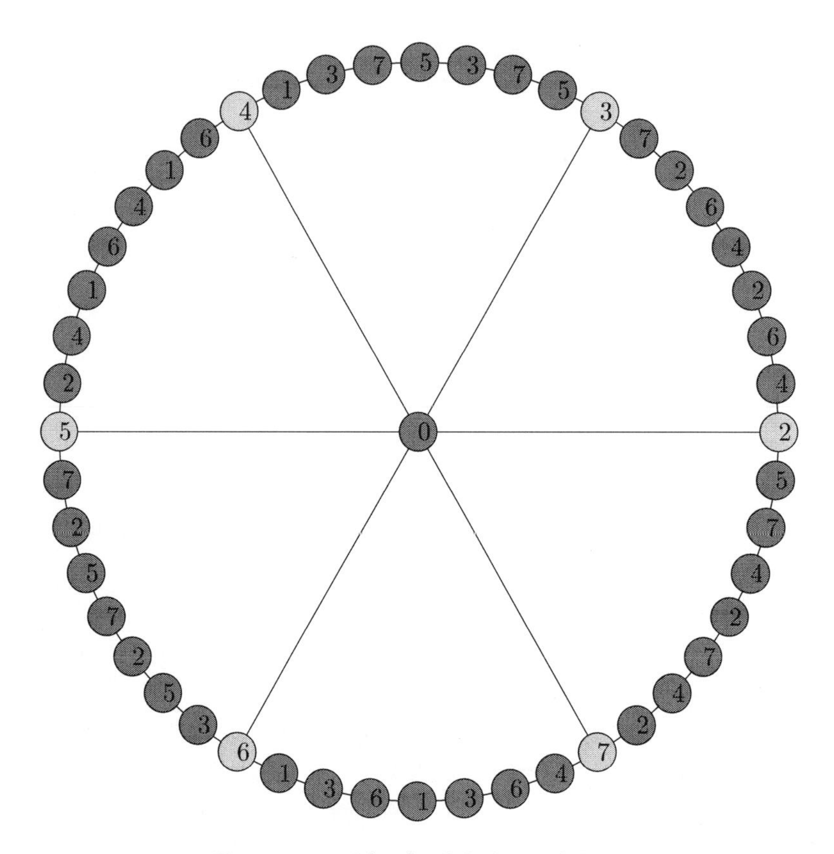

Figure 4.4: $L(2, 1)$ - labeling of $J_{8,6}$

4.4 CONCLUSION

Exact $L(2, 1)$ - labeling number for a generalized Jahangir graph $J_{n,m}$ is computed for $m \geq 5$ and $n \geq 3$. A lots of work has been done in this area and still work is being carried out for different graphs. Determination of an exact $L(2, 1)$-labeling number for other graph families is an open area of research.

REFERENCES

1. W. K. Hale. Frequency assignment: Theory and applications. *Proc. IEEE*, 68(12): 1497-1514, 1980.
2. J. R. Griggs and R. K. Yeh. Labeling graphs with condition at distance two. *SIAM J. Discrete Math.*, 5(4): 586-595, 1992.
3. J. A. Bondy and U. S. R. Murty. *Graph Theory with Applications* (2nd Edition). McMillan, New York, 1976.
4. J. A. Gallian. A dynamic survey of graph labeling. *The Electronic Journal of Combinatories*, #DS6, 1-518, 2019.

5. G. J. Chang and D. Kuo. The L(2,1)-labeling problem on graphs. *SIAM J. Discrete Math.*, 9(2): 309-316, 1996.
6. D. Kuo, and J. Yan. On L(2,1) - labelings of cartesian products of paths and cycles. *Discrete Math.* 283, 137-144, 2004.
7. S. K. Vaidya and D. D. Bantva. Labeling cacti with a condition at distance two. *Le Matematiche*, 66(1): 29-36, 2011.
8. Baby Smitha, K. M. and K. Thirusangu. Distance two labeling of certain snake graphs. *International Mathematical Forum*, 11(11): 503-512, 2016.
9. K. Ali, E. T. Baskoro and Tomescu I. On the Ramsey numbers for paths and generalized Jahangir graphs $J_{s,m}$. *Bull. Math. Soc. Sci. Math.*, 51: 177-182, 2008.
10. U. M. Prajapati and N. B. Patel. $L(2,1)$ labeling of line graph of some graphs. *Journal of Applied Science and Computations*, 6(5): 309-316, 2019.

5 V_4-Cordial Labeling of Some Ladder and Book related Graphs

Neha B. Rathod
Government Engineering College,
Bhavnagar, Gujarat (INDIA)
E-mail: rathodneha005@gmail.com

Kailas K. Kanani
Government Engineering College,
Rajkot, Gujarat (INDIA)
E-mail: kananikkk@yahoo.co.in

A-cordial labeling was introduced by Mark Hovey. He introduced A -cordial labeling for an abelian group as a simultaneous generalization of cordial and harmonious labeling. If $A = V_4$, it is known as V_4 -cordial labeling. Let V_4 be the Klein-four group. In this chapter we prove some ladder and book related graphs which admit V_4 -cordial labeling. We prove that Book Graph $B(5, n)$, Mobius Ladder M_n Open Ladder $O(L_n)$ and Mongolian Tent $MT(2, n)$ are V_4 -cordial graphs.

5.1 INTRODUCTION

A-cordial labeling was introduced by Mark Hovey[2]. He introduced A-cordial labeling for an abelian group as a simultaneous generalization of cordial and harmonious labeling. If $A = V_4$, it is known as V_4-cordial labeling. Let V_4 be the Klein-four group. In this work we discuss V_4-cordial labeling of some ladder and book related graphs. We prove that Book Graph $B(5, n)$, Möbius Ladder M_n, Open Ladder $O(L_n)$ and Mongolian Tent $MT(2, n)$ are V_4-cordial graphs. In this chapter, we consider finite, connected, undirected and simple graph $G = (V(G), E(G))$. We denote $|V(G)|$ = total number of vertices of graph G and $|E(G)|$ = total number of edges of graph G. A *graph labeling* is an assignment of numbers to the vertices or edges or both subject to certain condition(s). To understand more about graph labeling as well as bibliographic references we refer to Gallian[1].

5.2 BASIC DEFINITIONS

Let $V_4 = Z_2 \times Z_2 = \{0 =< 0,0 >, a =< 1,0 >, b =< 0,1 >, c =< 1,1 >\}$ be the Klein-four group in which the operation $*$ is defined as follows:

$*$	0	a	b	c
0	0	a	b	c
a	a	0	c	b
b	b	c	0	a
c	c	b	a	0

Definition. The graph $G = (V(G), E(G))$, with vertex set $V(G)$ and edge set $E(G)$, is said to be **V_4-cordial** if there exists a mapping $f : V(G) \to V_4$ which satisfies the following two conditions when the edge $e = uv$ is labelled as $f(u) * f(v)$

1. $|v_f(p) - v_f(q)| \leq 1$,
2. $|e_f(p) - e_f(q)| \leq 1$,

for all $p, q \in V_4$. Where
$v_f(p)$=the number of vertices with label p; $v_f(q)$=the number of vertices with label q;
$e_f(p)$=the number of edges with label p; $e_f(q)$=the number of edges with label q.

Definition. The *Book B(5,n)* with n-pages is defined as n copies of cycle C_5 sharing a common edge. The common edge is called the *spine or base* of the book.

Definition. A *Möbius Ladder graph* M_n is a graph obtained from the ladder $L_n = P_n \times P_2 (n \geq 2)$by joining the opposite end points of the two copies of P_n.

Definition. An *Open Ladder* $O(L_n)$, $n > 2$ is obtained from two paths of length $n-1$ with $V(G) = \{v_i, v_i' : 1 \leq i \leq n-1\}$ and $E(G) = \{v_i v_{i+1}, v_i' v_{i+1}' : 1 \leq i \leq n-1\} \cup \{v_i v_i' : 2 \leq i \leq n-1\}$.

Definition. A Mongolian Tent $MT(m,n)$ is defined as the graph obtained from $P_m \times P_n$ by adding a new vertex above the grid and joining every vertex of the top row to the new vertex. If we take $m = 2$ and $n \geq 2$ then it is Mongolian Tent $MT(2,n)$.

Definition. A graph is 1-factorable if the edges of the graph can be partitioned into disjoint perfect matchings.

5.3 EXISTING RESULTS

The concept of V_4-cordial labeling was introduced by Adrian Riskin[5] and proved the following results:

1. A complete graph K_n is V_4-cordial if and only if $n < 4$.
2. The n-star $K_{1,n}$ is V_4-cordial.

Seenivasan M and Lourdusamy A[6] proved the following results:

1. If G is an Eulerian graph with q edges, where $q \equiv 2(mod\ 4)$, then G has no V_4-cordial labeling.
2. If f is a V_4-cordial labeling of a graph G with $p \geq 4$ and uv is an edge of G such that $f(u) = 0$ and $f(u) \neq f(v)$, then the graph G' obtained from G by replacing the edge uv by a path of length five is V_4-cordial.
3. The paths P_4 and P_5 are not V_4-cordial.
4. All trees except P_4 and P_5 are V_4-cordial.
5. The cycle C_n is V_4-cordial if and only if $n \neq 4$ or 5 or $n \not\equiv 2(mod 4)$.

Oliver Pechenik and Jennifer Wise[3] proved the following results:

1. The complete bipartite graph $K_{m,n}$ is V_4-cordial if and only if m and n are not both congruent to $2(mod\ 4)$.
2. The path P_n is V_4-cordial unless $n \in \{4,5\}$.
3. The cycle C_n is V_4-cordial if and only if $n \notin \{4,5\}$ and $n \not\equiv 2(mod\ 4)$.
4. All ladders $P_2 \times P_k$ are V_4-cordial, except $P_2 \times P_2$.
5. The prism $P_2 \times C_k$ is V_4-cordial if and only if $k \not\equiv 2(mod\ 4)$.
6. The d-dimensional hypercube Q_d is V_4-cordial, unless $d = 2$.

Rathod and Kanani[4] proved the following results:

1. The crown graph $C_n \odot K_1$, armed crown AC_n, pan graph C_n^{+1} and corona graph $C_n \odot mK_1$ are V_4-cordial for all n.

5.4 MAIN RESULTS

Theorem 5.1

The Book graph $B(5,n)$ is V_4-cordial for all n. ■

Proof. Let $G = B(5,n)$ be the book graph. Let $v_1, v_1', v_1'', v_2, v_2', v_2'', \ldots, v_n, v_n', v_n''$ be the vertices of 1st page, 2nd page,...,n^{th} page respectively. Let u and v be the spine vertices. We note that $|V(G)| = 3n+2$ and $|E(G)| = 4n+1$.

Define V_4-cordial labeling $f : V(G) \to V_4$ as follows:

$f(u) = 0$ and $f(v) = c$,
$f(v_1) = a,\quad f(v_1') = b,\quad f(v_1'') = c,$
$f(v_2) = 0,\quad f(v_2') = b,\quad f(v_2'') = a.$

$f(v_i) = 0\ ;\quad i \equiv 1,3(mod\ 4),$
$f(v_i) = b\ ;\quad i \equiv 0,2(mod\ 4),\quad 3 \leq i \leq n.$

$f(v_i') = a\ ;\quad i \equiv 0,1,3(mod\ 4),$
$f(v_i') = c\ ;\quad i \equiv 2(mod\ 4),\quad 3 \leq i \leq n.$

$f(v_i'') = 0\ ;\quad i \equiv 2(mod\ 4),$
$f(v_i'') = b\ ;\quad i \equiv 0(mod\ 4),$
$f(v_i'') = c\ ;\quad i \equiv 1,3(mod\ 4),\quad 3 \leq i \leq n.$

In each possibility the graph under consideration satisfies the vertex conditions and edge conditions for V_4-cordial labeling is shown in Table 5.1. Hence, Book graph $B(5,n)$ is V_4-cordial for all n. □

Let $n = 4a + b$, where $a, b \in N \cup \{0\}$.

Table 5.1
V_4**-cordial Labeling of Book Graph** $B(5,n)$

b	Vertex conditions	Edge conditions
0	$v_f(0)+1 = v_f(a) = v_f(b) = v_f(c)+1$	$e_f(0)+1 = e_f(a)+1 = e_f(b)+1 = e_f(c)$
1	$v_f(0)+1 = v_f(a) = v_f(b)+1 = v_f(c)+1$	$e_f(0) = e_f(a)+1 = e_f(b)+1 = e_f(c)+1$
2	$v_f(0) = v_f(a) = v_f(b) = v_f(c)$	$e_f(0)+1 = e_f(a)+1 = e_f(b)+1 = e_f(c)$
3	$v_f(0) = v_f(a) = v_f(b)+1 = v_f(c)$	$e_f(0) = e_f(a)+1 = e_f(b)+1 = e_f(c)+1$

Illustration 5.4.1. The Book Graph $B(5,5)$ and its V_4-cordial labeling is shown in Figure 5.1.

Theorem 5.2

The Möbius Ladder M_n is V_4-cordial for all n except $n \equiv 2(mod\ 4)$. ■

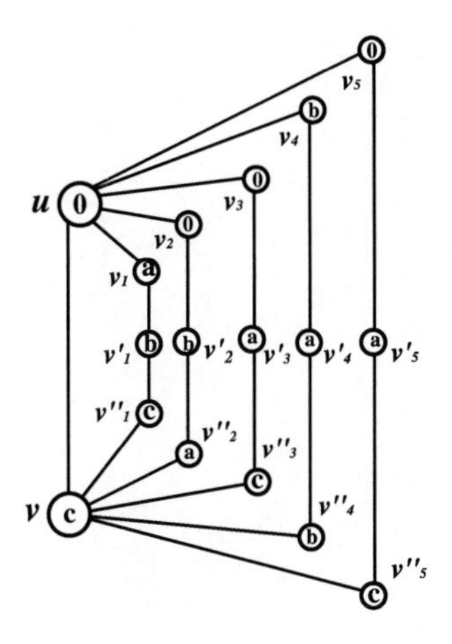

Figure 5.1: V_4-cordial labeling of Book Graph $B(5,5)$

Proof. Let $G = M_n$ be the Möbius Ladder obtained from the ladder $P_n \times P_2$ by joining the opposite end points of the two copies of P_n, where $v_1, v_2, v_3, \ldots, v_n$ are the vertices of the first path and $v_1', v_2', v_3', \ldots, v_n'$ are the vertices of the second path. We note that $|V(G)| = 2n$ and $|E(G)| = 3n$.

To define V_4-cordial labeling $f : V(G) \to V_4$ we consider the following cases:

Case 1: $n \equiv 0(mod\ 4)$.
$f(v_i) = 0$; $i \equiv 1(mod\ 4)$,
$f(v_i) = a$; $i \equiv 2,3(mod\ 4)$,
$f(v_i) = b$; $i \equiv 0(mod\ 4)$, $1 \le i \le n$.

$f(v_i') = 0$; $i \equiv 1(mod\ 4)$,
$f(v_i') = b$; $i \equiv 2(mod\ 4)$,
$f(v_i') = c$; $i \equiv 0,3(mod\ 4)$, $1 \le i \le n$.

Case 2: $n \equiv 1(mod\ 4)$.
$f(v_i) = 0$; $i \equiv 1(mod\ 4)$,
$f(v_i) = a$; $i \equiv 2,3(mod\ 4)$,
$f(v_i) = b$; $i \equiv 0(mod\ 4)$, $1 \le i \le n-5$,
$f(v_{n-4}) = 0$,
$f(v_{n-3}) = c$,
$f(v_{n-2}) = a$,
$f(v_{n-1}) = a$,
$f(v_n) = a$.

$f(v'_i) = 0$; $i \equiv 1 (mod\ 4)$,
$f(v'_i) = b$; $i \equiv 2 (mod\ 4)$,
$f(v'_i) = c$; $i \equiv 0,3 (mod\ 4)$, $1 \leq i \leq n-5$,
$f(v'_{n-4}) = 0$,
$f(v'_{n-3}) = b$,
$f(v'_{n-2}) = c$,
$f(v'_{n-1}) = b$,
$f(v'_n) = b$.

Case 3: $n \equiv 2 (mod\ 4)$.
Here, G is a 1-factorable graph with kN vertices and $lN \pm 2$ edges, where $k, l \in N$. Then G is not A-cordial[3].

Case 4: $n \equiv 3 (mod\ 4)$.
$f(v_i) = 0$; $i \equiv 1 (mod\ 4)$,
$f(v_i) = a$; $i \equiv 2,3 (mod\ 4)$,
$f(v_i) = b$; $i \equiv 0 (mod\ 4)$, $1 \leq i \leq n-3$,
$f(v_{n-2}) = 0$,
$f(v_{n-1}) = a$,
$f(v_n) = b$.

$f(v'_i) = 0$; $i \equiv 1 (mod\ 4)$,
$f(v'_i) = b$; $i \equiv 2 (mod\ 4)$,
$f(v'_i) = c$; $i \equiv 0,3 (mod\ 4)$, $1 \leq i \leq n-3$,
$f(v'_{n-2}) = 0$,
$f(v'_{n-1}) = c$,
$f(v'_n) = b$.

In each possibility the graph under consideration satisfies the vertex conditions and edge conditions for V_4-cordial labeling is shown in Table 5.2. Hence, Möbius Ladder M_n is V_4-cordial for all n except $n \equiv 2 (mod\ 4)$.

□

Let $n = 4a + b$, where $a, b \in N \cup \{0\}$.

Table 5.2
V_4-cordial Labeling of Möbius Ladder M_n

b	Vertex conditions	Edge conditions
0	$v_f(0) = v_f(a) = v_f(b) = v_f(c)$	$e_f(0) = e_f(a) = e_f(b) = e_f(c)$
1	$v_f(0) + 1 = v_f(a) = v_f(b) = v_f(c) + 1$	$e_f(0) = e_f(a) = e_f(b) = e_f(c) + 1$
3	$v_f(0) = v_f(a) + 1 = v_f(b) = v_f(c) + 1$	$e_f(0) + 1 = e_f(a) + 1 = e_f(b) = e_f(c) + 1$

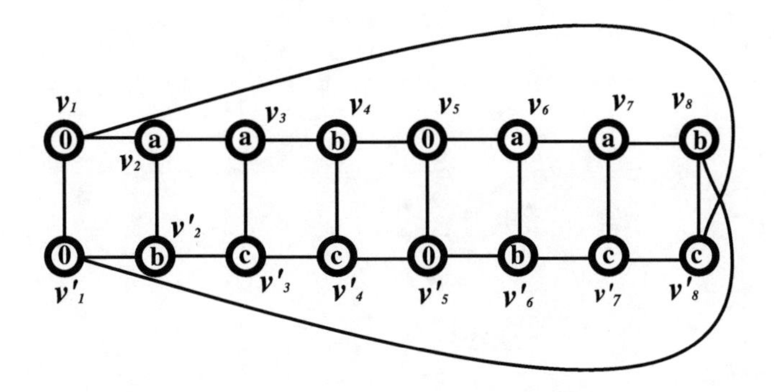

Figure 5.2: V_4-cordial labeling of Möbius Ladder M_8

Illustration 5.4.2. The Möbius Ladder M_8 and its V_4-cordial labeling is shown in Figure 5.2.

Theorem 5.3

The Open Ladder $O(L_n)$ is V_4-cordial for all n except $n \equiv 2(mod\ 4)$. ■

Proof. Let $G = O(L_n)$ be the Open Ladder obtained from two paths of length $n-1$ with $V(G) = \{v_i, v_i' : 1 \leq i \leq n-1\}$ and $E(G) = \{v_iv_{i+1}, v_i'v_{i+1}' : 1 \leq i \leq n-1\} \cup \{v_iv_i' : 2 \leq i \leq n-1\}$. We note that $|V(G)| = 2n$ and $|E(G)| = 3n-4$. To define V_4-cordial labeling $f : V(G) \to V_4$ we consider the following cases:

Case 1: $n \equiv 0(mod\ 4)$.
$f(v_i) = 0$; $i \equiv 1(mod\ 4)$,
$f(v_i) = a$; $i \equiv 2, 3(mod\ 4)$,
$f(v_i) = b$; $i \equiv 0(mod\ 4)$, $1 \leq i \leq n$.

$f(v_i') = 0$; $i \equiv 1(mod\ 4)$,
$f(v_i') = b$; $i \equiv 2(mod\ 4)$,
$f(v_i') = c$; $i \equiv 0, 3(mod\ 4)$, $1 \leq i \leq n$.

Case 2: $n \equiv 1(mod\ 4)$.
$f(v_i) = 0$; $i \equiv 1(mod\ 4)$,
$f(v_i) = a$; $i \equiv 2, 3(mod\ 4)$,
$f(v_i) = b$; $i \equiv 0(mod\ 4)$, $1 \leq i \leq n-5$,
$f(v_{n-4}) = 0$,
$f(v_{n-3}) = c$,

$f(v_{n-2}) = a,$
$f(v_{n-1}) = a,$
$f(v_n) = a.$

$f(v'_i) = 0\ ;\qquad i \equiv 1(mod\ 4),$
$f(v'_i) = b\ ;\qquad i \equiv 2(mod\ 4),$
$f(v'_i) = c\ ;\qquad i \equiv 0,3(mod\ 4),\qquad 1 \leq i \leq n-5,$
$f(v'_{n-4}) = 0,$
$f(v'_{n-3}) = b,$
$f(v'_{n-2}) = c,$
$f(v'_{n-1}) = b,$
$f(v'_n) = b.$

Case 3: $n \equiv 2(mod\ 4)$.
Here, G be a 1-factorable graph with kN vertices and $lN \pm 2$ edges, where $k, l \in N$. Then G is not A-cordial[3].

Case 4: $n \equiv 3(mod\ 4)$.
$f(v_i) = 0\ ;\qquad i \equiv 1(mod\ 4),$
$f(v_i) = a\ ;\qquad i \equiv 2,3(mod\ 4),$
$f(v_i) = b\ ;\qquad i \equiv 0(mod\ 4),\qquad 1 \leq i \leq n-3,$
$f(v_{n-2}) = 0,$
$f(v_{n-1}) = a,$
$f(v_n) = b.$

$f(v'_i) = 0\ ;\qquad i \equiv 1(mod\ 4),$
$f(v'_i) = b\ ;\qquad i \equiv 2(mod\ 4),$
$f(v'_i) = c\ ;\qquad i \equiv 0,3(mod\ 4),\qquad 1 \leq i \leq n-3,$
$f(v'_{n-2}) = 0,$
$f(v'_{n-1}) = c,$
$f(v'_n) = b.$

In each possibility the graph under consideration satisfies the vertex conditions and edge conditions for V_4-cordial labeling is shown in Table 5.3. Hence, Open Ladder $O(L_n)$ is V_4-cordial for all n except $n \equiv 2(mod\ 4)$.

□

Let $n = 4a + b$, where $a, b \in N \cup \{0\}$.

Illustration 5.4.3. The Open Ladder $O(L_5)$ and its V_4-cordial labeling is shown in Figure 5.3.

Theorem 5.4

The Mongolian Tent $MT(2, n)$ is V_4-cordial for all n. ■

Table 5.3
V_4-cordial Labeling of Open Ladder $O(L_n)$

b	Vertex conditions	Edge conditions
0	$v_f(0) = v_f(a) = v_f(b) = v_f(c)$	$e_f(0) = e_f(a) = e_f(b) = e_f(c)$
1	$v_f(0)+1 = v_f(a) = v_f(b) = v_f(c)+1$	$e_f(0) = e_f(a) = e_f(b) = e_f(c)+1$
3	$v_f(0) = v_f(a)+1 = v_f(b)+1 = v_f(c)$	$e_f(0)+1 = e_f(a)+1 = e_f(b)+1 = e_f(c)$

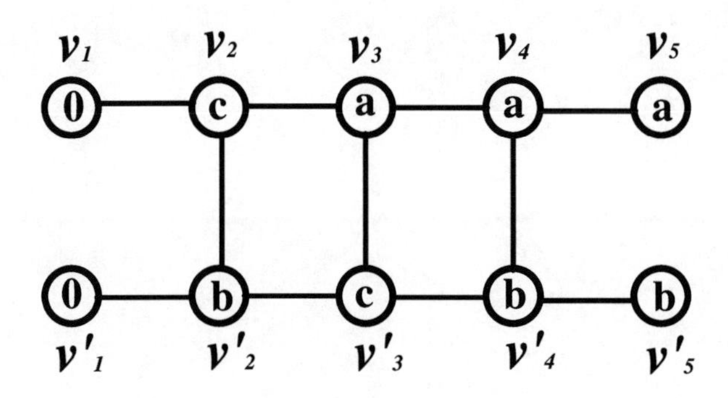

Figure 5.3: V_4-cordial labeling of Open Ladder $O(L_5)$

Proof. Let $G = MT(2,n)$ be the Mongolian Tent obtained from ladder $P_2 \times P_n$ by adding a new vertex above the ladder and joining every vertex of the first path to the new vertex u. Let $v_1, v_2, v_3, \ldots, v_n$ be the vertices of the first path and $v'_1, v'_2, v'_3, \ldots, v'_n$ be the vertices of the second path. We note that $|V(G)| = 2n+1$ and $|E(G)| = 4n-2$.

To define V_4-cordial labeling $f : V(G) \to V_4$ we consider the following cases:

Case 1: $n = 2, 3, 4, 5$.
$f(u) = 0.$
$f(v_1) = a,$ $\quad f(v'_1) = a,$
$f(v_2) = b,$ $\quad f(v'_2) = c,$
$f(v_3) = 0,$ $\quad f(v'_3) = c,$
$f(v_4) = c,$ $\quad f(v'_4) = b,$
$f(v_5) = 0,$ $\quad f(v'_5) = b.$

Case 2: $n = 6$.
$f(u) = 0.$
$f(v_1) = 0,$ $\quad f(v'_1) = b,$

$f(v_2) = a,$ $\quad f(v_2') = a,$
$f(v_3) = b,$ $\quad f(v_3') = c,$
$f(v_4) = 0,$ $\quad f(v_4') = c,$
$f(v_5) = c,$ $\quad f(v_5') = b,$
$f(v_6) = a,$ $\quad f(v_6') = b.$

Case 3: $n = 7$.
$f(u) = 0.$
$f(v_1) = a,$ $\quad f(v_1') = a,$
$f(v_2) = 0,$ $\quad f(v_2') = b,$
$f(v_3) = c,$ $\quad f(v_3') = c,$
$f(v_4) = 0,$ $\quad f(v_4') = a,$
$f(v_5) = b,$ $\quad f(v_5') = b,$
$f(v_6) = 0,$ $\quad f(v_6') = c,$
$f(v_7) = a,$ $\quad f(v_7') = c.$

Case 4: $n \geq 8$.
$f(u) = 0.$
$f(v_1) = 0,$ $\quad f(v_1') = b,$
$f(v_2) = a,$ $\quad f(v_2') = a,$
$f(v_3) = b,$ $\quad f(v_3') = c,$
$f(v_4) = c,$ $\quad f(v_4') = c,$
$f(v_5) = 0,$ $\quad f(v_5') = b,$
$f(v_6) = a,$ $\quad f(v_6') = a,$
$f(v_7) = b,$ $\quad f(v_7') = c,$
$f(v_8) = 0,$ $\quad f(v_8') = 0,$
$f(v_9) = b,$ $\quad f(v_9') = c,$

$f(v_i) = 0\ ;$ $\quad i \equiv 1,3,5(mod\ 6),$
$f(v_i) = a\ ;$ $\quad i \equiv 4(mod\ 6),$
$f(v_i) = b\ ;$ $\quad i \equiv 2(mod\ 6),$
$f(v_i) = c\ ;$ $\quad i \equiv 0(mod\ 6),$ $\quad 10 \leq i \leq n.$

$f(v_i') = a\ ;$ $\quad i \equiv 1,4(mod\ 4),$
$f(v_i') = b\ ;$ $\quad i \equiv 2,5(mod\ 4),$
$f(v_i') = c\ ;$ $\quad i \equiv 0,3(mod\ 4),$ $\quad 10 \leq i \leq n.$

In each possibility the graph under consideration satisfies the vertex conditions and edge conditions for V_4-cordial labeling is shown in Table 5.4. Hence, Mongolian Tent $MT(2,n)$ is V_4-cordial for all n.

□

Let $n = 6a + b$, where $a, b \in N \cup \{0\}$ and $n \geq 8$.

Illustration 5.4.4. The Mongolian Tent $MT(2,6)$ and its V_4-cordial labeling is shown in Figure 5.4.

Table 5.4
V_4-cordial Labeling of Mongolian Tent $MT(2,n)$

b	Vertex conditions	Edge conditions
0	$v_f(0)+1 = v_f(a)+1 = v_f(b)+1 = v_f(c)$	$e_f(0)+1 = e_f(a) = e_f(b)+1 = e_f(c)$
1	$v_f(0) = v_f(a) = v_f(b)+1 = v_f(c)$	$e_f(0)+1 = e_f(a) = e_f(b)+1 = e_f(c)$
2	$v_f(0)+1 = v_f(a)+1 = v_f(b) = v_f(c)+1$	$e_f(0)+1 = e_f(a)+1 = e_f(b) = e_f(c)$
3	$v_f(0) = v_f(a)+1 = v_f(b) = v_f(c)$	$e_f(0)+1 = e_f(a)+1 = e_f(b) = e_f(c)$
4	$v_f(0)+1 = v_f(a) = v_f(b)+1 = v_f(c)+1$	$e_f(0)+1 = e_f(a) = e_f(b) = e_f(c)+1$
5	$v_f(0) = v_f(a) = v_f(b) = v_f(c)+1$	$e_f(0)+1 = e_f(a) = e_f(b) = e_f(c)+1$

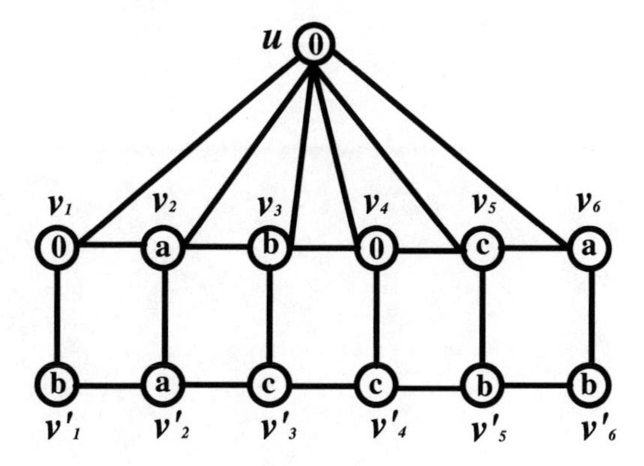

Figure 5.4: V_4-cordial labeling of Mongolian Tent $MT(2,6)$

5.5 CONCLUDING REMARKS

Graph labeling technique is a wide area of research. To investigate more graph families which admit V_4-cordial labeling is an open area of research.

REFERENCES

1. J A Gallian. A dynamic survey of graph labeling. *The Electronics Journal of Combinatorics*, 2019.
2. M Hovey. On V_4-cordial labeling of graphs. *Discrete Mathematics*, 93, 183-194, 1991.
3. O Pechenik and J Wise. Generalized graph cordiality. *Discussiones Mathematicae Graph Theory*, 32(3): 557-567, 2012.
4. N B Rathod and K K Kanani. On V_4-cordial labeling of graphs. *British Journal of Mathematics and Computer Science*, 13(4): 1-15, 2015.
5. A Riskin. Z_2^2-cordiality of K_n and $K_{m,n}$. *arXiv*:0709.0290, 2013.
6. M Seenivasan and A Lourdusamy. Some V_4-cordial graphs. *Sciencia Acta Xaveriana*, 1(1): 91-99, 2009.

6 SD-Prime Cordial Labeling of Double k-Polygonal Snake Graph

U. M. Prajapati
Department of Mathematics,
St. Xavier's College,
Ahmedabad, Gujarat (INDIA)
E-mail: udayan64@yahoo.com

A. V. Vantiya
Department of Mathematics,
Gujarat University,
Ahmedabad, Gujarat (INDIA)
E-mail: avantiya@yahoo.co.in

Let $f : V(G) \to \{1, 2, \ldots, |V(G)|\}$ be a bijection, and let us denote $S = f(u) + f(v)$ and $D = |f(u) - f(v)|$ for every edge uv in $E(G)$. Let f' be the induced edge labeling, induced by the vertex labeling f, defined as $f' : E(G) \to \{0, 1\}$ such that for any edge uv in $E(G), f'(uv) = 1$ if $\gcd(S, D) = 1$, and $f'(uv) = 0$ otherwise. Let $e_{f'}(0)$ and $e_{f'}(1)$ be the number of edges labeled with 0 and 1 respectively. f is SD-prime cordial labeling if $|e_{f'}(0) - e_{f'}(1)| \leq 1$ and G is an SD-prime cordial graph if it admits SD-prime cordial labeling. In this chapter, we discuss the SD-prime cordial labeling of double $k-$polygonal snake graph.

6.1 INTRODUCTION

Let $G = (V(G), E(G))$ be a finite, simple and undirected graph of order $|V(G)| = p$ and size $|E(G)| = q$. For standard terminology of Graph Theory, we used [1]. For all detailed surveys of graph labeling we refer to [2]. Lau, Chu, Suhadak, Foo, and Ng [3] have introduced SD-prime cordial labeling and they proved that path, complete bipartite graph, star, double star, wheel, fan, double fan, ladder etc. are SD-prime cordial graphs. Lourdusamy and Patrick [4] proved that $S'(K_{1,n}), D_2(K_{1,n}), S(K_{1,n}), DS(K_{1,n}), S'(B_{n,n})$, $D_2(B_{n,n}), TL_n, DS(B_{n,n}), S(B_{n,n}), K_{1,3} \star K_{1,n}, CH_n, Fl_n, P_n^2, T(P_n), T(C_n)$, $Q_n, A(T_n), P_n \odot K_1, C_n \odot K_1, J_n$ and the graph obtained by duplication of each vertex and cycle by an edge are SD-prime cordial. Lourdusamy, Wency and Patrick [5] proved that the union of star and path graphs, subdivision of comb graph, subdivision of ladder graph and the graph obtained by attaching a star graph at one end of the path are SD-prime cordial graphs. They proved that the union of two SD-prime cordial graphs need not be an SD-prime cordial graph. Also, they proved that given a positive integer n, there is an SD-prime cordial graph G with n vertices. Prajapati and Vantiya [7] proved that $T_n(n \neq 3), A(T_n), Q_n, A(Q_n), DT_n, DA(T_n), DQ_n$ and $DA(Q_n)$ are SD-prime cordial. Prajapati and Vantiya [8], [9] proved that $S(T_n)$, $S(A(T_n))$, $S(Q_n)$, $S(A(Q_n))$, $S(DT_n)$, $S(DA(T_n))$, $S(DQ_n)$ and $S(DA(Q_n))$ are SD-prime cordial. In this chapter, we investigate the SD-prime cordial labeling behavior of a double k-polygonal snake graph $D(S_nC_k)$.

Definition. If the vertices or edges or both of a graph are assigned values subject to certain conditions then it is known as *vertex or edge or total labeling* respectively.

Definition. [3] A bijection $f : V(G) \to \{1,2,\ldots,|V(G)|\}$ induces an edge labeling $f' : E(G) \to \{0,1\}$ such that for any edge uv in G, $f'(uv) = 1$ if $\gcd(S,D) = 1$, and $f'(uv) = 0$ otherwise, where $S = f(u) + f(v)$ and $D = |f(u) - f(v)|$ for every edge uv in $E(G)$. Let $e_{f'}(0)$ and $e_{f'}(1)$ be the number of edges labeled with 0 and 1 respectively. The labeling f is called *SD-prime cordial labeling* if $|e_{f'}(0) - e_{f'}(1)| \leq 1$. G is called an *SD-prime cordial graph* if it admits SD-prime cordial labeling.

Definition. [2] A *triangular snake* T_n is obtained from the path P_n by replacing every edge of a path by a triangle C_3. That is, it is obtained from a path $u_1,u_2,\ldots,u_n$ by joining u_i and u_{i+1} to a new vertex w_i for $i = 1,2,\ldots,n-1$.

Definition. [2] A *quadrilateral snake* Q_n is obtained from the path P_n by replacing every edge of a path by a cycle C_4, such that u_i, u_{1+1} remains adjacent for each $i = 1,2,\ldots,n-1$. That is, it is obtained from a path $u_1,u_2,\ldots,u_n$ by joining u_i and u_{i+1} to new vertices v_i and w_i respectively, and then joining v_i and w_i by an edge, for $i = 1,2,\ldots,n-1$.

Definition. [2] A *double triangular* DT_n snake is a graph formed by two triangular snakes having a common path. That is, it is obtained from a path $u_1,u_2,\ldots,u_n$ by joining u_i and u_{i+1} to two new vertices w_i and v_i for $i = 1,2,\ldots,n-1$.

Definition. [2] A *double quadrilateral snake* DQ_n is a graph formed by two quadrilateral snakes having a common path. That is, it is obtained from a path $u_1,u_2,\ldots,u_n$ by joining u_i and u_{i+1} to new vertices v_i, x_i and w_i, y_i respectively, and then adding the edges v_iw_i and x_i,y_i, for $i = 1,2,\ldots,n-1$.

The following graph is the generalization of T_n and Q_n.

Definition. [6] A *k-Polygonal Snake* is obtained by replacing every edge of a path $P_n : u_1, u_2, \ldots, u_n$ by k-cycle C_k for $k \geq 3$, such that each pair u_i, u_{i+1} remains adjacent for all $i = 1, 2, \ldots n$. It is denoted by $S_n(C_k)$.

Note that, $S_n(C_3)$ is the triangular snake graph and $S_n(C_4)$ is the quadrilateral snake graph.

We define the following graph as it is a natural extension of DT_n and DQ_n.

Definition. A *double k-polygonal snake graph* is obtained by replacing every edge of a path $P_n : u_1, u_2, \ldots, u_n$ by a double k-cycle C_k for $k \geq 3$ (the double k-cycle is obtained by taking two cycles C_k having exactly one common edge), such that each edge u_i, u_{i+1} is replaced by the common edge of a double cycle, for every $i = 1, 2, \ldots n$. We denote it by $D(S_nC_k)$.

Notation. Throughout this chapter, a path $P_n = u_1, u_2, \ldots, u_n$, where $n > 1$. Let $V(D(S_nC_k)) = V(P_n) \cup \{v_{i,j}, w_{i,j} : 1 \leq i \leq n-1, 1 \leq j \leq k-2\}$ and $E(D(S_nC_k)) = E(P_n) \cup \{u_iv_{i,1}, v_{i,k-2}u_{i+1}, u_iw_{i,1}, w_{i,k-2}u_{i+1} : 1 \leq i \leq n-1\} \cup \{v_{i,j}v_{i,j+1}, w_{i,j}w_{i,j+1} : 1 \leq i \leq n-1, 1 \leq j \leq k-3\}$. Then, the double k-polygonal Snake $D(S_nC_k)$ is of order $2nk - 3n - 2k + 4$ and size $(n-1)(2k-1)$. For example, see Figure 6.1.

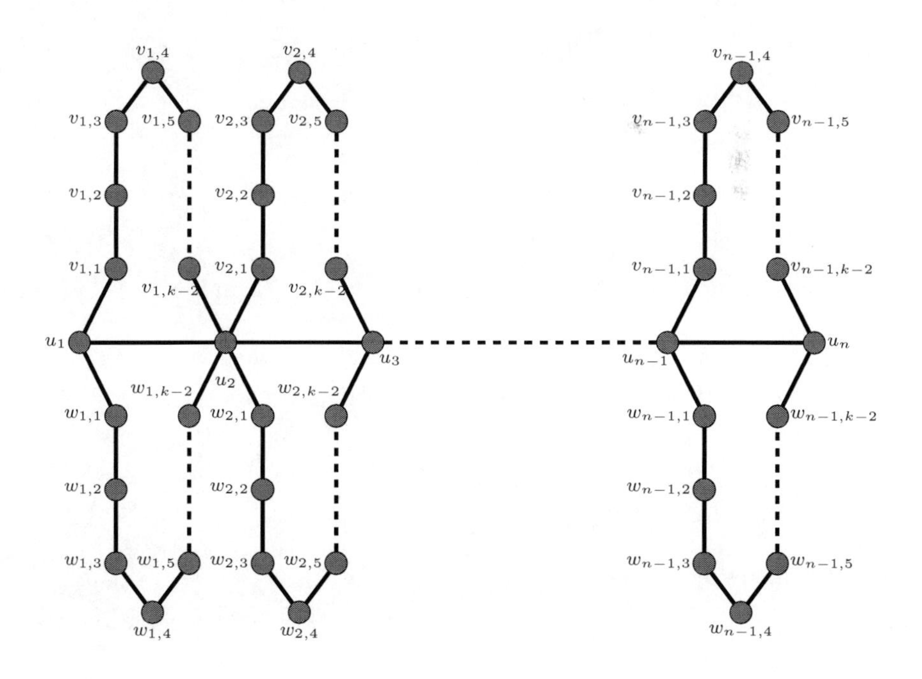

Figure 6.1: $D(S_nC_k)$

6.2 MAIN RESULTS

Prajapati and Vantiya [7] proved that double triangular snake DT_n, i.e. $D(S_nC_3)$ and double quadrilateral snake DQ_n i.e. $D(S_nC_4)$ are SD prime cordial. Thus for $k=3,4$; $D(S_nC_k)$ is SD-prime cordial. Here we consider the cases for remaining arbitrary values of $k \in N$.

Theorem 6.1

The graph $D(S_nC_k)$ is SD-prime cordial, for $k \equiv 1 (\text{mod } 4), k \geq 5$. ■

Proof. Define $f : V(D(S_nC_k)) \to \{1,2,\ldots,2nk-3n-2k+4\}$ as follows:

$$f(u_i) = (2k-3)i - \left(2k - \frac{7}{2} + \frac{(-1)^i}{2}\right), \quad \text{if } 1 \leq i \leq n;$$

$$f(v_{i,1}) = (2k-3)i - \left(2k - \frac{11}{2} + \frac{(-1)^i}{2}\right), \quad \text{if } 1 \leq i \leq n-1;$$

$$f(v_{i,2}) = (2k-3)i - \left(2k - \frac{15}{2} - \frac{(-1)^i}{2}\right), \quad \text{if } 1 \leq i \leq n-1;$$

$$f(v_{i,3}) = (2k-3)i - \left(2k - 10 + (-1)^i\right), \quad \text{if } 1 \leq i \leq n-1;$$

$$f(v_{i,j}) = (2k-3)i - \left(2k - 2j - 3 + (-1)^i\right), \quad \text{if } j \equiv 1 (\text{mod } 4), 5 \leq j \leq k-2, \ 1 \leq i \leq n-1;$$

$$f(v_{i,j}) = (2k-3)i - \left(2k - 2j - \frac{7}{2} - \frac{(-1)^i}{2}\right), \quad \text{if } j \equiv 2 (\text{mod } 4), 6 \leq j \leq k-2, \ 1 \leq i \leq n-1;$$

$$f(v_{i,j}) = (2k-3)i - \left(2k - 2j - 3 + 2(-1)^i\right), \quad \text{if } j \equiv 3 (\text{mod } 4), 7 \leq j \leq k-2, \ 1 \leq i \leq n-1;$$

$$f(v_{i,j}) = (2k-3)i - \left(2k - 2j - \frac{5}{2} - \frac{(-1)^i}{2}\right), \quad \text{if } j \equiv 0 (\text{mod } 4), 4 \leq j \leq k-2, \ 1 \leq i \leq n-1;$$

$$f(w_{i,1}) = (2k-3)i - \left(2k - \frac{11}{2} - \frac{(-1)^i}{2}\right), \quad \text{if } 1 \leq i \leq n-1;$$

$$f(w_{i,2}) = (2k-3)i - \left(2k - 8 + (-1)^i\right), \quad \text{if } 1 \leq i \leq n-1;$$

$$f(w_{i,j}) = (2k-3)i - (2k - 2j - 3), \quad \text{if } j \equiv 1 (\text{mod } 4), 5 \leq j \leq k-2, \ 1 \leq i \leq n-1;$$

$$f(w_{i,j}) = (2k-3)i - (2k - 2j - 5), \quad \text{if } j \equiv 2 (\text{mod } 4), 6 \leq j \leq k-2, \ 1 \leq i \leq n-1;$$

$$f(w_{i,j}) = (2k-3)i - \left(2k-2j-3-(-1)^i\right), \quad \text{if } j \equiv 3(\text{mod } 4), 3 \leq j \leq k-2, \ 1 \leq i \leq n-1;$$
$$f(w_{i,j}) = (2k-3)i - \left(2k-2j-5-(-1)^i\right), \quad \text{if } j \equiv 0(\text{mod } 4), 4 \leq j \leq k-2, \ 1 \leq i \leq n-1.$$

Therefore, $e_{f'}(0) = \lceil \frac{(n-1)(2k-1)}{2} \rceil, e_{f'}(1) = \lfloor \frac{(n-1)(2k-1)}{2} \rfloor$.

Thus $\left|e_{f'}(0) - e_{f'}(1)\right| \leq 1$.
Hence $D(S_nC_k)$ is SD-prime cordial, for $k \equiv 1(\text{mod } 4), k \geq 5$. □

Theorem 6.2

The graph $D(S_nC_k)$ is SD-prime cordial, for $k \equiv 2(\text{mod } 4), k \geq 6$. ■

Proof. Define $f : V(D(S_nC_k)) \to \{1, 2, \ldots, 2nk-3n-2k+4\}$ as follows:

$$f(u_i) = (2k-3)i - \left(2k - \frac{7}{2} + \frac{(-1)^i}{2}\right), \quad \text{if } 1 \leq i \leq n;$$
$$f(v_{i,1}) = (2k-3)i - (2k-5), \quad \text{if } 1 \leq i \leq n-1;$$
$$f(v_{i,2}) = (2k-3)i - \left(2k-7-(-1)^i\right), \quad \text{if } 1 \leq i \leq n-1;$$
$$f(v_{i,3}) = (2k-3)i - \left(2k-8-(-1)^i\right), \quad \text{if } 1 \leq i \leq n-1;$$
$$f(v_{i,4}) = (2k-3)i - \left(2k - \frac{19}{2} - \frac{3(-1)^i}{2}\right), \quad \text{if } 1 \leq i \leq n-1;$$
$$f(v_{i,j}) = (2k-3)i - \left(2k-2j-3+(-1)^i\right), \quad \text{if } j \equiv 1(\text{mod } 4, 5 \leq j \leq k-2, \ 1 \leq i \leq n-1;$$
$$f(v_{i,j}) = (2k-3)i - \left(2k-2j - \frac{5}{2} - \frac{3(-1)^i}{2}\right), \quad \text{if } j \equiv 2(\text{mod } 4), 6 \leq j \leq k-2, \ 1 \leq i \leq n-1;$$
$$f(v_{i,j}) = (2k-3)i - \left(2k-2j-2+(-1)^i\right), \quad \text{if } j \equiv 3(\text{mod } 4, 7 \leq j \leq k-2, \ 1 \leq i \leq n-1;$$
$$f(v_{i,j}) = (2k-3)i - \left(2k-2j - \frac{5}{2} - \frac{(-1)^i}{2}\right), \quad \text{if } j \equiv 0(\text{mod } 4), 8 \leq j \leq k-2, \ 1 \leq i \leq n-1;$$
$$f(w_{i,1}) = (2k-3)i - \left(2k-8+2(-1)^i\right), \quad \text{if } 1 \leq i \leq n-1;$$
$$f(w_{i,2}) = (2k-3)i - \left(2k-8+(-1)^i\right), \quad \text{if } 1 \leq i \leq n-1;$$

$$f(w_{i,j}) = (2k-3)i - \left(2k - 2j - \frac{9}{2} + \frac{3(-1)^i}{2}\right), \quad \text{if } j \equiv 1 (\text{mod } 4), 5 \le j \le k-2, \; 1 \le i \le n-1;$$

$$f(w_{i,j}) = (2k-3)i - \left(2k - 2j - 4 - (-1)^i\right), \quad \text{if } j \equiv 2 (\text{mod } 4), 6 \le j \le k-2, \; 1 \le i \le n-1;$$

$$f(w_{i,j}) = (2k-3)i - \left(2k - 2j - \frac{9}{2} + \frac{(-1)^i}{2}\right), \quad \text{if } j \equiv 3 (\text{mod } 4), 3 \le j \le k-2, \; 1 \le i \le n-1;$$

$$f(w_{i,j}) = (2k-3)i - \left(2k - 2j - 5 - (-1)^i\right), \quad \text{if } j \equiv 0 (\text{mod } 4), 4 \le j \le k-3, \; 1 \le i \le n-1;$$

$$f(w_{i,j}) = (2k-3)i + \left(\frac{1}{2} - \frac{(-1)^i}{2}\right), \quad \text{if } j = k-2, 1 \le i \le n-1.$$

Therefore, $e_{f'}(0) = \lfloor \frac{(n-1)(2k-1)}{2} \rfloor, e_{f'}(1) = \lceil \frac{(n-1)(2k-1)}{2} \rceil$.

Thus $\left| e_{f'}(0) - e_{f'}(1) \right| \le 1$.
Hence $D(S_n C_k)$ is SD-prime cordial, for $k \equiv 2 (\text{mod } 4), k \ge 6$. □

Theorem 6.3

The graph $D(S_n C_k)$ is SD-prime cordial, for $k \equiv 3 (\text{mod } 4), k \ge 7$. ■

Proof. Define $f : V(D(S_n C_k)) \to \{1, 2, \ldots, 2nk - 3n - 2k + 4\}$ as follows:

$$f(u_i) = (2k-3)i - \left(2k - \frac{7}{2} + \frac{(-1)^i}{2}\right), \quad \text{if } 1 \le i \le n;$$

$$f(v_{i,1}) = (2k-3)i - \left(2k - 6 + (-1)^i\right), \quad \text{if } 1 \le i \le n-1;$$

$$f(v_{i,2}) = (2k-3)i - \left(2k - \frac{11}{2} - \frac{(-1)^i}{2}\right), \quad \text{if } 1 \le i \le n-1;$$

$$f(v_{i,3}) = (2k-3)i - \left(2k - \frac{15}{2} + \frac{(-1)^i}{2}\right), \quad \text{if } 1 \le i \le n-1;$$

$$f(v_{i,4}) = (2k-3)i - \left(2k - 10 + 2(-1)^i\right), \quad \text{if } 1 \le i \le n-1;$$

$$f(v_{i,j}) = (2k-3)i - \left(2k - 2j - \frac{7}{2} - \frac{(-1)^i}{2}\right), \quad \text{if } j \equiv 1 (\text{mod } 4), 5 \le j \le k-2, \; 1 \le i \le n-1;$$

$$f(v_{i,j}) = (2k-3)i - \left(2k - 2j - 3 + 2(-1)^i\right), \quad \text{if } j \equiv 2 (\text{mod } 4), 6 \le j \le k-2, \; 1 \le i \le n-1;$$

$$f(v_{i,j}) = (2k-3)i - \left(2k-2j-\frac{5}{2}-\frac{(-1)^i}{2}\right), \quad \text{if } j \equiv 3(\text{mod } 4), 7 \leq j \leq k-2, \; 1 \leq i \leq n-1;$$

$$f(v_{i,j}) = (2k-3)i - \left(2k-2j-3+(-1)^i\right), \quad \text{if } j \equiv 0(\text{mod } 4), 8 \leq j \leq k-2, \; 1 \leq i \leq n-1;$$

$$f(w_{i,1}) = (2k-3)i - \left(2k-\frac{15}{2}-\frac{3(-1)^i}{2}\right), \quad \text{if } 1 \leq i \leq n-1;$$

$$f(w_{i,2}) = (2k-3)i - \left(2k-\frac{19}{2}-\frac{(-1)^i}{2}\right), \quad \text{if } 1 \leq i \leq n-1;$$

$$f(w_{i,j}) = (2k-3)i - (2k-2j-5), \quad \text{if } j\equiv 1(\text{mod } 4), 5\leq j\leq k-3, 1\leq i\leq n-1 \text{ or } j=k-2 \text{ and } i \text{ is odd}, 1\leq i\leq n-1;$$

$$f(w_{i,j}) = (2k-3)i - (2k-2j-3), \quad \text{if } j=k-2, \text{ and } i \text{ is even}, 1\leq i\leq n-1;$$

$$f(w_{i,j}) = (2k-3)i - \left(2k-2j-3-(-1)^i\right), \quad \text{if } j \equiv 2(\text{mod } 4), 6 \leq j \leq k-2, \; 1 \leq i \leq n-1;$$

$$f(w_{i,j}) = (2k-3)i - \left(2k-2j-5-(-1)^i\right), \quad \text{if } j \equiv 3(\text{mod } 4), 3 \leq j \leq k-2, \; 1 \leq i \leq n-1;$$

$$f(w_{i,j}) = (2k-3)i - (2k-2j-3), \quad \text{if } j \equiv 0(\text{mod } 4), 4 \leq j \leq k-2, \; 1 \leq i \leq n-1.$$

Therefore, $e_{f'}(0) = \lfloor \frac{(n-1)(2k-1)}{2} \rfloor, e_{f'}(1) = \lceil \frac{(n-1)(2k-1)}{2} \rceil$.

Thus $|e_{f'}(0) - e_{f'}(1)| \leq 1$.
Hence $D(S_nC_k)$ is SD-prime cordial, for $k \equiv 3(\text{mod } 4), k \geq 7$. □

Theorem 6.4

The graph $D(S_nC_k)$ is SD-prime cordial, for $k \equiv 0(\text{mod } 4), k \geq 8$. ■

Proof. Define $f : V(D(S_nC_k)) \to \{1, 2, \ldots, 2nk-3n-2k+4\}$ as follows:
Define $f(u_i), f(v_{i,j}), f(w_{i,1}), f(w_{i,2})$ for all i and j as per the previous theorem. And

$$f(w_{i,j}) = (2k-3)i - \left(2k-2j-\frac{9}{2}-\frac{(-1)^i}{2}\right), \quad \text{if } j \equiv 1(\text{mod } 4), 5 \leq j \leq k-2, \; 1 \leq i \leq n-1;$$

$$f(w_{i,j}) = (2k-3)i - \left(2k-2j-\frac{7}{2}-\frac{(-1)^i}{2}\right), \quad \text{if } j \equiv 2(\text{mod } 4), 6 \leq j \leq k-2, \; 1 \leq i \leq n-1;$$

$$f(w_{i,j}) = (2k-3)i - \left(2k - 2j - \frac{11}{2} - \frac{(-1)^i}{2}\right), \quad \text{if } j \equiv 3 (\text{mod } 4), 3 \leq j \leq k-2, \; 1 \leq i \leq n-1;$$

$$f(w_{i,j}) = (2k-3)i - \left(2k - 2j - \frac{5}{2} - \frac{(-1)^i}{2}\right), \quad \text{if } j \equiv 0 (\text{mod } 4), 4 \leq j \leq k-2, \; 1 \leq i \leq n-1.$$

Therefore, $e_{f'}(0) = \lceil \frac{(n-1)(2k-1)}{2} \rceil, e_{f'}(1) = \lfloor \frac{(n-1)(2k-1)}{2} \rfloor$.

Thus $|e_{f'}(0) - e_{f'}(1)| \leq 1$.
Hence $D(S_nC_k)$ is SD-prime cordial, for $k \equiv 0(\text{mod } 4), k \geq 8$. □

Theorem 6.5

The graph $D(S_nC_k)$ is SD-prime cordial, for all integers $k \geq 3, n \geq 2$. ■

Proof. The proof follows from Theorems 6.1 to 6.4. □

Illustration 6.2.1. SD-prime cordial labeling of the graph $D(S_7C_6)$ is shown in figure 6.2.

6.3 CONCLUSIONS

We have proved that the k-polygonal snake $D(S_nC_k)$ is SD-prime cordial, for all integers $k \geq 3, n \geq 2$. For instance, one can study the SD-prime cordial behaviour of cyclic ladder.

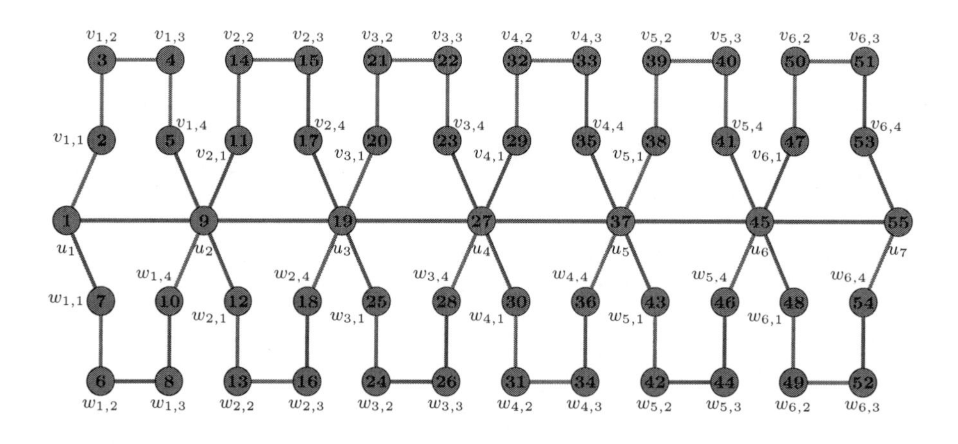

Figure 6.2: $D(S_7C_6), n = 7, k = 6, e_{f'}(1) = e_{f'}(0) = 33$

REFERENCES

1. J. A. Bondy and U. S. R. Murty. *Graph Theory with Applications.* 2nd Edition, MacMillan, New York, 1976.
2. J. A. Gallian. A dynamic survey of graph labeling. *The Electronic Journal of Combinatories, #DS6*, 1-502, 2018.
3. G. C. Lau, H. H. Chu, N. Suhadak, F. Y. Foo and H. K. Ng. On SD-prime cordial graphs. *International Journal of Pure and Applied Mathematics*, 106(4): 1017-1028, 2016.
4. A. Lourdusamy and F. Patrick. Some results on SD-prime cordial labeling. *Proyecciones Journal of Mathematics*, 36(4): 601-614, 2017.
5. A. Lourdusamy, S. J. Wency and F. Patrick. On SD-prime cordial labeling. *International Journal of Pure and Applied Mathematics*, 117(11): 221-228, 2017.
6. V. Ramachandran and C. Sekar. One modulo N gracefulness of $n-$polygonal snakes, $C_n^{(t)}$ and $P_{a,b}$. *International Journal of Engineering Research and Technology*, 2(10): 3514-3529, 2013.
7. U. M. Prajapati and A. V. Vantiya. SD-prime cordial labeling of some snake graphs. *Journal of Applied Science and Computations*, 6(4): 1857-1868, 2019.
8. U. M. Prajapati and A. V. Vantiya. SD-prime cordial labeling of subdivision of snake graphs. *International Journal of Scientific Research and Reviews*, 8(2): 2414-2423, 2019.
9. U. M. Prajapati and A. V. Vantiya. SD-prime cordial labeling of subdivision of some double snake graphs. *In Communication.*

7 Edge Product Cordial and Total Edge Product Cordial Labeling of Some Wheel related Graphs

Chirag M. Barasara
Department of Mathematics,
Hemchandracharya North Gujarat University,
Patan, Gujarat (INDIA)
E-mail: chirag.barasara@gmail.com

Variants of cordial labeling are known as equitable labeling, and edge product cordial labeling and total edge product cordial labeling are such equitable labelings.

The wheel graph is not an edge product cordial graph while it is a total edge product cordial graph. Here we have investigated edge product cordial labeling and total edge product cordial labeling for larger graphs obtained from wheel graphs by using some graph operations like the shadow graph, splitting graph, middle graph and total graph.

We found that the shadow graph of wheel and splitting graph of wheel are not edge product cordial graphs while the middle graph of wheel for $n > 5$ and total graph of wheel for $n > 8$ are edge product cordial graphs, while the shadow graph of wheel, splitting graph of wheel, middle graph of wheel and total graph of wheel are total edge product cordial graphs.

7.1 INTRODUCTION

In this chapter we consider simple, finite, connected and undirected graph $G = (V(G), E(G))$ with order $p = |V(G)|$ and size $q = |E(G)|$. The elements of sets $V(G)$ and $E(G)$ are commonly known as graph elements. For other graph theoretic terminologies and notations, we refer Clark and Holton [4].

The theory of graphs has a close interaction with many branches of mathematics and social sciences like discrete mathematics, network theory, information system and social network analysis. There are many interesting areas of research in graph theory. The labeling of graphs is one of such areas which cuts across a wide range of applications especially in coding theory, design of optimal circuit layout and X-ray crystallography as reported in Bloom and Golomb [2] and Yegnanarayanan and Vaidhyanathan [21].

The concept of graph labeling was originated by Rosa [10] in 1967. For an extensive survey and bibliographic references on graph labeling we refer to Gallian [5].

Definition (Graph Labeling). A *graph labeling* is an assignment of integers to the vertices or edges or both subject to certain condition(s). If the domain of mapping is the set of vertices(edges)(both) then the labeling is called a vertex(an edge)(total) labeling.

In 1987, Cahit [3] introduced the concept of cordial labeling which is defined as follows.

Definition (Cordial Labeling). For graph G, the vertex labeling function $f: V(G) \rightarrow \{0,1\}$ induces an edge labeling function $f^*: E(G) \rightarrow \{0,1\}$ defined by $f^*(e = uv) = |f(u) - f(v)|$. The function f is called a *cordial labeling* of graph g if the absolute difference of the number of vertices with label 1 and label 0 is at most 1 and the absolute difference of the number of edges with label 1 and label 0 is also at most 1. A graph G is called *cordial* if it admits a cordial labeling.

Some labelings like prime cordial labeling, A - cordial labeling, H - cordial labeling, product cordial labeling, edge product cordial labeling etc. were also introduced as variants of cordial labeling. Such labelings are commonly referred as equitable labelings.

In this chapter we consider two of the equitable labelings called edge product cordial labeling and total edge product cordial labeling and study them in the context of the larger graph obtained by graph operations.

Definition (Join of Graph). For graph G_1 and G_2, *join of graph* is defined to be the graph with vertex set $V(G_1) \bigcup V(G_2)$ and edge set $E(G_1) \bigcup E(G_2) \bigcup \{x_1x_2 / x_1 \in V(G_1)$ and $x_2 \in V(G_2)\}$ and is denoted by $G_1 + G_2$.

Definition (Wheel Graph). The *wheel graph* W_n is defined to be the join $K_1 + C_n$. The vertex corresponding to K_1 is known as the apex vertex and vertices corresponding to cycle are known as rim vertices while the edges corresponding to cycle are known as rim edges and edges joining the apex vertex and rim vertices are known as spoke edges.

7.2 EDGE PRODUCT CORDIAL LABELING

In cordial labeling the edge labels were induced by the absolute difference of vertex labels while in product cordial labeling the edge labels are induced by the product of vertex labels. The concept of product cordial labeling was introduced by Sundaram *et al.* [11] and defined as follows.

Definition (Product Cordial Labeling). For graph G, the vertex labeling function $f : V(G) \rightarrow \{0,1\}$ induces an edge labeling function $f^* : E(G) \rightarrow \{0,1\}$ defined by $f^*(e = uv) = f(u)f(v)$. The function f is called a *product cordial labeling* of graph G if the absolute difference of the number of vertices with label 1 and label 0 is at most 1 and the absolute difference of number of edges with label 1 and label 0 is also at most 1. A graph G is called a *product cordial graph* if it admits a product cordial labeling.

In 2012, Vaidya and Barasara [13] introduced the edge analogue of product cordial labeling and named it edge product cordial labeling which is defined as follows.

Definition (Edge Product Cordial Labeling). For graph G, the edge labeling function $f : E(G) \rightarrow \{0,1\}$ induces a vertex labeling function $f^* : V(G) \rightarrow \{0,1\}$ defined by $f^*(v) = \Pi f(e_i)$ for $\{e_i \in E(G)/e_i$ is incident to $v\}$. The function f is called an *edge product cordial labeling* of graph G if the absolute difference of the number of vertices with label 1 and label 0 is at most 1 and the absolute difference of number of edges with label 1 and label 0 is also at most 1. A graph G is called an *edge product cordial graph* if it admits an edge product cordial labeling.

Illustration 7.2.1. The cycle C_5 and its edge product cordial labeling is shown in Figure 7.1.

In [13, 16, 17, 18, 19] Vaidya and Barasara have investigated edge product cordial labeling for some standard graphs, a graph obtained by applying graph operations and different products of the graphs while in [20] they have looked into edge product cordial labeling of some degree splitting graphs. Prajapati

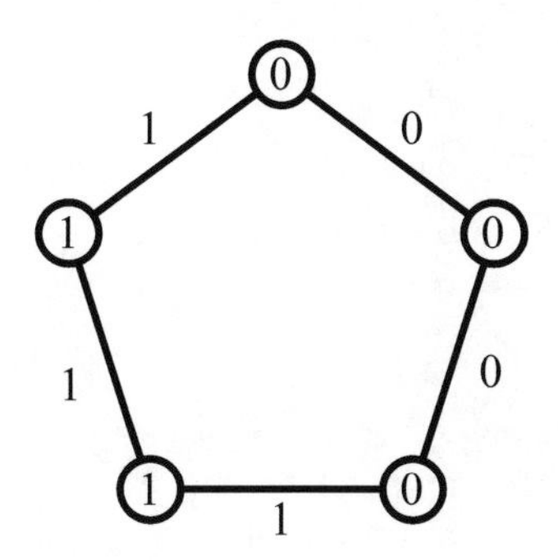

Figure 7.1: C_5 and its edge product cordial labeling

and Shah [7] have discussed edge product cordial labeling in the context of duplication of some graph elements.

Edge product cordial labeling for some cycle related graphs have been studied by Prajapati and Patel [8] while in [9] they have proved that $W_n^{(t)}$ for even t, PS_n and DPS_n for odd n are edge product cordial graphs. In [6] Ivančo has given some generalized results related to edge product cordial labeling.

For any graph G we introduce the following notations:

$v_f(0)$ = number of vertices having label 0.
$v_f(1)$ = number of vertices having label 1.
$e_f(0)$ = number of edges having label 0.
$e_f(1)$ = number of edges having label 1.

7.3 TOTAL EDGE PRODUCT CORDIAL LABELING

Another variant of product cordial labeling is total product cordial labeling which was introduced by Sundaram *et al.* [12] in 2006 and is defined as follows.

Definition (Total Product Cordial Labeling). For graph G, the vertex labeling function $f : V(G) \to \{0,1\}$ induces an edge labeling function $f^* : E(G) \to \{0,1\}$ defined by $f^*(e = uv) = f(u)f(v)$. The function f is called a *total product cordial labeling* of graph G if $|(v_f(0)+e_f(0)) - (v_f(1)+e_f(1))| \leq 1$. A graph G is called *total product cordial graph* if it admits a total product cordial labeling.

In 2013, Vaidya and Barasara [14] have introduced an edge analogue of total product cordial labeling and named it total edge product cordial labeling which is defined as follows.

Definition (Total Edge Product Cordial Labeling). For graph G, the edge labeling function $f : E(G) \to \{0,1\}$ induces a vertex labeling function $f^* : V(G) \to \{0,1\}$ defined by $f^*(v) = \Pi f(e_i)$ for $\{e_i \in E(G)/e_i$ is incident to $v\}$. The function f is called a *total edge product cordial labeling* of graph G if $|(v_f(0)+e_f(0)) - (v_f(1)+e_f(1))| \leq 1$. A graph G is called a *total edge product cordial graph* if it admits a total edge product cordial labeling.

Illustration 7.3.1. The graph W_6 and its total edge product cordial labeling is shown in Figure 7.2.

In [14] Vaidya and Barasara have investigated total edge product cordial labeling for some standard graphs while in [15] they have discussed total edge product cordial labeling in the context of some graph operations. Barasara [1] has investigated edge and total edge product cordial labeling for a cycle with one chord, a cycle with twin chords, a triangular ladder graph and a comb graph.

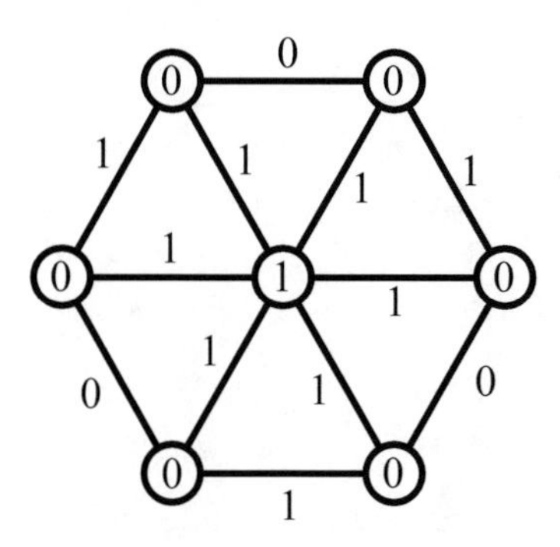

Figure 7.2: W_6 and its total edge product cordial labeling

7.4 RESULTS RELATED TO SHADOW GRAPH OF WHEEL

Definition (Shadow Graph). The *shadow graph* $D_2(G)$ of a connected graph G is constructed by taking two copies of G say G' and G''. Join each vertex u' in G' to the neighbours of the corresponding vertex u'' in G''.

Theorem 7.1

$D_2(W_n)$ is not an edge product cordial graph. ■

Proof. The graph $D_2(W_n)$ has order $2n+2$ and size $8n$. In order to satisfy the edge condition for a graph to be edge product cordial it is essential to assign label 0 to $4n$ edges. The edges with label 0 will give rise at least $n+3$ vertices with label 0 and at most $n-1$ vertices with label 1. Therefore $|v_f(0)-v_f(1)| \geq 4$. Thus the vertex condition for a graph to be edge product cordial is violated. Hence, $D_2(W_n)$ is not an edge product cordial graph. □

Theorem 7.2

$D_2(W_n)$ is a total edge product cordial graph. ■

Proof. For $D_2(W_n)$ let u be the apex vertex and $u_1, u_2, u_3, \ldots, u_n$ be the rim vertices of the first copy of wheel W_n and v be the apex vertex and $v_1, v_2, v_3, \ldots, v_n$ be the rim vertices of the second copy of wheel W_n. We define $f : E(D_2(W_n)) \rightarrow \{0,1\}$ as follows.

$$\begin{aligned}
&f(uu_i) = 1, && 1 \le i \le n;\\
&f(u_iu_{i+1}) = 1, && 1 \le i \le n-1;\\
&f(u_1u_n) = 1;\\
&f(vv_i) = 1, && 1 \le i \le n;\\
&f(v_iv_{i+1}) = 1, && 1 \le i \le n-1;\\
&f(v_1v_n) = 1;\\
&f(uv_i) = 1, && 1 \le i \le n;\\
&f(vu_i) = 0, && 1 \le i \le n;\\
&f(u_iv_j) = 0, && \text{for remaining } 2n \text{ edges.}
\end{aligned}$$

In view of the above defined labeling pattern we have $v_f(0) + e_f(0) = v_f(1) + e_f(1) = 5n + 1$. Therefore $|(v_f(0) + e_f(0)) - (v_f(1) + e_f(1))| = 0$. Hence, $D_2(W_n)$ is a total edge product cordial graph. □

Illustration 7.4.1. The graph $D_2(W_3)$ and its total edge product cordial labeling is shown in Figure 7.3. Here black colored elements are labeled with 1 while gray colored elements are labeled with 0.

7.5 RESULTS RELATED TO SPLITTING GRAPH OF WHEEL

Definition (Splitting Graph). The *splitting graph* $S'(G)$ of a graph G is obtained by adding to each vertex v a new vertex v' such that v' is adjacent to every vertex that is adjacent to v in G.

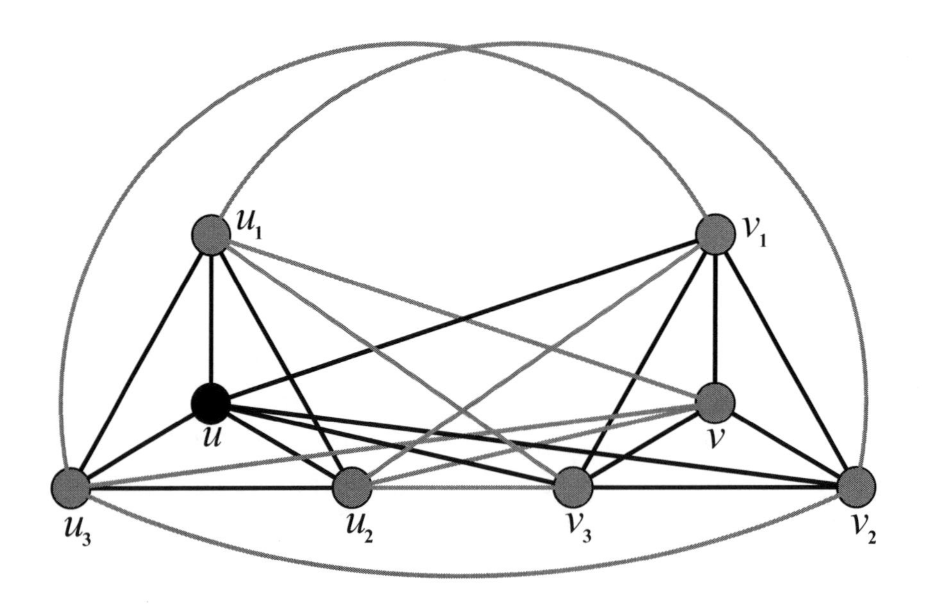

Figure 7.3: $D_2(W_3)$ and its total edge product cordial labeling

---

Theorem 7.3

$S'(W_n)$ is not an edge product cordial graph. ■

Proof. The graph $S'(W_n)$ has order $2n+2$ and size $6n$. In order to satisfy the edge condition for a graph to be edge product cordial it is essential to assign the label 0 to $3n$ edges. The edges with label 0 will give rise at least $n+2$ vertices with label 0 and at most n vertices with label 1. Therefore $|v_f(0)-v_f(1)| \geq 2$. Thus the vertex condition for a graph to be edge product cordial is violated.
Hence, $S'(W_n)$ is not an edge product cordial graph. □

Theorem 7.4

$S'(W_n)$ is a total edge product cordial graph. ■

Proof. Let u be the apex vertex and $u_1,u_2,u_3,\ldots,u_n$ be the rim vertices of wheel W_n. For graph $S'(W_n)$ added vertices corresponding to $u,u_1,u_2,u_3,\ldots,u_n$ are $v,v_1,v_2,v_3,\ldots,v_n$ respectively. We define $f: S'(W_n) \rightarrow \{0,1\}$ as follows.

$$
\begin{aligned}
&f(uv_i)=1, && 1 \leq i \leq n;\\
&f(uu_i)=1, && 1 \leq i \leq n;\\
&f(u_iu_{i+1})=1, && 1 \leq i \leq n-1;\\
&f(u_1u_n)=1; &&\\
&f(v_iu_{i+1})=1, && 1 \leq i \leq n-1;\\
&f(u_1v_n)=1; &&\\
&f(u_iv_{i+1})=0, && 1 \leq i \leq n-1;\\
&f(v_1u_n)=0; &&\\
&f(vu_i)=0, && 1 \leq i \leq n.
\end{aligned}
$$

In view of the above defined labeling pattern we have $v_f(0)+e_f(0)=v_f(1)+e_f(1)=4n+1$. Therefore $|\left(v_f(0)+e_f(0)\right)-\left(v_f(1)+e_f(1)\right)|=0$. Hence, $S'(W_n)$ is a total edge product cordial graph. □

Illustration 7.5.1. The graph $S'(W_3)$ and its total edge product cordial labeling is shown in Figure 7.4. Here black colored elements are labeled with 1 while gray colored elements are labeled with 0.

7.6 RESULTS RELATED TO MIDDLE GRAPH OF WHEEL

Definition (Middle Graph). The *middle graph* $M(G)$ of a graph G is the graph whose vertex set is $V(G)\bigcup E(G)$ and in which two vertices are adjacent

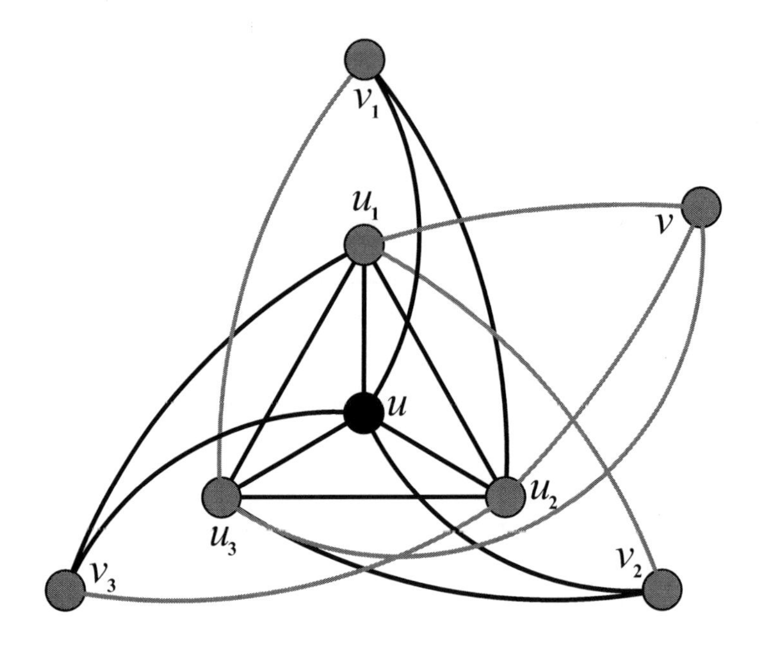

Figure 7.4: $S'(W_3)$ and its total edge product cordial labeling

if and only if either they are adjacent edges of G or one is a vertex of G and the other is an edge incident on it.

Theorem 7.5

$M(W_n)$ is an edge product cordial graph for $n > 5$ and not an edge product cordial graph for $n \leq 5$. ■

Proof. Let v be the apex vertex and $v_1, v_2, v_3, \ldots, v_n$ be the rim vertices of wheel W_n while $e_1, e_2, e_3, \ldots, e_n$ are rim edges and $e_{n+1}, e_{n+2}, e_{n+3}, \ldots, e_{2n}$ are spoke edges of wheel W_n. Then $M(W_n)$ has order $3n+1$ and size $\dfrac{n^2+13n}{2}$. To define $f : M(W_n) \to \{0,1\}$ we consider the following four cases.

Case - 1: For $n = 3$.
In order to satisfy the edge condition for a graph to be edge product cordial it is essential to assign label 0 to 12 edges. The edges with label 0 will give rise at least 6 vertices with label 0 and at most 4 vertices with label 1. Therefore $|v_f(0) - v_f(1)| \geq 2$. Thus the vertex condition for a graph to be edge product cordial is violated.

Case - 2: For $n=4$.
In order to satisfy the edge condition for a graph to be edge product cordial it is essential to assign label 0 to 17 edges. The edges with label 0 will give rise at least 8 vertices with label 0 and at most 5 vertices with label 1. Therefore $|v_f(0)-v_f(1)| \geq 3$. Thus the vertex condition for a graph to be edge product cordial is violated.

Case - 3: For $n=5$.
In order to satisfy the edge condition for a graph to be edge product cordial it is essential to assign label 0 to at least 22 edges. The edges with label 0 will give rise at least 9 vertices with label 0 and at most 7 vertices with label 1. Therefore $|v_f(0)-v_f(1)| \geq 2$. Thus the vertex condition for a graph to be edge product cordial is violated.

Case - 4: For $n>5$.

$$
\begin{array}{ll}
f(v_ie_{n+i})=1, & 1\leq i\leq n;\\
f(v_ie_i)=1, & 1\leq i\leq n;\\
f(v_ie_{i-1})=1, & 2\leq i\leq n;\\
f(v_1e_n)=1; & \\
f(e_1e_n)=1; & \\
f(e_ie_{i+1})=1, & 1\leq i\leq \left\lfloor \frac{3n+1}{2}\right\rfloor -n;\\
f(e_ie_{n+i})=1, & 1\leq i\leq \left\lfloor \frac{3n+1}{2}\right\rfloor -n;\\
f(e_ie_{n+i+1})=1, & 1\leq i\leq \left\lfloor \frac{3n+1}{2}\right\rfloor -n;\\
f(e_ie_{n+i})=0, & \left\lfloor \frac{3n+1}{2}\right\rfloor -n+1\leq i\leq n;\\
f(ve_{n+i})=0, & 1\leq i\leq n.
\end{array}
$$

For the remaining edges, give label 1 to any $\left(\left\lfloor \frac{n^2+13n}{4}\right\rfloor -3\left\lfloor \frac{3n+1}{2}\right\rfloor -1\right)$ edges and label 0 to other edges.
In view of the above defined labeling pattern we have $v_f(0)=\left\lceil \frac{3n+1}{2}\right\rceil$, $v_f(1)=\left\lfloor \frac{3n+1}{2}\right\rfloor$, $e_f(0)=\left\lceil \frac{n^2+13n}{4}\right\rceil$ and $e_f(1)=\left\lfloor \frac{n^2+13n}{4}\right\rfloor$. Therefore $|v_f(0)-v_f(1)|\leq 1$ and $|e_f(0)-e_f(1)|\leq 1$.
Hence, $M(W_n)$ is an edge product cordial graph for $n>5$ and not an edge product cordial graph for $n\leq 5$. □

Illustration 7.6.1. The graph $M(W_6)$ and its edge product cordial labeling is shown in Figure 7.5. Here black colored elements are labeled with 1 while gray colored elements are labeled with 0.

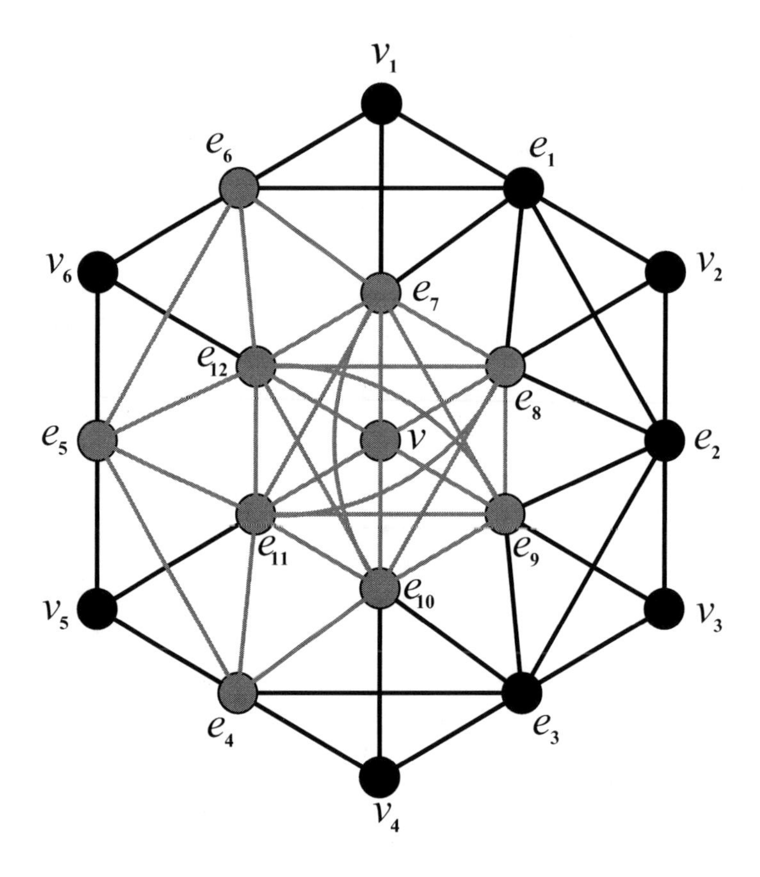

Figure 7.5: $M(W_6)$ and its edge product cordial labeling

Theorem 7.6

$M(W_n)$ is a total edge product cordial graph. ∎

Proof. Let v be the apex vertex and $v_1, v_2, v_3, \ldots, v_n$ be the rim vertices of wheel W_n while $e_1, e_2, e_3, \ldots, e_n$ are rim edges and $e_{n+1}, e_{n+2}, e_{n+3}, \ldots, e_{2n}$ are spoke edges of wheel W_n. Then $M(W_n)$ has order $3n+1$ and size $\dfrac{n^2+13n}{2}$. We define $f : M(W_n) \to \{0,1\}$ as follows.

$$
\begin{aligned}
f(v_i e_{n+i}) &= 0, \quad 1 \le i \le n; \\
f(e_i e_{i+1}) &= 0, \quad 1 \le i \le n-1; \\
f(e_1 e_n) &= 0; \\
f(v e_{n+1}) &= 0.
\end{aligned}
$$

For the remaining edges give label 1 to any $\left\lfloor \frac{n^2+19n+2}{4} \right\rfloor$ edges and label 0 to other edges.

In view of above defined labeling pattern we have $e_f(0) + v_f(0) = \left\lceil \frac{n^2+19n+2}{4} \right\rceil$ and $e_f(1)+v_f(1) = \left\lfloor \frac{n^2+19n+2}{4} \right\rfloor$.

Therefore $|\left(v_f(0)+e_f(0)\right) - \left(v_f(1)+e_f(1)\right)| \leq 1$. Hence, $M(W_n)$ is a total edge product cordial graph. □

Illustration 7.6.2. The graph $M(W_3)$ and its total edge product cordial labeling is shown in Figure 7.6. Here black colored elements are labeled with 1 while gray colored elements are labeled with 0.

7.7 RESULTS RELATED TO TOTAL GRAPH OF WHEEL

Definition (Total Graph). Let G be a graph with two or more vertices; then the *total graph* $T(G)$ of graph G is the graph whose vertex set is $V(G) \cup E(G)$ and two vertices are adjacent whenever they are either adjacent or incident in G.

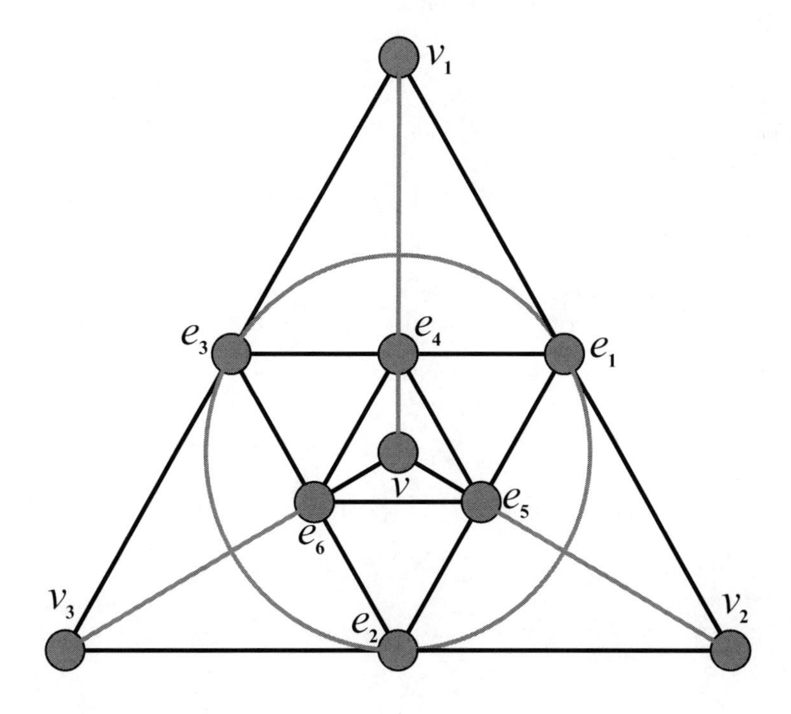

Figure 7.6: $M(W_3)$ and its total edge product cordial labeling

Theorem 7.7

$T(W_n)$ is an edge product cordial graph for $n > 8$ and not an edge product cordial graph for $n \leq 8$. ■

Proof. Let v be the apex vertex and $v_1, v_2, v_3, \ldots, v_n$ be the rim vertices of wheel W_n while $e_1, e_2, e_3, \ldots, e_n$ are rim edges and $e_{n+1}, e_{n+2}, e_{n+3}, \ldots, e_{2n}$ are spoke edges of wheel W_n. Then $T(W_n)$ has order $3n+1$ and size $\frac{n^2+17n}{2}$. To define $f: T(W_n) \rightarrow \{0,1\}$ we consider the following six cases.

Case - 1: For $n=3$ or $n=5$ or $n=7$.
In order to satisfy the edge condition for a graph to be edge product cordial it is essential to assign label 0 to at least $\left\lfloor \frac{n^2+17n}{4} \right\rfloor$ edges. The edges with label 0 will give rise to at least $\frac{3n+5}{2}$ vertices with label 0 and at most $\frac{3n-3}{2}$ vertices with label 1. Therefore $|v_f(0) - v_f(1)| \geq 4$. Thus the vertex condition for a graph to be edge product cordial is violated.

Case - 2: For $n=4$.
In order to satisfy the edge condition for a graph to be edge product cordial it is essential to assign label 0 to 21 edges. The edges with label 0 will give rise to at least 9 vertices with label 0 and at most 4 vertices with label 1. Therefore $|v_f(0) - v_f(1)| \geq 5$. Thus the vertex condition for a graph to be edge product cordial is violated.

Case - 3: For $n=6$ or $n=8$.
In order to satisfy the edge condition for a graph to be edge product cordial it is essential to assign label 0 to at least $\left\lfloor \frac{n^2+17n}{4} \right\rfloor$ edges. The edges with label 0 will give rise to at least $\left\lceil \frac{3n+1}{2} \right\rceil + 1$ vertices with label 0 and at most $\left\lfloor \frac{3n+1}{2} \right\rfloor - 1$ vertices with label 1. Therefore $|v_f(0) - v_f(1)| \geq 3$. Thus the vertex condition for a graph to be edge product cordial is violated.

Case - 4: For $n > 8$ and $n \equiv 0 (\text{mod } 4)$.

$$f(v_1 v_2) = 1;$$
$$f(v_i v_{i+1}) = 1, \quad 2 \leq i \leq \left\lfloor \frac{3n+1}{4} \right\rfloor + 1;$$
$$f(v_i v) = 1, \quad 2 \leq i \leq \left\lfloor \frac{3n+1}{4} \right\rfloor + 1;$$

$$
\begin{aligned}
&f(v_i e_{i-1}) = 1, && 2 \le i \le \left\lfloor \frac{3n+1}{4} \right\rfloor + 1;\\
&f(v_i e_i) = 1, && 2 \le i \le \left\lfloor \frac{3n+1}{4} \right\rfloor + 1;\\
&f(v_i v_{n+i}) = 1, && 2 \le i \le \left\lfloor \frac{3n+1}{4} \right\rfloor + 1;\\
&f(e_1 e_2) = 1; &&\\
&f(e_i e_{i+1}) = 1, && 2 \le i \le \left\lfloor \frac{3n+1}{4} \right\rfloor + 1;\\
&f(e_i e_{n+i}) = 1, && 2 \le i \le \left\lfloor \frac{3n+1}{4} \right\rfloor + 1;\\
&f(e_i e_{n+i+1}) = 1, && 2 \le i \le \left\lfloor \frac{3n+1}{4} \right\rfloor + 1;\\
&f(e_i v_{i+1}) = 1, && i = \left\lfloor \frac{3n+1}{4} \right\rfloor + 1;
\end{aligned}
$$

For the remaining edges, give label 0 to $\frac{n^2+17n}{4}$ edges such that all the vertices have at least one incident edge with label 0 and label 1 to other edges. In view of the above defined labeling pattern we have $v_f(0) = \left\lceil \frac{3n+1}{2} \right\rceil$, $v_f(1) = \left\lfloor \frac{3n+1}{2} \right\rfloor$, $e_f(0) = \frac{n^2+17n}{4}$ and $e_f(1) = \frac{n^2+17n}{4}$. Therefore $|v_f(0) - v_f(1)| \le 1$ and $|e_f(0) - e_f(1)| = 0$.

Case - 5: For $n > 8$ and $n \equiv 1(\text{mod } 4)$.

$$
\begin{aligned}
&f(v_1 v_2) = 1; &&\\
&f(v_i v_{i+1}) = 1, && 2 \le i \le \frac{3n+5}{4};\\
&f(v_i v) = 1, && 2 \le i \le \frac{3n+5}{4};\\
&f(v_i e_{i-1}) = 1, && 2 \le i \le \frac{3n+5}{4};\\
&f(v_i e_i) = 1, && 2 \le i \le \frac{3n+5}{4};\\
&f(v_i v_{n+i}) = 1, && 2 \le i \le \frac{3n+5}{4};\\
&f(e_1 e_2) = 1; &&\\
&f(e_i e_{i+1}) = 1, && 2 \le i \le \frac{3n+5}{4};\\
&f(e_i e_{n+i}) = 1, && 2 \le i \le \frac{3n+5}{4};\\
&f(e_i e_{n+i+1}) = 1, && 2 \le i \le \frac{3n+5}{4};\\
&f(e_i v_{i+1}) = 1, && i = \frac{3n+5}{4};
\end{aligned}
$$

For the remaining edges, give label 0 to $\left\lfloor \frac{n^2+17n}{4} \right\rfloor$ edges such that all the vertices have at least one incident edge with label 0 and label 1 to other edges. In view of the above defined labeling pattern we have $v_f(0) = \frac{3n+1}{2}$, $v_f(1) = \frac{3n+1}{2}$, $e_f(0) = \left\lfloor \frac{n^2+17n}{4} \right\rfloor$ and $e_f(1) = \left\lceil \frac{n^2+17n}{4} \right\rceil$. Therefore $|v_f(0) - v_f(1)| = 0$ and $|e_f(0) - e_f(1)| \leq 1$.

Case - 6: For $n > 8$ and $[n \equiv 2 (\mathrm{mod}\ 4)$ or $n \equiv 3 (\mathrm{mod}\ 4)]$.

$$
\begin{aligned}
& f(v_1v_2) = 1; && \\
& f(v_iv_{i+1}) = 1, && 2 \leq i \leq \left\lceil \frac{3n+1}{4} \right\rceil + 1; \\
& f(v_iv) = 1, && 2 \leq i \leq \left\lceil \frac{3n+1}{4} \right\rceil + 1; \\
& f(v_ie_{i-1}) = 1, && 2 \leq i \leq \left\lceil \frac{3n+1}{4} \right\rceil + 1; \\
& f(v_ie_i) = 1, && 2 \leq i \leq \left\lceil \frac{3n+1}{4} \right\rceil + 1; \\
& f(v_iv_{n+i}) = 1, && 2 \leq i \leq \left\lceil \frac{3n+1}{4} \right\rceil + 1; \\
& f(e_1e_2) = 1; && \\
& f(e_ie_{i+1}) = 1, && 2 \leq i \leq \left\lfloor \frac{3n+1}{4} \right\rfloor + 1; \\
& f(e_ie_{n+i}) = 1, && 2 \leq i \leq \left\lfloor \frac{3n+1}{4} \right\rfloor + 1; \\
& f(e_ie_{n+i+1}) = 1, && 2 \leq i \leq \left\lfloor \frac{3n+1}{4} \right\rfloor + 1;
\end{aligned}
$$

For the remaining edges, give label 0 to $\left\lfloor \frac{n^2+17n}{4} \right\rfloor$ edges such that all the vertices have at least one incident edge with label 0 and label 1 to other edges. In view of the above defined labeling pattern we have $v_f(0) = \left\lceil \frac{3n+1}{2} \right\rceil$, $v_f(1) = \left\lfloor \frac{3n+1}{2} \right\rfloor$, $e_f(0) = \left\lfloor \frac{n^2+17n}{4} \right\rfloor$ and $e_f(1) = \left\lceil \frac{n^2+17n}{4} \right\rceil$. Therefore $|v_f(0) - v_f(1)| \leq 1$ and $|e_f(0) - e_f(1)| \leq 1$.
Hence, $T(W_n)$ is an edge product cordial graph for $n > 8$ and not an edge product cordial graph for $n \leq 8$. □

Theorem 7.8

$T(W_n)$ is a total edge product cordial graph. ■

Proof. Let v be the apex vertex and $v_1, v_2, v_3, \ldots, v_n$ be the rim vertices of wheel W_n while $e_1, e_2, e_3, \ldots, e_n$ are rim edges and $e_{n+1}, e_{n+2}, e_{n+3}, \cdots, e_{2n}$ are spoke edges of wheel W_n. Then $T(W_n)$ has order $3n+1$ and size $\frac{n^2+17n}{2}$. We define $f : T(W_n) \rightarrow \{0,1\}$ as follows.

$$\begin{aligned} &f(v_i e_{n+i}) = 0, \quad 1 \leq i \leq n; \\ &f(e_i e_{i+1}) = 0, \quad 1 \leq i \leq n-1; \\ &f(e_1 e_n) = 0; \\ &f(v e_{n+1}) = 0. \end{aligned}$$

For the remaining edges give label 1 to any $\left\lfloor \frac{n^2+23n+2}{4} \right\rfloor$ edges and label 0 to other edges.

In view of above defined labeling pattern we have $e_f(0) + v_f(0) = \left\lceil \frac{n^2+23n+2}{4} \right\rceil$ and $e_f(1) + v_f(1) = \left\lfloor \frac{n^2+23n+2}{4} \right\rfloor$.

Therefore $|\left(v_f(0)+e_f(0)\right) - \left(v_f(1)+e_f(1)\right)| \leq 1$. Hence, $T(W_n)$ is a total edge product cordial graph. □

Illustration 7.7.1. The graph $T(W_3)$ and its total edge product cordial labeling is shown in Figure 7.7. Here black colored elements are labeled with 1 while gray colored elements are labeled with 0.

7.8 CONCLUDING REMARKS

The wheel graph is not an edge product cordial graph as shown in [13] while it is a total edge product cordial graph as proved in [14]. Here I have investigated edge product cordial labeling and total edge product cordial labeling for larger graphs obtained from a wheel graph by using graph operations like shadow graph, splitting graph, middle graph and total graph.

7.9 OPEN PROBLEMS

To investigate some characterization(s) or necessary and sufficient condition(s) for the graph to be edge product cordial or total edge product cordial and identify linkages between cordial labeling, edge product cordial labeling and total edge product cordial labeling are open areas of research.

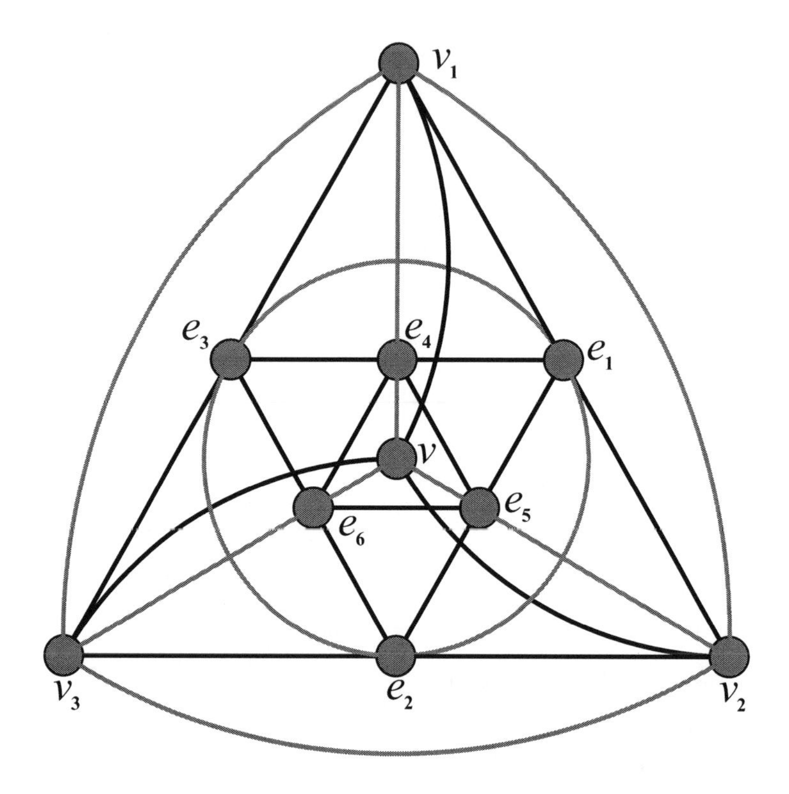

Figure 7.7: $T(W_3)$ and its total edge product cordial labeling

REFERENCES

1. C. M. Barasara. Edge and total edge product cordial labeling of some new graphs. *International Journal of Engineering, Science and Mathematics*, 7(2): 263-273, 2018.
2. G. S. Bloom and S. W. Golomb. Application of Numbered Undirected Graphs. *Proceedings of IEEE*, 65, 562-570, 1977.
3. I. Cahit. Cordial graphs: A weaker version of graceful and harmonious graphs. *Ars Combinatoria*, 23, 201-207, 1987.
4. J. Clark and D. A. Holton. *A First Look at Graph Theory.* Allied Publisher Ltd., 1991.
5. J. A. Gallian. A dynamic survey of graph labeling. *The Electronic Journal of Combinatorics*, 22, #DS6, 2019.
6. J. Ivančo. On edge product cordial graphs. *Opuscula Mathematica*, 39(5): 691-703, 2019.
7. U. M. Prajapati and P. D. Shah. Some edge product cordial graphs in the context of duplication of some graph elements. *Open Journal of Discrete Mathematics*, 6(4): 248-258, 2016.

8. U. M. Prajapati and N. B. Patel. Edge product cordial labeling of some cycle related graphs. *Open Journal of Discrete Mathematics*, 6(4): 268-278, 2016.
9. U. M. Prajapati and N. B. Patel. Edge product cordial labeling of some graphs. *Journal of Applied Mathematics and Computational Mechanics*, 18(1): 69-76, 2019.
10. A. Rosa. On certain valuations of the vertices of a graph. in: *Theory of Graphs (International Symposium, Rome, July 1966)*, Gordon and Breach, New York and Dunod Paris, 349-355, 1967.
11. M. Sundaram, R. Ponraj and S. Somasundaram. Product cordial labeling of graphs. *Bulletin of Pure and Applied Sciences (Mathematics and Statistics)*, 23E, 155-163, 2004.
12. M. Sundaram, R. Ponraj and S. Somasundaram. Total product cordial labeling of graphs. *Bulletin of Pure and Applied Sciences (Mathematics and Statistics)*, 25E(1): 199-203, 2006.
13. S. K. Vaidya and C. M. Barasara. Edge product cordial labeling of graphs. *Journal of Mathematical and Computational Science*, 2(5): 1436-1450, 2012.
14. S. K. Vaidya and C. M. Barasara. Total edge product cordial labeling of graphs. *Malaya Journal of Matematik*, 3(1): 55-63, 2013.
15. S. K. Vaidya and C. M. Barasara. On total edge product cordial labeling. *International Journal of Mathematics and Scientific Computing*, 3(2): 12-16, 2013.
16. S. K. Vaidya and C. M. Barasara. Some edge product cordial graphs. *International Journal of Mathematics and Soft Computing*, 3(3): 49-53 , 2013.
17. S. K. Vaidya and C. M. Barasara. Edge product cordial labeling in the context of some graph operations. *International Journal of Mathematics and Scientific Computing*, 3(1): 4-7, 2013.
18. S. K. Vaidya and C. M. Barasara. Some new families of edge product cordial graphs. *Advanced Modeling and Optimization*, 15(1): 103-111, 2013.
19. S. K. Vaidya and C. M. Barasara. On edge product cordial labeling of some product related graphs. *International Journal of Mathematics And Its Applications*, 2(2): 15-22, 2014
20. S. K. Vaidya and C. M. Barasara. Product and edge product cordial labeling of degree splitting graph of some graphs. *Advances and Applications in Discrete Mathematics*, 15(1): 61-74, 2015.
21. V. Yegnanarayanan and P. Vaidhyanathan. Some interesting applications of graph labellings. *Journal of Mathematical and Computational Science*, 2(5): 1522-1531, 2012.

8 Product Cordial Labeling for the Line Graph of Bistar

M I Bosmia
Department of Mathematics,
Government Engineering College,
Gandhinagar, Gujarat (INDIA)
E-mail: cosmicmohit@gmail.com

Kailas K. Kanani
Department of Mathematics,
Government Engineering College,
Rajkot, Gujarat (INDIA)
E-mail: kananikkk@yahoo.co.in

The line graph $L(G)$ of a graph G is the graph whose vertex set is the edge set of G and two vertices are adjacent in $L(G)$ whenever they are incident in G. In this chapter an investigation of the line graph of bistars $B_{n,n+1}$ and $B_{n,n+2}$ for product cordial labeling has been done.

8.1 INTRODUCTION

In this chapter, all graphs under consideration are finite, connected, undirected and simple $G = (V(G), E(G))$ of order $|V(G)|$ and size $|E(G)|$. For any undefined notation and terminology related to graph theory Gross and Yellen[5] is referred while for number theory Burton[2] is referred.

Many graph labeling techniques have been introduced so far and explored as well by many researchers. Graph labelings have massive applications not only in mathematics but in several areas of computer science and communication networks. A dynamic survey on various graph labeling problems with a wide-ranging bibliography can be found in Gallian[4].

Definition. If the vertices or edges or both are assigned numbers subject to certain condition(s) then it is known as *graph labeling.*

Definition. A mapping $f : V(G) \longrightarrow \{0,1\}$ is called *binary vertex labeling* of G and $f(v)$ is called the label of the vertex v of G under f.

Notations:
If for an edge $e = uv$, the induced edge labeling $f^* : E(G) \longrightarrow \{0,1\}$ is given by $f^*(e = uv) = |f(u) - f(v)|$, then

$v_f(i)$=number of vertices of G having label i under f,
$e_f(i)$=number of edges of G having label i under f^*.

Definition. A binary vertex labeling f of a graph G is called a *cordial labeling* if $|v_f(0) - v_f(1)| \leq 1$ and $|e_f(0) - e_f(1)| \leq 1$. A graph which admits cordial labeling is called a *cordial graph.*

The concept of cordial labeling was introduced by Cahit[3] as a weaker version of graceful and harmonious labeling in 1987. In the same paper Cahit proved the following results:

1. All trees are cordial.
2. The complete graph K_n is cordial if and only if $n \leq 3$.
3. The complete bipartite graph $K_{m,n}$ is cordial.

Vaidya and Shah[8] proved the following results:

1. The shadow graph $D_2(B_{n,n})$ of bistar $B_{n,n}$ is a cordial graph.
2. The splitting graph $S'(B_{n,n})$ of bistar $B_{n,n}$ is a cordial graph.
3. The degree splitting graph $DS(B_{n,n})$ of bistar $B_{n,n}$ is a cordial graph.

Kanani and Bosmia[1] discussed cordial labeling for the line graph $L(B_{n,n})$ of bistar $B_{n,n}$.

Definition. For a graph $G = (V(G); E(G))$, a vertex labeling function $f : V(G) \longrightarrow \{0, 1\}$ induces an edge labeling function $f^* : E(G) \longrightarrow \{0, 1\}$ defined as $f^*(e = uv) = f(u)f(v)$. Then the function f is called a *product cordial labeling* of graph G if $|v_f(0) - v_f(1)| \leq 1$ and $|e_f(0) - e_f(1)| \leq 1$. A graph which admits product cordial labeling is called a *product cordial graph.*

The concept of product cordial labeling was introduced by Sundaram et al.[6] in 2004. In the same paper they proved the following results:

1. All trees are product cordial.
2. Helm H_n is product cordial.
3. $P_m \cup P_n$ is product cordial.

Vaidya and Barasara[7] proved the following results:

1. The graph $L(M(P_n))$ is product cordial for odd n and not product cordial for even n.
2. The graph $L(T_n)$ is product cordial for even n and not product cordial for odd n.
3. The graph $L(ACr_n)$ is a product cordial graph.
4. The graph $L(P_n^2)$ is not product cordial for odd $n > 3$.
5. The graph $L(S'(P_n))$ is not product cordial for even $n > 2$.
6. The graph $L(T(P_n))$ is not a product cordial graph.

Kanani and Bosmia[1] discussed product cordial labeling for the line graph $L(B_{n,n})$ of bistar $B_{n,n}$.

Definition. *Bistar* $B_{m,n}$ is the graph obtained by joining the center(apex) vertices of $K_{1,m}$ and $K_{1,n}$ by an edge.

Definition. The line graph $L(G)$ of a graph G is the graph whose vertices are the edges of G, with $ef \in E(L(G))$ when $e = uv$ and $f = vw$ in G.

Illustration 8.1.1. Bistar $B_{5,7}$ is shown in the following Figure 8.1.

Line graph $L(B_{5,7})$ of bistar $B_{5,7}$ is shown in the Figure 8.2.
Some characteristic features of line graph $L(B_{m,n})$ of bistar $B_{m,n}$ are:

1. $L(B_{m,n})$ is isomorphic to $(K_m \cup K_n) + K_1$.
2. The vertex e_0 in $L(B_{m,n})$ is the apex vertex with degree $d(e_0) = m + n$.
3. If the apex vertex e_0 is removed in $L(B_{m,n})$ then two complete graphs of order m and n as two components of a vertex deleted subgraph are obtained.

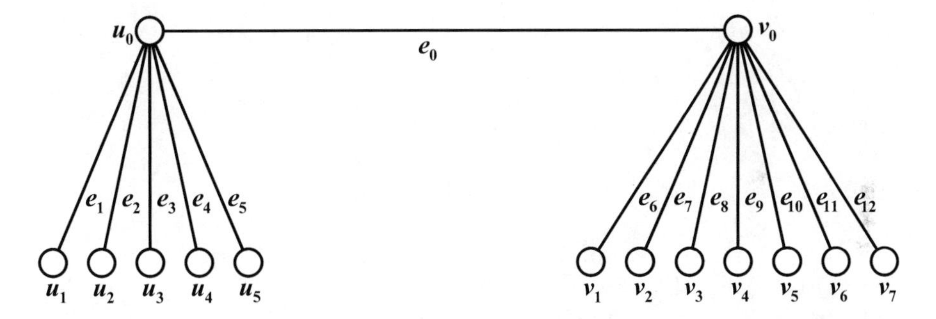

Figure 8.1: Bistar $B_{5,7}$

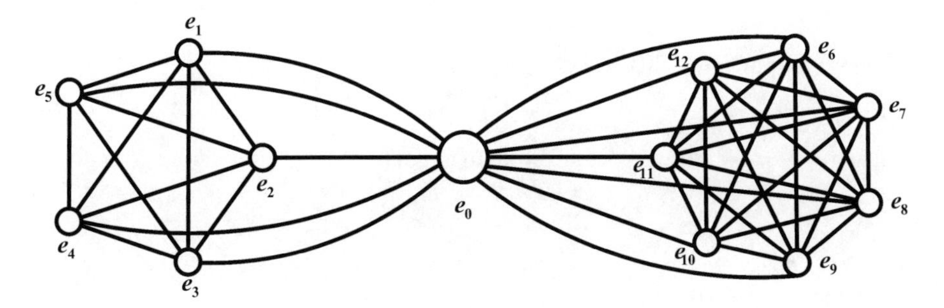

Figure 8.2: $L(B_{5,7})$ Line Graph of Bistar $B_{5,7}$

8.2 MAIN RESULTS

Theorem 8.1

$L(B_{n,n+1})$ is not product cordial for any n. ■

Proof. Let $B_{n,n+1}$ be the bistar with vertex set $\{u_0, v_0, u_i, v_i, v_{n+1} : 1 \leq i \leq n\}$, where u_0, v_0 are apex vertices and u_i, v_i, v_{n+1} are pendent vertices for all $1 \leq i \leq n$. Let $\{e_0 = u_0v_0, e_i = u_0u_i, e_{n+i} = v_0v_i, e_{2n+1} = v_0v_{n+1} : 1 \leq i \leq n\}$ be the edge set of $B_{n,n+1}$. Then $V(L(B_{n,n+1})) = \{e_0, e_1, e_2, \ldots, e_n, e_{n+1}, e_{n+2}, \ldots, e_{2n+1}\}$. Hence, $|V(L(B_{n,n+1}))| = 2n+2$ and $|E(L(B_{n,n+1}))| = (n+1)^2$.

To define vertex labeling $f : V(L(B_{n,n+1})) \longrightarrow \{0,1\}$ the following two cases are considered:

Case 1: When $f(e_0) = 1$.

In order to satisfy vertex condition $|v_f(0) - v_f(1)| \leq 1$ it must be $v_f(0) = v_f(1) = n+1$.

Define $f : V(L(B_{n,n+1})) \longrightarrow \{0,1\}$ as follows:

$f(e_0) = 1$.

Now, $n+1$ vertices must be labeled with label 0 and the remaining n vertices with label 1. To consider all different possibilities of labeling the variable j is defined as $j =$ Number of vertices having label 0 from the vertices $e_{n+1}, e_{n+2}, \ldots, e_{2n+1} = 1 +$ Number of vertices having label 1 from the vertices $e_1, e_2, \ldots, e_n$. It is noted that $1 \leq j \leq n+1$.

Without loss of generality $f(e_i)$ is defined as $f(e_i) = \begin{cases} 1; & 1 \leq i \leq j-1 \\ 0; & j \leq i \leq n \\ 0; & n+1 \leq i \leq n+j \\ 1; & n+j+1 \leq i \leq 2n+1 \end{cases}$

In view of the above define labeling pattern $e_f(1) = n + \frac{(j-1)(j-2)}{2} + \frac{(n+1-j)(n-j)}{2}$.

Now, $|E(L(B_{n,n+1}))| = (n+1)^2$ is either even or odd. Therefore, the following two subcases are considered.

Subcase 1.1: When n is even.

$(n+1)^2$ is odd.

$e_f(1) = \frac{(n+1)^2+1}{2}$ or $\frac{(n+1)^2-1}{2}$. Therefore, the following two subsubcases are considered.

Subsubcase 1.1.1: When $e_f(1) = \frac{(n+1)^2+1}{2}$.

$L(B_{n,n+1})$ is product cordial if and only if $|e_f(0) - e_f(1)| \leq 1$
is product cordial if and only if $e_f(1) = \frac{(n+1)^2+1}{2} = 1 + e_f(0)$
is product cordial if and only if $e_f(1) = n + \frac{(j-1)(j-2)}{2} + \frac{(n+1-j)(n-j)}{2}$
is product cordial if and only if $2j^2 + n - (2n+4)j = 0$
is product cordial if and only if $n = \frac{2j(j-2)}{2j-1}$
But, $n = \frac{2j(j-2)}{2j-1}$ is not a positive integer for any j, $1 \leq j \leq n+1$.
Hence, $L(B_{n,n+1})$ is not product cordial.

Subsubcase 1.1.2: When $e_f(1) = \frac{(n+1)^2-1}{2}$.

$L(B_{n,n+1})$ is product cordial if and only if $|e_f(0) - e_f(1)| \leq 1$
is product cordial if and only if $1 + e_f(1) = 1 + \frac{(n+1)^2-1}{2} = e_f(0)$
is product cordial if and only if $e_f(1) = n + \frac{(j-1)(j-2)}{2} + \frac{(n+1-j)(n-j)}{2}$
is product cordial if and only if $2j^2 + n - (2n+4)j + 2 = 0$
is product cordial if and only if $n = \frac{2(j-1)^2}{2j-1}$
But, $n = \frac{2(j-1)^2}{2j-1}$ is not a positive integer for any j, $1 \leq j \leq n+1$.
Hence, $L(B_{n,n+1})$ is not product cordial.

Subcase 1.2: When n is odd.

$(n+1)^2$ is even.
$e_f(1) = \frac{(n+1)^2}{2}$.

$L(B_{n,n+1})$ is product cordial if and only if $|e_f(0) - e_f(1)| \leq 1$
is product cordial if and only if $e_f(1) = \frac{(n+1)^2}{2} = e_f(0)$
is product cordial if and only if $e_f(1) = n + \frac{(j-1)(j-2)}{2} + \frac{(n+1-j)(n-j)}{2}$
is product cordial if and only if $2j^2 + n - (2n+4)j + 1 = 0$
is product cordial if and only if $n = \frac{2(j-1)^2-1}{2j-1}$
But, $n = \frac{2(j-1)^2-1}{2j-1}$ is not a positive integer for any j, $1 \leq j \leq n+1$.
Hence, $L(B_{n,n+1})$ is not product cordial.

Case 2: When $f(e_0) = 0$.

In order to satisfy vertex condition $|v_f(0) - v_f(1)| \leq 1$ it must be $v_f(0) = v_f(1) = n+1$.

Define $f : V(L(B_{n,n+1})) \longrightarrow \{0, 1\}$ as follows:

$f(e_0) = 0$.

Now, $n+1$ vertices must be labeled with label 1 and the remaining n vertices with label 0. To consider all different possibilities of labeling the variable j is defined as j = Number of vertices having label 1 from the vertices $e_{n+1}, e_{n+2}, \ldots, e_{2n+1}$ = 1 + Number of vertices having label 0 from the vertices $e_1, e_2, \ldots, e_n$. It is noted that $1 \leq j \leq n+1$.

Without loss of generality $f(e_i)$ is defined as $f(e_i) = \begin{cases} 0; \ 1 \leq i \leq j-1 \\ 1; \ j \leq i \leq n \\ 1; \ n+1 \leq i \leq n+j \\ 0; \ n+j+1 \leq i \leq 2n+1 \end{cases}$

In view of the above define labeling pattern $e_f(1) = \frac{(n-(j-1))(n-j)}{2} + \frac{j(j-1)}{2}$.

Now, $|E(L(B_{n,n+1}))| = (n+1)^2$ is either even or odd. Therefore, the following two subcases are considered.

Subcase 2.1: When n is even.

$(n+1)^2$ is odd.
$e_f(1) = \frac{(n+1)^2+1}{2}$ or $\frac{(n+1)^2-1}{2}$. Therefore, the following two subsubcases are considered.

Subsubcase 2.1.1: When $e_f(1) = \frac{(n+1)^2+1}{2}$.

$$\begin{aligned} L(B_{n,n+1}) &\text{ is product cordial if and only if } |e_f(0) - e_f(1)| \leq 1 \\ &\text{is product cordial if and only if } e_f(1) = \frac{(n+1)^2+1}{2} = 1 + e_f(0) \\ &\text{is product cordial if and only if } e_f(1) = \frac{(n-(j-1))(n-j)}{2} + \frac{j(j-1)}{2} \\ &\text{is product cordial if and only if } 2j^2 - n - (2n+2)j - 2 = 0 \\ &\text{is product cordial if and only if } n = \frac{2(j^2-j-1)}{2j+1} \end{aligned}$$

But, $n = \frac{2(j^2-j-1)}{2j+1}$ is not a positive integer for any j, $1 \leq j \leq n+1$.
Hence, $L(B_{n,n+1})$ is not product cordial.

Subsubcase 2.1.2: When $e_f(1) = \frac{(n+1)^2-1}{2}$.

$$\begin{aligned} L(B_{n,n+1}) &\text{ is product cordial if and only if } |e_f(0) - e_f(1)| \leq 1 \\ &\text{is product cordial if and only if } 1 + e_f(1) = 1 + \frac{(n+1)^2-1}{2} = e_f(0) \\ &\text{is product cordial if and only if } e_f(1) = \frac{(n-(j-1))(n-j)}{2} + \frac{j(j-1)}{2} \\ &\text{is product cordial if and only if } 2j^2 - n - (2n+2)j = 0 \\ &\text{is product cordial if and only if } n = \frac{2j(j-1)}{2j+1} \end{aligned}$$

But, $n = \frac{2j(j-1)}{2j+1}$ is not a positive integer for any j, $1 \leq j \leq n+1$.
Hence, $L(B_{n,n+1})$ is not product cordial.

Subcase 1.2: When n is odd.

$(n+1)^2$ is even.
$e_f(1) = \frac{(n+1)^2}{2}$.

$$\begin{aligned}L(B_{n,n+1}) &\text{ is product cordial if and only if } |e_f(0) - e_f(1)| \leq 1\\ &\text{ is product cordial if and only if } e_f(1) = \tfrac{(n+1)^2}{2} = e_f(0)\\ &\text{ is product cordial if and only if } e_f(1) = \tfrac{(n-(j-1))(n-j)}{2} + \tfrac{j(j-1)}{2}\\ &\text{ is product cordial if and only if } 2j^2 - n - (2n+2)j - 1 = 0\\ &\text{ is product cordial if and only if } n = \tfrac{2j^2-2j-1}{2j+1}\end{aligned}$$

But, $n = \frac{2j^2-2j-1}{2j+1}$ is not a positive integer for any j, $1 \leq j \leq n+1$.
Hence, $L(B_{n,n+1})$ is not product cordial.

Therefore, $L(B_{n,n+1})$ is not product cordial for any n. □

Theorem 8.2

$L(B_{n,n+2})$ is product cordial for every n. ■

Proof. Let $B_{n,n+2}$ be the bistar with vertex set $\{u_0, v_0, u_i, v_i, v_{n+1}, v_{n+2} : 1 \leq i \leq n\}$, where u_0, v_0 are apex vertices and $u_i, v_i, v_{n+1}, v_{n+2}$ are pendent vertices for all $1 \leq i \leq n$. Let $\{e_0 = u_0v_0, e_i = u_0u_i, e_{n+i} = v_0v_i, e_{2n+1} = v_0v_{n+1}, e_{2n+2} = v_0v_{n+2} : 1 \leq i \leq n\}$ be the edge set of $B_{n,n+2}$. Then $V(L(B_{n,n+2})) = \{e_0, e_1, e_2, \ldots, e_n, e_{n+1}, e_{n+2}, \ldots, e_{2n+2}\}$. Hence, $|V(L(B_{n,n+1}))| = 2n+3$ and $|E(L(B_{n,n+1}))| = n^2+3n+3$.

Define vertex labeling $f : V(L(B_{n,n+2})) \longrightarrow \{0,1\}$ as follows:

$f(e_0) = 0$,
$f(e_i) = 0; 1 \leq i \leq n$,
$f(e_{n+i}) = 1$,
$f(e_{2n+1}) = 1$,
$f(e_{2n+2}) = 1$,

In view of the above define labeling pattern $v_f(0) = n+1$ and $v_f(1) = n+2$. Now, $e_f(0) = \frac{n(n-1)}{2} + n + n + 2 = \frac{n^2+3n+4}{2}$ and $e_f(1) = \frac{(n+2)(n+1)}{2} = \frac{n^2+3n+2}{2}$.
Thus, $|v_f(0) - v_f(1)| \leq 1$ and $|e_f(0) - e_f(1)| \leq 1$.

Hence, $L(B_{n,n+2})$ is product cordial for every n.

□

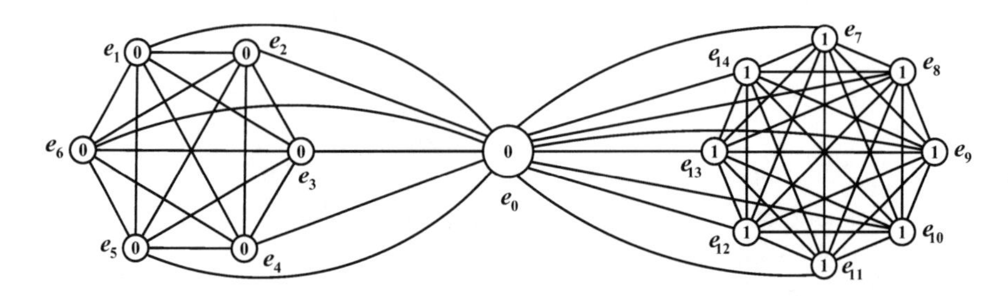

Figure 8.3: Product cordial labeling of $L(B_{6,8})$

Illustration 8.2.1. Product cordial labeling of the graph $L(B_{6,8})$ is shown in the Figure 8.3.

Remark. $B_{n,n+1}$ and $B_{n,n+2}$ are trees and every tree is product cordial. Hence, $B_{n,n+1}$ and $B_{n,n+2}$ are product cordial.

8.3 CONCLUDING REMARK

Bistars $B_{n,n+1}$ and $B_{n,n+2}$ are product cordial as proved by Sundaram et al.[6] while it is proved that the line graph of bistar $B_{n,n+1}$ is not product cordial for any n but the line graph of bistar $B_{n,n+2}$ is product cordial for every n. Thus, product cordiality is not invariant under the line graph for bistar $B_{n,n+1}$ but product cordiality is invariant under the line graph for bistar $B_{n,n+2}$.

REFERENCES

1. M. I. Bosmia and K. K. Kanani. Various graph labeling techniques for the line graph of bistar. *International Journal of Technical Innovation in Modern Engineering & Science*, 4(9): 851-858, 2018.
2. D. M. Burton. *Elementary Number Theory*. Seventh Edition, McGraw-Hill Publisher, 2010.
3. I. Cahit. Cordial Graphs: A weaker version of graceful and harmonious graphs. *Ars Combinatoria*, 23, 201-207. 1987.
4. J. A. Gallian. A dynamic Survey of graph labeling. *The Electronic Journal of Combinatorics*, 20, # D56, 2017.
5. J. Gross and J. Yellen. *Graph Theory and its Applications*. CRC Press, 2005.
6. M. Sundaram, R. Ponraj and S. Somasundaram. Product cordial labeling of graphs. *Bulletin Pure and Applied Sciences (Mathematics & Statistics)*, 23, 155-163, 2004.
7. S. K. Vaidya and C. M. Barasara. Product cordial labeling of line graph of some graphs. *Kragujevac Journal of Mathematics*, 40(2): 290-297, 2016.
8. S. K. Vaidya and N. H. Shah. Cordial labeling of some bistar related graphs. *International Journal of Mathematics and Soft Computing*, 4(2): 33-39, 2014.

9 Sum Divisor Cordial Labeling for Vertex Switching of Cycle related Graphs

D. G. Adalja
Marwadi Education Foundation,
Rajkot, Gujarat (INDIA)
E-mail: divya.adalja@marwadieducation.edu.in

G. V. Ghodasara
H. & H. B. Kotak Institute of Science,
Rajkot, Gujarat (INDIA)
E-mail: gaurang_enjoy@yahoo.co.in

A sum divisor cordial labeling of a graph G with vertex set V (G) is a bijection f from $V(G)$ to $\{1,2,3,\ldots,|V(G)|\}$ such that an edge $e=uv$ is assigned the label 1 if $2/[f(u)+f(v)]$ and 0 otherwise, then the number of edges labeled with 0 and the number of edges labeled with 1 differ by at most 1. A graph which admits sum divisor cordial labeling is called a sum divisor cordial graph. In this research article we prove that the graphs obtained by switching of a vertex in C_n ($n \geq 4$) with one chord, $\mathbf{C}_{n,3}(n \geq 5)$ cycle with twin chords, $C_n(1,1,n-5)(n \geq 6)$, with triangle, crown $C_n \odot P_1$, armed crown $C_n \odot P_2$ are sum divisor cordial.

9.1 INTRODUCTION

Throughout this work, by a graph we mean a simple, finite, undirected graph $G=(V(G),E(G))$ of order $|V(G)|$ and size $|E(G)|$. For terms and notations related to graph theory which are not defined here, we refer to Gross and Yellen[5] and for standard terminology and notations related to number theory we refer to Burton[2]. We will provide a brief summary of definitions and other information which are necessary for the present investigations.

Definition (Graph labeling). If the vertices or edges or both of the graph are assigned values subject to certain conditions it is known as *graph labeling*.

For a dynamic survey on various graph labeling problems along with an extensive bibliography we refer to Gallian[3].

Definition (Divisor cordial labeling). Let $G=(V,E)$ be a simple, finite, connected and undirected graph. A bijection $f: V(G) \rightarrow \{1,2,\ldots,|V(G)|\}$ is said to be *divisor cordial labeling* if the induced function $f^*: E(G) \rightarrow \{0,1\}$ defined by

$$f^*(e=uv) = \begin{cases} 1; & \text{if } f(u) \mid f(v) \text{ or } f(v) \mid f(u) \\ 0; & \text{otherwise} \end{cases}$$

satisfies the condition $|e_f(0) - e_f(1)| \leq 1$.
A graph with divisor cordial labeling is called a *divisor cordial graph.*

By combining the divisibility concept in number theory and the cordial labeling concept in graph labeling, Varatharajan, Navaneethakrishnan and Nagarajan originated the notion called divisor cordial labeling in 2011. They proved that the graphs such as path, cycle, wheel, star and some complete bipartite graphs are divisor cordial graphs. In[8], the same authors proved that some special classes of graphs such as full binary tree, dragon, corona, $G*K_{2,n}$ and $G*K_{3,n}$ are divisor cordial.
Ghodasara and Adalja[4] derived divisor cordial labeling for the ringsum of some standard graphs with star graphs.

Definition (Sum divisor cordial labeling)**.** Let $G = (V,E)$ be a simple graph, $f: V(G) \rightarrow \{1,2,3,\ldots,|V(G)|\}$ be a bijection and the induced function $f^*: E(G) \rightarrow \{0,1\}$ be defined as

$$f^*(e=uv) = \begin{cases} 1; & \text{if } 2 \mid [f(u)+f(v)] \\ 0; & \text{otherwise.} \end{cases}$$

Then f is called *sum divisor cordial labeling* if $|e_f(0) - e_f(1)| \leq 1$.
A graph which admits sum divisor cordial labeling is called a *sum divisor cordial graph.*

Inspired by the idea of divisor cordial labeling in 2016, A. Lourdusamy and F. Patrick originated the concept of one of the variants of divisor cordial labeling called sum divisor cordial labeling. In[7] they proved that the shadow graph and splitting graph of $K_{1,n}$, shadow graph, subdivision graph, splitting graph and degree splitting graph of $B_{n,n}$, subdivision graph of ladder, corona of ladder, triangular ladder with K_1 and closed helm are sum divisor cordial. In[1] Adalja and Ghodasara proved the sum divisor cordial labeling of some new graphs.

Definition (Vertex switching)**.** The *vertex switching* G_v of a graph G is the graph obtained by taking a vertex v of G, removing all the edges incident to v and adding edges joining v to every other vertex which is not adjacent to v in G.

9.2 SUM DIVISOR CORDIAL LABELING FOR VERTEX SWITCHING OF CYCLE RELATED GRAPHS

Definition (A chord of a cycle)**.** *A chord of a cycle* C_n is an edge joining two non-adjacent vertices.

Theorem 9.1

Let G be a graph acquired by vertex switching of cycle $C_n (n \geq 4)$ with one chord such that the chord forms a triangle with two edges of C_n. Then G is a sum divisor cordial graph. ■

Proof. Let G be the cycle C_n with one chord. Let $v_1, v_2, \ldots, v_n$ be the successive vertices of C_n and $e = v_2 v_n$ be the chord of C_n.
The edges $e = v_2 v_n, e_1 = v_1 v_2, e_2 = v_1 v_n$ form a triangle.
Without loss of generality let the switched vertex be v_1 (of degree 2 or degree 3) and let G_{v_1} denote the vertex switching of G with respect to v_1.
Corresponding to the vertices of different degree in C_n with one chord, it is required to discuss the following two cases.

Case 1: $\deg(v_1) = 2$.
Then by the effect of switching operation, the edge set of G_{v_1} is

$$E(G_{v_1}) = \{v_i v_{i+1} \mid 2 \leq i \leq n-1\} \bigcup \{v_2 v_n\} \bigcup \{v_1 v_i \mid 3 \leq i \leq n-1\}.$$

It is to be noted that, $|V(G_{v_1})| = n$ and $|E(G_{v_1})| = 2n-4$.
Here we define labeling function $f : V(G_{v_1}) \to \{1, 2, 3, \ldots, n\}$ as per the following cases.

Subcase 1: $n \equiv 0 (mod\ 4)$.

$$\begin{aligned} f(v_1) &= 1. \\ f(v_2) &= 2. \\ f(v_i) &= \begin{cases} i & ; i \equiv 3 (mod\ 4) \\ i+1 & ; i \equiv 1, 0 (mod\ 4) \\ i+2 & ; i \equiv 2 (mod\ 4); \quad 3 \leq i \leq n-1. \end{cases} \\ f(v_n) &= 4. \end{aligned}$$

Subcase 2: $n \equiv 1 (mod\ 4)$.

$$f(v_i) = \begin{cases} i & ; i \equiv 1, 2 (mod\ 4) \\ i+1 & ; i \equiv 3 (mod\ 4) \\ i-1 & ; i \equiv 0 (mod\ 4); \quad 1 \leq i \leq n. \end{cases}$$

Subcase 3: $n \equiv 2 (mod\ 4)$.

$$\begin{aligned} f(v_1) &= n. \\ f(v_2) &= 2. \end{aligned}$$

$$f(v_i) = \begin{cases} i+1 & ; i \equiv 3(mod\ 4) \\ i-1 & ; i \equiv 0(mod\ 4) \\ i & ; i \equiv 1,2(mod\ 4); \quad 3 \leq i \leq n-1. \end{cases}$$
$$f(v_n) = 1.$$

Subcase 4: $n \equiv 3(mod\ 4)$.

$$f(v_1) = 1.$$
$$f(v_2) = 2.$$
$$f(v_i) = \begin{cases} i & ; i \equiv 3(mod\ 4) \\ i+1 & ; i \equiv 0,1(mod\ 4) \\ i+2 & ; i \equiv 2(mod\ 4); \quad 3 \leq i \leq n-2. \end{cases}$$
$$f(v_{n-1}) = n.$$
$$f(v_n) = 4.$$

By looking into the above prescribed pattern,

$$e_f(1) = e_f(0) = n-2.$$

Case 2 $\deg(v_1) = 3$.
Then by the effect of switching operation, the edge set of G_{v_1} is

$$E(G_{v_1}) = \{v_i v_{i+1} \mid 2 \leq i \leq n-1\} \bigcup \{v_1 v_i \mid 4 \leq i \leq n-1\}.$$

Also it is to be noted that, $|V(G_{v_1})| = n$ and $|E(G_{v_1})| = 2n-6$.
Consider a bijection $f : V(G_{v_1}) \to \{1, 2, 3, \ldots, n\}$ defined as below.
Subcase 1: $n \equiv 0, 1, 2(mod\ 4)$.

$$f(v_i) = \begin{cases} i & ; i \equiv 1,2(mod\ 4) \\ i+1 & ; i \equiv 3(mod\ 4) \\ i-1 & ; i \equiv 0(mod\ 4); \quad 1 \leq i \leq n. \end{cases}$$

Subcase 2: $n \equiv 3(mod\ 4)$.

$$f(v_i) = \begin{cases} i & ; i \equiv 1,2(mod\ 4) \\ i+1 & ; i \equiv 3(mod\ 4) \\ i-1 & ; i \equiv 0(mod\ 4); \quad 1 \leq i \leq n-1. \end{cases}$$
$$f(v_n) = n.$$

By looking into the above prescribed pattern, $e_f(1) = e_f(0) = n-3$.
Then we get, $|e_f(0) - e_f(1)| \leq 1$ in each case.
That is, G_v is sum divisor cordial, where G is cycle C_n with one chord. □

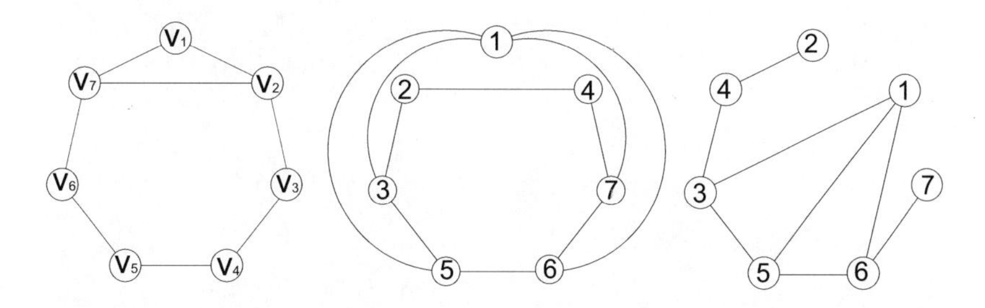

Figure 9.1: Sum divisor cordial labeling in the graph obtained by vertex switching of cycle C_7 with one chord

Illustration 9.2.1. Cycle graph C_7 with one chord and sum divisor cordial labeling for the graph obtained by vertex switching of cycle C_7 with one chord with respect to vertex of degree 2 and degree 3 respectively are shown in Figure 9.1.

Definition (Two chords of a cycle). C_n are said to be twin chords if they form a triangle with an edge of C_n.

For positive integers n and p with $5 \leq p+2 \leq n$, $C_{n,p}$ is the graph consisting of a cycle C_n with twin chords where chords form cycles C_p, C_3 and C_{n+1-p} without chords on the edges of C_n.

Theorem 9.2

Let G be a graph acquired by vertex switching of cycle $C_n (n \geq 5)$ with twin chords such that the chord forms two triangles with edges of C_n. Then G is a sum divisor cordial graph. ■

Proof. Let $V(C_{n,3}) = \{v_1, v_2, \ldots, v_n\}$, where $v_i (1 \leq i \leq n)$ are the vertices of C_n.

Let $E(C_{n,3}) = \{v_i v_{i+1} \mid 1 \leq i \leq n-1\} \bigcup \{v_n v_1\} \bigcup \{v_2 v_n\} \bigcup \{v_2 v_{n-1}\}$, where $v_2 v_n$, $v_2 v_{n-1}$ are chords.

Without loss of generality let v_1 be the switched vertex and let $(C_{n,3})_{v_1}$ denote the graph constructed from switching of vertex v_1 in $C_{n,3}$.

Corresponding to the vertices of different degree in $C_{n,3}$, it is required to discuss the following three cases.

Case 1: $\deg(v_1) = 2$.

Then by the effect of switching operation, the edge set of $(C_{n,3})_{v_1}$ is

$$E((C_{n,3})_{v_1}) = \{v_i v_{i+1} \mid 2 \leq i \leq n-1\} \bigcup \{v_2 v_n\} \bigcup \{v_2 v_{n-1}\} \bigcup \{v_1 v_i \mid 3 \leq i \leq n-1\}.$$

In this case it is to be noted that, $|V((C_{n,3})_{v_1})| = n$ and $|E((C_{n,3})_{v_1})| = 2n-3$. Here we define labeling function $f : V((C_{n,3})_{v_1}) \rightarrow \{1,2,3,\ldots,n\}$ as per the following cases.

Subcase 1: $n \equiv 0,1,2(mod\ 4)$.

$$f(v_i) = \begin{cases} i & ; i \equiv 1,2(mod\ 4) \\ i+1 & ; i \equiv 3(mod\ 4) \\ i-1 & ; i \equiv 0(mod\ 4); \quad 1 \leq i \leq n. \end{cases}$$

Subcase 2: $n \equiv 3(mod\ 4)$.

$$f(v_i) = \begin{cases} i & ; i \equiv 1,2(mod\ 4) \\ i+1 & ; i \equiv 3(mod\ 4) \\ i-1 & ; i \equiv 0(mod\ 4); \quad 1 \leq i \leq n-1. \end{cases}$$
$$f(v_n) = n.$$

By looking into the above prescribed pattern,

Cases of n	Edge label conditions
$n \equiv 0,1,3(mod\ 4)$	$e_f(1) = n-2, e_f(0) = n-1$
$n \equiv 2(mod\ 4)$	$e_f(1) = n-1, e_f(0) = n-2$

Case 2: $\deg(v_1) = 3$.

Then by the effect of switching operation, the edge set of $(C_{n,3})_{v_1}$ is

$$E((C_{n,3})_{v_1}) = \{v_iv_{i+1} \mid 2 \leq i \leq n-1\} \bigcup \{v_3v_n\} \bigcup \{v_1v_i \mid 4 \leq i \leq n-1\}.$$

In this case it is to be noted that, $|V((C_{n,3})_{v_1})| = n$ and $|E((C_{n,3})_{v_1})| = 2n-5$. Consider a bijection $f : V((C_{n,3})_{v_1}) \rightarrow \{1,2,3,\ldots,n\}$ defined as below.

Subcase 1: $n \equiv 0,1,2(mod\ 4)$.

$$f(v_i) = \begin{cases} i & ; i \equiv 1,2(mod\ 4) \\ i+1 & ; i \equiv 3(mod\ 4) \\ i-1 & ; i \equiv 0(mod\ 4); \quad 1 \leq i \leq n. \end{cases}$$

Subcase 2: $n \equiv 3(mod\ 4)$.

$$f(v_i) = \begin{cases} i & ; i \equiv 1,2(mod\ 4) \\ i+1 & ; i \equiv 3(mod\ 4) \\ i-1 & ; i \equiv 0(mod\ 4); \quad 1 \leq i \leq n-1. \end{cases}$$
$$f(v_n) = n.$$

By looking into the above prescribed pattern,

Cases of n	Edge label conditions
$n \equiv 0,2,3(mod\ 4)$	$e_f(1) = \lfloor \frac{2n-5}{2} \rfloor, e_f(0) = \lceil \frac{2n-5}{2} \rceil$
$n \equiv 1(mod\ 4)$	$e_f(0) = \lfloor \frac{2n-5}{2} \rfloor, e_f(1) = \lceil \frac{2n-5}{2} \rceil$

Case 3: $\deg(v_1) = 4$.
Then by the effect of switching operation, the edge set of $(C_{n,3})_{v_1}$ is

$$E((C_{n,3})_{v_1}) = \{v_i v_{i+1} \mid 2 \le i \le n-1\} \bigcup \{v_1 v_i \mid 3 \le i \le n-3\}.$$

In this case it is to be noted that, $|V((C_{n,3})_{v_1})| = n$ and $|E((C_{n,3})_{v_1})| = 2n-7$.
Consider a bijection f from $f : V((C_{n,3})_{v_1}) \to \{1,2,3,\ldots,n\}$ defined as below.
Subcase 1: $n \equiv 0,1,2(mod\ 4)$.

$$f(v_i) = \begin{cases} i & ; i \equiv 1,2(mod\ 4) \\ i+1 & ; i \equiv 3(mod\ 4) \\ i-1 & ; i \equiv 0(mod\ 4); \quad 1 \le i \le n. \end{cases}$$

Subcase 2: $n \equiv 3(mod\ 4)$.

$$f(v_n) = n.$$

$$f(v_i) = \begin{cases} i & ; i \equiv 1,2(mod\ 4) \\ i+1 & ; i \equiv 3(mod\ 4) \\ i-1 & ; i \equiv 0(mod\ 4); \quad 1 \le i \le n-1. \end{cases}$$

By looking into the above prescribed pattern,

Cases of n	Edge label conditions
$n \equiv 0,1,2(mod\ 4)$	$e_f(0) = \left\lfloor \frac{2n-7}{2} \right\rfloor, e_f(1) = \left\lceil \frac{2n-7}{2} \right\rceil$
$n \equiv 3(mod\ 4)$	$e_f(1) = \left\lfloor \frac{2n-7}{2} \right\rfloor, e_f(0) = \left\lceil \frac{2n-7}{2} \right\rceil$

Then we get, $|e_f(0) - e_f(1)| \le 1$ in each case.
That is, $(C_{n,3})_v$ is a sum divisor cordial graph. □

Illustration 9.2.2. $C_{8,3}$ graph and sum divisor cordial labeling for the graph obtained by vertex switching of $C_{8,3}$ with respect to vertex of degree 2, degree 3 and degree 4 respectively are shown in Figure 9.2.

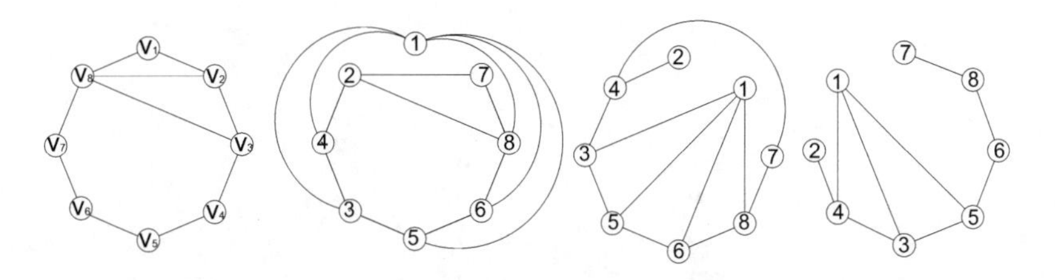

Figure 9.2: Sum divisor cordial labeling in the graph obtained by vertex switching of $C_{8,3}$

Definition (A cycle with triangle). *A cycle with triangle* is a cycle with three chords which by themselves form a triangle.

For positive integers p,q,r and $n \geq 6$ with $p+q+r+3=n$, $C_n(p,q,r)$ denotes a cycle with triangle whose edges form the edges of cycles C_{p+2}, C_{q+2}, C_{r+2} without chords.

Theorem 9.3

Let G be a graph acquired by vertex switching of cycle $C_n (n \geq 6)$ with triangle such that three chords by themselves form a triangle. Then G is a sum divisor cordial graph. ■

Proof. Let $G = C_n(1,1,n-5)$. Let $V(G) = \{v_i \mid 1 \leq i \leq n\} = V(C_n)$.
Let $E(G) = \{v_i v_{i+1} \mid 1 \leq i \leq n-1\} \bigcup \{v_n v_1\} \bigcup \{v_1 v_3\} \bigcup \{v_3 v_{n-1}\} \bigcup \{v_{n-1} v_1\}$, where $v_1 v_3$, $v_3 v_{n-1}, v_{n-1} v_1$ are chords.
Without loss of generality let v_1 be the switched vertex.
Let G_{v_1} denote the graph constructed from switching of arbitrary vertex v_1 of G.
Corresponding to the vertices of different degree in $C_n(1,1,n-5)$, it is required to discuss the following two cases.
Case 1: $\deg(v_1) = 2$.
Then by the effect of switching operation, the edge set of G_{v_1} is $E(G_{v_1}) = \{v_i v_{i+1} \mid 2 \leq i \leq n-1\} \bigcup \{v_2 v_4\} \bigcup \{v_4 v_{n-1}\} \bigcup \{v_{n-1} v_2\} \bigcup \{v_1 v_i \mid 3 \leq i \leq n-1\}$.
In this case it is to be noted that, $|V(G_{v_1})| = n$ and $|E(G_{v_1})| = 2n-2$.
Here we define labeling function $f : V(G_{v_1}) \to \{1,2,3,\ldots,n\}$ as per the following cases.
Subcase 1: $n \equiv 1,2 (mod\ 4)$.

$$f(v_i) = \begin{cases} i & ; i \equiv 1,2(mod\ 4) \\ i+1 & ; i \equiv 3(mod\ 4) \\ i-1 & ; i \equiv 0(mod\ 4); \quad 1 \leq i \leq n. \end{cases}$$

Subcase 2: $n \equiv 3(mod\ 4)$.

$$f(v_1) = n-1.$$

$$f(v_i) = \begin{cases} i-1 & ; i \equiv 1,2(mod\ 4) \\ i & ; i \equiv 3(mod\ 4) \\ i-2 & ; i \equiv 0(mod\ 4); \quad 2 \leq i \leq n. \end{cases}$$

By looking into the above prescribed pattern,

$$e_f(1) = e_f(0) = n-1.$$

Case 2: $\deg(v_1) = 4$.
Then by the effect of switching operation, the edge set of G_{v_1} is

$$E(G_{v_1}) = \{v_iv_{i+1} \mid 2 \leq i \leq n-1\} \bigcup \{v_3v_{n-1}\} \bigcup \{v_1v_i \mid 4 \leq i \leq n-2\}.$$

In this case it is to be noted that, $|V(G_{v_1})| = n$ and $|E(G_{v_1})| = 2n-6$.
Consider a bijection $f : V(G_{v_1}) \rightarrow \{1,2,3,\ldots,n\}$ defined as below.
Subcase 1: $n \equiv 1,2(mod\ 4)$.

$$f(v_i) = \begin{cases} i & ; i \equiv 1,2(mod\ 4) \\ i+1 & ; i \equiv 3(mod\ 4) \\ i-1 & ; i \equiv 0(mod\ 4); \quad 1 \leq i \leq n. \end{cases}$$

Subcase 2: $n \equiv 3(mod\ 4)$.

$$\begin{aligned} f(v_1) &= n-1. \\ f(v_i) &= \begin{cases} i & ; i \equiv 1,2(mod\ 4) \\ i+1 & ; i \equiv 3(mod\ 4) \\ i-1 & ; i \equiv 0(mod\ 4); \quad 2 \leq i \leq n-2. \end{cases} \\ f(v_{n-1}) &= n. \\ f(v_n) &= 1. \end{aligned}$$

By looking into the above prescribed pattern,

$$e_f(1) = e_f(0) = n-3.$$

Then we get, $|e_f(0) - e_f(1)| \leq 1$ in each case.
That is, $(C_n(1,1,n-5))_v$ is a sum divisor cordial graph. □

Illustration 9.2.3. $C_7(1,1,2)$ graph and sum divisor cordial labeling for the graph obtained by vertex switching of $C_7(1,1,2)$ with respect to the vertex of degree 2 and degree 4 respectively are shown in Figure 9.3.

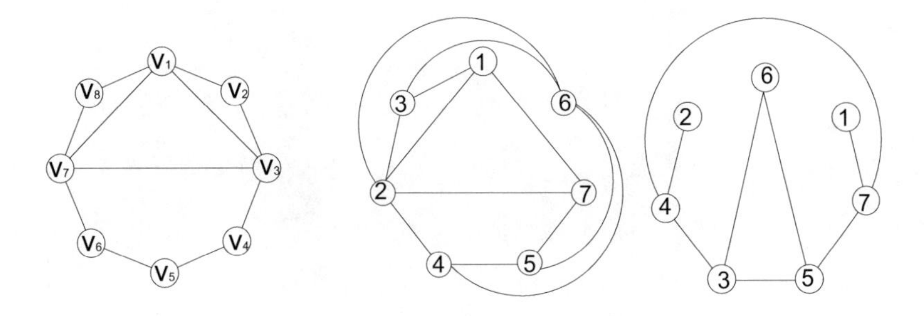

Figure 9.3: Sum divisor cordial labeling in the graph obtained by vertex switching of $C_7(1,1,2)$

Definition (Crown graph). *Crown graph* $C_n \odot K_1$ is the graph obtained by joining a pendent edge to each vertex of cycle C_n. It is denoted by C_n, where n is the number of vertices in cycle C_n.

Theorem 9.4

Let G be a graph acquired by vertex switching of the crown $C_n \odot K_1$. Then G is a sum divisor cordial graph. ∎

Proof. Let $G = C_n \odot K_1$ be the crown with the vertex set $V(C_n \odot K_1) = \{v_i, u_i \mid 1 \leq i \leq n\}$, where v_i are pendent vertices and u_i are vertices of degree 3.

$$E(C_n \odot K_1) = \{u_i u_{i+1}, u_n u_1 \mid 1 \leq i \leq n-1\}, \bigcup \{u_i v_i \mid 1 \leq i \leq n\}.$$

Let $(C_n \odot K_1)_v$ be the graph constructed from switching of an arbitrary vertex v in $C_n \odot K_1$.
Corresponding to the vertices of different degree in $C_n \odot K_1$, it is required to discuss the following two cases.
Case 1: $\deg(v_1) = 1$.
Without loss of generality let us assume that the switched pendent vertex is v_1.
In this case it is to be noted that, $|V(C_n \odot K_1)_{v_1}| = 2n$ and $|E(C_n \odot K_1)_{v_1}| = 4n-3$.
We define vertex labeling $f : V(G_v) \to \{1, 2, \ldots 2n\}$ as follows.

$$\begin{aligned} f(v_i) &= 2i-1; \quad 1 \leq i \leq n. \\ f(u_i) &= 2i; \qquad 1 \leq i \leq n. \end{aligned}$$

By looking into the above prescribed pattern,

$$e_f(1) = 2n-1, e_f(1) = 2n-2.$$

Case 2: $\deg(u_1) = 3$.
Without loss of generality let us assume that the switched vertex is u_1.
In this case it is to be noted that, $|V(C_n \odot K_1)_{u_1}| = 2n$ and $|E(C_n \odot K_1)_{u_1}| = 4n-7$.
Consider a bijection $f : V(G_v) \to \{1, 2, \ldots 2n\}$ defined as below.

$$\begin{aligned} f(u_1) &= 1. \\ f(v_1) &= 2. \\ f(u_i) &= 2i; \qquad 2 \leq i \leq n. \\ f(v_i) &= 2i-1; \quad 2 \leq i \leq n. \end{aligned}$$

By looking into the above prescribed pattern,

$$e_f(1) = 2n-3, e_f(0) = 2n-4.$$

Then we get, $|e_f(0) - e_f(1)| \leq 1$ in each case.
That is, the graph G_v obtained by switching of a vertex in the crown $C_n \odot K_1$ is sum divisor cordial. □

Illustration 9.2.4. The crown graph $C_7 \odot K_1$ and sum divisor cordial labeling for the graph obtained by vertex switching in $C_7 \odot K_1$ with respect to vertex v_1 of degree 1 are shown in Figure 9.4.

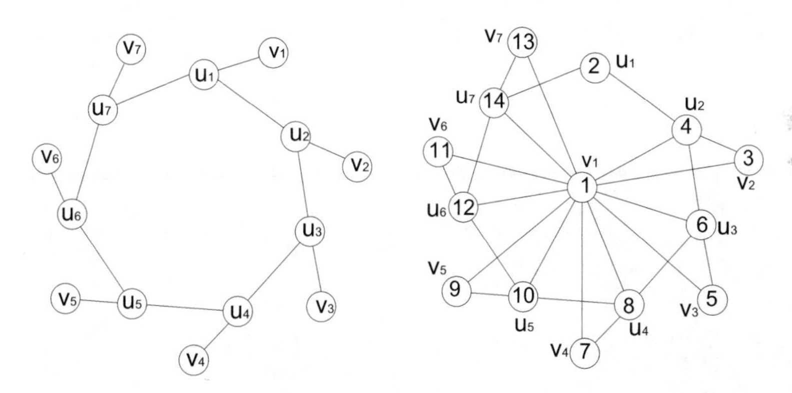

Figure 9.4: Sum divisor cordial labeling in the graph $(C_7 \odot K_1)_{v_1}$

Illustration 9.2.5. The crown graph $C_7 \odot K_1$ and sum divisor cordial labeling for the graph obtained by vertex switching in $C_7 \odot K_1$ with respect to vertex u_1 of degree 3 are shown in Figure 9.5.

Definition (The armed crown graph)**.** AC_n is the graph obtained by attaching a path P_2 at each vertex of cycle C_n. It is denoted by AC_n, where n is the number of vertices in cycle C_n.

Theorem 9.5

Let G be a graph acquired by vertex switching of the armed crown AC_n. Then G is a sum divisor cordial graph. ■

Proof. Let AC_n be the armed crown graph with the vertex set $V(G) = \{v_i, w_i, u_i \mid 1 \leq i \leq n\}$, where v_i, w_i and u_i are vertices of degree one, two and three respectively.

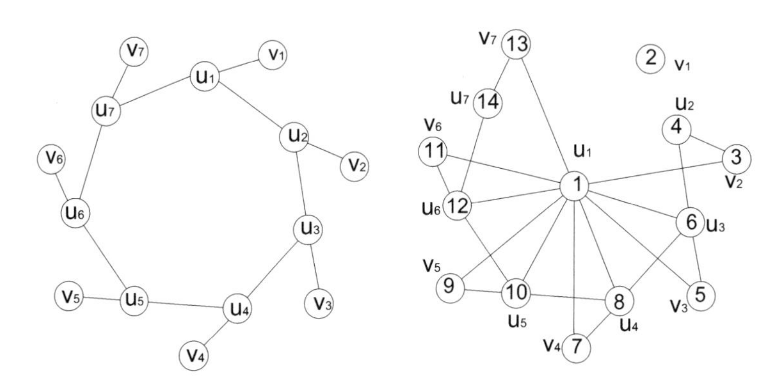

Figure 9.5: Sum divisor cordial labeling in $(C_7 \odot K_1)_{u_1}$

$$E(AC_n) = \{u_iu_{i+1}, u_nu_1 \mid 1 \leq i \leq n-1\} \bigcup \{u_iw_i, w_iv_i \mid 1 \leq i \leq n\}.$$

Let $(AC_n)_v$ denote the graph constructed from switching of an arbitrary vertex v in AC_n.
According to different degrees of vertices of the graph $(AC_n)_v$, it is required to discuss the following three cases.
Case 1: $\deg(v_1) = 1$.
Without loss of generality let us assume that the switched pendent vertex is v_1.
In this case it is to be noted that, $|V((AC_n)_{v_1})| = 3n$ and $|E((AC_n)_{v_1})| = 6n-3$.
We define vertex labeling $f : V(G_v) \to \{1, 2, \ldots 3n\}$ as follows.

$$\begin{aligned} f(v_1) &= 1. \\ f(u_i) &= 2i+1; \quad 1 \leq i \leq n. \\ f(w_i) &= 2i; \qquad 1 \leq i \leq n. \\ f(v_i) &= 2n+i; \quad 2 \leq i \leq n. \end{aligned}$$

By looking into the above prescribed pattern,

$$e_f(1) = 3n-1, \; e_f(1) = 3n-2.$$

Case 2: $\deg(w_1) = 2$.
Without loss of generality let us assume that the switched vertex is w_1.
In this case it is to be noted that, $|V((AC_n)_{w_1})| = 3n$ and $|E((AC_n)_{w_1})| = 6n-5$.
Consider a bijection $f : V(G_v) \to \{1, 2, \ldots 3n\}$ defined as below.

$$\begin{aligned} f(v_1) &= 2n. \\ f(w_1) &= 1. \\ f(u_i) &= 2i+1; \quad 1 \leq i \leq n. \end{aligned}$$

$$\begin{aligned} f(w_i) &= 2(i-1); \quad 2 \leq i \leq n. \\ f(v_i) &= 2n+i; \quad 2 \leq i \leq n. \end{aligned}$$

By looking into the above prescribed pattern,

$$e_f(1) = 3n-2, \ e_f(0) = 3n-3.$$

Case 3: $\deg(u_1) = 3$.
Without loss of generality let us assume that the switched vertex is u_1.
In this case it is to be noted that, $|V((AC_n)_{u_1})| = 3n$ and $|E((AC_n)_{u_1})| = 6n-7$.
Consider a bijection $f : V(G_v) \to \{1,2,\ldots 3n\}$ defined as below.
For $n \leq 7$:
Subcase 1: $n \equiv 0,2(mod\ 4)$

$$\begin{aligned} f(u_1) &= 1. \\ f(u_i) &= 2n+i; \qquad 2 \leq i \leq n. \\ f(w_{2i-1}) &= 4i-2; \qquad 1 \leq i \leq \frac{n}{2}. \\ f(w_{2i}) &= 4i-1; \qquad 1 \leq i \leq \frac{n}{2}. \\ f(v_{2i-1}) &= 4i; \qquad 1 \leq i \leq \frac{n}{2}. \\ f(v_{2i}) &= 4i+1; \qquad 1 \leq i \leq \frac{n}{2}. \end{aligned}$$

Subcase 2: $n \equiv 1,3(mod\ 4)$

$$\begin{aligned} f(u_1) &= 1. \\ f(u_2) &= 2n+1. \\ f(u_3) &= 2n+3. \\ f(u_i) &= 2n+i; \qquad 4 \leq i \leq n. \\ f(w_{2i-1}) &= 4i-2; \qquad 1 \leq i \leq \left\lceil \frac{n}{2} \right\rceil. \\ f(w_{2i}) &= 4i-1; \qquad 1 \leq i \leq \left\lfloor \frac{n}{2} \right\rfloor. \\ f(v_{2i-1}) &= 4i; \qquad 1 \leq i \leq \left\lceil \frac{n}{2} \right\rceil. \\ f(v_{2i}) &= 4i+1; \qquad 1 \leq i \leq \left\lfloor \frac{n}{2} \right\rfloor. \end{aligned}$$

By looking into the above prescribed pattern,

$$e_f(1) = \left\lfloor \frac{6n-7}{2} \right\rfloor, \quad e_f(0) = \left\lceil \frac{6n-7}{2} \right\rceil.$$

For $n > 7$:
Subcase 1: $n \equiv 0,2(mod\ 4)$

$$f(u_1) = 1.$$

$$f(u_2) = 2n+3.$$
$$f(u_3) = 2n+2.$$
$$f(u_i) = 2n+i; \quad 4 \le i \le n.$$
$$f(w_{2i-1}) = 4i-2; \quad 1 \le i \le \frac{n}{2}.$$
$$f(w_{2i}) = 4i-1; \quad 1 \le i \le \frac{n}{2}.$$
$$f(v_{2i-1}) = 4i; \quad 1 \le i \le \frac{n}{2}.$$
$$f(v_{2i}) = 4i+1; \quad 1 \le i \le \frac{n}{2}.$$

Subcase 2: $n \equiv 1,3(mod\ 4)$

$$f(u_1) = 1.$$
$$f(u_2) = 2n+1.$$
$$f(u_3) = 2n+4.$$
$$f(u_4) = 2n+3.$$
$$f(u_i) = 2n+i; \quad 4 \le i \le n.$$
$$f(w_{2i-1}) = 4i-2; \quad 1 \le i \le \left\lceil \frac{n}{2} \right\rceil.$$
$$f(w_{2i}) = 4i-1; \quad 1 \le i \le \left\lfloor \frac{n}{2} \right\rfloor.$$
$$f(v_{2i-1}) = 4i; \quad 1 \le i \le \left\lceil \frac{n}{2} \right\rceil.$$
$$f(v_{2i}) = 4i+1; \quad 1 \le i \le \left\lfloor \frac{n}{2} \right\rfloor.$$

By looking into the above prescribed pattern,

$$e_f(0) = \left\lfloor \frac{6n-7}{2} \right\rfloor, \quad e_f(1) = \left\lceil \frac{6n-7}{2} \right\rceil.$$

Then we get, $|e_f(0) - e_f(1)| \le 1$ in each case.
That is, the graph G_v obtained by switching of a vertex in the armed crown AC_n is sum divisor cordial. □

Illustration 9.2.6. The armed crown graph AC_5 and sum divisor cordial labeling for the graph obtained by vertex switching in AC_5 with respect to vertex v_1 of degree 1 are shown in Figure 9.6.

Illustration 9.2.7. The armed crown graph AC_5 and sum divisor cordial labeling for the graph obtained by vertex switching in AC_5 with respect to vertex w_1 of degree 1 are shown in Figure 9.7.

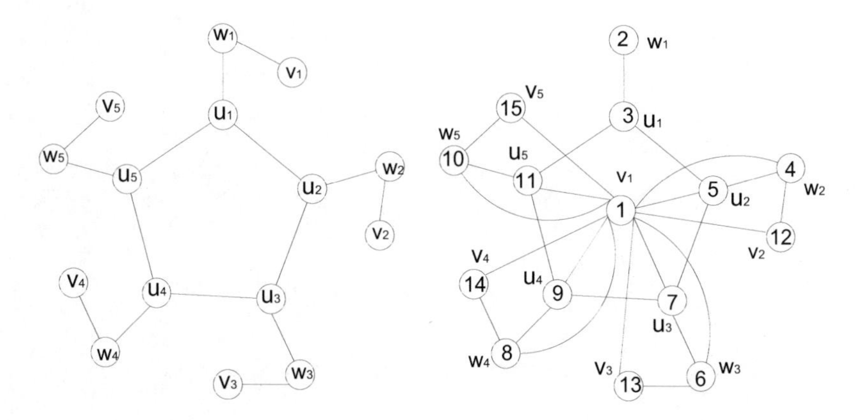

Figure 9.6: Sum divisor cordial labeling in the graph $(AC_5)_{v_1}$

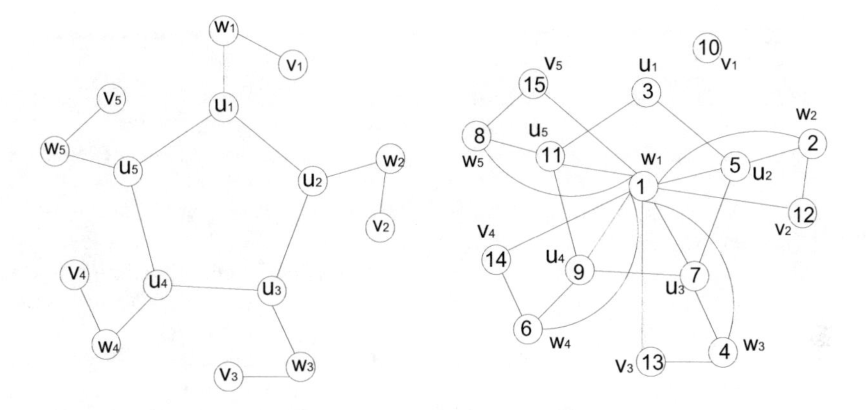

Figure 9.7: Sum divisor cordial labeling in the graph $(AC_5)_{w_1}$

Illustration 9.2.8. The armed crown graph AC_5 and sum divisor cordial labeling for the graph obtained by vertex switching in AC_5 with respect to vertex u_1 of degree 1 are shown in Figure 9.8.

9.3 CONCLUDING REMARKS

Here, we have investigated some new results related to the graph operation vertex switching for the sum divisor cordial labeling technique. To explore some new sum divisor cordial graphs in the context of other graph operations is an open area of research.

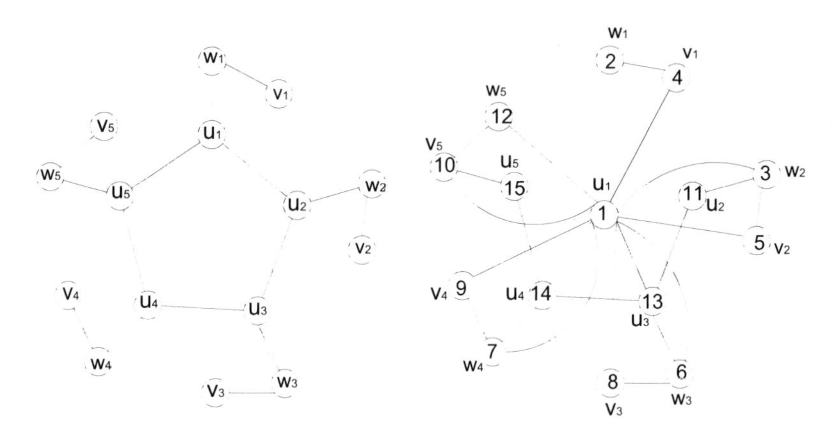

Figure 9.8: Sum divisor cordial labeling in the graph $(AC_5)_{u_1}$

REFERENCES

1. D. G. Adalja and G. V. Ghodasara. Some new sum divisor cordial graphs. *International Journal of Applied Graph Theory*, 2(1): 19-33, 2018.
2. D. M. Burton. *Elementary Number Theory.* Second Edition, Brown Publishers, 1990.
3. J. A. Gallian. A dynamic survey of graph labeling. *The Electronic Journal of Combinatorics*, # DS6, 20, 2017.
4. G. V. Ghodasara and D. G. Adalja. Divisor cordial labeling in context of ring sum of graphs. *International Journal of Mathematics and Soft Computing*, 7(1): 23-31, 2017.
5. J. Gross and J. Yellen. *Graph Theory and Its Applications.* CRC Press, 2004.
6. A. Lourdusamy, F. Patrick and J. Shiama. Sum divisor cordial graphs. *Proyecciones Journal of Mathematics*, 35(1): 119-136, 2016.
7. A. Lourdusamy and F. Patrick. Sum divisor cordial labeling for star and ladder related graphs. *Proyecciones Journal of Mathematics*, 35(4): 437-455, 2016.
8. R. Varatharajan, S. Navanaeethakrishnan and K. Nagarajan. Divisor cordial graphs. *International J. Math. Combin.*, 4, 15-25, 2011.
9. V. Yegnanaryanan and P. Vaidhyanathan. Some interesting applications of graph labellings. *J. Math. Comput. Sci.*, 2(5):1522-1531, 2012.

10 A Few Results on Fibonacci Cordial Labeling

U. M. Prajapati
St. Xavier's College,
Ahmedabad, Gujarat (INDIA)
E-mail: udayan64@yahoo.com

K. K. Raval
Department of Mathematics,
Gujarat University,
Ahmedabad, Gujarat (INDIA)
E-mail: karishma.raval13@gmail.com

An injective function $f : V(G) \rightarrow \{F_0, F_1, F_2, \dots, F_n\}$, where F_j is the jth Fibonacci number is called Fibonacci cordial labeling if the induced function $f^* : E(G) \rightarrow \{0,1\}$ defined by $f^*(uv) = (f(u) + f(v))\ (mod\ 2)$ satisfies the condition that the total number of edges with label 1 and total number of edges with label 0 differ by at most 1. A graph which admits Fibonacci cordial labeling is called a Fibonacci cordial graph. We show that a comb graph and helm graph are Fibonacci cordial graphs. The cycle admits Fibonacci cordial labeling if and only if $n \equiv 0,1,3\ (mod\ 4)$ and $n \geq 3$. A friendship graph is a Fibonacci cordial graph. C_n^2 is a Fibonacci cordial graph if n is even and $n \geq 4$.

10.1 INTRODUCTION

We begin with simple, finite, undirected graph G, where $V(G)$ and $E(G)$ are the vertex set and edge set of G respectively. For all other terminology we follow Gross [4]. Now we provide a brief summary of definitions and other information which are necessary for the present investigations.

10.1.1 DEFINITION

Definition. The Fibonacci sequence can be defined by the linear recurrence relation satisfying:

$$F_n = \begin{cases} 0 & , \text{if } n = 0; \\ 1 & , \text{if } n = 1; \\ F_{n-1} + F_{n-2} & , \text{if } n > 1. \end{cases}$$

Definition. Let $f: V(G) \to \{0,1\}$ for each edge uv assigned the label $|f(u) - f(v)|$, then f is said to be a cordial labeling [2] of G if the number of vertices with label 0 and the number of vertices with label 1 differ at most by 1, and the number of edges with label 0 and the number of edges with label 1 differ by at most 1.

Definition. An injective function $f: V(G) \to \{F_0, F_1, F_2, \ldots, F_n\}$, where F_j is the jth Fibonacci number, is said to be Fibonacci cordial labeling [5] if the induced function $f^*: E(G) \to \{0,1\}$ defined by $f^*(uv) = (f(u)+f(v))(mod 2)$ satisfies the condition $|e_f(0) - e_f(1)| \leq 1$, where $e_f(0)$ and $e_f(1)$ are the total number of edges with label 0 and 1 respectively. A graph which admits Fibonacci cordial labeling is called a Fibonacci cordial graph. It was first introduced by Rokad and Ghodasara in [6].
Karthikeyan et. al. showed that the wheel graph and bistar graph are Fibonacci cordial graphs [5].
We also note that F_n is even if and only if $n \equiv 0 \ (mod\ 3)$.

Definition. The graph $W_n = C_n + K_1$ is called a wheel graph [3]. The vertex corresponding to K_1 is called the apex vertex and the vertices corresponding to C_n are called rim vertices.

Definition. The circulant graph [7] $C_n(a_1, a_2, \ldots, a_m)$ is the graph with vertex set $\{v_1, v_2, \ldots, v_m\}$ and the edge set $\{v_i v_{i+a_j}, 1 \leq i \leq n, 1 \leq j \leq m\}$ where addition of indices in modulo n and $m, n, a_1, a_2, \ldots, a_m$ are positive integers such that $1 \leq a_i \leq \left\lfloor \frac{n}{2} \right\rfloor$ and the a_i's are distinct. C_n^k is the generalization of the circulant graphs $C_n(1,2,3,\ldots,k)$ given by Anholcer and Palmer [1].
In this chapter, we use the notation C_n^2 for the circulant graph $C_n(1,2)$. C_n^2 can also be considered as square of cycle graph C_n.

Definition. A helm graph [3] is obtained from the wheel graph W_n by adding a pendent edge on each rim vertex of the wheel.

Definition. The friendship graph [3] is one point union of C_3 which is denoted by $F_n = C_3^n$, where n stands for the total number of C_3.

10.2 MAIN RESULTS

Theorem 10.1

C_n is a Fibonacci cordial graph if and only if $n \equiv 0,1,3 \ (mod\ 4)$ and $n \geq 3$. ■

Proof. Case 1: $n \equiv 0,1,3 (mod\ 4)$.
We label all the consecutive vertices of cycle C_n as $v_1, v_2, v_3, \ldots, v_n$. Thus, $|V(C_n)| = |E(C_n)| = n$.

Sub case 1: $n \equiv 1\ (mod\ 6)$.
We define an injective function $f : V(G) \to \{F_0, F_1, F_2, \ldots, F_n\}$ such that, $f(v_1) = F_0$, $f(v_2) = F_1$, $f(v_3) = F_2$, $f(v_4) = F_4$, $f(v_5) = F_3$, $f(v_6) = F_6$ and $f(v_7) = F_5$ and for $n > 7$:

$$f(v_n) = \begin{cases} F_{n-1} & , \text{if } n \equiv 2,3\ (mod\ 6); \\ F_n & , \text{if } n \equiv 0\ (mod\ 6); \\ F_{n-2} & , \text{if } n \equiv 1\ (mod\ 6); \\ F_n & , \text{if } n \equiv 10\ (mod\ 12); \\ F_{n-2} & , \text{if } n \equiv 11\ (mod\ 12); \\ F_{n-1} & , \text{if } n \equiv 4,5\ (mod\ 12). \end{cases}$$

Then $e_f(1) = \frac{n+1}{2}$ and $e_f(0) = \frac{n-1}{2}$.
Sub case 2: $n \equiv 8\ (mod\ 12)$.
We define an injective function $f : V(G) \to \{F_0, F_1, F_2, \ldots, F_n\}$ such that, $f(v_1) = F_0$, $f(v_2) = F_1$ and for $n > 2$:

$$f(v_n) = \begin{cases} F_{n-1} & , \text{if } n \equiv 2,3\ (mod\ 6); \\ F_n & , \text{if } n \equiv 0\ (mod\ 6); \\ F_{n-2} & , \text{if } n \equiv 1\ (mod\ 6); \\ F_n & , \text{if } n \equiv 4\ (mod\ 12); \\ F_{n-2} & , \text{if } n \equiv 5\ (mod\ 12); \\ F_{n-1} & , \text{if } n \equiv 10,11\ (mod\ 12). \end{cases}$$

Then $e_f(1) = \frac{n}{2}$ and $e_f(0) = \frac{n}{2}$.
Sub case 3: $n \equiv 3\ (mod\ 6)$.
We define an injective function $f : V(G) \to \{F_0, F_1, F_2, \ldots, F_n\}$ such that, $f(v_1) = F_0$, $f(v_2) = F_1$, $f(v_3) = F_2$ and for $n > 3$:

$$f(v_n) = \begin{cases} F_n & , \text{if } n \equiv 0\ (mod\ 6); \\ F_{n-2} & , \text{if } n \equiv 1\ (mod\ 6); \\ F_{n-1} & , \text{if } n \equiv 2,3\ (mod\ 6); \\ F_{n-1} & , \text{if } n \equiv 10,11\ (mod\ 12); \\ F_n & , \text{if } n \equiv 4\ (mod\ 12); \\ F_{n-2} & , \text{if } n \equiv 5\ (mod\ 12). \end{cases}$$

Then $e_f(0) = \frac{n-1}{2}$ and $e_f(1) = \frac{n+1}{2}$, for $n = 3$ and $e_f(1) = \frac{n-1}{2}$ and $e_f(0) = \frac{n+1}{2}$, for $n > 3$.
Sub case 4: $n \equiv 4\ (mod\ 12)$.

We define an injective function $f : V(G) \to \{F_0, F_1, F_2, \ldots, F_n\}$ such that, $f(v_1) = F_0$, $f(v_2) = F_3$, $f(v_3) = F_1$, $f(v_4) = F_2$ and for $n > 4$:

$$f(v_n) = \begin{cases} F_{n+2} & \text{, if } n \equiv 1 \ (mod\ 6); \\ F_{n-3} & \text{, if } n \equiv 2 \ (mod\ 6); \\ F_{n-2} & \text{, if } n \equiv 3,4 \ (mod\ 6); \\ F_{n-1} & \text{, if } n \equiv 5 \ (mod\ 12); \\ F_{n+1} & \text{, if } n \equiv 11 \ (mod\ 12); \\ F_n & \text{, if } n \equiv 6 \ (mod\ 12); \\ F_{n-2} & \text{, if } n \equiv 0 \ (mod\ 12). \end{cases}$$

Then $e_f(0) = \frac{n}{2}$ and $e_f(1) = \frac{n}{2}$.
Sub case 5: $n \equiv 5 \ (mod\ 6)$.
We define an injective function $f : V(G) \to \{F_0, F_1, F_2, \ldots, F_n\}$ such that, $f(v_1) = F_0$, $f(v_2) = F_3$, $f(v_3) = F_1$, $f(v_4) = F_2$, $f(v_5) = F_4$ and for $n > 5$:

$$f(v_n) = \begin{cases} F_{n+1} & \text{, if } n \equiv 2 \ (mod\ 6); \\ F_{n-2} & \text{, if } n \equiv 3,4 \ (mod\ 6); \\ F_{n-1} & \text{, if } n \equiv 5 \ (mod\ 6); \\ F_n & \text{, if } n \equiv 6 \ (mod\ 12); \\ F_{n-1} & \text{, if } n \equiv 0,1 \ (mod\ 12); \\ F_{n-2} & \text{, if } n \equiv 7 \ (mod\ 12). \end{cases}$$

Then $e_f(0) = \frac{n+1}{2}$ and $e_f(1) = \frac{n-1}{2}$, for $n = 5$ and $e_f(0) = \frac{n-1}{2}$ and $e_f(1) = \frac{n+1}{2}$, for $n > 5$.
Sub case 6: $n \equiv 0 \ (mod\ 12)$.
We define an injective function $f : V(G) \to \{F_0, F_1, F_2, \ldots, F_n\}$ such that,

$$f(v_n) = \begin{cases} F_{n-1} & \text{, if } n \equiv 1,2,3,10,11 \ (mod\ 12); \\ F_n & \text{, if } n \equiv 7,8,4 \ (mod\ 12); \\ F_{n-3} & \text{, if } n \equiv 9 \ (mod\ 12); \\ F_{n-2} & \text{, if } n \equiv 5 \ (mod\ 12); \\ F_{n-1} & \text{, if } n \equiv 0 \ (mod\ 6). \end{cases}$$

Then $e_f(1) = \frac{n}{2}$ and $e_f(0) = \frac{n}{2}$.
Thus, for all the above cases, $|e_f(0) - e_f(1)| \leq 1$. Hence, C_n admits Fibonacci cordial labeling for $n \equiv 0,1,3 \ (mod\ 4)$ and $n \geq 3$.
Case 2: $n \equiv 2 \ (mod\ 4)$. Hence, $n = 4k + 2$ for some positive integer k. Thus, $|V(C_n)| = |E(C_n)| = 2(2k+1)$. Clearly, to satisfy the condition of Fibonacci cordial labeling $e_f(0) = e_f(1) = 2k+1$, which is an odd number. Suppose, we label the consecutive vertices of C_n with even Fibonacci

numbers $\{F_{3i},\ 0 \le 3i \le n\}$ and the remaining vertices with odd Fibonacci numbers. In this case, $e_f(0) = 4k$ and $e_f(1) = 2$. Now, if we interchange the labels of two vertices having odd and even Fibonacci numbers respectively then $e_f(0) = 4k - 2$ and $e_f(1) = 4$. Continuing in such a manner we obtain $e_f(0) = 2k + 2$ and $e_f(1) = 2k$. So $e_f(0)$ and $e_f(1)$ will always be even. Hence, $|e_f(0) - e_f(1)| \ge 2$. So C_n is not a Fibonacci cordial graph for $n \equiv 2\ (mod\ 4)$.

□

Theorem 10.2

C_n^2 is a Fibonacci cordial graph if n is even and $n \ge 4$. ■

Proof. Let $G = C_n^2$, we label all the consecutive vertices of the cycle C_n as $v_1, v_2, v_3, \ldots, v_n$. Thus, $|V(G)| = n$ and $|E(G)| = 2n$.

Case 1: $n \equiv 0\ (mod\ 6)$.
We define an injective function $f : V(G) \to \{F_0, F_1, F_2, \ldots, F_n\}$ such that,

$$f(v_n) = \begin{cases} F_{n-1} & \text{, if } n \equiv 1,2,5,0\ (mod\ 6); \\ F_n & \text{, if } n \equiv 3\ (mod\ 6); \\ F_{n-2} & \text{, if } n \equiv 4\ (mod\ 6). \end{cases}$$

Thus, $e_f(1) = e_f(0) = n$.
Case 2: $n \equiv 2\ (mod\ 6)$.
We define an injective function $f : V(G) \to \{F_0, F_1, F_2, \ldots, F_n\}$ such that, $f(v_1) = F_0$, $f(v_2) = F_1$, $f(v_3) = F_3$ and for $n > 3$:

$$f(v_n) = \begin{cases} F_{n-1} & \text{, if } n \equiv 2\ (mod\ 6); \\ F_{n+1} & \text{, if } n \equiv 5\ (mod\ 6); \\ F_n & \text{, if } n \equiv 3\ (mod\ 6); \\ F_{n-2} & \text{, if } n \equiv 0,1,4\ (mod\ 6). \end{cases}$$

Thus, $e_f(1) = e_f(0) = n$.
Case 3: $n \equiv 4\ (mod\ 6)$.
We define an injective function $f : V(G) \to \{F_0, F_1, F_2, \ldots, F_n\}$ such that, $f(v_1) = F_0$, $f(v_2) = F_1$, $f(v_3) = F_3$ and for $n > 3$:

$$f(v_n) = \begin{cases} F_{n-2} & \text{, if } n \equiv 0,4,3\ (mod\ 6); \\ F_{n+1} & \text{, if } n \equiv 5\ (mod\ 6); \\ F_{n+2} & \text{, if } n \equiv 1\ (mod\ 6); \\ F_{n-3} & \text{, if } n \equiv 2\ (mod\ 6). \end{cases}$$

Thus $e_f(1) = e_f(0) = n$.

Thus, from all the above cases, $|e_f(0)-e_f(1)| \leq 1$. Hence, C_n^2 admits Fibonacci cordial labeling. □

Theorem 10.3

Comb graph $P_n \odot K_1$ is a Fibonacci cordial graph. ■

Proof. We label all the consecutive vertices of the path P_n in $G = P_n \odot K_1$ as $v_1, v_2, v_3, \ldots, v_n, v_{n+1}$. We label all the vertices as $\{u_i, 1 \leq i \leq n+1\}$ corresponding to each of the vertices $\{v_i,\ 1 \leq i \leq n+1\}$. Clearly, $|V(G)| = 2n+2$ and $|E(G)| = 2n+1$.
In all the below cases let S be a set of unlabeled vertices and g be an injective function form S to $F_{3i+1}, F_{3i+2}, 0 \leq 3i+1, 3i+2 \leq 2n+2$.

Case 1: $1 \leq n \leq 6$.

Sub case 1: n is odd.
We define an injective function $f : V(G) \to \{F_0, F_1, F_2, \ldots, F_{2n+2}\}$ such that,

$$f(x) = \begin{cases} F_{3i} & \text{, if } x = v_{i+1},\ i = 0,1,2,\ldots,\dfrac{n+1}{2}; \\ g(x) & \text{, if } x \in S. \end{cases}$$

Sub case 2: $n = 2$.
We define an injective function $f : V(G) \to \{F_0, F_1, F_2, \ldots, F_{2n+2}\}$ such that,

$$f(x) = \begin{cases} F_{3i} & \text{, if } x = v_{i+1},\ i = 0,1; \\ g(x) & \text{, if } x \in S. \end{cases}$$

Sub case 3: $n = 4, 6$.
We define an injective function $f : V(G) \to \{F_0, F_1, F_2, \ldots, F_{2n+2}\}$ such that,

$$f(x) = \begin{cases} F_{3i} & \text{, if } x = v_{i+1},\ i = 0,1,2,\ldots,\dfrac{n}{2}+1; \\ g(x) & \text{, if } x \in S. \end{cases}$$

Case 2: $n > 6$.
Sub case 1: $n \equiv 1\ (mod\ 6)$.
We define an injective function $f : V(G) \to \{F_0, F_1, F_2, \ldots, F_{2n+2}\}$ such that,

$$f(x) = \begin{cases} F_{3(i-1)} & \text{, if } x = v_{2i-1},\ i = 1,2,\ldots,\left\lfloor \dfrac{n}{6} \right\rfloor; \\ F_{3\left\lfloor \frac{n}{6} \right\rfloor + 3i} & \text{, if } x = u_{n-i+1},\ i = 0,1,2,\ldots,\dfrac{n+1}{2}; \\ g(x) & \text{, if } x \in S. \end{cases}$$

Sub case 2: $n \equiv 2\ (mod\ 6)$.
We define an injective function $f: V(G) \to \{F_0, F_1, F_2, \ldots, F_{2n+2}\}$ such that,

$$f(x) = \begin{cases} F_{3(i-1)} & \text{, if } x = v_{2i-1},\ i = 1,2,\ldots,\left\lceil \frac{n}{6} \right\rceil; \\ F_{3(\left\lceil \frac{n}{6} \right\rceil + 3i)} & \text{, if } x = u_{n-i+1},\ i = 0,1,2,\ldots,\frac{n-2}{2}; \\ g(x) & \text{, if } x \in S. \end{cases}$$

Sub case 3: $n \equiv 3, 5\ (mod\ 6)$.
We define an injective function $f: V(G) \to \{F_0, F_1, F_2, \ldots, F_{2n+2}\}$ such that,

$$f(x) = \begin{cases} F_{3(i-1)} & \text{, if } x = v_{2i-1},\ i = 1,2,\ldots,\left\lceil \frac{n}{6} \right\rceil; \\ F_{3(\left\lceil \frac{n}{6} \right\rceil + 3i)} & \text{, if } x = u_{n-i+1},\ i = 0,1,2,\ldots,\frac{n-1}{2}; \\ g(x) & \text{, if } x \in S. \end{cases}$$

Sub case 4: $n \equiv 0, 4\ (mod\ 6)$.
We define an injective function $f: V(G) \to \{F_0, F_1, F_2, \ldots, F_{2n+2}\}$ such that,

$$f(x) = \begin{cases} F_{3(i-1)} & \text{, if } x = v_{2i-1},\ i = 1,2,\ldots,\left\lceil \frac{n}{6} \right\rceil; \\ F_{3(\left\lceil \frac{n}{6} \right\rceil) + 3i} & \text{, if } x = u_{n-i+1},\ i = 0,1,2,\ldots,\frac{n}{2}; \\ g(x) & \text{, if } x \in S. \end{cases}$$

Thus, for all the above cases $|e_f(0) - e_f(1)| \leq 1$. Hence, the comb graph is Fibonacci cordial graph. □

Theorem 10.4

A helm graph is a Fibonacci cordial graph. ■

Proof. We label the apex vertex as v_0 and all the consecutive vertices of the cycle C_n in H_n as $v_1, v_2, v_3, \ldots, v_n$. We label all the vertices as $\{w_i, 1 \leq i \leq n\}$ corresponding to each of the vertices $\{v_i,\ 1 \leq i \leq n\}$. Clearly, $|V(H_n)| = 2n+1$ and $|E(H_n)| = 3n$.

Case 1: $n \equiv 1\ (mod\ 6)$.
Let g be any injection from the set
$S = \left\{ v_1, v_2, \ldots, v_{\frac{n-3}{2}} \right\} \cup \left\{ v_{\frac{n+1}{2}}, v_{\frac{n+3}{2}}, \ldots, v_n \right\} \cup \left\{ w_{\frac{n+1}{2}}, w_{\frac{n+3}{2}}, \ldots, w_n \right\}$ to the set

$\left\{F_{3i+1}, F_{3i+2},\ i=0,1,2,\ldots,\left\lceil\frac{2n+1}{3}\right\rceil\right\}$. We define an injective function $f: V(G) \rightarrow \{F_0, F_1, F_2, \ldots, F_{2n+1}\}$ such that,

$$f(v_n)=\begin{cases} F_0 & \text{, if } x=v_0; \\ F_{3i} & \text{, if } x=w_i, i=1,2,3,\ldots,\frac{n-1}{2}; \\ F_{3(i+1)} & \text{, if } x=v_i,\ i=\frac{n-1}{2}+k,\ k=0,1,2,\ldots,\left\lfloor\frac{n}{6}\right\rfloor-1; \\ g(x) & \text{, if } x\in S. \end{cases}$$

Then $e_f(0)=\left\lceil\frac{3n}{2}\right\rceil$ and $e_f(1)=\left\lfloor\frac{3n}{2}\right\rfloor$.

Case 2: $n \equiv 3,5$ $(mod\ 6)$.

Let g be any bijection from the set $S=\left\{v_1,v_2,\ldots,v_{\frac{n-3}{2}}\right\}\bigcup\left\{v_{\frac{n+1}{2}},v_{\frac{n+3}{2}},\ldots,v_n\right\}$ $\bigcup\left\{w_{\frac{n+1}{2}},w_{\frac{n+3}{2}},\ldots,w_n\right\}$ to the set $\left\{F_{3i+1}, F_{3i+2},\ i=0,1,2,\ldots,\frac{2n+1}{3}\right\}$. We define an injective function $f: V(G) \rightarrow \{F_0, F_1, F_2, \ldots, F_{2n+1}\}$ such that,

$$f(v_n)=\begin{cases} F_0 & \text{, if } x=v_0; \\ F_{3i} & \text{, if } x=w_i, i=1,2,3,\ldots,\frac{n-1}{2}; \\ F_{3(i+1)} & \text{, if } x=v_i,\ i=\frac{n-1}{2}+k,\ k=0,1,2,\ldots,\left\lfloor\frac{n}{6}\right\rfloor; \\ g(x) & \text{, if } x\in S. \end{cases}$$

Then $e_f(0)=\left\lceil\frac{3n}{2}\right\rceil$ and $e_f(1)=\left\lfloor\frac{3n}{2}\right\rfloor$.

Case 3: $n \equiv 0,2,4$ $(mod\ 6)$.

Sub case 1: $n=4$.

Let g be an injection from the set $S=\{v_1,v_3,v_4\}\bigcup\{w_2,w_3,w_4\}$ to the set $\{F_1,F_2,F_4,F_5,F_7,F_8,F_9\}$. We define an injective function $f: V(G) \rightarrow \{F_0, F_1, F_2, \ldots, F_8\}$ such that,

$$f(v_n)=\begin{cases} F_0 & \text{, if } x=v_0; \\ F_3 & \text{, if } x=w_1; \\ F_6 & \text{, if } x=v_2; \\ g(x) & \text{, if } x\in S. \end{cases}$$

Then $e_f(1)=e_f(0)=6$.

Sub case 2: $n\geq 6$.

Let g be an injection from the set $S=\left\{v_{\frac{n}{2}+\left\lfloor\frac{n}{6}\right\rfloor}, v_{\frac{n}{2}+\left\lfloor\frac{n}{6}\right\rfloor+1},\ldots,v_n\right\}$

$\bigcup \left\{ v_1, v_2, \ldots, v_{\frac{n}{2}-1} \right\} \bigcup \left\{ w_{\frac{n}{2}+1}, w_{\frac{n}{2}+2}, \ldots, w_n \right\}$ to the set $\left\{ F_{3i+1}, F_{3i+2},\ i = 0,1,2,\ldots, \left\lceil \frac{2n+1}{3} \right\rceil \right\}$. We define an injective function $f : V(G) \to \{F_0, F_1, F_2, \ldots, F_{2n+1}\}$ such that,

$$f(v_n) = \begin{cases} F_0 & \text{, if } x = v_0; \\ F_{3i} & \text{, if } x = w_i, i = 1,2,3,\ldots,\frac{n}{2}; \\ F_{3(i+1)} & \text{, if } x = v_i,\ i = \frac{n}{2} + k,\ k = 0,1,2,\ldots,\left\lfloor \frac{n}{6} \right\rfloor - 1; \\ g(x) & \text{, if } x \in S. \end{cases}$$

Then $e_f(1) = e_f(0) = \frac{3n}{2}$.

Thus, from all the above cases $|e_f(0) - e_f(1)| \leq 1$. Hence, H_n admits Fibonacci cordial labeling. H_n is a Fibonacci cordial graph. □

Theorem 10.5

A friendship graph is a Fibonacci cordial graph. ■

Proof. A friendship graph is one point union of C_3 which is denoted by $F_n = C_3^n$, where n stands for the total number of C_3. We label the center vertex as v_0. All other vertices are labeled with $\{v_i,\ i = 1,2,3,\ldots,2n\}$ consecutively. Thus, $|V(F_n)| = 2n+1$ and $|E(F_n)| = 3n$.

Case 1: $n \equiv 1\ (mod\ 6)$.

Sub case 1: $n = 1$.

We define an injective function $f : V(G) \to \{F_0, F_1, F_2\}$ such that, $f(v_0) = F_0, f(v_1) = F_1, f(v_2) = F_2$. Thus, $e_f(0) = 1$ and $e_f(1) = 2$.

Sub case 2: $n \equiv 1\ (mod\ 12)$ and $n > 1$.

Let g be a injection from the set $S = \left\{ v_i,\ i = \frac{n+3}{2} + 2k,\ k = 0,1,2,\ldots,\frac{3(n-1)}{4} \right\}$ $\bigcup \left\{ v_i,\ i = \frac{5n+1}{6} + 2k,\ k = 0,1,2,\ldots,\frac{7(n-1)}{12} \right\}$ to the set $\left\{ F_{3i+1}, F_{3i+2},\ i = 0,1,2,\ldots, \left\lceil \frac{2n+1}{3} \right\rceil \right\}$. We define an injective function $f : V(G) \to \{F_0, F_1, F_2, \ldots, F_{2n+1}\}$ such that,

$$f(v_n) = \begin{cases} F_0 & \text{, if } x = v_0; \\ F_{3i} & \text{, if } x = v_i,\ i = 1,2,3,\ldots,\frac{n-1}{2}; \\ F_{\frac{3(n+2k+1)}{2}} & \text{, if } x = v_i,\ i = \frac{n+1}{2} + 2k,\ k = 0,1,2,\ldots,\frac{n-7}{6}; \\ g(x) & \text{, if } x \in S. \end{cases}$$

By the above induced labeling we obtain, $e_f(0) = \left\lfloor \frac{3n}{2} \right\rfloor$ and $e_f(1) = \left\lceil \frac{3n}{2} \right\rceil$.

Sub case 3: $n = 7$.

Let g be an injection from the set $S = \{v_i,\ i = 5, 6, \ldots, 14\}$ to the set $\{F_{3i+1}, F_{3i+2},\ i = 0, 1, 2, 3, 4\}$. We define an injective function $f : V(G) \to \{F_0, F_1, F_2, \ldots, F_{2n+1}\}$ such that,

$$f(v_n) = \begin{cases} F_0 & \text{, if } x = v_0; \\ F_{3i} & \text{, if } x = v_i,\ i = 1, 2, 3, \ldots, \frac{n+1}{2}; \\ g(x) & \text{, if } x \in S. \end{cases}$$

By the above induced labeling we obtain, $e_f(1) = 10$ and $e_f(0) = 11$.

Sub case 4: $n \equiv 7\ (mod\ 12)$ and $n > 7$.

Let g be an injection from the set $S = \left\{ v_i,\ i = \frac{n-7}{2} + 2k,\ k = 0, 1, 2, \ldots, \frac{3n-5}{4} \right\}$ $\cup \left\{ v_i,\ i = \frac{5(n-1)}{6} + 2k,\ k = 0, 1, 2, \ldots, \frac{7n-1}{12} \right\}$ to the set $\left\{ F_{3i+1}, F_{3i+2},\ i = 0, 1, 2, \ldots, \left\lceil \frac{2n+1}{3} \right\rceil \right\}$. We define an injective function $f : V(G) \to \{F_0, F_1, F_2, \ldots, F_{n-1}\}$ such that,

$$f(v_n) = \begin{cases} F_0 & \text{, if } x = v_0; \\ F_{3i} & \text{, if } x = v_i,\ i = 1, 2, 3, \ldots, \frac{n+1}{2}; \\ F_{\frac{3(n+2k+3)}{2}} & \text{, if } x = v_i,\ i = \frac{n+3}{2} + 2k,\ k = 0, 1, 2, \ldots, \frac{n-13}{6}; \\ g(x) & \text{, if } x \in S. \end{cases}$$

By the above induced labeling we obtain, $e_f(1) = \left\lfloor \frac{3n}{2} \right\rfloor$ and $e_f(0) = \left\lceil \frac{3n}{2} \right\rceil$.

Case 2: $n \equiv 0\ (mod\ 12)$.

Let g be an injection from the set $S = \left\{ v_i,\ i = \frac{n+4}{2} + 2k,\ k = 0, 1, 2, \ldots, \frac{3n-4}{4} \right\}$ $\cup \left\{ v_i,\ i = \frac{5n+6}{6} + 2k,\ k = 0, 1, 2, \ldots, \frac{7n-12}{12} \right\}$ to the set $\left\{ F_{3i+1}, F_{3i+2},\ i = 0, 1, 2, \ldots, \left\lceil \frac{2n+1}{3} \right\rceil \right\}$. We define an injective function

$f: V(G) \to \{F_0, F_1, F_2, \ldots, F_{2n+1}\}$ such that,

$$f(v_n) = \begin{cases} F_0 & , \text{ if } x = v_0; \\ F_{3i} & , \text{ if } x = v_i,\ i = 1,2,3,\ldots,\frac{n}{2}; \\ F_{\frac{3(n+2k+2)}{2}} & , \text{ if } x = v_i,\ i = \frac{n+2}{2} + 2k,\ k = 0,1,2,\ldots,\frac{n-6}{6}; \\ g(x) & , \text{ if } x \in S. \end{cases}$$

By the above induced labeling we obtain, $e_f(1) = e_f(0) = \frac{3n}{2}$.

Case 3: $n \equiv 8 \ (mod\ 12)$.

Let g be an injection from the set $S = \{v_i,\ i = \frac{n+4}{2} + 2k,\ k = 0,1,2,\ldots,\frac{3n-4}{4}\} \cup \left\{v_i,\ i = \frac{5n+2}{6} + 2k,\ k = 0,1,2,\ldots,\frac{7n-8}{12}\right\}$ to the set $\left\{F_{3i+1}, F_{3i+2},\ i = 0,1,2,\ldots,\left\lceil\frac{2n+1}{3}\right\rceil\right\}$. We define an injective function $f: V(G) \to \{F_0, F_1, F_2, \ldots, F_{2n+1}\}$ such that,

$$f(v_n) = \begin{cases} F_0 & , \text{ if } x = v_0; \\ F_{3i} & , \text{ if } x = v_i,\ i = 1,2,3,\ldots,\frac{n}{2}; \\ F_{\frac{3(n+2k+2)}{2}} & , \text{ if } x = v_i,\ i = \frac{n+2}{2} + 2k,\ k = 0,1,2,\ldots,\frac{n-8}{6}; \\ g(x) & , \text{ if } x \in S. \end{cases}$$

By the above induced labeling we obtain, $e_f(1) = e_f(0) = \frac{3n}{2}$.

Case 4: $n \equiv 3(\ mod\ 6)$.

Sub case 1: $n = 3$.

Let g be an injection from the set $S = \{v_i,\ i = 3,4,5,6\}$ to the set $\{F_{3i+1}, F_{3i+2},\ i = 0,1\}$. We define an injective function $f: V(G) \to \{F_0, F_1, F_2, \ldots, F_{2n+1}\}$ such that,

$$f(v_n) = \begin{cases} F_0 & , \text{ if } x = v_0; \\ F_{3i} & , \text{ if } x = v_i,\ i = 1,2; \\ g(x) & , \text{ if } x \in S. \end{cases}$$

By the above induced labeling we obtain, $e_f(1) = 4$ and $e_f(0) = 5$.

Sub case 2: $n \equiv 3 \ (mod\ 12)$ and $n > 3$.

Let g be an injection from the set $S = \left\{v_i,\ i = \frac{n+5}{2} + 2k,\ k = 0,1,2,\ldots,\frac{3n-5}{4}\right\}$ $\cup \left\{v_i,\ i = \frac{5n+3}{6} + 2k,\ k = 0,1,2,\ldots,\frac{7n-9}{12}\right\}$ to the set

$\left\{F_{3i+1}, F_{3i+2},\ i=0,1,2,\ldots,\left\lceil\frac{2n+1}{3}\right\rceil\right\}$. We define an injective function $f: V(G) \rightarrow \{F_0, F_1, F_2, \ldots, F_{n-1}\}$ such that,

$$f(v_n)=\begin{cases} F_0 & \text{, if } x=v_0; \\ F_{3i} & \text{, if } x=v_i,\ i=1,2,3,\ldots,\left\lceil\frac{n}{2}\right\rceil; \\ F_{\frac{3(n+2k+3)}{2}} & \text{, if } x=v_i,\ i=\frac{n+3}{2}+2k,\ k=0,1,2,\ldots,\left\lfloor\frac{n}{6}\right\rfloor-1; \\ g(x) & \text{, if } x\in S. \end{cases}$$

By the above induced labeling we obtain, $e_f(1)=\left\lfloor\frac{3n}{2}\right\rfloor$ and $e_f(0)=\left\lceil\frac{3n}{2}\right\rceil$.

Sub case 3: $n\equiv 9\ (mod\ 12)$.

Let g be an injection from the set $S=\left\{v_i,\ i=\frac{5n-9}{6}+2k,\ k=0,1,2,\ldots,\frac{7n+9}{12}\right\}\cup\left\{v_i,\ i=\frac{5n+9}{6}+2k,\ k=0,1,2,\ldots,\frac{7n-15}{12}\right\}$ to the set $\left\{F_{3i+1}, F_{3i+2},\ i=0,1,2,\ldots,\left\lceil\frac{2n+1}{3}\right\rceil\right\}$. We define an injective function $f: V(G)\rightarrow\{F_0, F_1, F_2, \ldots, F_{2n+1}\}$ such that,

$$f(v_n)=\begin{cases} F_0 & \text{, if } x=v_0; \\ F_{3i} & \text{, if } x=v_i,\ i=1,2,3,\ldots,\left\lfloor\frac{n}{2}\right\rfloor; \\ F_{\frac{3(n+2k+1)}{2}} & \text{, if } x=v_i,\ i=\frac{n+1}{2}+2k,\ k=0,1,2,\ldots,\left\lfloor\frac{n}{6}\right\rfloor; \\ g(x) & \text{, if } x\in S. \end{cases}$$

By the above induced labeling we obtain, $e_f(0)=\left\lfloor\frac{3n}{2}\right\rfloor$ and $e_f(1)=\left\lceil\frac{3n}{2}\right\rceil$.

Case 5: $n\equiv 4\ (mod\ 6)$.

Sub case 1: $n=4$.

Let g be an injection from the set $S=\{v_i,\ i=3,4,\ldots,8\}$ to the set $\{F_{3i+1}, F_{3i+2},\ i=0,1,2\}$. We define an injective function $f: V(G)\rightarrow\{F_0, F_1, F_2, \ldots, F_{2n+1}\}$ such that,

$$f(v_n)=\begin{cases} F_0 & \text{, if } x=v_0; \\ F_{3i} & \text{, if } x=v_i,\ i=1,2; \\ g(x) & \text{, if } x\in S. \end{cases}$$

By the above induced labeling we obtain, $e_f(1)=e_f(0)=6$. Thus, $|e_f(0)-e_f(1)|\leq 1$.
Sub case 2: $n>4$.
Let g be an injection from the set $S=\{v_i,\ i=\frac{n+4}{2}+2k,\ k=0,1,2,\ldots,\frac{3n-4}{4}\}$ $\bigcup\left\{v_i,\ i=\dfrac{5n-2}{6}+2k,\ k=0,1,2,\ldots,\dfrac{n-10}{6}\right\}$ to the set $\left\{F_{3i+1},F_{3i+2},\ i=0,1,2,\ldots,\left\lceil\dfrac{2n+1}{3}\right\rceil\right\}$. We define an injective function $f:V(G)\rightarrow\{F_0,F_1,F_2,\ldots,F_{2n+1}\}$ such that,

$$f(v_n)=\begin{cases}F_0 & ,\text{ if } x=v_0;\\ F_{3i} & ,\text{ if } x=v_i,\ i=1,2,3,\ldots,\dfrac{n}{2};\\ F_{\frac{3(n+2k+2)}{2}} & ,\text{ if } x=v_i,\ i=\dfrac{n+2}{2}+2k,\ k=0,1,2,\ldots,\dfrac{n-10}{6};\\ g(x) & ,\text{ if } x\in S.\end{cases}$$

By the above induced labeling we obtain, $e_f(1)=e_f(0)=\dfrac{3n}{2}$.
Case 6: $n\equiv 5\ (mod\ 6)$.
Sub case 1: $n=5$.
Let g be an injection from the set $S=\{v_i,\ i=4,5,\ldots,10\}$ to the set $\{F_{3i+1}, i=0,1,2,3\}\bigcup\{F_{3i+2},\ i=0,1,2\}$. We define an injective function $f:V(G)\rightarrow\{F_0,F_1,F_2,\ldots,F_{2n+1}\}$ such that,

$$f(v_n)=\begin{cases}F_0 & ,\text{ if } x=v_0;\\ F_{3i} & ,\text{ if } x=v_i,\ i=1,2,3;\\ g(x) & ,\text{ if } x\in S.\end{cases}$$

By the above induced labeling we obtain, $e_f(1)=8$ and $e_f(0)=7$.
Sub case 2: $n\equiv\ 5\ (mod\ 12)$ and $n>5$.
Let g be an injection from the set $S=\{v_i,\ i=\frac{n+3}{2}+2k,\ k=0,1,2,\ldots,\frac{3n-3}{4}\}$ $\bigcup\left\{v_i,\ i=\dfrac{5n+5}{6}+2k,\ k=0,1,2,\ldots,\dfrac{7n-11}{12}\right\}$ to the set $\left\{F_{3i+1},F_{3i+2},\ i=0,1,2,\ldots,\left\lceil\dfrac{2n+1}{3}\right\rceil\right\}$. We define an injective function $f:V(G)\rightarrow\{F_0,F_1,F_2,\ldots,F_{2n+1}\}$ such that,

$$f(v_n)=\begin{cases}F_0 & ,\text{ if } x=v_0;\\ F_{3i} & ,\text{ if } x=v_i,\ i=1,2,3,\ldots,\dfrac{n+1}{2};\\ F_{\frac{3n+6k+9}{2}} & ,\text{ if } x=v_i,\ i=\dfrac{n+5}{2}+2k,\ k=0,1,2,\ldots,\dfrac{n-11}{6};\\ g(x) & ,\text{ if } x\in S.\end{cases}$$

By the above induced labeling we obtain, $e_f(0) = \left\lfloor \frac{3n}{2} \right\rfloor$ and $e_f(1) = \left\lceil \frac{3n}{2} \right\rceil$.

Sub case 3: $n \equiv 11 \ (mod\ 12)$.

Let g be an injection from the set $S = \{v_i,\ i = \frac{n+5}{2} + 2k,\ k = 0,1,2,\ldots,\frac{3n-5}{4}\}$ $\bigcup \left\{ v_i,\ i = \frac{5n-1}{6} + 2k,\ k = 0,1,2,\ldots,\frac{7n-5}{12} \right\}$ to the set $\left\{ F_{3i+1}, F_{3i+2},\ i = 0,1,2,\ldots,\left\lceil \frac{2n+1}{3} \right\rceil \right\}$. We define an injective function $f : V(G) \to \{F_0, F_1, F_2, \ldots, F_{2n+1}\}$ such that,

$$f(v_n) = \begin{cases} F_0 & ,\text{ if } x = v_0; \\ F_{3i} & ,\text{ if } x = v_i,\ i = 1,2,3,\ldots,\frac{n+1}{2}; \\ F_{\frac{3(n+2k+3)}{2}} & ,\text{ if } x = v_i,\ i = \frac{n+3}{2} + 2k,\ k = 0,1,2,\ldots,\frac{n-11}{6}; \\ g(x) & ,\text{ if } x \in S. \end{cases}$$

By the above induced labeling we obtain, $e_f(1) = \left\lfloor \frac{3n}{2} \right\rfloor$ and $e_f(0) = \left\lceil \frac{3n}{2} \right\rceil$.

Thus, by all the above cases $|e_f(0) - e_f(1)| \leq 1$. Hence, F_n is a Fibonacci cordial graph. □

10.3 CONCLUSION

We conclude that a cycle admits Fibonacci cordial labeling if and only if $n \geq 3$ and $n \equiv 0,1,3 \ (mod\ 4)$. Some other graphs like the comb graph, helm graph and friendship graph are also Fibonacci cordial graphs. C_n^2 is a Fibonacci cordial graph if n is even and $n \geq 4$. For the future, one can explore more Fibonacci cordial graph families. Also, different patterns of labeling can be discovered such that the graph admits Fibonacci cordial labeling.

REFERENCES

1. M. Anholcer and C. Palmer. Irregular labelings of circulant graphs. *Discrete Mathematics*, 312(23): 3461-3466, 2012.
2. I. Cahit. Cordial graphs: A weaker version of graceful and harmonious graphs. *Ars Combinatoric*, 23, 201-207, 2018.
3. J. A. Gallian. A dynamic survey of graph labeling. *The Electronic Journal of Combinatorics*, #DS6, 2018.
4. J. Gross and J. Yellen. *Handbook of Graph Theory*, CRC Press, 2004.

5. C. Karthikeyan, M. Abinaya, S. Arthi, V. Surya and K. V. Sreelakshmi. Fibonacci cordial labeling of some special graphs. *International Journal for Scientific Research and Development*, 5(12): 348-350, 2018.
6. A. H. Rokad and G. V. Ghodasara. Fibonacci cordial labeling of some special graphs. *Annals of Pure and Applied Mathematics*, 11(1): 133-144, 2016.
7. A. Semaničová. On magic and supermagic circulant graphs. *Discrete Mathematics*, 306, 2263-2269, 2006.

11 Some More Parity Combination Cordial Graphs

U. M. Prajapati
St. Xavier's College,
Ahmedabad, Gujarat (INDIA)
E-mail: udayan64@yahoo.com

K. P. Shah
Department of Mathematics,
Gujarat University,
Ahmedabad, Gujarat (INDIA)
E-mail: kinjal.shah09@yahoo.com

For a graph $G(V,E)$, a bijective function $f : V \to \{1, 2, \ldots, |V|\}$ is called a parity combination cordial labeling (PCC-labeling) if for each edge $uv \in E$, assigned with label $\binom{f(u)}{f(v)}$ or $\binom{f(v)}{f(u)}$ according as $f(u) > f(v)$ or $f(v) > f(u)$ satisfying $|e_f(0) - e_f(1)| \leq 1$, where $e_f(0)$ denotes the number of edges labeled with even numbers and $e_f(1)$ denotes the number of edges labeled with odd numbers. The graph satisfying PCC-labeling is called a parity combination cordial graph or PCC-graph. In this chapter, some standard graphs like coconut tree, m-star graph, triangular book, friendship graph, cycle cactus, and double fan are proved to have PCC-labeling.

11.1 INTRODUCTION

$G(V,E)$ is an undirected, connected, simple, finite and non-trivial graph throughout the chapter with V or $V(G)$ and E or $E(G)$ respectively as the vertex set and the edge set of G. $|V|$ denotes the number of vertices and $|E|$ denotes the number of edges respectively of G. Clark and Holton [5] is referred to for graph theoretical terminologies and notations whereas Burton [1] is referred to for number theoretical notations that are not defined in this chapter. Subject to certain conditions, the assignment of integers to vertices or edges or both for a graph is called *graph labeling* [3]. One such labeling, called parity combination cordial labeling has been studied here. Certain standard graphs obtained using different graph operations on the path, cycle and star graph are proved here to have parity combination cordial labeling. The next section gives a brief on notations and definitions required in this regard.

11.2 NOTATIONS AND DEFINITIONS

Notation. [6] $[n]$ is denoted as the set of first n natural numbers for any natural number n. i. e. $[n] = \{1,2,3,\ldots,n\}$.

Definition (Coconut Tree). *Coconut Tree* is obtained by identifying the apex vertex of the star graph $K_{1,m}$ with a pendent vertex of path P_n. It is denoted by $T(n,m)$ where n stands for number of vertices of the path and m stands for number of pendent edges of the star graph.

Note: Another notation for coconut tree is $CT(n,m)$. It is sometimes also known as Broom and denoted as $B_{n,t}$.

Definition (Y-tree). *Y-tree* is a graph obtained from a path by adding an edge to a vertex of a path adjacent to an end point.

Note: Y-tree is a special case of coconut tree $T(2,n)$.

Definition (m-star). *m-star* is a graph obtained by subdividing each pendent edge of a star graph $K_{1,n}$ m times. We denote it as $K_{1,n}^m$.

Definition (Triangular book). *Triangular book* with n-pages is obtained by identifying an edge of n copies of cycle C_3. It is denoted by $B(3,n)$.

Definition. One point union of n cycles C_k is called one point union of cycles or cycle cactus. It is denoted by C_k^n.

Definition (friendship graph). A *friendship graph* F_n is a one point union of n cycles C_3.

Note: A friendship graph is a special case of cycle cactus with $k = 3$, i.e. $C_3^n = F_n$.

Definition. Double fan is defined to be $P_n + \overline{K_2}$.

Definition. [7] Let $G(V,E)$ be a graph. A bijective function $f : V \to \{1,2,\ldots,|V|\}$ is called *parity combination cordial labeling (PCC-labeling)* if for each edge $uv \in E$, the induced edge labeling is given by:

$$f^*(uv) = \begin{cases} \binom{f(u)}{f(v)}, & \text{if } f(u) > f(v); \\ \\ \binom{f(v)}{f(u)}, & \text{if } f(v) > f(u), \end{cases}$$

satisfying $|e(0) - e(1)| \leq 1$, where $e(0)$ denotes the number of edges labeled with even numbers and $e(1)$ denotes the number of edges labeled with odd numbers.
The graph satisfying PCC-labeling is called a *parity combination cordial graph* or *PCC-graph.*

Ponraj, Sathish Narayan and Ramasamy [7] introduced parity combination cordial labeling, motivated by two labelings: cordial labeling and combination labeling introduced by Cahit [2] and Hedge and Shetty [4] respectively. They [7] introduced this labeling in 2015 and proved that some standard graphs like paths, cycles, stars, triangular snakes, alternate triangular snakes, olive trees, comb $P_n \bigodot K_1$, crown graph $C_n \bigodot K_1$, fan graph, wheel graph W_n iff $n \geq 4$ and umbrella graph $U_{m,n}$ with $m \geq 2$ are parity combination cordial graphs. Ponraj, Rajpal Singh and Satish Narayan [8] investigated the PCC-labeling of helms, P_n^2, dragon and bistar $B_{m,n}$. They [8] also showed that identifying any arbitrary vertex of the cycle and star graph is a parity combination cordial graph. Further they [8] proved that the disjoint union of the parity combination cordial graph G and path P_n is also a parity combination cordial graph if $n \neq 2,4$. Soni, Khanna and Bhathawala [9] showed that standard graphs like the path, cycle and star graph admit PCC-labeling under duplication of graph elements.

The next section shows that graphs obtained with the help of graph operations like identifying vertices/edges, subdividing edges and joining of two graphs on a cycle, path and star graph have parity combination labeling or PCC-labeling.

11.3 MAIN RESULTS

Theorem 11.1

Coconut tree T(n, m) is a PCC graph. ■

Proof. Consider $\{v_i \,|\, i \in [n] \cup \{0\}\}$ and $\{v_0 v_i \,|\, i \in [n]\}$ as the vertex and edge set respectively for star graph $K_{1,n}$, where v_0 is the apex vertex. Also consider $\{u_i \,|\, i \in [m-1] \cup \{0\}\}$ and $\{u_i u_{i+1} \,|\, i \in [m-1]\}$ as the vertex and edge set respectively for the path P_n. Let the coconut tree $T(n,m)$ be obtained by identifying the apex vertex v_0 of the star graph and the pendent vertex u_0 of the path graph. Assume $v_0 = u_0 = w$. Thus the vertex set and the edge set of the coconut tree can be given by $V = \{v_i, u_j, w \,|\, i \in [n], j \in [m-1]\}$ and $E = \{wv_i, wu_1, u_j u_{j+1} \,|\, i \in [n], j \in [m-1]\}$ with $|V| = m+n$ and $|E| = m+n-1$. It is easy to see that $m \geq 3$ and $n \geq 2$.

Define $f : V \to [m+n]$ as $f(x) = \begin{cases} 1, & \text{if } x = w; \\ i+1, & \text{if } x = v_i, i \in [n]; \\ j+n+1, & \text{if } x = u_j, j \in [m-1]. \end{cases}$

The corresponding induced edge labeling is given by:

$$f^*(e) = \begin{cases} i+1, & \text{if } e = wv_i, i \in [n]; \\ n+2, & \text{if } e = wu_1; \\ j+n+2, & \text{if } e = u_j u_{j+1}, j \in [m-2]. \end{cases}$$

Depending on the nature of m,n we have Table 11.1:

Table 11.1

		n is even			n is odd		
Edges		wv_i	wu_i, u_ju_{j+1}	Total	wv_i	wu_i, u_ju_{j+1}	Total
m is even	$e_f(0)$	$\frac{n}{2}$	$\frac{m}{2}$	$\frac{m+n}{2}$	$\frac{n+1}{2}$	$\frac{m}{2}-1$	$\frac{m+n-1}{2}$
	$e_f(1)$	$\frac{n}{2}$	$\frac{m}{2}-1$	$\frac{m+n}{2}-1$	$\frac{n-1}{2}$	$\frac{m}{2}$	$\frac{m+n-1}{2}$
m is odd	$e_f(0)$	$\frac{n}{2}$	$\frac{m-3}{2}+1$	$\frac{m+n-1}{2}$	$\frac{n+1}{2}$	$\frac{m-1}{2}$	$\frac{m+n}{2}$
	$e_f(1)$	$\frac{n}{2}$	$\frac{m-1}{2}$	$\frac{m+n-1}{2}$	$\frac{n-1}{2}$	$\frac{m-3}{2}+1$	$\frac{m+n}{2}-1$

This concludes that $|c_f(1) - c_f(0)| \leq 1$ in all the cases proving that the function satisfies PCC labeling and hence the graph is a PCC-graph. □

Corollary 11.3.1. *$Y-tree$ is a PCC graph.*

Proof. $Y-tree$ is a special case of coconut tree with $n=2$. Thus it is a PCC graph. □

Theorem 11.2

$m-star$ is a PCC graph. ■

Proof. Let an $m-$ star graph be obtained from a star graph $K_{1,n}$ by subdividing each pendent vertex m times.
Consider the vertex set as $V = \{v_0, v_{i,j} \,|\, i \in [n], j \in [m+1]\}$, where v_0 is the apex vertex of the star graph, $v_{i,m+1} = v_i$ for each $i \in [n]$ is the pendent vertex and for each edge v_0v_i, vertices $v_{i,j}$, $j \in [m]$ subdivide the edge. Here the $m-$ star graph has $mn+n+1$ vertices and $mn+n$ edges.
Define $f: V \to [mn+n+1]$ as $f(x)$ where

$$f(x) = \begin{cases} 1, & \text{if } x = v_0; \\ (i-1)(m+1)+j+1, & \text{if } x = v_{i,j}, i \in [n], j \in [m+1]. \end{cases}$$

The induced edge labeling is given by:

$$f^*(e) = \begin{cases} (m+1)i+(1-m), & \text{if } e = v_0v_{i,1}, i \in [n]; \\ (m+1)i+(j+1-m), & \text{if } e = v_{i,j}v_{i,j+1}, i \in [n], j \in [m-1]. \end{cases}$$

Depending on the nature of m, n we have Table 11.2:

Table 11.2

		n is even			n is odd		
Edges		$v_0v_{i,1}$	$v_{i,j}, v_{i,j+1}$	Total	$v_0v_{i,1}$	$v_{i,j}, v_{i,j+1}$	Total
m is even	$e_f(0)$	$\frac{n}{2}$	$\frac{mn}{2}$	$\frac{mn+n}{2}$	$\frac{n+1}{2}$	$\frac{mn}{2}$	$\frac{mn+n+1}{2}$
	$e_f(1)$	$\frac{n}{2}$	$\frac{mn}{2}$	$\frac{mn+n}{2}$	$\frac{n-1}{2}$	$\frac{mn}{2}$	$\frac{mn+n-1}{2}$
m is odd	$e_f(0)$	n	$\frac{mn-n}{2}+1$	$\frac{mn+n}{2}$	n	$\frac{mn-n}{2}+1$	$\frac{mn+n}{2}$
	$e_f(1)$	0	$\frac{mn+n}{2}$	$\frac{mn+n}{2}$	0	$\frac{mn+n}{2}$	$\frac{mn+n}{2}$

This concludes that $|e_f(1)-e_f(0)| \leq 1$ in all the cases proving that the function satisfies PCC labeling and hence the graph is a PCC-graph. □

Theorem 11.3

Triangular book is a PCC graph. ■

Proof. Consider a triangular book with n-pages obtained by identifying an edge of each of the n copies of cycle C_3. Let the identified edge be u_1u_2 and the n cycles C_3 be $u_1u_2v_i$ for each $i \in [n]$. Here the graph has $n+2$ vertices and $2n+1$ edges. Defining $f : V \to [n+2]$ as $f(x) = \begin{cases} i, & \text{if } x = u_i, i \in [2]; \\ i+2, & \text{if } x = v_i, i \in [n]. \end{cases}$ the induced edge labeling is given by: $f^*(e) = \begin{cases} 2, & \text{if } e = u_1u_2; \\ i+2, & \text{if } e = u_1v_i, i \in [n]; \\ \frac{(i+2)(i+1)}{2}, & \text{if } e = u_2v_i, i \in [n]. \end{cases}$

Depending on the nature of n we have Table 11.3:
This concludes that $|e_f(1)-e_f(0)| \leq 1$ in all the cases proving that the function satisfies PCC labeling and hence the graph is a PCC-graph. □

Theorem 11.4

One point union of cycles or cycle cactus C_k^n, $k \geq 4$ is a PCC graph. ■

Table 11.3

Edges		u_1u_2	u_1v_i	u_2v_i	Total
n is odd	$e_f(0)$	1	$\frac{n+1}{2}$	$\frac{n-1}{2}$	$n+1$
	$e_f(1)$	0	$\frac{n-1}{2}$	$\frac{n+1}{2}$	n
n is even	$e_f(0)$	1	$\frac{n}{2}$	$\frac{n}{2}$	$n+1$
	$e_f(1)$	0	$\frac{n}{2}$	$\frac{n}{2}$	n

Proof. Let there be n cycles C_k of length k. Let these n cycles be denoted by $C_k^{(i)}$, $i \in [n]$ with vertex set $\{v_{i,j} \,|\, j \in [k]\}$. Consider the one point union of cycles where the vertices $v_{i,1}$ from each cycle $C_k^{(j)}$ are identified. Let this vertex be denoted by v. Thus, $v_{i,1} = v$, $i \in [n]$. Here $|V| = nk - n + 1$ and $|E| = nk$. Define $f : V \to [nk - n + 1]$ as follows:

(1) Let k be even, then

(a) if n is even, define $f(x) = \begin{cases} 1, & \text{if } x = v; \\ (i-1)(k-1)+j, & \text{if } x = v_{i,j}, j \in [k] - \{1\}, \\ & i \in [n]. \end{cases}$

The induced edge labeling given by:

$$f^*(e) = \begin{cases} (i-1)k - (i-2) + j, & \text{if } e = v_{i,j}v_{i,j+1}, i \in [n], j \in [k-1]; \\ ik - (i-1), & \text{if } e = v_{i,k}v, i \in [n]. \end{cases}$$

gives us Table 11.4:

Table 11.4

Edges	$v_{2i+1,j}, v_{2i+1,j+1}$	$v_{2i,j}v_{2i,j+1}$	Total
$e_f(0)$	$\frac{nk+n}{4}$	$\frac{nk-n}{4}$	$\frac{nk}{2}$
$e_f(1)$	$\frac{nk-n}{4}$	$\frac{nk+n}{4}$	$\frac{nk}{2}$

(b) if n is odd,

$$f(x)=\begin{cases}1, & \text{if } x=v;\\ 3, & \text{if } x=v_{1,2};\\ 2, & \text{if } x=v_{1,3};\\ j, & \text{if } x=v_{1,j}, j\in[k]-\{1,2,3\};\\ (i-1)(k-1)+j, & \text{if } e=v_{i,j}, i\in[n]-\{1\}, j\in[k]-\{1\}.\end{cases}$$

The induced edge labeling:

$$f^*(e)=\begin{cases}3, & \text{if } e=vv_{1,2};\\ 3, & \text{if } e=v_{1,2}v_{1,3};\\ 6, & \text{if } e=v_{1,3}v_{1,4};\\ j, & \text{if } e=v_{1,j}v_{1,j+1}, j\in[k-1]-\{1,2,3\};\\ k, & \text{if } e=v_{1,k}v;\\ (i-1)k-(i-2)+j, & \text{if } e=v_{i,j}v_{i,j+1}, i\in[n]-\{1\},\\ & \qquad j\in[k-1]-\{1\};\\ ik-(i-1), & \text{if } e=v_{i,k}v, i\in[n].\end{cases}$$

gives us Table 11.5:

Table 11.5

Edges	$v_{2i+1,j}, v_{2i+1,j+1}$	$v_{2i,j}v_{2i,j+1}$	Total
$e_f(0)$	$\left(\frac{n-1}{2}\right)\left(\frac{k}{2}+1\right)+\frac{k}{2}$	$\left(\frac{n-1}{2}\right)\left(\frac{k}{2}-1\right)+\frac{k}{2}$	$\frac{nk}{2}$
$e_f(1)$	$\left(\frac{n-1}{2}\right)\left(\frac{k}{2}-1\right)+\frac{k}{2}$	$\left(\frac{n-1}{2}\right)\left(\frac{k}{2}+1\right)+\frac{k}{2}$	$\frac{nk}{2}$

1. Let k be odd, then
   (a) if n is even,

$$f(x)=\begin{cases}1, & \text{if } x=v;\\ (2i-2)(k-1)+j, & \text{if } x=v_{2i-1,j}, j\in[k]-\{1\}, i\in\left[\frac{n}{2}\right];\\ (2i-1)(k-1)+j, & \text{if } x=v_{2i,j}, j\in[k-2]-\{1\}, i\in\left[\frac{n}{2}\right];\\ 2i(k-1)+1, & \text{if } x=v_{2i,k-1}, i\in\left[\frac{n}{2}\right];\\ 2i(k-1), & \text{if } x=v_{2i,k}, i\in\left[\frac{n}{2}\right].\end{cases}$$

The induced edge labeling:

$$f^*(e)=\begin{cases}(2i-2)k-(2i-3)+j, & \text{if } e=v_{2i-1,j}v_{2i-1,j+1}, i\in\left[\frac{n}{2}\right], j\in[k-1];\\ (2i-1)k-(2i-2), & \text{if } e=v_{2i-1,k}v, i\in\left[\frac{n}{2}\right];\\ (2i-1)k-(2i-2)+j, & \text{if } e=v_{2i,j}v_{2i,j+1}, i\in\left[\frac{n}{2}\right], j\in[k-3];\\ (2ik-2i+1)(ik-i), & \text{if } e=v_{2i,k-2}v_{2i,k-1}, i\in\left[\frac{n}{2}\right];\\ 2ik-(2i-1), & \text{if } e=v_{2i,k-1}v_{2i,k}, i\in\left[\frac{n}{2}\right];\\ 2ik-2i, & \text{if } e=v_{2i,k}v, i\in\left[\frac{n}{2}\right];\end{cases}$$

gives us Table 11.6:

Table 11.6

Edges	$v_{2i+1,j}, v_{2i+1,j+1}$	$v_{2i,j}v_{2i,j+1}$	Total
$e_f(0)$	$\frac{nk-n}{4}$	$\frac{nk+n}{4}$	$\frac{nk}{2}$
$e_f(1)$	$\frac{nk+n}{4}$	$\frac{nk-n}{4}$	$\frac{nk}{2}$

(b) if n is odd,

$$f(x)=\begin{cases}1, & \text{if } x=v;\\ (2i-2)(k-1)+j, & \text{if } x=v_{2i-1,j}, j\in[k]-\{1\}, i\in\left[\frac{n+1}{2}\right];\\ (2i-1)(k-1)+j, & \text{if } x=v_{2i,j}, j\in[k-2]-\{1\}, i\in\left[\frac{n-1}{2}\right];\\ 2i(k-1)+1, & \text{if } x=v_{2i,k-1}, i\in\left[\frac{n-1}{2}\right];\\ 2i(k-1), & \text{if } x=v_{2i,k}, i\in\left[\frac{n-1}{2}\right].\end{cases}$$

The induced edge labeling is given by:

$$f^*(e) = \begin{cases} (2i-2)k-(2i-3)+j, & \text{if } e = v_{2i-1,j}v_{2i-1,j+1}, i \in \left[\dfrac{n+1}{2}\right], \\ & \quad j \in [k-1]; \\ (2i-1)k-(2i-2), & \text{if } e = v_{2i-1,k}v, i \in \left[\dfrac{n+1}{2}\right]; \\ (2i-1)k-(2i-2)+j, & \text{if } e = v_{2i,j}v_{2i,j+1}, i \in \left[\dfrac{n-1}{2}\right], j \in [k-3]; \\ (2ik-2i+1)(ik-i), & \text{if } e = v_{2i,k-2}v_{2i,k-1}, i \in \left[\dfrac{n-1}{2}\right]; \\ 2ik-(2i-1), & \text{if } e = v_{2i,k-1}v_{2i,k}, i \in \left[\dfrac{n-1}{2}\right]; \\ 2ik-2i, & \text{if } e = v_{2i,k}v, i \in \left[\dfrac{n-1}{2}\right]; \end{cases}$$

This gives us Table 11.7:

Table 11.7

Edges	$v_{2i+1,j}, v_{2i+1,j+1}$	$v_{2i,j}v_{2i,j+1}$	Total
$e_f(0)$	$\dfrac{(n+1)(k-1)}{4}$	$\dfrac{(n-1)(k+1)}{4}$	$\dfrac{nk-1}{2}$
$e_f(1)$	$\dfrac{(n+1)(k+1)}{4}$	$\dfrac{(n-1)(k-1)}{4}$	$\dfrac{nk+1}{2}$

This concludes that $|e_f(1) - e_f(0)| \leq 1$ in all the cases proving that the function satisfies PCC labeling and hence the graph is a PCC-graph. □

Theorem 11.5

Friendship graph is a PCC graph. ■

Proof. Let there by $n \geq 2$ cycles C_3, where each one is denoted as C_3^j, $j \in [n]$ with vertex set as $\{v_{i,j} \mid i \in [3], j \in [n]\}$. Consider the one point union of these cycles where the vertices $v_{1,j}$ from each cycle C_3^j are identified. Let this vertex be denoted by v. Thus, $v_{1,j} = v$, $j \in [n]$. The so obtained graph is the friendship graph F_n with $2n+1$ vertices and $3n$ edges. For $f : V \to [2n+1]$:

1. If $n \equiv 0 \pmod 4$, define $f(x)$ as:

$$f(x) = \begin{cases} 1, & \text{if } x = v; \\ 4j-2, & \text{if } x = v_{2,2j-1}, j \in \left[\frac{n}{4}\right]; \\ 4j-1, & \text{if } x = v_{2,2j}, j \in \left[\frac{n}{4}\right]; \\ f(v_{2,j})+2, & \text{if } x = v_{3,j}, j \in \left[\frac{n}{2}\right]; \\ 2j, & \text{if } x = v_{2,j}, j \in [n] - \left[\frac{n}{2}\right]; \\ 2j+1, & \text{if } x = v_{3,j}, j \in [n] - \left[\frac{n}{2}\right], \end{cases}$$

The induced edge labeling is given by:

$$f^*(e) = \begin{cases} 4j-2, & \text{if } e = vv_{2,2j-1}, j \in \left[\frac{n}{4}\right]; \\ 4j-1, & \text{if } e = vv_{2,2j}, j \in \left[\frac{n}{4}\right]; \\ 2j, & \text{if } e = vv_{2,j}, j \in [n] - \left[\frac{n}{2}\right]; \\ 4j, & \text{if } e = vv_{3,2j-1}, j \in \left[\frac{n}{4}\right]; \\ 4j+1, & \text{if } e = vv_{3,2j}, j \in \left[\frac{n}{4}\right]; \\ 2j+1, & \text{if } e = vv_{3,j}, j \in [n] - \left[\frac{n}{2}\right], \\ 2j(4j-1), & \text{if } e = v_{2,2j-1}v_{3,2j-1}, j \in \left[\frac{n}{4}\right], \\ 2j(4j+1), & \text{if } e = v_{2,2j}v_{3,2j}, j \in \left[\frac{n}{4}\right], \\ 2j+1, & \text{if } e = v_{2,j}v_{3,j}, j \in [n] - \left[\frac{n}{2}\right], \end{cases}$$

This gives us Table 11.8:

Table 11.8

Edges	$vv_{2,j}$	$vv_{3,j}$	$v_{2,j}v_{3,j}$	Total
$e_f(0)$	$\frac{3n}{4}$	$\frac{n}{4}$	$\frac{n}{2}$	$\frac{3n}{2}$
$e_f(1)$	$\frac{n}{4}$	$\frac{3n}{4}$	$\frac{n}{2}$	$\frac{3n}{2}$

2. If $n \equiv 1 (\text{mod } 4)$, define $f(x)$ as:

$$f(x) = \begin{cases} 1, & \text{if } x = v; \\ 4j-2, & \text{if } x = v_{2,2j-1}, j \in \left[\frac{n-1}{4}\right]; \\ 4j-1, & \text{if } x = v_{2,2j}, j \in \left[\frac{n-1}{4}\right]; \\ f(v_{2,j})+2, & \text{if } x = v_{3,j}, j \in \left[\frac{n-1}{2}\right]; \\ 2j, & \text{if } x = v_{2,j}, j \in [n] - \left[\frac{n-1}{2}\right]; \\ 2j+1, & \text{if } x = v_{3,j}, j \in [n] - \left[\frac{n-1}{2}\right], \end{cases}$$

The induced edge labeling is given by:

$$f^*(e) = \begin{cases} 4j-2, & \text{if } e = vv_{2,2j-1}, j \in \left[\frac{n-1}{4}\right]; \\ 4j-1, & \text{if } e = vv_{2,2j}, j \in \left[\frac{n-1}{4}\right]; \\ 2j, & \text{if } e = vv_{2,j}, j \in [n] - \left[\frac{n-1}{2}\right]; \\ 4j, & \text{if } e = vv_{3,2j-1}, j \in \left[\frac{n-1}{4}\right]; \\ 4j+1, & \text{if } e = vv_{3,2j}, j \in \left[\frac{n-1}{4}\right]; \\ 2j+1, & \text{if } e = vv_{3,j}, j \in [n] - \left[\frac{n-1}{2}\right], \\ 2j(4j-1), & \text{if } e = v_{2,2j-1}v_{3,2j-1}, j \in \left[\frac{n-1}{4}\right], \\ 2j(4j+1), & \text{if } e = v_{2,2j}v_{3,2j}, j \in \left[\frac{n-1}{4}\right], \\ 2j+1, & \text{if } e = v_{2,j}v_{3,j}, j \in [n] - \left[\frac{n-1}{2}\right], \end{cases}$$

This gives us Table 11.9:

Table 11.9

Edges →	$vv_{2,j}$	$vv_{2,j}$	$v_{2,j}v_{3,j}$	Total
$e_f(0)$	$\frac{3n+1}{4}$	$\frac{n-1}{4}$	$\frac{n-1}{2}$	$\frac{3n-1}{2}$
$e_f(1)$	$\frac{n-1}{4}$	$\frac{3n+1}{4}$	$\frac{n+1}{2}$	$\frac{3n+1}{2}$

3. If $n \equiv 2 \pmod 4$, for $n > 2$ define $f(x)$ as:

$$f(x) = \begin{cases} 1, & \text{if } x = v; \\ 4j-2, & \text{if } x = v_{2,2j-1}, j \in \left[\frac{n-2}{4}\right]; \\ 4j-1, & \text{if } x = v_{2,2j}, j \in \left[\frac{n-2}{4}\right]; \\ f(v_{2,j})+2, & \text{if } x = v_{3,j}, j \in \left[\frac{n}{2}-1\right]; \\ 2j, & \text{if } x = v_{2,j}, j \in [n] - \left[\frac{n}{2}\right]; \\ 2j+1, & \text{if } x = v_{3,j}, j \in [n] - \left[\frac{n}{2}+2\right] \cup \left\{\frac{n}{2}\right\}, \\ n+5, & \text{if } x = v_{3,\frac{n}{2}+1}; \\ n+3, & \text{if } x = v_{3,\frac{n}{2}+2}, \end{cases}$$

The induced edge labeling is given by:

$$f^*(e) = \begin{cases} 4j-2, & \text{if } e = vv_{2,2j-1}, j \in \left[\frac{n-2}{4}\right]; \\ 4j-1, & \text{if } e = vv_{2,2j}, j \in \left[\frac{n-2}{4}\right]; \\ 2j, & \text{if } e = vv_{2,j}, j \in [n] - \left[\frac{n-2}{2}\right]; \\ 4j, & \text{if } e = vv_{3,2j-1}, j \in \left[\frac{n-2}{4}\right]; \\ 4j+1, & \text{if } e = vv_{3,2j}, j \in \left[\frac{n-2}{4}\right]; \\ 2j+1, & \text{if } e = vv_{3,j}, j \in [n] - \left[\frac{n}{2}+2\right] \cup \left\{\frac{n}{2}\right\}; \\ n+5, & \text{if } e = vv_{3,\frac{n}{2}+1}; \\ n+3, & \text{if } e = vv_{3,\frac{n}{2}+2}; \\ 2j(4j-1), & \text{if } e = v_{2,2j-1}v_{3,2j-1}, j \in \left[\frac{n-2}{4}\right]; \\ 2j(4j+1), & \text{if } e = v_{2,2j}v_{3,2j}, j \in \left[\frac{n-2}{4}\right]; \\ 2j+1, & \text{if } e = v_{2,j}v_{3,j}, j \in [n] - \left[\frac{n}{2}+2\right] \cup \left\{\frac{n}{2}\right\}; \\ (n+5)\left(\frac{n}{2}+2\right), & \text{if } e = v_{2,\frac{n}{2}+1}v_{3,\frac{n}{2}+1}; \\ n+4, & \text{if } e = v_{2,\frac{n}{2}+2}v_{3,\frac{n}{2}+2}. \end{cases}$$

This gives us Table 11.10:

Table 11.10

Edges	$vv_{2,j}$	$vv_{3,j}$	$v_{2,j}v_{3,j}$	Total
$e_f(0)$	$\frac{3n+2}{4}$	$\frac{n-2}{4}$	$\frac{n}{2}$	$\frac{3n}{2}$
$e_f(1)$	$\frac{n-2}{4}$	$\frac{3n+2}{4}$	$\frac{n}{2}$	$\frac{3n}{2}$

For $n = 2$, the Figure 11.1 proves the result.

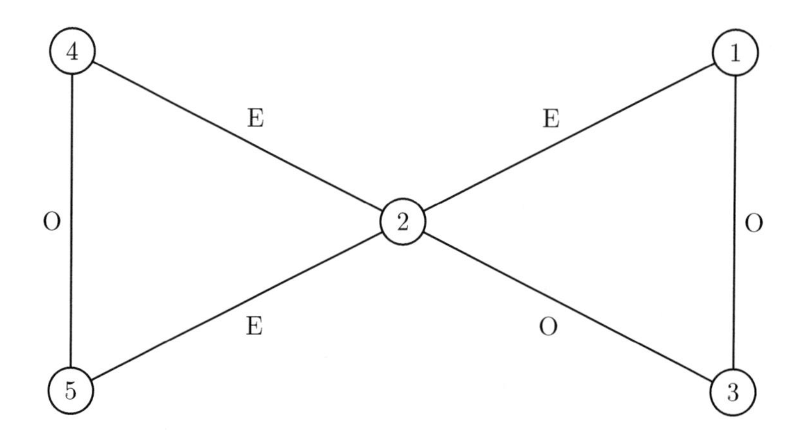

Figure 11.1: Friendship graph F_2 satisfying PCC-labeling

4. If $n \equiv 3 \pmod 4$, define $f(x)$ as:

$$f(x) = \begin{cases} 1, & \text{if } x = v; \\ 4j-2, & \text{if } x = v_{2,2j-1}, j \in \left[\frac{n+1}{4}\right]; \\ 4j-1, & \text{if } x = v_{2,2j}, j \in \left[\frac{n+1}{4}\right]; \\ f(v_{2,j})+2, & \text{if } x = v_{3,j}, j \in \left[\frac{n+1}{2}\right]; \\ 2j, & \text{if } x = v_{2,j}, j \in [n] - \left[\frac{n+1}{2}\right]; \\ 2j+1, & \text{if } x = v_{3,j}, j \in [n] - \left[\frac{n+1}{2}\right], \end{cases}$$

The induced edge labeling is given by:

$$f^*(e) = \begin{cases} 4j-2, & \text{if } e = vv_{2,2j-1}, j \in \left[\frac{n+1}{4}\right]; \\ 4j-1, & \text{if } e = vv_{2,2j}, j \in \left[\frac{n+1}{4}\right]; \\ 2j, & \text{if } e = vv_{2,j}, j \in [n] - \left[\frac{n+1}{2}\right]; \\ 4j, & \text{if } e = vv_{3,2j-1}, j \in \left[\frac{n+1}{4}\right]; \\ 4j+1, & \text{if } e = vv_{3,2j}, j \in \left[\frac{n+1}{4}\right]; \\ 2j+1, & \text{if } e = vv_{3,j}, j \in [n] - \left[\frac{n+1}{2}\right], \\ 2j(4j-1), & \text{if } e = v_{2,2j-1}v_{3,2j-1}, j \in \left[\frac{n+1}{4}\right], \\ 2j(4j+1), & \text{if } e = v_{2,2j}v_{3,2j}, j \in \left[\frac{n+1}{4}\right], \\ 2j+1, & \text{if } e = v_{2,j}v_{3,j}, j \in [n] - \left[\frac{n+1}{2}\right], \end{cases}$$

This gives us Table 11.11:

Table 11.11

Edges	$vv_{2,j}$	$vv_{3,j}$	$v_{2,j}v_{3,j}$	Total
$e_f(0)$	$\frac{3n-1}{4}$	$\frac{n+1}{4}$	$\frac{n+1}{2}$	$\frac{3n+1}{2}$
$e_f(1)$	$\frac{n+1}{4}$	$\frac{3n-1}{4}$	$\frac{n-1}{2}$	$\frac{3n-1}{2}$

This concludes that $|e_f(1) - e_f(0)| \leq 1$ in all the cases proving that the function satisfies PCC labeling and hence the graph is a PCC-graph. □

Theorem 11.6

Double fan $P_n + \overline{K_2}$ is a PCC graph. ■

Proof. Let the vertex and edge set of double fan graph $G = P_n + \overline{K_2}$ be $V = \{u_i, v_j \mid i \in [2], j \in [n]\}$ and $E = \{u_i v_j, v_j v_{j+1}, \mid i \in [2], j \in [n-1]\} \cup \{u_1 v_n, u_2 v_n\}$. Thus, $|V| = n+2$ and $|E| = 3n-1$. Define $f : V \rightarrow [n+2]$ as follows:

1. If $n \not\equiv 1 \,(\text{mod } 4)$, $f(x) = \begin{cases} i, & \text{if } x = u_i, i \in [2]; \\ i+2, & \text{if } x = v_i, i \in [n]. \end{cases}$

   The induced edge labeling is given by:

$$f^*(e) = \begin{cases} i+2, & \text{if } e = u_1 v_i, i \in [n]; \\ \dfrac{(i+1)(i+2)}{2}, & \text{if } e = u_2 v_i, i \in [n]; \\ i+3, & \text{if } e = v_i v_{i+1}, i \in [n-1]. \end{cases}$$

   Depending on the nature of n, we have Table 11.12:

Table 11.12

Edges		u_1v_i	u_2v_i	v_iv_{i+1}	Total
n is odd	$e_f(0)$	$\frac{n-1}{2}$	$\frac{n+1}{2}$	$\frac{n-1}{2}$	$\frac{3n-1}{2}$
	$e_f(1)$	$\frac{n+1}{2}$	$\frac{n-1}{2}$	$\frac{n-1}{2}$	$\frac{3n-1}{2}$
n is even	$e_f(0)$	$\frac{n}{2}$	$\frac{n}{2}$	$\frac{n}{2}$	$\frac{3n}{2}$
	$e_f(1)$	$\frac{n}{2}$	$\frac{n}{2}$	$\frac{n}{2}-1$	$\frac{3n}{2}-1$

2. If $n \equiv 1 \,(\text{mod } 4)$, $f(x) = \begin{cases} i, & \text{if } x = u_i, i \in [2]; \\ 4, & \text{if } x = v_1; \\ 3, & \text{if } x = v_2; \\ i+2, & \text{if } x = v_i, i \in [n] - \{1,2\}. \end{cases}$

   The induced edge labeling is given by:

$$f^*(e) = \begin{cases} i+2, & \text{if } e = u_1 v_i, i \in [n] - \{1,2\}; \\ 4, & \text{if } e = u_1 v_1; \\ 3, & \text{if } e = u_1 v_2; \\ \dfrac{(i+1)(i+2)}{2}, & \text{if } e = u_2 v_i, i \in [n] - \{1,2\}; \\ 6, & \text{if } e = u_2 v_1; \\ 3, & \text{if } e = u_2 v_2; \\ i+3, & \text{if } e = v_i v_{i+1}, i \in [n-1] - \{1,2\}; \\ 4, & \text{if } e = v_1 v_2; \\ 10, & \text{if } e = v_2 v_3. \end{cases}$$

This gives us Table 11.13:

Table 11.13

	u_1v_i	u_2v_i	v_iv_{i+1}	Total
$e_f(0)$	$\frac{n-1}{2}$	$\frac{n-1}{2}$	$\frac{n+1}{2}$	$\frac{3n-1}{2}$
$e_f(1)$	$\frac{n+1}{2}$	$\frac{n+1}{2}$	$\frac{n-3}{2}$	$\frac{3n-1}{2}$

This concludes that $|e_f(1) - e_f(0)| \leq 1$ in all the cases proving that the function satisfies PCC labeling and hence the graph is a PCC-graph. □

11.4 CONCLUSION

In this chapter, graphs obtained by graph operations like identifying vertices, identifying edges, subdividing edges and joining two graphs on standard graphs like path, cycle, star graph have been proved to be parity combination cordial graphs.

REFERENCES

1. D. M. Burton. *Elementary Number Theory.* 6th edition, Tata Macgraw-Hill, 2007.
2. I. Cahit. Codial graphs: a weaker version of graceful and harmonious graphs. *Ars Combinatoria*, 23, 201-208,1987
3. J. A. Gallian. A dynamic survey of graph labeling. *The Electronic Journal of Combinatorics*, # DS6, 2016
4. S. M. Hegde and S. Shetty. Combinatorial labelings of graphs. *Applied Mathematics E-Notes*, 6, 251-258, 2006
5. John Clark and Derek Allan Holton. *A First Look at Graph Theory*, World Scientific, 1995.
6. O. Pikhurko. Trees are almost prime. *Discrete Math.*, 307 1455-1462, 2007.
7. R. Ponraj, S. Sathish Narayanan and A. M. S. Ramasamy. Parity combination cordial labeling of graphs. *Jordan J. Math. and Stat.*, 8(4): 293-308, 2015.
8. R. Ponraj, R. Singh, and S. Sathish Narayanan. On parity combination cordial graphs. *Palestine J. Math.*, 1825.
9. Soni Devyani Vinodkumar, Ritu Khanna and P. H. Bhathawala. Parity combination cordial labeling in the context of duplication of graph elements. *IOSR Journal of Mathematics*, 15(2): 39-45, 2019.

12 Total Neighborhood Prime Labeling of Join Graphs

N P Shrimali
Department of Mathematics,
Gujarat University,
Ahmedabad, Gujarat (INDIA)
E-mail: narenp05@gmail.com

A K Rathod
Department of Mathematics,
Gujarat University,
Ahmedabad, Gujarat (INDIA)
E-mail: ashwin.rathodmaths@gmail.com

Let G be a graph with $V(G)$ as the vertex set and $E(G)$ as the edge set. A bijective function $f: V(G) \cup E(G) \to \{1,2,3,\ldots,|V(G) \cup E(G)|\}$ is said to be a total neighborhood prime labeling, if for each vertex in G having degree greater than 1, the gcd of the labels of its neighborhood vertices is 1 and the gcd of the labels of its induced edges is 1. A graph which admits a total neighborhood prime labeling is called a total neighborhood prime graph. In this chapter, we investigate total neighborhood prime labeling for some join graphs.

12.1 INTRODUCTION AND DEFINITIONS

In this chapter, we consider the simple, finite, connected and undirected graph $G = (V(G), E(G))$ with $V(G)$ as vertex set and $E(G)$ as edge set. For various notations and terminology of graph theory, we follow Gross and Yellen [3] and for some results of number theory, we follow Burton [2].

Let G be a graph with n vertices. A bijective function $f : V(G) \to \{1,2,3,\ldots,n\}$ is said to be a **neighborhood-prime labeling** if for every vertex u in $V(G)$ with $deg(u) > 1$, gcd $\{f(p) | p \in N(u)\} = 1$, where $N(u) = \{w \in V(G) | uw \in E(G)\}$. A graph which admits neighborhood-prime labeling is called a neighborhood-prime graph.

The notion of neighborhood-prime labeling was introduced by Patel and Shrimali [4]. In [5] they proved the union of some graphs are neighborhood-prime graphs. They also proved that the product of some graphs are neighborhood-prime [6]. For a further list of results regarding neighborhood-prime graphs the reader may refer to [1].

In a neighborhood-prime labeling, edges are not considered. Kumar and Varkey extended the condition on edges and they defined total neighborhood prime labeling[7].

Definition (Total neighborhood prime labeling). Let G be a graph. For $u \in V(G)$,

$$N_V(u) = \{w \in V(G) | uw \in E(G)\}$$
$$N_E(u) = \{e \in E(G) | e = uv, \text{for some } v \in V(G)\}$$

A bijective function $f : V(G) \cup E(G) \rightarrow \{1,2,3,\ldots,|V(G) \cup E(G)|\}$ is said to be a total neighborhood prime labeling, if for each vertex u with degree greater than 1, gcd $\{f(w)|w \in N_V(u)\} = 1$ and $gcd\{f(e)|e \in N_E(u)\} = 1$. A graph which admits total neighborhood prime labeling is called a total neighborhood prime graph.

Kumar and Varkey proved that path, cycle C_{4k} and comb graphs admit total neighborhood prime labeling[7]. Shrimali and Pandya proved comb, disjoint union of paths, disjoint union of sunlet graphs, disjoint union of wheel graphs, graph obtained by one copy of path P_n and n copies of $K_{1,m}$ and joining i^{th} vertex of P_n with an edge to fixed vertex in the ith copy of $K_{1,m}$, corona product of cycle with m copy of K_1 and subdivision of bistar are total neighborhood prime graphs[8].

Definition (Armed crown). Armed crown is the graph obtained by attaching a path P_2 at each vertex of cycle C_n. It is denoted by ACn.

Definition (Join graph). Let $G_1 = (V(G_1), E(G_1))$ and $G_2 = (V(G_2), E(G_2))$ be two graphs with no vertex in common. The join graph of G_1 and G_2 denoted by $G_1 + G_2$ is the graph with vertex set $V(G_1 + G_2) = V(G_1) \bigcup V(G_2)$ and edge set $E(G_1 + G_2) = E(G_1) \bigcup E(G_2) \bigcup \{uv : u \in V(G_1), v \in V(G_2)\}$.

12.2 MAIN RESULTS

Theorem 12.1

$AC_n + K_1$ is a total neighborhood prime graph. ■

Proof. Let $G = AC_n + K_1$. Let $u_1, u_2, \ldots, u_n$ be the vertices of C_n , $u_{i,1}$ and $u_{i,2}$ be the vertices of ith copy of path P_2 for each i in AC_n and u be the vertex corresponding to K_1. By the definition of join graph, u is adjacent to each vertex of a graph AC_n, $u_{i,1}$ is adjacent to $u_{i,2}$ and u_i for $i = 1,2,\ldots,n$. Let $d_i = u_i\ u_{i+1}$(value of i is taken modulo n), $e_{i,j} = u\ u_{i,j}$ and $e_i = uu_i$ for $i = 1,2,\ldots,n$ and $j = 1,2$ and $g_{i,1} = u_iu_{i,1}, g_{i,2} = u_{i,1}u_{i,2}$ for $i = 1,2,\ldots,n$.

So, vertex set $V(G) = \{u, u_i, u_{i,j} / i = 1, 2, \ldots, n \text{ and } j = 1, 2\}$ and edge set $E(G) = \{e_i, d_i / i = 1, 2, \ldots, n\} \cup \{e_{i,j}, g_{i,j} / i = 1, 2, \ldots, n \text{ and } j = 1, 2\}$
We define $f : V(G) \cup E(G) \longrightarrow \{1, 2, 3, \ldots, |V(G) \cup E(G)|\}$ as follows.

$f(u) = 1,$
$f(u_i) = 8i - 1, \quad 1 \le i \le n$
$f(u_{i,j}) = \begin{cases} 8i, & j = 1 \\ 8i + 1, & j = 2. \end{cases}, 1 \le i \le n$
$f(e_i) = 8(i-1) + 4, \quad 1 \le i \le n$
$f(e_{i,j}) = \begin{cases} 8(i-1) + 6, & j = 1 \\ 8(i-1) + 3, & j = 2. \end{cases}, 1 \le i \le n$
$f(g_{i,j}) = \begin{cases} 8(i-1) + 5, & j = 1 \\ 8(i-1) + 2, & j = 2. \end{cases}, 1 \le i \le n$
$f(d_i) = 8n + 1 + i, \quad 1 \le i \le n$

In order to prove f is a total neighborhood prime labeling, we have to show that both the conditions are satisfied by f. Let w be an arbitrary vertex in $V(G)$. For $w \ne u$, since $f(u) = 1$ and $u \in N_V(w)$, gcd $\{f(p)/p \in N_V(w)\} = 1$ and for $w = u$, $\{f(p)/p \in N_V(w)\}$ contains at least two consecutive numbers, so gcd $\{f(p)/p \in N_V(w)\} = 1$. For any vertex w, $\{f(e)/e \in N_E(w)\}$ contains at least two consecutive numbers or consecutive odd numbers, so gcd $\{f(e)/e \in N_E(w)\} = 1$. Therefore, f is a total neighborhood prime labeling. Hence, G is a total neighborhood prime graph. □

Illustration 12.2.1. Total neighborhood prime labeling of $AC_4 + K_1$ is shown in Figure 12.1.

Theorem 12.2

$(\bigcup_{i=1}^{p} C_{n_i}) + K_1$ is a total neighborhood prime graph. ■

Proof. Let $G = (\bigcup_{i=1}^{p} C_{n_i}) + K_1$. Let $u_{i,1}, u_{i,2}, u_{i,3}, \ldots, u_{i,n_i}$ be the vertices of ith cycle C_{n_i} where $1 \le i \le p$ and u be the vertex corresponding to K_1. By the definition of join graph, u is adjacent to each vertex of $\bigcup_{i=1}^{p} C_{n_i}$. Let $e_{i,j} = u_{i,j}u_{i,j+1}$ and $d_{i,j} = uu_{i,j}$ for $1 \le i \le p$, $1 \le j \le n_i$ where the value of j is taken modulo n_i. Vertex set $V(G) = \{u_{i,j}, u / i = 1, 2, \ldots, p \text{ and } j = 1, 2, \ldots, n_i\}$ and edge set $E(G) = \{e_{i,j}, d_{i,j} / i = 1, 2, \ldots, p \text{ and } j = 1, 2, \ldots, n_i\}$.
Now we define the function $f : V(G) \cup E(G) \longrightarrow \{1, 2, 3, \ldots, |V(G) \cup E(G)|\}$ as follows.

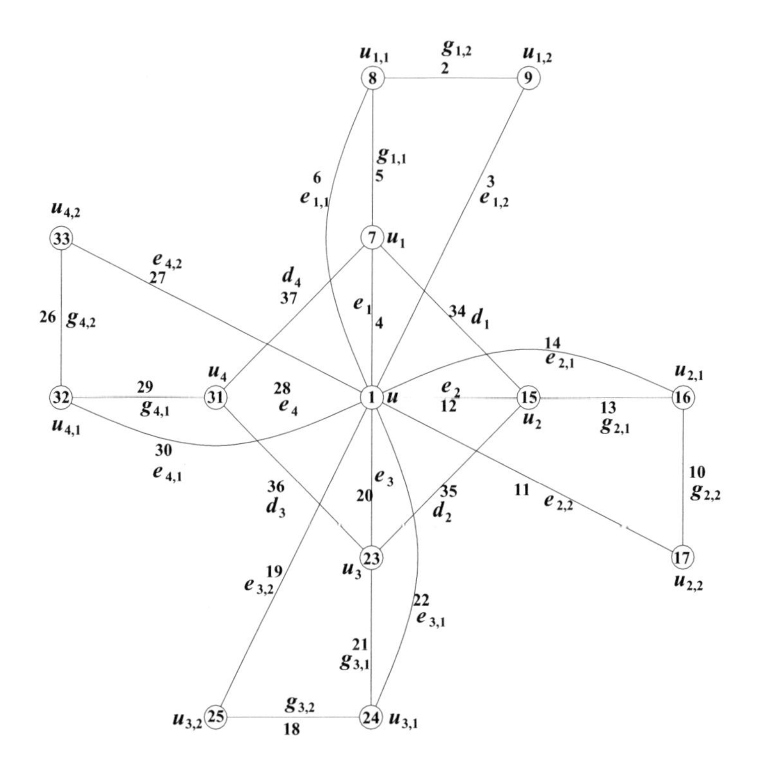

Figure 12.1: Total neighborhood prime labeling of $AC_4 + K_1$

$f(u) = 1,$

$$f(u_{i,j}) = \begin{cases} 2n_1+1+j, & i=1, 1 \le j \le n_1 \\ 1+3(n_1+n_2+,\dots,+n_{i-1})+2n_i+j, & 2 \le i \le p, 1 \le j \le n_i \end{cases}$$

$$f(e_{i,j}) = \begin{cases} 1+j, & i=1, 1 \le j \le n_1 \\ 1+3(n_1+n_2+,\dots,+n_{i-1})+j, & 2 \le i \le p, 1 \le j \le n_i \end{cases}$$

$$f(d_{i,j}) = \begin{cases} n_1+j+1, & i=1, 1 \le j \le n_1 \\ 1+3(n_1+n_2+,\dots,+n_{i-1})+n_i+j, & 2 \le i \le p, 1 \le j \le n_i \end{cases}$$

We will show that both the conditions for total neighborhood prime labeling are satisfied by f. Let w be an arbitrary vertex in $V(G)$. For $w \neq u$, gcd $\{f(p)/p \in N_V(w)\} = 1$ because $f(u) = 1$ and $u \in N_V(w)$. For $w = u$, $\{f(p)/p \in N_V(w)\}$ contains at least two consecutive numbers, so gcd $\{f(p)/p \in N_V(w)\} = 1$. For any vertex w, $\{f(e)/e \in N_E(w)\}$ contains at least two consecutive numbers or consecutive odd numbers, so gcd $\{f(e)/e \in N_E(w)\} = 1$. Therefore, f is a total neighborhood prime labeling and G is a total neighborhood prime graph. □

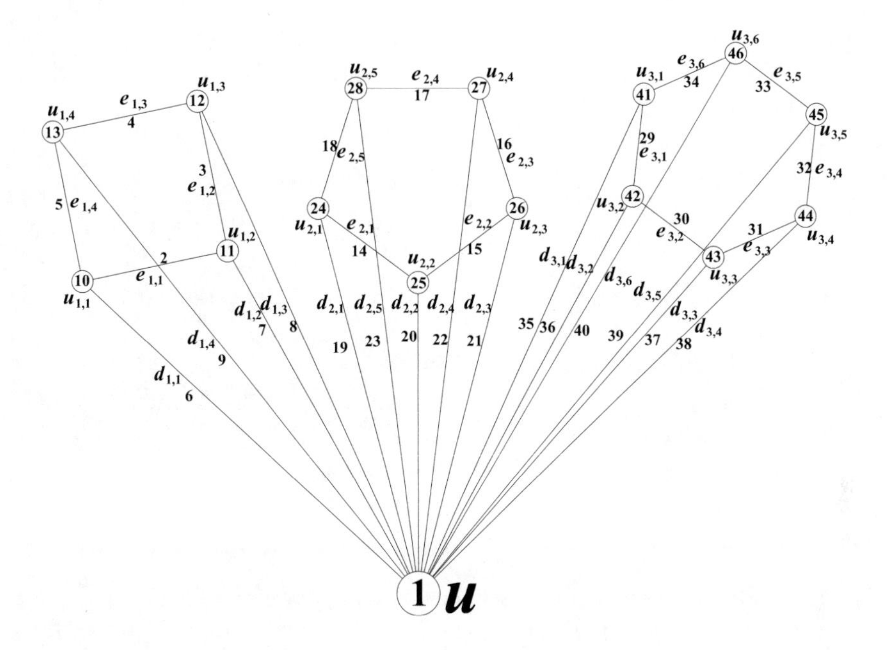

Figure 12.2: Total neighborhood prime labeling of $(C_4 \bigcup C_5 \bigcup C_6) + K_1$

Illustration 12.2.2. Total neighborhood prime labeling of $(C_4 \bigcup C_5 \bigcup C_6) + K_1$ is shown in Figure 12.2.

Theorem 12.3

$[P_n \bigcup (\bigcup_{i=1}^{p} C_{n_i})] + K_1$ is a total neighborhood prime graph. ∎

Proof. Let $G = [P_n \bigcup (\bigcup_{i=1}^{p} C_{n_i})] + K_1$. Let $v_1, v_2, \ldots, v_n$ be the consecutive vertices of path P_n and $u_{i,1}, u_{i,2}, u_{i,3}, \ldots, u_{i,n_i}$ be the vertices of ith cycle C_{n_i} where $1 \leq i \leq p$. Let u be the vertex corresponding to K_1. So by the definition of join graph, u is adjacent to each vertex of $P_n \bigcup (\bigcup_{i=1}^{p} C_{n_i})$. Let $e_{i,j} = u_{i,j}u_{i,j+1}$ and $d_{i,j} = uu_{i,j}$ for $1 \leq i \leq p$, $1 \leq j \leq n_i$ where value of j is taken modulo n_i. Let $g_i = v_i v_{i+1}$ for $i = 1, 2, \ldots, n-1$ and $g'_i = uv_i$ for $i = 1, 2, \ldots, n$.

Vertex set $V(G) = \{u_{i,j}, u / i = 1, 2, \ldots, p \text{ and } j = 1, 2, \ldots, n_i\} \bigcup \{v_1, v_2, \ldots, v_n\}$ and edge set $E(G) = \{e_{i,j}, d_{i,j} / i = 1, 2, \ldots, p \text{ and } j = 1, 2, \ldots, n_i\} \bigcup \{g_1, g_2, \ldots, g_{n-1}\} \bigcup \{g'_1, g'_2, \ldots, g'_n\}$.
Now we define the function $f : V(G) \cup E(G) \longrightarrow \{1, 2, 3, \ldots, |V(G) \cup E(G)|\}$ as follows.

$$f(u) = 1,$$
$$f(u_{i,j}) = \begin{cases} 2n_1 + 1 + j, & i = 1, 1 \le j \le n_1 \\ 1 + 3(n_1 + n_2 +, \ldots, + n_{i-1}) + 2n_i + j, & 2 \le i \le p, 1 \le j \le n_i \end{cases}$$
$$f(e_{i,j}) = \begin{cases} 1 + j, & i = 1, 1 \le j \le n_1 \\ 1 + 3(n_1 + n_2 +, \ldots, + n_{i-1}) + j, & 2 \le i \le p, 1 \le j \le n_i \end{cases}$$
$$f(d_{i,j}) = \begin{cases} n_1 + j + 1, & i = 1, 1 \le j \le n_1 \\ 1 + 3(n_1 + n_2 +, \ldots, + n_{i-1}) + n_i + j, & 2 \le i \le p, 1 \le j \le n_i \end{cases}$$
$$f(v_i) = 3(n_1 + n_2 +, \ldots, + n_p) + 2n + i, \quad 1 \le i \le n,$$
$$f(g_i) = 3(n_1 + n_2 +, \ldots, + n_p) + 2i + 1, \quad 1 \le i \le n - 1,$$
$$f(g'_i) = 3(n_1 + n_2 +, \ldots, + n_p) + 2i, \quad 1 \le i \le n.$$

Let w be an arbitrary vertex in $V(G)$. For $w \neq u$, since $f(u) = 1$ and $u \in N_V(w)$, gcd $\{f(p)/p \in N_V(w)\} = 1$ and for $w = u$, gcd $\{f(p)/p \in N_V(w)\} = 1$ because $\{f(p)/p \in N_V(w)\}$ contains at least two consecutive numbers. For any vertex w, gcd $\{f(e)/e \in N_E(w)\} = 1$ because $\{f(e)/e \in N_E(w)\}$ contains at least two consecutive numbers or consecutive odd numbers. Since both the conditions for total neighborhood prime labeling are satisfied, f is a total neighborhood prime labeling and hence G is a total neighborhood prime graph. □

Illustration 12.2.3. Total neighborhood prime labeling of $(P_9 \bigcup C_3 \bigcup C_5 \bigcup C_6) + K_1$ is shown in Figure 12.3.

Theorem 12.4

$[K_{1,n} \bigcup (\bigcup_{i=1}^{p} C_{n_i})] + K_1$ is a total neighborhood prime graph. ■

Proof. Let $G = [K_{1,n} \bigcup (\bigcup_{i=1}^{p} C_{n_i})] + K_1$. Let $v, v_1, v_2, \ldots, v_n$ be the vertices of star graph $K_{1,n}$ where v is the apex vertex. Let $u_{i,1}, u_{i,2}, u_{i,3}, \ldots, u_{i,n_i}$ be the vertices of ith cycle C_{n_i} where $1 \le i \le p$. Let u be the vertex corresponding to K_1. So by the definition of join graph, u is adjacent to each vertex of $K_{1,n} \bigcup (\bigcup_{i=1}^{p} C_{n_i})$. Let $e_{i,j} = u_{i,j}u_{i,j+1}$ and $d_{i,j} = uu_{i,j}$ for $1 \le i \le p$, $1 \le j \le n_i$

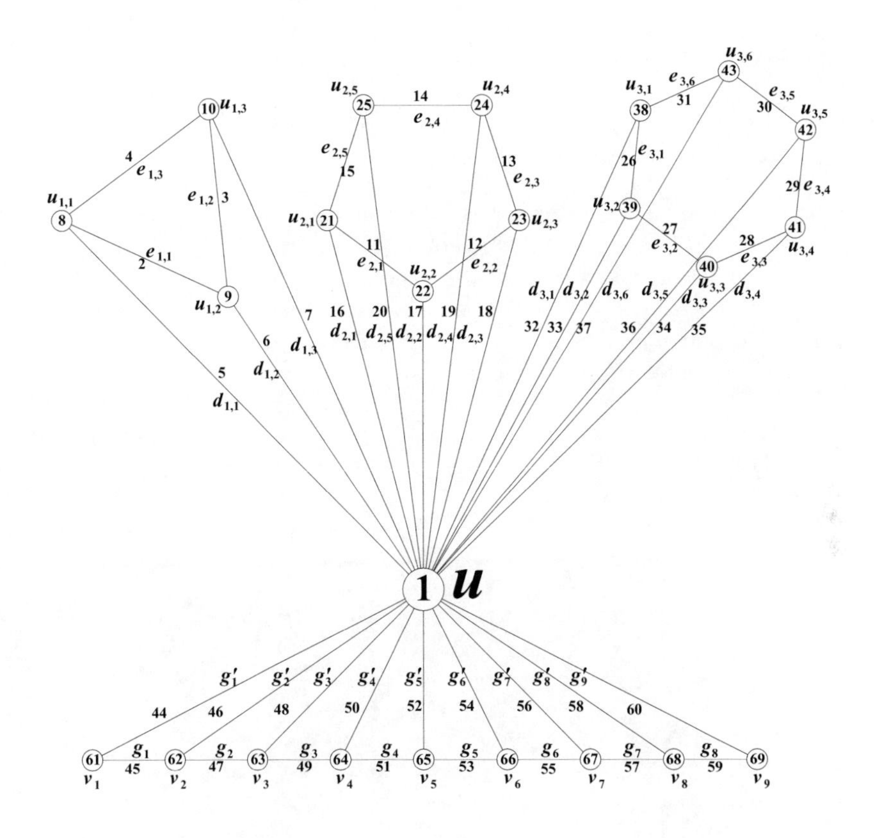

Figure 12.3: Total neighborhood prime labeling of $(P_9 \bigcup C_3 \bigcup C_5 \bigcup C_6) + K_1$

where value of j is modulo n_i. Let $g_i = uv_i$, $h_i = vv_i$ for $i = 1,2,\ldots,n$ and $e = uv$.
Vertex set $V(G) = \{u_{i,j}/i = 1,2,\ldots,p$ and $j = 1,2,\ldots,n_i\} \bigcup \{v_1, v_2, \ldots, v_n\}$ $\bigcup \{u,v\}$ and edge set $E(G) = \{e_{i,j}, d_{i,j}/i = 1,2,\ldots,p$ and $j = 1,2,\ldots,n_i\} \bigcup \{g_1, g_2, \ldots, g_n\} \bigcup \{h_1, h_2, \ldots, h_n\} \bigcup \{x, x_1, x_2\}$.
Now we define the function $f: V(G) \cup E(G) \longrightarrow \{1,2,3,\ldots,|V(G) \cup E(G)|\}$ as follows.

$f(u) = 1,$

$$f(u_{i,j}) = \begin{cases} 2n_1 + 1 + j, & i = 1, 1 \le j \le n_1 \\ 1 + 3(n_1 + n_2 +, \ldots, +n_{i-1}) + 2n_i + j, & 2 \le i \le p, 1 \le j \le n_i \end{cases}$$

$$f(e_{i,j}) = \begin{cases} 1 + j, & i = 1, 1 \le j \le n_1 \\ 1 + 3(n_1 + n_2 +, \ldots, +n_{i-1}) + j, & 2 \le i \le p, 1 \le j \le n_i \end{cases}$$

$$f(d_{i,j}) = \begin{cases} n_1 + j + 1, & i = 1, 1 \le j \le n_1 \\ 1 + 3(n_1 + n_2 +, \ldots, +n_{i-1}) + n_i + j, & 2 \le i \le p, 1 \le j \le n_i \end{cases}$$

$f(v_i) = 3(n_1 + n_2 +, \ldots, +n_p) + 2n + 1 + i, \quad 1 \le i \le n,$

$f(g_i)=3(n_1+n_2+,\ldots,+n_p)+2i+1, \quad 1\le i\le n,$
$f(h_i)=3(n_1+n_2+,\ldots,+n_p)+2i, \quad 1\le i\le n,$
$f(e)=3(n_1+n_2+,\ldots,+n_p)+3n+3,$
$f(v)=3(n_1+n_2+,\ldots,+n_p)+3n+2.$

Let w be an arbitrary vertex in $V(G)$. For $w\neq u$, gcd $\{f(p)/p\in N_V(w)\}=1$ because $f(u)=1$ and $u\in N_V(w)$. For $w=u$, since $\{f(p)/p\in N_V(w)\}$ contains at least two consecutive numbers, gcd $\{f(p)/p\in N_V(w)\}=1$. For any vertex w, $\{f(e)/e\in N_E(w)\}$ contains at least two consecutive numbers or consecutive odd numbers, so gcd $\{f(e)/e\in N_E(w)\}=1$. Therefore, f is a total neighborhood prime labeling. Thus, G is a total neighborhood prime graph. □

Illustration 12.2.4. Total neighborhood prime labeling of $(K_{1,8}\bigcup C_4\bigcup C_5 \bigcup C_6)+K_1$ is shown in Figure 12.4.

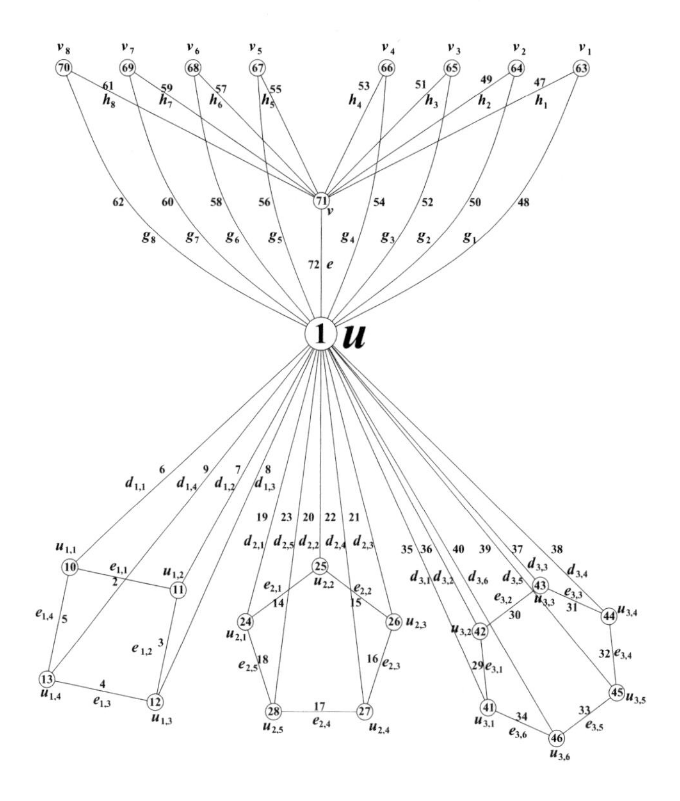

Figure 12.4: Total neighborhood prime labeling of $(K_{1,8}\bigcup C_4\bigcup C_5\bigcup C_6)+K_1$

Theorem 12.5

$[B_{m,n} \bigcup (\bigcup_{i=1}^{p} C_{n_i})] + K_1$ is a total neighborhood prime graph. ■

Proof. Let $G = [B_{m,n} \bigcup (\bigcup_{i=1}^{p} C_{n_i})] + K_1$. Let $v_1, v_2, v_{1,1}, v_{1,2}, \ldots, v_{1,m}, v_{2,1}, v_{2,2}, \ldots, v_{2,n}$ be the vertices of bistar graph $B_{m,n}$ where $v_{1,1}, v_{1,2}, \ldots, v_{1,m}$ are adjacent to v_1 and $v_{2,1}, v_{2,2}, \ldots, v_{2,n}$ are adjacent to v_2. Let $u_{i,1}, u_{i,2}, u_{i,3}, \ldots, u_{i,n_i}$ be the vertices of ith cycle C_{n_i} where $1 \leq i \leq p$. Let u be the vertex corresponding to K_1. So by the definition of join graph, u is adjacent to each vertex of $B_{m,n} \bigcup (\bigcup_{i=1}^{p} C_{n_i})$. Let $e_{i,j} = u_{i,j}u_{i,j+1}$ and $d_{i,j} = uu_{i,j}$ for $1 \leq i \leq p$, $1 \leq j \leq n_i$ where the value of j is taken modulo n_i. Let $g_{i,j} = v_i v_{i,j}$ and $h_{i,j} = uv_{i,j}$ for $i = 1$, $j = 1, 2, \ldots, m$ and for $i = 2$, $j = 1, 2, \ldots, n$. let $t = v_1 v_2$ and $t_i = uv_i$ for $i = 1, 2$.
Vertex set $V(G) = \{u_{i,j} / i = 1, 2, \ldots, p \text{ and } j = 1, 2, \ldots, n_i\} \bigcup \{v_{1,1}, v_{1,2}, \ldots, v_{1,m}\} \bigcup \{v_{2,1}, v_{2,2}, \ldots, v_{2,n}\} \bigcup \{u, v_1, v_2\}$ and edge set $E(G) = \{e_{i,j}, d_{i,j} / i = 1, 2, \ldots, p \text{ and } j = 1, 2, \ldots, n_i\} \bigcup \{g_{1,1}, g_{1,2}, \ldots, g_{1,m}, g_{2,1}, g_{2,2}, \ldots, g_{2,n}\} \bigcup \{h_{1,1}, h_{1,2}, \ldots, h_{1,m}, h_{2,1}, h_{2,2}, \ldots, h_{2,n}\}$.
Now we define the function $f : V(G) \cup E(G) \longrightarrow \{1, 2, 3, \ldots, |V(G) \cup E(G)|\}$ as follows.

$$f(u) = 1,$$

$$f(u_{i,j}) = \begin{cases} 2n_1 + 1 + j, & i = 1, 1 \leq j \leq n_1 \\ 1 + 3(n_1 + n_2 +, \ldots, +n_{i-1}) + 2n_i + j, & 2 \leq i \leq p, 1 \leq j \leq n_i \end{cases}$$

$$f(e_{i,j}) = \begin{cases} 1 + j, & i = 1, 1 \leq j \leq n_1 \\ 1 + 3(n_1 + n_2 +, \ldots, +n_{i-1}) + j, & 2 \leq i \leq p, 1 \leq j \leq n_i \end{cases}$$

$$f(d_{i,j}) = \begin{cases} n_1 + j + 1, & i = 1, 1 \leq j \leq n_1 \\ 1 + 3(n_1 + n_2 +, \ldots, +n_{i-1}) + n_i + j, & 2 \leq i \leq p, 1 \leq j \leq n_i \end{cases}$$

$$f(v_i) = 3(n_1 + n_2 +, \ldots, +n_p) + 3(m + n) + 4 + i, \quad i = 1, 2,$$

$$f(v_{i,j}) = \begin{cases} 3(n_1 + n_2 +, \ldots, +n_p) + 2(m + n) + 4 + j, & i = 1, 1 \leq j \leq m \\ 3(n_1 + n_2 +, \ldots, +n_p) + 3m + 2n + 4 + j, & i = 2, 1 \leq j \leq n \end{cases}$$

$$f(g_{i,j}) = \begin{cases} 3(n_1 + n_2 +, \ldots, +n_p) + 2j + 1, & i = 1, 1 \leq j \leq m \\ 3(n_1 + n_2 +, \ldots, +n_p) + 2m + 2j + 1, & i = 2, 1 \leq j \leq n \end{cases}$$

$$f(h_{i,j}) = \begin{cases} 3(n_1 + n_2 +, \ldots, +n_p) + 2j, & i = 1, 1 \leq j \leq m \\ 3(n_1 + n_2 +, \ldots, +n_p) + 2m + 2j, & i = 2, 1 \leq j \leq n \end{cases}$$

$$f(t_i) = 3(n_1 + n_2 +, \ldots, +n_p) + 2(m + n) + 2 + i, \quad i = 1, 2,$$
$$f(t) = 3(n_1 + n_2 +, \ldots, +n_p) + 2(m + n) + 2.$$

Let w be an arbitrary vertex in $V(G)$. For $w \neq u$, since $f(u) = 1$ and $u \in N_V(w)$, gcd $\{f(p) / p \in N_V(w)\} = 1$ and for $w = u$, gcd $\{f(p) / p \in N_V(w)\} = 1$

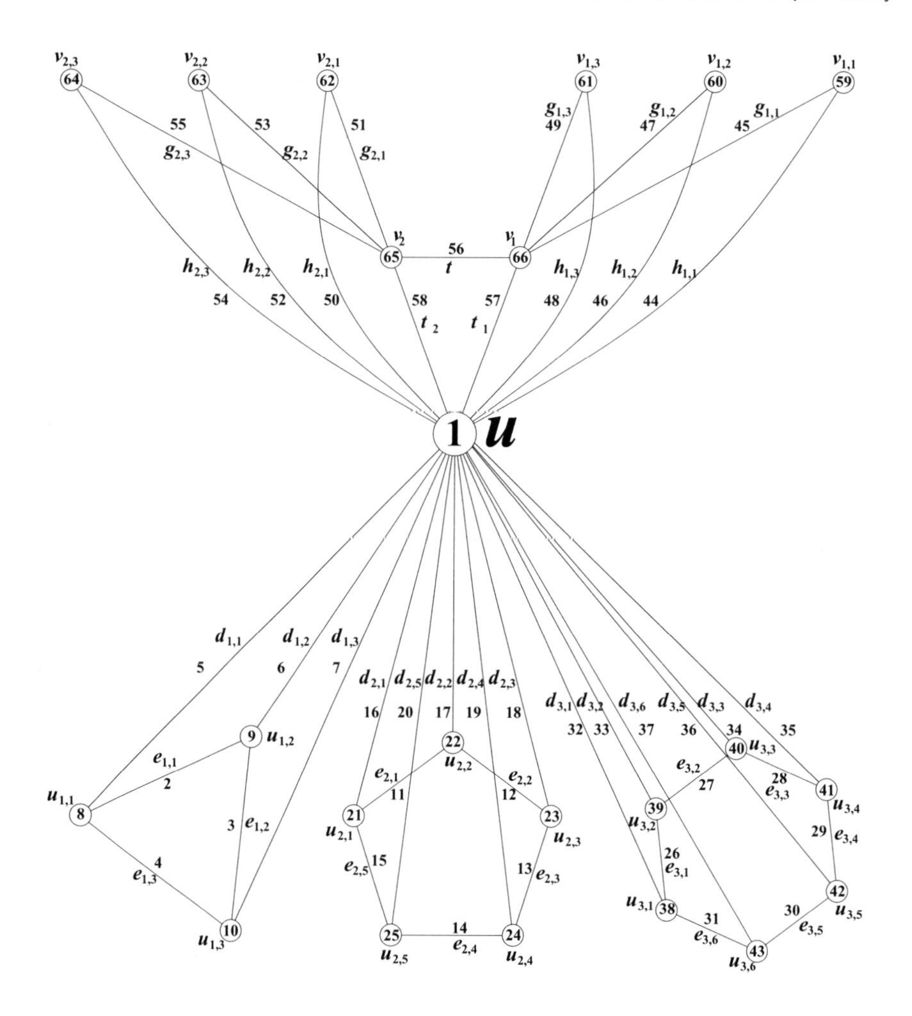

Figure 12.5: Total neighborhood prime labeling of $(B_{3,3} \bigcup C_3 \bigcup C_5 \bigcup C_6) + K_1$

because $\{f(p)/p \in N_V(w)\}$ contains at least two consecutive numbers. For any vertex w, gcd $\{f(e)/e \in N_E(w)\} = 1$ because $\{f(e)/e \in N_E(w)\}$ contains at least two consecutive numbers or consecutive odd numbers. Since both the conditions for total neighborhood prime labeling are satisfied, f is a total neighborhood prime labeling and G is a total neighborhood prime graph. □

Illustration 12.2.5. Total neighborhood prime labeling of $(B_{3,3} \bigcup C_3 \bigcup C_5 \bigcup C_6) + K_1$ is shown in Figure 12.5.

Theorem 12.6

$[(P_n \odot K_1)\bigcup(\bigcup_{i=1}^{p} C_{n_i})] + K_1$ is a total neighborhood prime graph. ■

Proof. Let $G = [(P_n \odot K_1) \bigcup_{i=1}^{p} C_{n_i})] + K_1$. Let $v_{1,j}, v_{2,j}, \ldots, v_{n,j}$ be the vertices of comb graph $(P_n \odot K_1)$ for $j = 1,2$. $v_{i,1}$ is adjacent to $v_{i-1,1}$ and $v_{i+1,1}$ for $i = 2,3,\ldots,n-1$. $v_{i,1}$ is also adjacent $v_{i,2}$ for $i = 1,2,\ldots,n$. Let $u_{i,1}, u_{i,2}, u_{i,3}, \ldots, u_{i,n_i}$ be the vertices of ith cycle C_{n_i} where $1 \leq i \leq p$. Let u be the vertex corresponding to K_1. So by the definition of join graph, u is adjacent to each vertex of $[(P_n \odot K_1) \bigcup_{i=1}^{p} C_{n_i})]$. Let $e_{i,j} = u_{i,j}u_{i,j+1}$ and $d_{i,j} = uu_{i,j}$ for $1 \leq i \leq p$, $1 \leq j \leq n_i$ where the value of j is taken modulo n_i. Let $g_i = v_{i,1}v_{i+1,1}$ for $i = 1,2,\ldots,n-1$, $g_i' = v_{i,1}v_{i,2}$ for $i = 1,2,\ldots,n$ and $h_{i,j} = uv_{i,j}$ for $i = 1,2,\ldots,n$ and $j = 1,2$.
Vertex set $V(G) = \{u_{i,j}/i = 1,2,\ldots,p \text{ and } j = 1,2,\ldots,n_i\}\bigcup\{v_{i,1}, v_{i,2}, \ldots, v_{i,n}/ i = 1,2\}$ and edge set $E(G) = \{e_{i,j}, d_{i,j}/i = 1,2,\ldots,p \text{ and } j = 1,2,\ldots,n_i\}\bigcup \{g_1, g_2, \ldots, g_{n-1}\}\bigcup\{g_1', g_2', \ldots, g_n'\}\bigcup\{h_{i,1}, h_{i,2}, \ldots, h_{i,n}/i = 1,2\}$.
Now we define the function $f : V(G) \cup E(G) \longrightarrow \{1,2,3,\ldots,|V(G) \cup E(G)|\}$ as follows.

$$f(u) = 1,$$
$$f(u_{i,j}) = \begin{cases} 2n_1 + 1 + j, & i = 1, 1 \leq j \leq n_1 \\ 1 + 3(n_1 + n_2 +, \ldots, +n_{i-1}) + 2n_i + j, & 2 \leq i \leq p, 1 \leq j \leq n_i \end{cases}$$
$$f(e_{i,j}) = \begin{cases} 1 + j, & i = 1, 1 \leq j \leq n_1 \\ 1 + 3(n_1 + n_2 +, \ldots, +n_{i-1}) + j, & 2 \leq i \leq p, 1 \leq j \leq n_i \end{cases}$$
$$f(d_{i,j}) = \begin{cases} n_1 + j + 1, & i = 1, 1 \leq j \leq n_1 \\ 1 + 3(n_1 + n_2 +, \ldots, +n_{i-1}) + n_i + j, & 2 \leq i \leq p, 1 \leq j \leq n_i \end{cases}$$
$$f(v_{i,j}) = \begin{cases} 3(n_1 + n_2 +, \ldots, +n_p) + 4n + i, & 1 \leq i \leq n, j = 1 \\ 3(n_1 + n_2 +, \ldots, +n_p) + 5n + i, & 1 \leq i \leq n, j = 2 \end{cases}$$
$$f(h_{i,j}) = \begin{cases} 3(n_1 + n_2 +, \ldots, +n_p) + 2n + 2i, & 1 \leq i \leq n, j = 1 \\ 3(n_1 + n_2 +, \ldots, +n_p) + 2i, & 1 \leq i \leq n, j = 2 \end{cases}$$
$$f(g_i) = 3(n_1 + n_2 +, \ldots, +n_p) + 2n + 2i + 1, \quad i = 1,2,\ldots,n-1,$$
$$f(g_i') = 3(n_1 + n_2 +, \ldots, +n_p) + 2i + 1, \quad i = 1,2,\ldots,n$$

Let w be an arbitrary vertex in $V(G)$. For $w \neq u$, gcd $\{f(p)/p \in N_V(w)\} = 1$ because $f(u) = 1$ and $u \in N_V(w)$. For $w = u$, since $\{f(p)/p \in N_V(w)\}$ contains at least two consecutive numbers, gcd $\{f(p)/p \in N_V(w)\} = 1$. For any vertex w, $\{f(e)/e \in N_E(w)\}$ contains at least two consecutive numbers or consecutive odd numbers, so gcd $\{f(e)/e \in N_E(w)\} = 1$. Since both the conditions for total

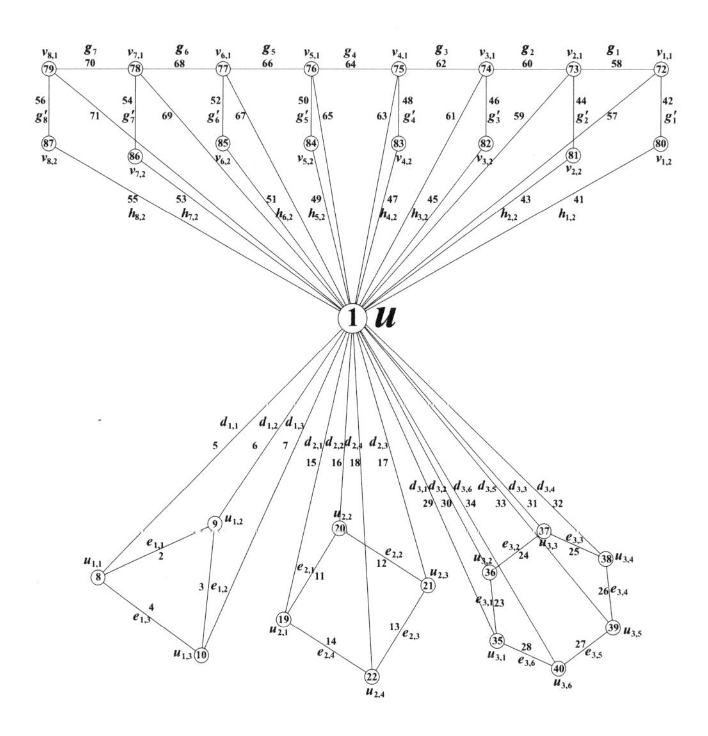

Figure 12.6: Total neighborhood prime labeling of $[(P_8 \odot K_1)\bigcup C_3 \bigcup C_4 \bigcup C_6]+K_1$

neighborhood prime labeling are satisfied, f is a total neighborhood prime labeling and G is a total neighborhood prime graph. □

Illustration 12.2.6. Total neighborhood prime labeling of $[(P_8 \odot K_1)\bigcup C_3 \bigcup C_4 \bigcup C_6]+K_1$ is shown in Figure 12.6.

REFERENCES

1. J. A. Gallian. A dynamic survey of graph labeling. *The Electronic Journal of Combinatorics*, #DS6, 2018.
2. D. M. Burton. *Elementary Number Theory.* Seventh Edition, McGraw-Hill Publisher, 2010.
3. J. Gross and J. Yellen. *Graph Theory and its Applications.* CRC Press, 2005.
4. S. K. Patel and N. P. Shrimali. Neighborhood-prime labeling. *International Journal of Mathematics and Soft Computing*, 5(2): 135-143, 2015.
5. S. K. Patel and N. P. Shrimali. Neighborhood-prime labeling of some union graphs. *International Journal of Mathematics and Soft Computing*, 6(1): 39-47, 2016.
6. S. K. Patel and N. P. Shrimali. Neighborhood-prime labeling of some product graphs. *Algebra and Discrete Mathematics*, 25(1): 118-129, 2018.

7. Rajesh Kumar and Mathew Varkey. A note on Total neighborhood prime labeling. *International Journal of Pure and Applied Mathematics*, 118(4): 1007-1013, 2018.
8. N. P. Shrimali and Parul B. Pandya. Total neighborhood prime labeling of some graphs. *International Journal of Scientific Research in Mathematical and Statistical Sciences*, 5(6): 157-163, 2018.

13 Gaussian Vertex Prime Labeling of Some Graphs Obtained from Origami Models

N P Shrimali
Department of Mathematics,
Gujarat University,
Ahmedabad, Gujarat (INDIA)
E-mail: narenp05@gmail.com

S K Singh
Department of Mathematics,
Gujarat University,
Ahmedabad, Gujarat (INDIA)
E-mail: sachinkumar28singh@gmail.com

A graph with edge set E has a Gaussian vertex prime labeling if its edges can be labeled with first $|E|$ Gaussian integers $\gamma_1, \gamma_2, \ldots, \gamma_{|E|}$ such that for each vertex of degree at least 2 the greatest common divisor of the labels on its incident edges is unit. A graph which admits Gaussian vertex prime labeling is known as a Gaussian vertex prime graph. In this chapter, we investigate Gaussian vertex prime labeling for a boreale star graph, holiday star graph, kusudama flower graph, christmas star graph, braided star graph, and cherry blossom graph.

13.1 INTRODUCTION

We consider here only undirected, connected and simple graph $G = (V(G), E(G))$ with the vertex set $V(G)$ and edge set $E(G)$. For various graph theoretic notations and terminology we follow Gross and Yellen [4] and D. M. Burton[1] for number theoretic results.

The notion of a prime labeling originated with Roger Entringer and was introduced in a paper by Tout et al.[12]. Many researchers have studied prime labeling for a good number of graphs listed in [3].

In [2, 5], Hunter Lehmann et al. defined a beautiful ordering in Gaussian integers and named it as "spiral ordering in Gaussian integers". Motivated by prime labeling they introduced Gaussian prime labeling of graphs with respect to spiral ordering. So, Gaussian prime labeling is an extension of prime labeling.

We will start our discussion by giving some definitions starting from spiral ordering of Gaussian integers and its properties given by Steven Klee et al. [5].

13.1.1 SPIRAL ORDERING OF THE GAUSSIAN INTEGERS

Gaussian integers are complex numbers of the form $\gamma = x + iy$ where x and y are integers and $i^2 = -1$. The set of Gaussian integers is usually denoted by $\mathbb{Z}[i]$. A Gaussian integer γ is said to be an even Gaussian integer if γ is divisible by $1+i$ and otherwise is called an odd Gaussian integer. The norm on $\mathbb{Z}[i]$ is defined as $d(x+iy) = x^2 + y^2$. One can easily see that the only units of $\mathbb{Z}[i]$ are $\pm 1, \pm i$. The associates of Gaussian integer γ are unit multiples of γ. In $\mathbb{Z}[i]$, two Gaussian integers are relatively prime if their common divisors are the only units of $\mathbb{Z}[i]$. A Gaussian integer γ is said to be a prime Gaussian integer if and only if $\pm 1, \pm i, \pm\gamma, \pm i\gamma$ are the only divisors of γ.

In [2, 5], the recursion relation of spiral ordering of the Gaussian integers starting with $\gamma_1 = 1$ is defined as follows:

$$\gamma_{n+1} = \begin{cases} \gamma_n + i & \text{if } Re(\gamma_n) \equiv 1 \pmod 2, \quad Re(\gamma_n) > Im(\gamma_n) + 1 \\ \gamma_n - 1 & \text{if } Im(\gamma_n) \equiv 0 \pmod 2, \quad Re(\gamma_n) \leq Im(\gamma_n) + 1, \quad Re(\gamma_n) > 1 \\ \gamma_n + 1 & \text{if } Im(\gamma_n) \equiv 1 \pmod 2, \quad Re(\gamma_n) < Im(\gamma_n) + 1 \\ \gamma_n + i & \text{if } Im(\gamma_n) \equiv 0 \pmod 2, \quad Re(\gamma_n) = 1 \\ \gamma_n - i & \text{if } Re(\gamma_n) \equiv 0 \pmod 2, \quad Re(\gamma_n) \geq Im(\gamma_n) + 1, \quad Im(\gamma_n) > 0 \\ \gamma_n - i & \text{if } Re(\gamma_n) \equiv 0 \pmod 2, \quad Im(\gamma_n) = 0 \end{cases}$$

The notation γ_n is used to denote the nth Gaussian integer under the above ordering. We symbolically write first $'n'$ Gaussian integers by $[\gamma_n]$. The index of Gaussian integer $x + iy$ in the spiral ordering is denoted by $I(x+iy)$.

In [5] Steven Klee et al. established some interesting basic properties about Gaussian integers with the above ordering like:

1. Any two consecutive Gaussian integers are relatively prime.
2. Any two consecutive odd Gaussian integers are relatively prime.
3. γ and $\gamma + \mu$ are relatively prime, if γ is a Gaussian integer and μ is a unit.
4. γ and $\gamma + \mu(1+i)^k$ are relatively prime, if γ is an odd Gaussian integer and μ is a unit. where k is a positive integer.
5. γ and $\gamma + \pi$ are relatively prime if and only if π does not divide γ where γ is a Gaussian integer and π is a prime Gaussian integer.

Definition. [2, 5] Let G be a graph having n vertices. A bijective function $g : V(G) \to [\gamma_n]$ is called Gaussian prime labeling, if the images of adjacent vertices are relatively prime. A graph which admits Gaussian prime labeling is known as a Gaussian prime graph.

Hunter Lehmann and Andrew Park proved that any tree with ≤ 72 vertices is a Gaussian prime tree in [2]. In [5] Steven Lee et al. proved that n-centipede tree, $(n, 2)-$centipede tree, path graph, spider tree, star graph, $(n, 3)$ firecracker tree, (n, k, m) double star tree are Gaussian prime graphs.
Deretsky, Lee and Mitchem defined a dual of prime labeling named vertex prime labeling as follows.

Definition. [13] A graph with edge set E has a vertex prime labeling if its edges can be labeled with distinct integers $1, 2, \ldots, |E|$ such that for each vertex of degree at least 2 the greatest common divisor of the labels on its incident edges is 1.

Deretsky, Lee and Mitchem[13] proved that forests, all connected graphs, $C_{2k} \bigcup C_n$, $C_{2m} \bigcup C_{2n} \bigcup C_{2k+1}$, $C_{2m} \bigcup C_{2n} \bigcup C_{2t} \bigcup C_k$, and $5C_{2m}$ are vertex prime graphs. They further proved that a graph with exactly two components, one of which is not an odd cycle, has a vertex prime labeling and a 2-regular graph with at least two odd cycles does not have a vertex prime labeling. Here, we introduce Gaussian vertex prime labeling with respect to spiral ordering motivated by Gaussian prime labeling and vertex prime labeling.

Definition. A graph with edge set E has a Gaussian vertex prime labeling if its edges can be labeled with first $|E|$ Gaussian integers $\gamma_1, \gamma_2, \cdots, \gamma_{|E|}$ such that for each vertex of degree at least 2 the greatest common divisor of the labels on its incident edges is unit. A graph which admits Gaussian vertex prime labeling is called a Gaussian vertex prime graph.

13.1.2 ORIGAMI MODELS

Origami is a paper folding art associated with Japanese culture. The word origami originates from two Japanese words "ori" and "kami" where "ori" represents folding and "kami" represents paper, here kami changes to gami due to rendaku. An origami practitioner's goal is to transform a flat square sheet of paper into a finished sculpture through folding and sculpting techniques without use of glue, cuts and markings on the paper. The Japanese word "kirigami" refers to designs which use cuts.
The Miura fold is a form of rigid origami(a branch of origami which is concerned with folding structures using flat rigid sheets joined by hinges). The Miura fold has a lot of practical applications. Let us see some of them for examples. Large solar panel arrays for space satellites in the Japanese space program have been Miura folded before launch and then spread out in space. Flat foldable furniture is another application. Miura fold has applications in surgical devices such as stents. Also University of Fribourg used this fold to stack hydrogel films, generating electricity similarly to electric eels.

The following definitions of graphs, inspired from origami models is given by U. M. Prajapati and R. M. Gajjar. And they have studied various labeling techniques for these graphs. We denote the set $\{1, 2, \ldots, n\}$ by $[n]$.

Definition. [6] Let a_0 be the apex vertex and $a_1, a_2, \ldots, a_{n-1}, a_n$ be consecutive n rim vertices of wheel graph W_n, $n \geq 3$; let $b_1, b_2, b_3, \ldots, b_{2n-1}, b_{2n}$ be consecutive $2n$ vertices of cycle C_{2n}; let $c_1, c_2, c_3, \ldots, c_{2n-1}, c_{2n}$ be consecutive $2n$ vertices of cycle C_{2n}. Join each a_i to b_{2i-1} by an edge and b_{2i} to c_{2i} by an edge. Take a new vertex d_i; Join each d_i to c_{2i-1} and c_{2i+1} by an edge, for each $i \in [n]$, subscripts are taken modulo n. The resulting graph is called a braided star graph BRS_n.

Definition. [7] Let $v_1, v_2, \ldots, v_{4n-1}, v_{4n}$ be consecutive $4n$ vertices of cycle C_{4n}, $n \geq 3$. Let u_0 be the central vertex and $u_1, u_2, \ldots, u_{2n-1}, u_{2n}$ be end vertices of star $K_{1,2n}$. Now join u_0 to v_{4i-3} by an edge; join u_{2i-1} to v_{4i-2} and u_{2i} by an edge, for each $i \in [n]$. The resulting graph is called a holiday star graph HS_n.

Definition. [8] Let v_0 be the apex vertex and $v_1, v_2, v_3, \ldots, v_{2n-1}, v_{2n}$ be consecutive $2n$ rim vertices of wheel graph W_{2n}, $n \geq 3$. Subdivide spoke edge v_0v_{2i-1} with vertex w_i and at each w_i, join two copies of the path of length 2; $P_2^l = v_0, u_{2i-1}, w_i$ and $P_2^r = v_0, u_{2i}, w_i$ for each $i \in [n]$. The resulting graph is called the kusudama flower graph KF_n.

Definition. [9] Let u_0 be the apex vertex and $v_1, v_2, v_3, \ldots, v_{4n-1}, v_{4n}$ be consecutive $4n$ rim vertices of wheel graph W_{4n}, $n \geq 3$. Subdivide the edge u_0v_{2i-1} by a vertex u_i, for $i \in [2n]$. The resulting graph is called the Christmas star graph CS_n.

Definition. [10] Let a_0 be the central vertex and $a_1, a_2, \ldots, a_{n-1}, a_n$ be end vertices of star $K_{1,n}$; let $b_1, b_2, b_3, \ldots, b_{2n-1}, b_{2n}$ be consecutive $2n$ vertices of cycle C_{2n}, $n \geq 3$; let $c_1, c_2, c_3, \ldots, c_{2n-1}, c_{2n}$ be consecutive $2n$ vertices of cycle C_{2n}. Join each a_i to b_{2i-1} and b_{2i+1} by an edge; join each b_{2i} to c_{2i} by an edge. Take a new vertex d_i; join each d_i to c_{2i-1} and c_{2i+1} by an edge, for each $i \in [n]$, subscripts are taken modulo n. The resulting graph is called the Boreale star graph BLS_n.

Definition. [11] Let u_0 be the apex vertex and $v_1, v_2, v_3, \ldots, v_{4n-1}, v_{4n}$ be consecutive $4n$ rim vertices of wheel graph $W_{4n}, n \geq 3$. A new vertex u_i; join each u_i to v_{2i} and u_0 by an edge, for each $i \in [2n]$. The resulting graph is called the cherry blossom graph CB_n.

Throughout this chapter, we will understand that the graph that is Gaussian vertex prime means that it is a Gaussian vertex prime graph with respect to the spiral ordering of Gaussian integers.

13.2 MAIN RESULTS

Theorem 13.1

Boreale star graph BLS_n is Gaussian vertex prime. ■

Proof. For the graph BLS_n, $V(BLS_n) = \{a_0\} \cup \{a_i, d_i | i \in [n]\} \cup \{b_i, c_i | i \in [2n]\}$ and
$E(BLS_n) = \{a_0a_i, a_ib_{2i-1}, b_{2i}c_{2i}, c_{2i-1}d_i | i \in [n]\} \cup \{b_ib_{i+1}, c_ic_{i+1} | i \in [2n-1]\} \cup \{a_ib_{2i+1}, d_ic_{2i+1} | i \in [n-1]\} \cup \{a_nb_1, c_{2n}c_1, b_{2n}b_1, d_nc_1\}$
Here, $|V(BLS_n)| = 6n+1$ and $|E(BLS_n)| = 10n$
Define a bijection $f : E(BLS_n) \to [\gamma_{10n}]$ as follows:

$$f(e) = \begin{cases} \gamma_{10i-9} & \text{if } e \in \{a_0a_i | i \in [n]\} \\ \gamma_{10i-8} & \text{if } e \in \{a_ib_{2i-1} | i \in [n]\} \\ \gamma_{10i-7} & \text{if } e \in \{a_ib_{2i+1} | i \in [n-1]\} \\ \gamma_{10i-6} & \text{if } e \in \{b_{2i-1}b_{2i} | i \in [n]\} \\ \gamma_{10i-5} & \text{if } e \in \{b_{2i+1}b_{2i} | i \in [n]\} \\ \gamma_{10i-4} & \text{if } e \in \{b_{2i}c_{2i} | i \in [n]\} \\ \gamma_{10i-3} & \text{if } e \in \{c_{2i-1}c_{2i} | i \in [n]\} \\ \gamma_{10i-2} & \text{if } e \in \{c_{2i+1}c_{2i} | i \in [n]\} \\ \gamma_{10i-1} & \text{if } e \in \{d_ic_{2i+1} | i \in [n-1]\} \\ \gamma_{10i} & \text{if } e \in \{d_ic_{2i-1} | i \in [n]\} \\ \gamma_{10n-1} & \text{if } e = d_nc_1 \\ \gamma_{10n-2} & \text{if } e = c_{2n}c_1 \\ \gamma_{10n-5} & \text{if } e = b_{2n}b_1 \\ \gamma_{10n-7} & \text{if } e = a_nb_1 \end{cases}$$

Let v be an arbitrary vertex of BLS_n. We prove f is a Gaussian vertex prime labeling of BLS_n by considering the following cases:
If $v = a_0$, $gcd(f(a_0a_1), f(a_0a_2), \ldots, f(a_0a_n)) = gcd(\gamma_1, \gamma_{11}, \ldots, \gamma_{10n-1}) = 1$
If $v = a_i$,
$gcd(f(a_ia_0), f(a_ib_{2i+1}), f(a_ib_{2i-1})) = gcd(\gamma_{10i-9}, \gamma_{10i-7}, \gamma_{10i-8}) = 1$, $i \in [n-1]$
If $v = a_n$,
$gcd(f(a_na_0), f(a_nb_1), f(a_nb_{2n-1})) = gcd(\gamma_{10n-9}, \gamma_{10n-7}, \gamma_{10n-8}) = 1$
If $v = b_{2i}$,
$gcd(f(b_{2i}b_{2i+1}), f(b_{2i}b_{2i-1}), f(b_{2i}c_{2i})) = gcd(\gamma_{10i-5}, \gamma_{10i-6}, \gamma_{10i-4}) = 1$, $i \in [n-1]$
If $v = b_1$,
$gcd(f(b_1b_2), f(b_1b_{2n}), f(a_1b_1), f(a_nb_1)) = gcd(\gamma_4, \gamma_{10n-5}, \gamma_2, \gamma_{10n-7}) = 1$
If $v = b_{2n}$,
$gcd(f(b_{2n}b_1), f(b_{2n}b_{2n-1}), f(b_{2n}c_{2n})) = gcd(\gamma_{10n-5}, \gamma_{10n-6}, \gamma_{10n-4}) = 1$

If $v = c_{2i-1}$,
$gcd(f(c_{2i-1}c_{2i-2}), f(c_{2i-1}c_{2i}), f(c_{2i-1}d_{i-1}), f(c_{2i-1}d_i))$
$= gcd(\gamma_{10i-12}, \gamma_{10i-3}, \gamma_{10i-11}, \gamma_{10i}) = 1, i \in [n] - \{1\}$
If $v = c_{2i}$,
$gcd(f(c_{2i}c_{2i-1}), f(c_{2i+1}c_{2i}), f(c_{2i}b_{2i})) = gcd(\gamma_{10i-3}, \gamma_{10i-2}, \gamma_{10i-4}) = 1$, $i \in [n-1]$
If $v = c_1$,
$gcd(f(c_1c_2), f(c_1c_{2n}), f(c_1d_1), f(c_1d_n)) = gcd(\gamma_7, \gamma_{10n-2}, \gamma_{10}, \gamma_{10n-1}) = 1$
If $v = c_{2n}$,
$gcd(f(c_{2n}c_1), f(c_{2n}c_{2n-1}), f(b_{2n}c_{2n})) = gcd(\gamma_{10n-2}, \gamma_{10n-3}, \gamma_{10n-4}) = 1$
If $v = d_i$,
$gcd(f(d_ic_{2i+1}), f(d_ic_{2i-1})) = gcd(\gamma_{10i-1}, \gamma_{10i}) = 1, i \in [n-1]$
If $v = d_n$,
$gcd(f(d_nc_1), f(d_nc_{2n-1})) = gcd(\gamma_{10n}, \gamma_{10n-1}) = 1$
If $v = b_{2i-1}$,
$gcd(f(b_{2i-1}b_{2i}), f(b_{2i-2}b_{2i-1}), f(b_{2i-1}a_i), f(b_{2i-1}a_{i-1})) - gcd(\gamma_{10i-6}, \gamma_{10i-15}, \gamma_{10i-8}, \gamma_{10i-7}) = 1, i \in [n] - \{1\}$
Hence, BLS_n is a Gaussian vertex prime graph.

□

Figure 13.1 illustrates a Gaussian vertex prime labeling of BLS_8.

Theorem 13.2

Holiday star graph HS_n is Gaussian vertex prime. ■

Proof. For the graph HS_n, $V(HS_n) = \{u_i | 0 \leq i \leq 2n\} \cup \{v_i | 1 \leq i \leq 4n\}$ and $E(HS_n) = \{u_0u_{2i-1}, u_0u_{2i}, u_{2i-1}v_{4i-2}, u_{2i}v_{4i}, u_{2i-1}u_{2i}, u_0v_{4i-3} | i \in [n]\} \cup \{v_iv_{i+1} | i \in [4n-1]\} \cup \{v_{4n}v_1\}$
Here, $|V(HS_n)| = 6n + 1$ and $|E(HS_n)| = 10n$
Let us define a bijection $f : E(HS_n) \to [\gamma_{10n}]$ as follows:

$$f(e) = \begin{cases} \gamma_{10i-9} & \text{if } e \in \{u_0v_{4i-3} | i \in [n]\} \\ \gamma_{10i-8} & \text{if } e \in \{v_{4i-2}v_{4i-3} | i \in [n]\} \\ \gamma_{10i-7} & \text{if } e \in \{u_0u_{2i-1} | i \in [n]\} \\ \gamma_{10i-6} & \text{if } e \in \{u_{2i-1}v_{4i-2} | i \in [n]\} \\ \gamma_{10i-5} & \text{if } e \in \{v_{4i-2}v_{4i-1} | i \in [n]\} \\ \gamma_{10i-4} & \text{if } e \in \{v_{4i-1}v_{4i} | i \in [n]\} \\ \gamma_{10i-3} & \text{if } e \in \{v_{4i}u_{2i} | i \in [n]\} \\ \gamma_{10i-2} & \text{if } e \in \{u_0u_{2i} | i \in [n]\} \\ \gamma_{10i-1} & \text{if } e \in \{u_{2i}u_{2i-1} | i \in [n]\} \\ \gamma_{10i} & \text{if } e \in \{v_{4i}v_{4i+1} | i \in [n-1]\} \\ \gamma_{10n} & \text{if } e = v_{4n}v_1 \end{cases}$$

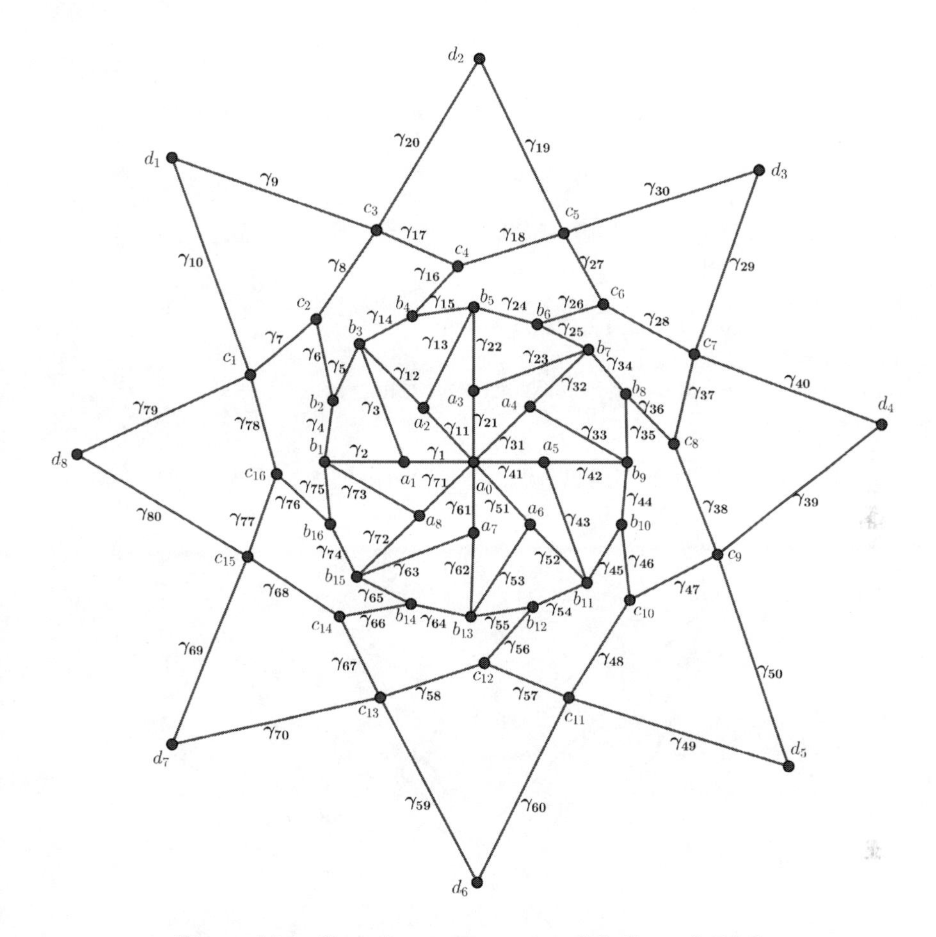

Figure 13.1: Gaussian vertex prime labeling of BLS_8

Let v be an arbitrary vertex of HS_n. To prove f is Gaussian vertex prime labeling, let us consider the following cases:
If $v = u_0$,
$gcd(f(u_0u_{2i-1}), f(u_0u_{2i}), f(u_0v_{4i-3})) = gcd(\gamma_{10i-7}, \gamma_{10i-2}, \gamma_{10i-9}) = 1, i \in [n]$
If $v = u_{2i-1}$,
$gcd(f(u_0u_{2i-1}), f(v_{4i-2}u_{2i-1}), f(u_{2i}u_{2i-1})) = gcd(\gamma_{10i-7}, \gamma_{10i-6}, \gamma_{10i-1}) = 1,$
$i \in [n]$
If $v = u_{2i}$,
$gcd(f(u_0u_{2i}), f(v_{4i}u_{2i}), f(u_{2i}u_{2i-1})) = gcd(\gamma_{10i-2}, \gamma_{10i-3}, \gamma_{10i-1}) = 1, i \in [n]$
If $v = v_{4i-1}$,
$gcd(f(v_{4i-1}v_{4i}), f(v_{4i-2}v_{4i-1})) = gcd(\gamma_{10i-4}, \gamma_{10i-5}) = 1, i \in [n]$
If $v = v_{4i}$,
$gcd(f(v_{4i}v_{4i+1}), f(v_{4i}u_{2i}), f(v_{4i}v_{4i-1})) = gcd(\gamma_{10i}, \gamma_{10i-3}, \gamma_{10i-4}) = 1,$

$i \in [n-1]$
If $v = v_{4n}$,
$gcd(f(v_{4n}v_1), f(v_{4n}u_{2n}), f(v_{4n}v_{4n-1})) = gcd(\gamma_{10n}, \gamma_{10n-3}, \gamma_{10n-4}) = 1$
If $v = v_1$,
$gcd(f(v_1v_2), f(v_1v_{4n}), f(v_1u_0)) = gcd(\gamma_2, \gamma_{10n}, \gamma_{10n-1}) = 1$
If $v = v_{4i-2}$,
$gcd(f(v_{4i-2}v_{4i-1}), f(v_{4i-2}v_{4i-3}), f(v_{4i-2}u_{2i-1})) = gcd(\gamma_{10i-5}, \gamma_{10i-8}, \gamma_{10i-6}) = 1, i \in [n]$
If $v = v_{4i-3}$,
$gcd(f(v_{4i-3}v_{4i-2}), f(v_{4i-3}v_{4i-4}), f(u_0v_{4i-3})) = gcd(\gamma_{10i-8}, \gamma_{10i-10}, \gamma_{10i-9}) = 1, i \in [n] - \{1\}$
Hence, HS_n is a Gaussian vertex prime graph.

□

Figure 13.2 illustrates a Gaussian vertex prime labeling of HS_4.

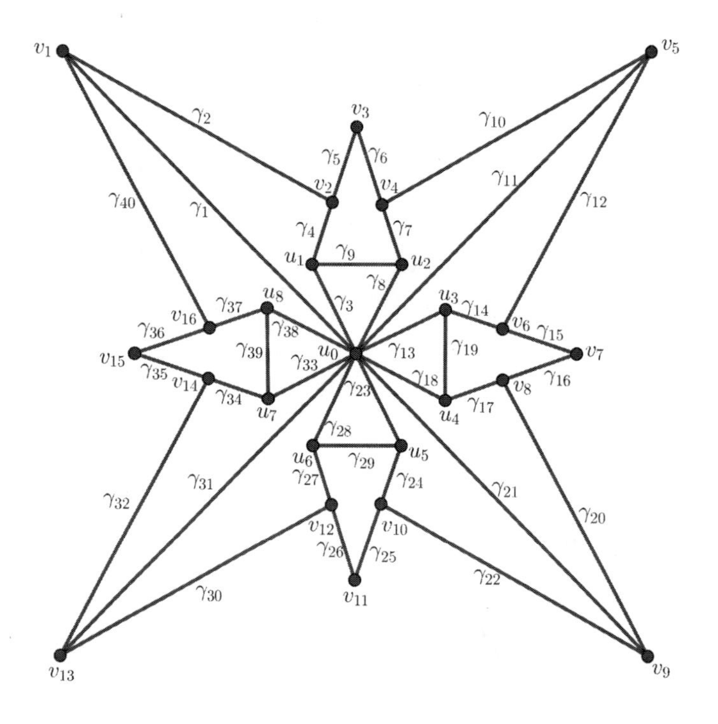

Figure 13.2: Gaussian vertex prime labeling of HS_4

Theorem 13.3

Kusudama flower graph KF_n is Gaussian vertex prime. ■

Proof. For the graph KF_n, $V(KF_n) = \{v_0\} \cup \{w_i | i \in [n]\} \cup \{u_i, v_i | i \in [2n]\}$ and $E(KF_n) = \{v_0v_{2i}, v_0w_i, v_0u_{2i-1}, v_0u_{2i}, w_iv_{2i-1}, w_iu_{2i-1}, w_iu_{2i} | i \in [n]\} \cup \{v_iv_{i+1} | i \in [2n-1]\} \cup \{v_{2n}v_1\}$
Here, $|V(KF_n)| = 5n+1$ and $|E(KF_n)| = 9n$
Define a bijection $f : E(KF_n) \to [\gamma_{9n}]$ as follows:

$$f(e) = \begin{cases} \gamma_{9i-8} & \text{if } e \in \{v_0u_{2i-1} | i \in [n]\} \\ \gamma_{9i-6} & \text{if } e \in \{v_0w_i | i \in [n]\} \\ \gamma_{9i-5} & \text{if } e \in \{v_0u_{2i} | i \in [n]\} \\ \gamma_{9i-3} & \text{if } e \in \{w_iv_{2i-1} | i \in [n]\} \\ \gamma_{9i-2} & \text{if } e \in \{v_{2i-1}v_{2i} | i \in [n]\} \\ \gamma_{9i-1} & \text{if } e \in \{v_{2i}v_0 | i \in [n]\} \\ \gamma_{9i} & \text{if } e \in \{v_{2i}v_{2i+1} | i \in [n]\} \\ \gamma_{9i-(3j+4)} & \text{if } e \in \{u_{2i-j}w_i | i \in [n], j \in \{0,1\}\} \\ \gamma_{9n} & \text{if } e = v_1v_{2n} \end{cases}$$

Let $v \in V(KF_n)$ be an arbitrary vertex of KF_n. Now, f is proved to be a Gaussian vertex prime labeling of KF_n by considering the following cases:

If $v = v_0$, $gcd(f(v_0u_{2i-1}), f(v_0u_{2i}), f(v_0w_i), f(v_0v_{2i})) = gcd(\gamma_{9i-8}, \gamma_{9i-5}, \gamma_{9i-6}, \gamma_{9i-1}) = 1, i \in [n]$
If $v = u_{2i}$, $gcd(f(u_{2i}w_i), f(v_0u_{2i})) = gcd(\gamma_{9i-4}, \gamma_{9i-5}) = 1, 1 \leq i \leq n$
If $v = u_{2i-1}$, $gcd(f(u_{2i-1}w_i), f(v_0u_{2i-1})) = gcd(\gamma_{9i-7}, \gamma_{9i-8}) = 1, 1 \leq i \leq n$
If $v = v_{2i-1}$, $gcd(f(v_{2i-1}w_i), f(v_{2i-1}v_{2i}), f(v_{2i-1}v_{2i-2})) = gcd(\gamma_{9i-3}, \gamma_{9i-2}, \gamma_{9i-9}) = 1, 2 \leq i \leq n$
If $v = v_1$, $gcd(f(v_1w_1), f(v_1v_2), f(v_1v_{2n})) = gcd(\gamma_6, \gamma_7, \gamma_{9n}) = 1$
If $v = v_{2n}$, $gcd(f(v_{2n}v_0), f(v_{2n-1}v_{2n}), f(v_{2n}v_1)) = gcd(\gamma_{9n-1}, \gamma_{9n-2}, \gamma_{9n}) = 1$
If $v = v_{2i}$, $gcd(f(v_{2i}v_0), f(v_{2i}v_{2i-1}), f(v_{2i+1}v_{2i})) = gcd(\gamma_{9i-1}, \gamma_{9i-2}, \gamma_{9i}) = 1, 1 \leq i \leq n-1$
If $v = w_i$, $gcd(f(u_{2i-1}w_i), f(w_iv_0), f(w_iu_{2i}), f(w_iv_{2i-1})) = gcd(\gamma_{9i-7}, \gamma_{9i-6}, \gamma_{9i-4}, \gamma_{9i-3}) = 1, i \in [n]$
Hence, KF_n is a Gaussian vertex prime graph.

□

Figure 13.3 illustrates a Gaussian vertex prime labeling of KF_5.

Theorem 13.4

Christmas Star graph CS_n is Gaussian vertex prime. ■

Proof. For the graph CS_n, $V(CS_n) = \{u_i | 0 \leq i \leq 2n\} \cup \{v_i | 1 \leq i \leq 4n\}$ and $E(CS_n) = \{u_0u_i | i \in [2n]\} \cup \{u_{2i-1}v_{4i-3}, u_{2i}v_{4i-1}, u_0v_{4i-2}, u_0v_{4i} | i \in [n]\} \cup \{v_iv_{i+1} | i \in [4n-1]\} \cup \{v_{4n}v_1\}$
Here, $|V(CS_n)| = 6n+1$ and $|E(CS_n)| = 10n$

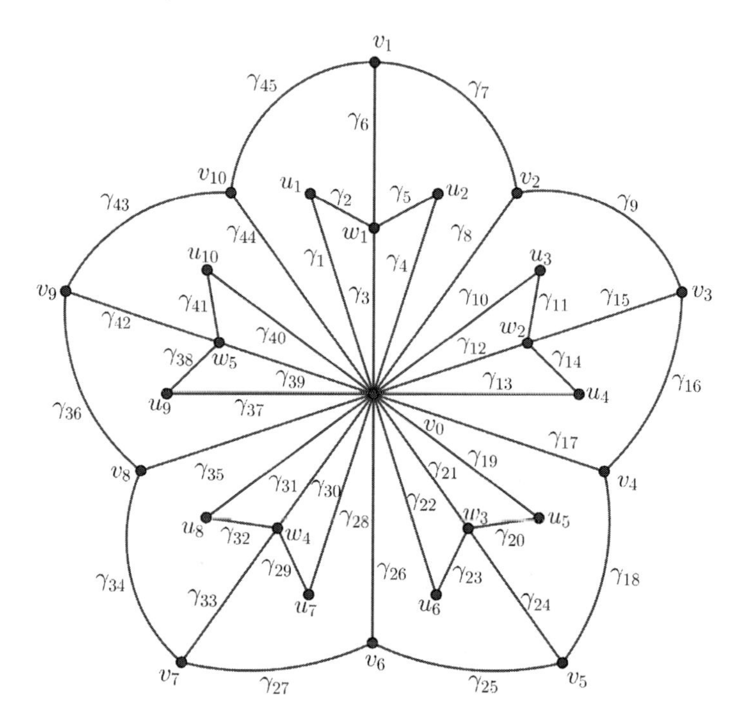

Figure 13.3: Gaussian vertex prime labeling of KF_5

Define a bijection $f : E(CS_n) \to [\gamma_{10n}]$ as follows:

$$f(e) = \begin{cases} \gamma_{10i-9} & \text{if } e \in \{u_0u_{2i-1} | i \in [n]\} \\ \gamma_{10i-8} & \text{if } e \in \{u_{2i-1}v_{4i-3} | i \in [n]\} \\ \gamma_{10i-7} & \text{if } e \in \{v_{4i-3}v_{4i-2} | i \in [n]\} \\ \gamma_{10i-6} & \text{if } e \in \{u_0v_{4i-2} | i \in [n]\} \\ \gamma_{10i-5} & \text{if } e \in \{u_0u_{2i} | i \in [n]\} \\ \gamma_{10i-4} & \text{if } e \in \{v_{4i-1}u_{2i} | i \in [n]\} \\ \gamma_{10i-3} & \text{if } e \in \{v_{4i-2}v_{4i-1} | i \in [n]\} \\ \gamma_{10i-2} & \text{if } e \in \{v_{4i}v_{4i-1} | i \in [n]\} \\ \gamma_{10i-1} & \text{if } e \in \{u_0v_{4i} | i \in [n]\} \\ \gamma_{10i} & \text{if } e \in \{v_{4i}v_{4i+1} | i \in [n-1]\} \\ \gamma_{10n} & \text{if } e = v_{4n}v_1 \end{cases}$$

Let v be an arbitrary vertex of CS_n. We prove f is a Gaussian vertex prime labeling of CS_n by considering the following cases:
If $v = u_0$, $gcd(f(u_0u_{2i-1}), f(u_0u_{2i}), f(u_0v_{4i-2}), f(u_0v_{4i})) = gcd(\gamma_{10i-9}, \gamma_{10i-5}, \gamma_{10i-6}, \gamma_{10i-1}) = 1, i \in [n]$
If $v = u_{2i-1}$, $gcd(f(u_0u_{2i-1}), f(v_{4i-3}u_{2i-1})) = gcd(\gamma_{10i-9}, \gamma_{10i-8}) = 1, i \in [n]$
If $v = u_{2i}$, $gcd(f(u_0u_{2i}), f(v_{4i-1}u_{2i})) = gcd(\gamma_{10i-5}, \gamma_{10i-4}) = 1, i \in [n]$

If $v = v_{4i-1}$, $gcd(f(v_{4i-1}v_{4i}), f(v_{4i-1}u_{2i}), f(v_{4i-2}v_{4i-1})) = gcd(\gamma_{10i-2}, \gamma_{10i-4}, \gamma_{10i-3}) = 1, i \in [n]$
If $v = v_{4i}$, $gcd(f(v_{4i}v_{4i+1}), f(v_{4i}u_0), f(v_{4i}v_{4i-1})) = gcd(\gamma_{10i}, \gamma_{10i-1}, \gamma_{10i-2}) = 1, i \in [n-1]$
If $v = v_{4n}$, $gcd(f(v_{4n}v_1), f(v_{4n}u_0), f(v_{4n}v_{4n-1})) = gcd(\gamma_{10n}, \gamma_{10n-1}, \gamma_{10n-2}) = 1$
If $v = v_1$, $gcd(f(v_1v_2), f(v_1v_{4n}), f(v_1u_1)) = gcd(\gamma_3, \gamma_{10n}, \gamma_2) = 1$
If $v = v_{4i-2}$, $gcd(f(v_{4i-2}v_{4i-1}), f(v_{4i-2}v_{4i-3}), f(v_{4i-2}u_0)) = gcd(\gamma_{10i-3}, \gamma_{10i-7}, \gamma_{10i-6}) = 1, i \in [n]$
If $v = v_{4i-3}$, $gcd(f(v_{4i-3}v_{4i-2}), f(v_{4i-3}v_{4i-4}), f(u_{2i-1}v_{4i-3})) = gcd(\gamma_{10i-7}, \gamma_{10i-10}, \gamma_{10i-8}) = 1, i \in [n] - \{1\}$
Hence, CS_n is a Gaussian vertex prime graph.

□

Figure 13.4 illustrates a Gaussian vertex prime labeling of CS_4.

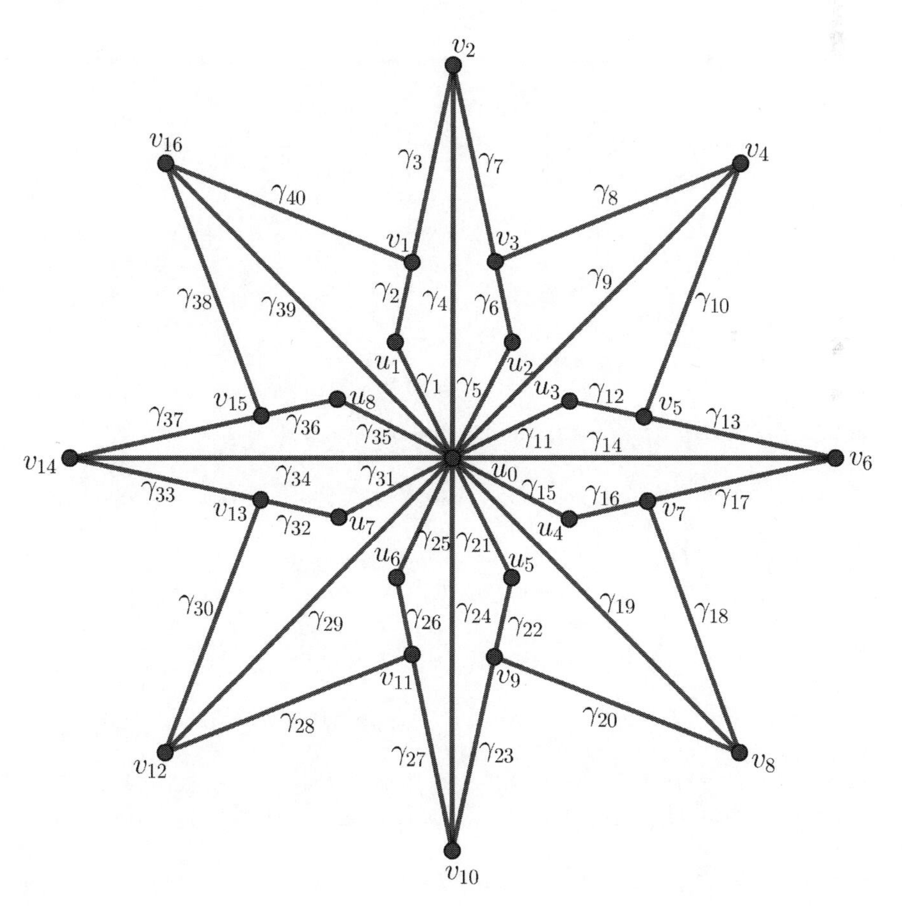

Figure 13.4: Gaussian vertex prime labeling of CS_4

Theorem 13.5

Braided star graph BRS_n is Gaussian vertex prime. ■

Proof. For the graph BRS_n, $V(BRS_n) = \{a_0\} \cup \{a_i, d_i | i \in [n]\} \cup \{b_i, c_i | i \in [2n]\}$ and
$E(BRS_n) = \{a_0a_i, a_ib_{2i-1}, b_{2i}c_{2i}, c_{2i-1}d_i | i \in [n]\} \cup \{b_ib_{i+1}, c_ic_{i+1} | i \in [2n-1]\} \cup \{a_na_1, c_{2n}c_1, b_{2n}b_1, d_nc_1\} \cup \{a_ia_{i+1}, d_ic_{2i+1} | i \in [n-1]\}$
Here, $|V(BRS_n)| = 6n+1$ and $|E(BRS_n)| = 10n$
Define a bijection $f : E(BRS_n) \to [\gamma_{10n}]$ as follows:

$$f(e) = \begin{cases} \gamma_{10i-9} & \text{if } e \in \{a_0a_i | i \in [n]\} \\ \gamma_{10i-8} & \text{if } e \in \{a_ia_{i+1} | i \in [n-1]\} \\ \gamma_{10i-7} & \text{if } e \in \{a_ib_{2i-1} | i \in [n]\} \\ \gamma_{10i-6} & \text{if } e \in \{b_{2i-1}b_{2i} | i \in [n]\} \\ \gamma_{10i-5} & \text{if } e \in \{b_{2i+1}b_{2i} | i \in [n-1]\} \\ \gamma_{10i-4} & \text{if } e \in \{b_{2i}c_{2i} | i \in [n]\} \\ \gamma_{10i-3} & \text{if } e \in \{c_{2i-1}c_{2i} | i \in [n]\} \\ \gamma_{10i-2} & \text{if } e \in \{c_{2i+1}c_{2i} | i \in [n-1]\} \\ \gamma_{10i-1} & \text{if } e \in \{d_ic_{2i+1} | i \in [n-1]\} \\ \gamma_{10i} & \text{if } e \in \{d_ic_{2i-1} | i \in [n]\} \\ \gamma_{10n-1} & \text{if } e = d_nc_1 \\ \gamma_{10n-2} & \text{if } e = c_{2n}c_1 \\ \gamma_{10n-5} & \text{if } e = b_{2n}b_1 \\ \gamma_{10n-8} & \text{if } e = a_na_1 \end{cases}$$

Let v be an arbitrary vertex of BRS_n. We prove f is a Gaussian vertex prime labeling of BRS_n by considering the following cases:

If $v = a_n$, $gcd(f(a_na_0), f(a_na_1), f(a_nb_{2n-1}), f(a_na_{n-1})) = gcd(\gamma_{10n-9}, \gamma_{10n-8}, \gamma_{10n-7}, \gamma_{10n-18}) = 1$
If $v = a_0$, $gcd(f(a_0a_1), f(a_0a_2), \ldots, f(a_0a_n)) = gcd(\gamma_1, \gamma_{11}, \ldots, \gamma_{10n-9}) = 1$
If $v = a_1$, $gcd(f(a_1a_0), f(a_2a_1), f(a_1b_1), f(a_na_1)) = gcd(\gamma_1, \gamma_2, \gamma_3, \gamma_{10n-8}) = 1$
If $v = b_{2i}$, $gcd(f(b_{2i}b_{2i+1}), f(b_{2i}b_{2i-1}), f(b_{2i}c_{2i})) = gcd(\gamma_{10i-5}, \gamma_{10i-6}, \gamma_{10i-4}) = 1, i \in [n-1]$
If $v = b_{2n}$, $gcd(f(b_{2n}b_1), f(b_{2n}b_{2n-1}), f(b_{2n}c_{2n})) = gcd(\gamma_{10n-5}, \gamma_{10n-6}, \gamma_{10n-4}) = 1$
If $v = b_1$, $gcd(f(b_1b_2), f(b_1b_{2n}), f(a_1b_1)) = gcd(\gamma_4, \gamma_{10n-5}, \gamma_3) = 1$
If $v = c_{2i}$, $gcd(f(c_{2i}c_{2i-1}), f(c_{2i+1}c_{2i}), f(c_{2i}b_{2i})) = gcd(\gamma_{10i-3}, \gamma_{10i-2}, \gamma_{10i-4}) = 1, i \in [n-1]$
If $v = c_1$, $gcd(f(c_1c_2), f(c_1c_{2n}), f(c_1d_1), f(c_1d_n)) = gcd(\gamma_7, \gamma_{10n-2}, \gamma_{10}, \gamma_{10n-1}) = 1$

If $v = c_{2n}$, $gcd(f(c_{2n}c_1), f(c_{2n}c_{2n-1}), f(b_{2n}c_{2n})) = gcd(\gamma_{10n-2}, \gamma_{10n-3}, \gamma_{10n-4}) = 1$
If $v = d_i$, $gcd(f(d_ic_{2i+1}), f(d_ic_{2i-1})) = gcd(\gamma_{10i-1}, \gamma_{10i}) = 1, i \in [n-1]$
If $v = d_n$, $gcd(f(d_nc_1), f(d_nc_{2n-1})) = gcd(\gamma_{10n}, \gamma_{10n-1}) = 1$
If $v = a_i$, $gcd(f(a_ia_0), f(a_ia_{i+1}), f(a_ib_{2i-1}), f(a_ia_{i-1})) = gcd(\gamma_{10i-9}, \gamma_{10i-8}, \gamma_{10i-7}, \gamma_{10i-18}) = 1$, $i \in [n-1] - \{1\}$
If $v = b_{2i-1}$, $gcd(f(b_{2i-1}b_{2i}), f(b_{2i-2}b_{2i-1}), f(b_{2i-1}a_i)) = gcd(\gamma_{10i-6}, \gamma_{10i-15}, \gamma_{10i-7}) = 1, i \in [n] - \{1\}$
If $v = c_{2i-1}$, $gcd(f(c_{2i-1}c_{2i-2}), f(c_{2i-1}c_{2i}), f(c_{2i-1}d_{i-1}), f(c_{2i-1}d_i)) = gcd(\gamma_{10i-12}, \gamma_{10i-3}, \gamma_{10i-11}, \gamma_{10i}) = 1, i \in [n] - \{1\}$
Hence, BRS_n is a Gaussian vertex prime graph.

□

Figure 13.5 illustrates a Gaussian vertex prime labeling of BRS_8.

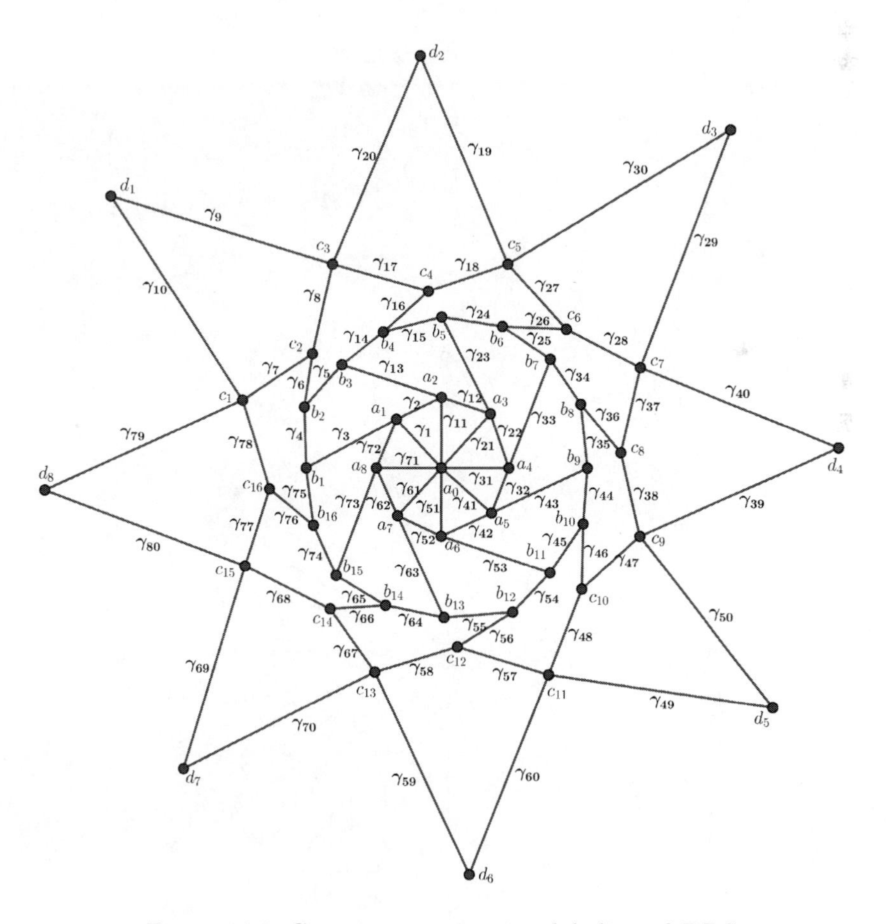

Figure 13.5: Gaussian vertex prime labeling of BRS_8

Theorem 13.6

Cherry blossom graph CB_n is Gaussian vertex prime. ■

Proof. For the graph CB_n, $V(CB_n) = \{u_0\} \cup \{v_i | i \in [4n]\} \cup \{u_i | i \in [2n]\}$ and $E(CB_n) = \{u_i v_{2i}, u_0 u_i | i \in [2n]\} \cup \{v_i v_{i+1} | i \in [4n-1]\} \cup \{u_0 v_i | i \in [4n]\} \cup \{v_{4n} v_1\}$
Here, $|V(CB_n)| = 6n + 1$ and $|E(CB_n)| = 12n$
Define a bijection $f : E(CB_n) \to [\gamma_{12n}]$ as follows:

$$f(e) = \begin{cases} \gamma_{12i-11} & \text{if } e \in \{u_0 v_{4i-3} | 1 \leq i \leq n\} \\ \gamma_{12i-10} & \text{if } e \in \{v_{4i-3} v_{4i-2} | 1 \leq i \leq n\} \\ \gamma_{12i-8} & \text{if } e \in \{u_0 u_{2i-1} | 1 \leq i \leq n\} \\ \gamma_{12i-7} & \text{if } e \in \{u_{2i-1} v_{4i-2} | 1 \leq i \leq n\} \\ \gamma_{12i-6} & \text{if } e \in \{v_{4i-2} v_{4i-1} | 1 \leq i \leq n\} \\ \gamma_{12i-4} & \text{if } e \in \{v_{4i-1} v_{4i} | 1 \leq i \leq n\} \\ \gamma_{12i-3} & \text{if } e \in \{u_0 u_{2i} | 1 \leq i \leq n\} \\ \gamma_{12i-2} & \text{if } e \in \{u_{2i} v_{4i} | 1 \leq i \leq n\} \\ \gamma_{12i} & \text{if } e \in \{v_{4i} v_{4i+1} | 1 \leq i < n\} \\ \gamma_{12i-(4j+1)} & \text{if } e \in \{u_0 v_{4i-j} | 1 \leq i \leq n, j \in \{0,1,2\}\} \\ \gamma_{12n} & \text{if } e = v_{4n} v_1 \end{cases}$$

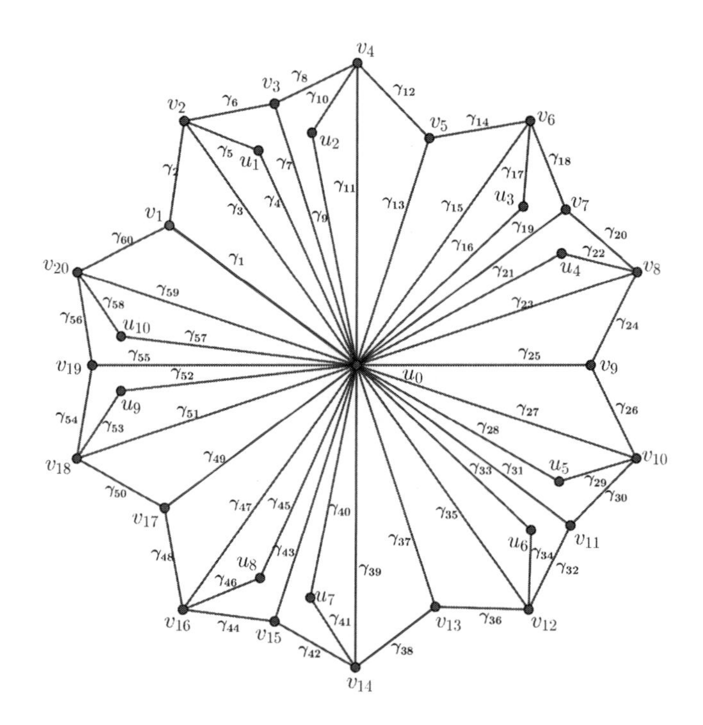

Figure 13.6: Gaussian vertex prime labeling of CB_5

The set of labels of incident edges of each vertex of CB_n contains at least one pair of consecutive Gaussian integers and hence f is a Gaussian vertex prime labeling and CB_n is a Gaussian vertex prime graph.

□

Figure 13.6 illustrates a Gaussian vertex prime labeling of CB_5.

13.3 CONCLUDING REMARK

In this chapter we have studied Gaussian vertex prime labeling of some graphs originated from origami models. One can also study other labeling techniques for these graphs.

REFERENCES

1. D. M. Burton. *Elementary Number Theory.* Tata McGraw Hill, 1990.
2. H. Lehmann and A. Park. Prime labeling of small trees with Gaussian integers. *Rose-Hulman Undergraduate Mathematics Journal*, 17(1): 72-97, 2016.
3. J. Gallian. A dynamic survey of graph labeling. *The Electronic Journal of Combinatorics*, #DS6, 2018.
4. J. Gross and J. Yellen. *Graph Theory and its Applications.* CRC Press, 1999.
5. S. Klee, H. Lehmann and A. Park. Prime labeling of families of trees with Gaussian integers. *AKCE International Journal of Graphs and Combinatorics*, 13(2): 165-176, 2016.
6. U. M. Prajapati and R. M. Gajjar. Some labeling techniques of braided star graph. *International Journal of Mathematics And its Applications*, 5(4-C): 361-369, 2017.
7. U. M. Prajapati and R. M. Gajjar. Labeling techniques of holiday star graph. *International Journal of Mathematics Trends and Technology*, 49(6): 339-344, 2017.
8. R. M. Gajjar and U. M. Prajapati. Some labeling techniques of kusudama flower graph. *Mathematics Today*, 34(A): 211-217, 2018.
9. U. M. Prajapati and R. M. Gajjar. Some labeling techniques of Christmas star graph. *International Journal of Mathematics and Soft Computing*, 8(2): 01-09, 2018.
10. R. M. Gajjar and U. M. Prajapati. Some labeling techniques of boreale star graph. *International Journal of Research and Analytical Reviews*, 6(1): 668-671, 2019.
11. R. M. Gajjar. Study of Various Techniques of Graph Labeling. Ph.D. Thesis, Gujarat University, 2019.
12. A. Tout, A. N. Dabboucy and K. Howalla. Prime labeling of graphs. *Nat. Acad. Sci. Letters*, 11, 365-368, 1982.
13. T. Deretsky, S. M. Lee, and J. Mitchem, On vertex prime labelings of graphs. *Graph Theory, Combinatorics and Applications*, J. Alavi, G. Chartrand, O. Oellerman, and A. Schwenk, eds., *Proceedings 6th International Conference Theory and Applications of Graphs*, 1, 359-369, 1991.

14 Vertex Magic Total Labeling of Tensor Product of Cycles

N P Shrimali
Department of Mathematics,
Gujarat University,
Ahmedabad, Gujarat (INDIA)
E-mail: narenp05@gmail.com

S T Trivedi
Department of Mathematics,
Gujarat University,
Ahmedabad, Gujarat (INDIA)
E-mail: siddh.trivedi@gmail.com

A vertex magic total labeling of a graph $G = (V, E)$ assigns to all vertices and edges of G labels from the set $\{1, 2, 3, \ldots, |V| + |E|\}$ so that the sum of the vertex label and labels of all incident edges does not depend on the vertex. In this chapter, we present a construction of vertex magic total labeling of the tensor product of $C_n \otimes C_n$ for odd n. The construction is based on a vertex magic total labeling of Cartesian product $C_n \times C_n$ for odd n.

14.1 INTRODUCTION

In 1990, MacDougall, Miller, Slamin and Wallis[2] introduced the vertex magic total labeling. In this paper, MacDougall *et.al.* defined vertex magic total labeling as follows:
Let $G = (V, E)$ be a graph. A bijective function $f : V(G) \cup E(G) \rightarrow \{1, 2, \ldots, |V| + |E|\}$ is called vertex magic total labeling(VMTL) of G if there exists a constant k such that

$$f(v) + \sum_{u \in N(v)} f(uv) = k$$

for every vertex $v \in V(G)$. Further, the graph having such a labeling is known as a vertex magic total graph. They have proved that the graphs $P_n(n > 1), K_n, C_n, K_{n,n} - e(n > 2)$ and $K_{n,n}(n > 1)$ for odd natural number n have vertex magic total labeling. For the detailed list of vertex magic graphs one may see the dynamic survey of graph labeling[5] written by J. A. Gallian. Furthermore, The construction of VMTL of Cartesian product $C_n \times C_n$ was given by Dalibor Fronček et al. in [1].

14.2 CALCULATION OF MAGIC CONSTANT

In [4], W. D. Wallis derived one concrete inequality for vertex magic total labeling, which gives us the bounds of the magic constant for any given graph $G = (V,E)$.

$$\binom{|E|+1}{2} + \binom{|V|+|E|+1}{2} \leq |V|k \leq 2\binom{|V|+|E|+1}{2} - \binom{|V|+1}{2} \quad (14.1)$$

For the tensor product $G = C_n \otimes C_n$, where $|V(G)| = n^2$ and $E(G) = 2n^2$, inequality (14.1) gives

$$\frac{13n^2+5}{2} \leq k \leq \frac{17n^2+5}{2} \quad (14.2)$$

14.3 THE PRODUCT $C_N \otimes C_N$ POSSESSES N DIFFERENT CYCLES OF LENGTH N

Before going to the main result, we will find the cyclic grid form of tensor product $C_n \otimes C_n$ with the help of different n cycles of length n in $C_n \otimes C_n$. To get in-depth understanding, let us start with a definition of tensor product $G \otimes H$:

If G and H are two graphs then the tensor product $G \otimes H$ is a graph such that

(1) The vertices of $G \otimes H$ is the cartesian product $V(G) \times V(H)$.
(2) The distinct vertices (u,v) and (u',v') are adjacent in $G \otimes H$ if and only if u is adjacent to u' and v is adjacent to v'.

In a particular case, take $G = H = C_5$. Now, we denote the vertices of two copies of cycles by u_0, u_1, u_2, u_3, u_4. (see Figure 14.1).
The tensor product of these two cycle is shown in Figure 14.2 as follows:

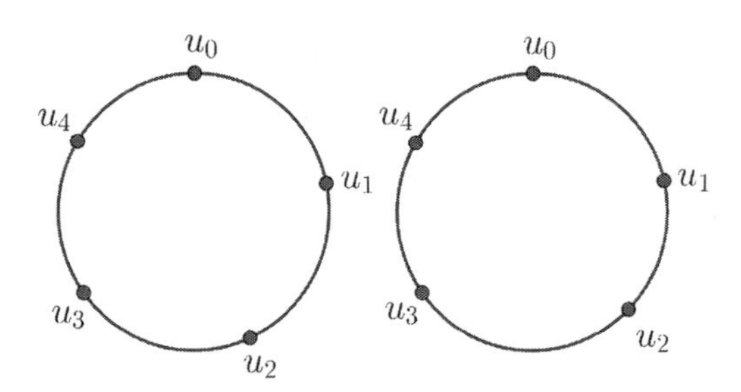

Figure 14.1: Two copies of Cycle C_5

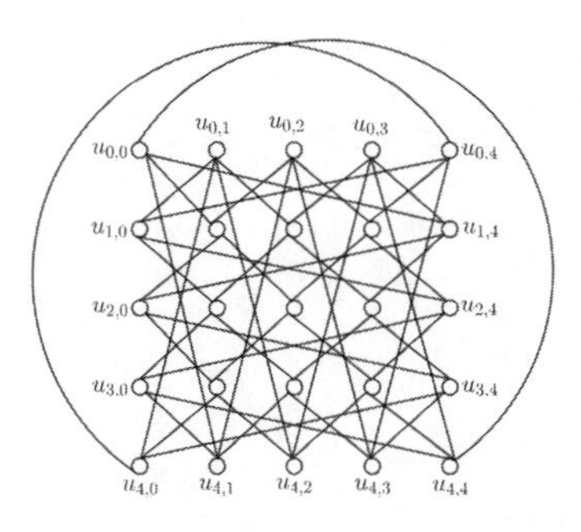

Figure 14.2: Tensor product $C_5 \otimes C_5$

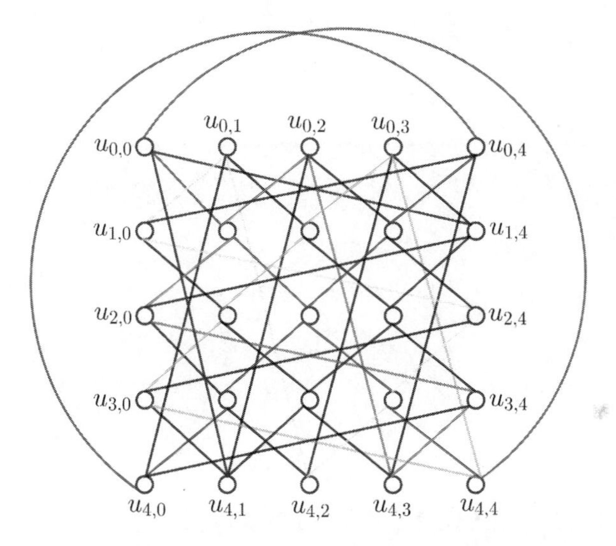

Figure 14.3: The tensor product $C_5 \otimes C_5$ possesses 5 different cycles

Now, one can easily see the different cycles in Figure 14.3.
These different cycles of length 5 are:

$$u_{0,0} - u_{1,4} - u_{2,3} - u_{3,2} - u_{4,1}$$

$$u_{1,1} - u_{2,0} - u_{3,4} - u_{4,3} - u_{0,2}$$

$$u_{2,2} - u_{3,1} - u_{4,0} - u_{0,4} - u_{1,3}$$

$$u_{3,3} - u_{4,2} - u_{0,1} - u_{1,0} - u_{2,4}$$

$$u_{4,4} - u_{0,3} - u_{1,2} - u_{2,1} - u_{3,0}$$

$$u_{0,0} - u_{1,1} - u_{2,2} - u_{3,3} - u_{4,4}$$

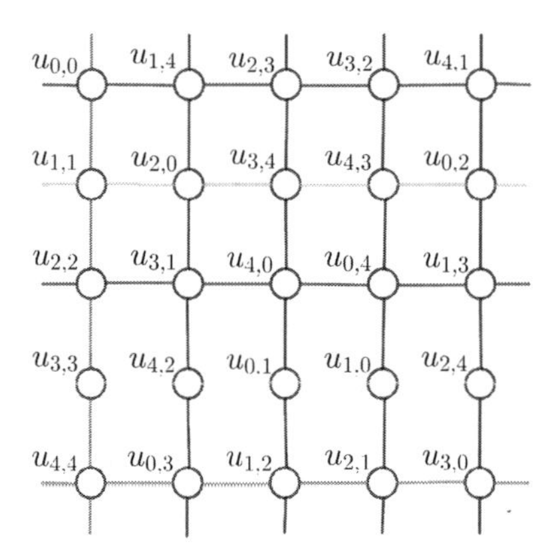

Figure 14.4: Cyclic grid form of tensor product $C_5 \otimes C_5$

Inductively, cycles occurring in tensor product $C_n \otimes C_n$ are given by:

$$u_{i,i+(n-1)} - u_{i+1,i+(n-1)-1} - \cdots - u_{i+(n-1),i} \tag{14.3}$$

for all $i = 0, 1, 2, .., n-1$, where the subscript of u_i's is calculated under modulo n.
Next we arrange first five cycles as horizontally and last cycle as vertically. The structure shown in Figure 14.4, is called cyclic grid form. We will use this cyclic grid form in our main result.

14.4 KNOWN RESULTS AND METHODS

In [3], M. Bača et al. proved the following result to get another VMT labeling through existing VMT labeling:

Theorem 14.1

Let λ be a vertex magic total labeling of an r-regular graph $G(V, E)$ with v vertices and e edges and a magic constant k. The labeling λ' given by

$$\lambda'(x) = v + e + 1 - \lambda(x), x \in V$$

$$\lambda'(xy) = v + e + 1 - \lambda(xy), xy \in E$$

which is also a vertex magic total labeling of G with the magic constant $(r+1)(v+e+1) - k$. ■

In [1], Dalibor Fronček et al. proved the following result for cartesian product $C_m \times C_n$ of cycles C_m and C_n

---

Theorem 14.2

For each $m,n \geq 3$ and m,n are odd, there exists a vertex magic total labeling of $C_m \times C_n$, given by

$$\lambda(v_{i,j}) = jm+1+i,$$

$$\lambda(v_{i,j}v_{i+1,j}) = \begin{cases} (2n-(j+1))m+1+\frac{i}{2} & \text{if } i \text{ even} \\ (2n-(j+1))m+1+\frac{n+i}{2} & \text{if } i \text{ odd,} \end{cases}$$

$$\lambda(v_{i,j}v_{i,j+1}) = \begin{cases} (2n+\frac{j}{2}+1)m-i & \text{if } j \text{ even} \\ (2n+\frac{n+j}{2}+1)m-i & \text{if } j \text{ odd} \end{cases}$$

with magic constant $\frac{17mn+5}{2}$, where the $v_{i,j}$ denotes vertices, $v_{i,j}v_{i+1,j}$ vertical edges and $v_{i,j}v_{i,j+1}$ horizontal edges in the graph $C_m \times C_n$. ■

Vertex magic total labeling of $C_3 \times C_3$ is shown in Figure 14.5.

Theorem 14.3

For odd natural number n, the tensor product $C_n \otimes C_n$ has vertex magic total labeling with magic constant $k = \frac{17n^2+5}{2}$. ■

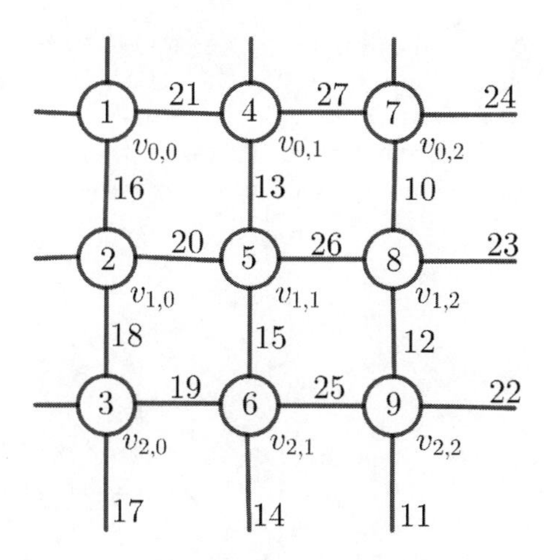

Figure 14.5: Vertex magic total labeling of $C_3 \times C_3$

Proof. We denote the graphs $C_n \otimes C_n$ and $C_n \times C_n$ by G and H respectively. The vertex sets of G and H are defined by $V(G) = \{u_{i,j} | i,j = 0,1,\cdots,n-1\}$ and $V(H) = \{v_{i,j} | i,j = 0,1,\cdots,n-1\}$ respectively. By substituting $m = n$ in Theorem 14.2, the vertex magic total labeling $g : V(H) \cup E(H) \to \{1,2,3,\ldots,n^2+1\}$ is defined as follows:

$$g(v_{i,j}) = jn + 1 + i,$$

For edges of vertical cycles:

$$g(v_{i,j}v_{i+1,j}) = \begin{cases} (2n-(j+1))n+1+\frac{i}{2} & \text{if } i \text{ even} \\ (2n-(j+1))n+1+\frac{n+i}{2} & \text{if } i \text{ odd,} \end{cases}$$

For edges of horizontal cycles:

$$g(v_{i,j}v_{i,j+1}) = \begin{cases} (2n+\frac{j}{2}+1)n-i & \text{if } j \text{ even} \\ (2n+\frac{n+j}{2}+1)n-i & \text{if } j \text{ odd} \end{cases}$$

Now, we define a bijection $f : V(G) \cup E(G) \to V(H) \cup E(H)$ as follows:

$$f(u_{i,j}) = \begin{cases} v_{i,i-j} & \text{if } i-j \equiv 0 \pmod n \\ v_{i-1,i-j-1} & \text{if } i-j \equiv 2 \pmod n \\ \vdots & \vdots \\ v_{i-(\frac{n-1}{2}),i-j-(\frac{n-1}{2})} & \text{if } i-j \equiv (n-1) \pmod n \\ v_{i-(\frac{n+1}{2})+1,i-j-(\frac{n+1}{2})} & \text{if } i-j \equiv 1 \pmod n \\ \vdots & \vdots \\ v_{i-(n-1),i-j-(n-1)} & \text{if } i-j \equiv (n-2) \pmod n \end{cases}$$

Similarly, each edge $e = u_{i,j}u_{k,l}$ of G joining two vertices $u_{i,j}$ and $u_{k,l}$ will be mapped onto $e' = v_{p,q}v_{r,s}$ for some p,q,r,s under f. Further, note that the subscripts of vertices and edges are calculated under modulo n.
Now, we claimed that the composition of f and g, that is $g \circ f : V(G) \cup E(G) \to \{1,2,\cdots,n^2+1\}$ will become a vertex magic total labeling of G with magic constant $k = \frac{17n^2+5}{2}$.

Let $u_{i,j}$ be the arbitrary vertex of $V(G)$.

Case-I: $i-j \equiv 0 \pmod n$

In this case, the edges incident on the vertex $u_{i,j}$ are $u_{i,j}u_{i+1,j-1}$, $u_{i,j}u_{i-1,j+1}$, $u_{i,j}u_{i-1,j-1}$ and $u_{i,j}u_{i+1,j+1}$, where $u_{i,j}u_{i+1,j-1}$ and $u_{i,j}u_{i-1,j+1}$ are edges of vertical cycle and $u_{i,j}u_{i-1,j-1}$ and $u_{i,j}u_{i+1,j+1}$ are edges of horizontal cycle in the cyclic grid form of $C_n \otimes C_n$. Let us see the images of vertex $u_{i,j}$ and incident edges on it under f:

$f(u_{i,j}) = v_{i,i-j} = v_{i,0}$, $f(u_{i,j}u_{i+1,j-1}) = v_{i,0}v_{i,1}$, $f(u_{i,j}u_{i-1,j+1}) = v_{i,0}v_{i,n-1}$,
$f(u_{i,j}u_{i-1,j-1}) = v_{i,0}v_{i-1,0}$, $f(u_{i,j}u_{i+1,j+1}) = v_{i,0}v_{i+1,0}$

Now, we apply g on images of f to get the labels of $u_{i,j}$ and incident edges on it under $g \circ f$:

$g(f(u_{i,j})) = g(v_{i,0}) = 1+i$,
$g(f(u_{i,j}u_{i+1,j-1})) = g(v_{i,0}v_{i,1}) = (2n+1)n-i$,
$g(f(u_{i,j}u_{i-1,j+1})) = g(v_{i,n-1}v_{i,0}) = \left(2n+\frac{n-1}{2}+1\right)n-i$,
$g(f(u_{i,j}u_{i-1,j-1})) = g(v_{i-1,0}v_{i,0}) = (2n-1)n+1+\frac{n+i-1}{2}$,
$g(f(u_{i,j}u_{i+1,j+1})) = g(v_{i,0}v_{i+1,0}) = \left(2n+(j+1)\right)n+1+\frac{i}{2}$

The sum of the label of vertex $u_{i,j}$ and labels of edges incident on it under $g \circ f$:

$$\begin{aligned}
&g \circ f(u_{i,j}) + \sum_{v \in N(u_{i,j})} g \circ f(vu_{i,j}) \\
&= g(f(u_{i,j})) + g(f(u_{i,j}u_{i+1,j-1})) + g(f(u_{i,j}u_{i-1,j+1})) \\
&\quad + g(f(u_{i,j}u_{i-1,j-1})) + g(f(u_{i,j}u_{i+1,j+1})) \\
&= g(v_{i,0}) + g(v_{i,0}v_{i,1}) + g(v_{i,0}v_{i,n-1}) + g(v_{i,0}v_{i-1,0}) \\
&\quad + g(v_{i,0}v_{i+1,0}) \\
&= 1+i+(2n+1)n-i+\left(2n+\frac{n-1}{2}+1\right)n-i \\
&\quad + (2n-1)n+1+\frac{n+i-1}{2} \\
&\quad + \left(2n+(j+1)\right)n+1+\frac{i}{2} \\
&= \frac{17n^2+5}{2}
\end{aligned}$$

<u>**Case-II**</u>: $i-j \equiv 2 \pmod{n}$

$g(f(u_{i,j})) = g(v_{i-1,i-j-1}) = g(v_{i-1,1}) = n+i$,
$g(f(u_{i,j}u_{i+1,j-1})) = g(v_{i-1,1}v_{i-1,2}) = \left(2n+\frac{n+1}{2}+1\right)n-(i-1)$,
$g(f(u_{i,j}u_{i-1,j+1})) = g(v_{i-1,0}v_{i-1,1}) = \left(2n+1\right)n-(i-1)$,

Subcase-I: i and j both are odd

$g(f(u_{i,j}u_{i-1,j-1})) = g(v_{i-2,1}v_{i-1,1}) = (2n-2)n+1+\frac{n+(i-2)}{2}$,
$g(f(u_{i,j}u_{i+1,j+1})) = g(v_{i-1,1}v_{i,1}) = (2n-2)n+1+\frac{i-1}{2}$

The sum of the label of vertex $u_{i,j}$ and labels of edges incident on it under $g \circ f$:

$$g \circ f(u_{i,j}) + \sum_{v \in N(u_{i,j})} g \circ f(vu_{i,j}) = g(f(u_{i,j})) + g(f(u_{i,j}u_{i+1,j-1})) + g(f(u_{i,j}u_{i-1,j+1})) + g(f(u_{i,j}u_{i-1,j-1})) + g(f(u_{i,j}u_{i+1,j+1}))$$

$$\begin{aligned} g \circ f(u_{i,j}) + \sum_{v \in N(u_{i,j})} g \circ f(vu_{i,j}) &= g(v_{i-1,1}) + g(v_{i-1,1}v_{i-1,2}) + g(v_{i-1,0}v_{i-1,1}) \\ &\quad + g(v_{i-2,1}v_{i-1,1}) + g(v_{i-1,1}v_{i,1}) \\ &= n+i+\left(2n+\tfrac{n+1}{2}+1\right)n-(i-1)+\left(2n+1\right)n-(i-1) \\ &\quad +\left(2n-2\right)n+1+\tfrac{n+(i-2)}{2}+\left(2n-2\right)n+1+\tfrac{i-1}{2} \\ &= \frac{17n^2+5}{2} \end{aligned}$$

Subcase-II: i and j both are even

$$g(f(u_{i,j}u_{i-1,j-1})) = g(v_{i-2,1}v_{i-1,1}) = \left(2n-2\right)n+1+\tfrac{i-2}{2},$$
$$g(f(u_{i,j}u_{i+1,j+1})) = g(v_{i-1,1}v_{i,1}) = \left(2n-2\right)n+1+\tfrac{n+(i-1)}{2}$$

The sum of the label of vertex $u_{i,j}$ and labels of edges incident on it under $g \circ f$:

$$\begin{aligned} g \circ f(u_{i,j}) + \sum_{v \in N(u_{i,j})} g \circ f(vu_{i,j}) &= g(f(u_{i,j})) + g(f(u_{i,j}u_{i+1,j-1})) + g(f(u_{i,j}u_{i-1,j+1})) \\ &\quad + g(f(u_{i,j}u_{i-1,j-1})) + g(f(u_{i,j}u_{i+1,j+1})) \\ &= g(v_{i-1,1}) + g(v_{i-1,1}v_{i-1,2}) + g(v_{i-1,0}v_{i-1,1}) \\ &\quad + g(v_{i-2,1}v_{i-1,1}) + g(v_{i-1,1}v_{i,1}) \\ &= n+i+\left(2n+\tfrac{n+1}{2}+1\right)n-(i-1)+\left(2n+1\right)n-(i-1) \\ &\quad +\left(2n-2\right)n+1+\tfrac{i-2}{2}+\left(2n-2\right)n+1+\tfrac{n+(i-1)}{2} \\ &= \frac{17n^2+5}{2} \end{aligned}$$

The proof is similar in the remaining cases, which completes the proof. □

Now, we illustrate the method of constructing vertex magic total labeling of $C_5 \otimes C_5$ for $n = 5$.

Illustration 14.4.1. Let $G = C_5 \otimes C_5$ and $H = C_5 \times C_5$. We define bijection f from $V(G) \cup E(G)$ to $V(H) \cup E(H)$ as follows:

$$f(u_{i,j}) = \begin{cases} v_{i,i-j} & \text{if } i-j=0 \\ v_{i-1,i-j-1} & \text{if } i-j=2 \\ v_{i-2,i-j-2} & \text{if } i-j=4 \\ v_{i-3,i-j-3} & \text{if } i-j=1 \\ v_{i-4,i-j-4} & \text{if } i-j=3 \end{cases}$$

Similarly, each edge $e = u_{i,j}u_{k,l}$ of G joining two vertices $u_{i,j}$ and $u_{k,l}$ will be mapped onto $e' = v_{p,q}v_{r,s}$ for some p,q,r,s under f. Further, note that the subscripts of vertices and edges are calculated under modulo 5.

Note that the function $g \circ f$ defined in Theorem 14.3 admits vertex magic total labeling of $C_5 \otimes C_5$ where g is vertex magic total labeling of $C_5 \times C_5$(See Theorem 14.2).
Consider a vertex $u_{3,1}$ in the cyclic grid form of $C_5 \otimes C_5$. The vertices adjacent to $u_{3,1}$ are $u_{2,2}, u_{4,0}, u_{2,0}$ and $u_{4,2}$. The edges incident to the vertex $u_{3,1}$ are $u_{2,2}u_{3,1}$, $u_{4,0}u_{3,1}$, $u_{2,0}u_{3,1}$ and $u_{4,2}u_{3,1}$. Now, we find the labels assigned to the vertex and edges incident on it using g defined in Theorem 14.2 and f defined in proof of Theorem 14.3. At last, we will show that the sum of these labels is constant 215.
The label of vertex of $u_{3,1}$ and labels of edges incident on it under $g \circ f$ in tensor product $C_5 \otimes C_5$ are:

$g(f(u_{3,1})) = g(v_{2,1}) = 8,$
$g(f(u_{3,1}u_{2,2})) = g(v_{2,1}v_{2,0}) = 53,$
$g(f(u_{3,1}u_{4,0})) = g(v_{2,1}v_{2,2}) = 68,$
$g(f(u_{3,1}u_{2,0})) = g(v_{2,1}v_{1,1}) = 44,$
$g(f(u_{3,1}u_{4,2})) = g(v_{2,1}v_{3,1}) = 42$

The sum of label of vertex $u_{3,1}$ and labels of incident edges on it under $g \circ f$ is:

$$\begin{aligned} g \circ f(u_{3,1}) + \sum_{v \in N(u_{3,2})} g \circ f(vu_{3,1}) &= g(f(u_{3,1})) + g(f(u_{3,1}u_{2,2})) + g(f(u_{3,1}u_{4,0})) \\ &\quad + g(f(u_{3,1}u_{2,0})) + g(f(u_{3,1}u_{4,2})) \\ &= 8 + 53 + 68 + 44 + 42 \\ &= 215 \end{aligned}$$

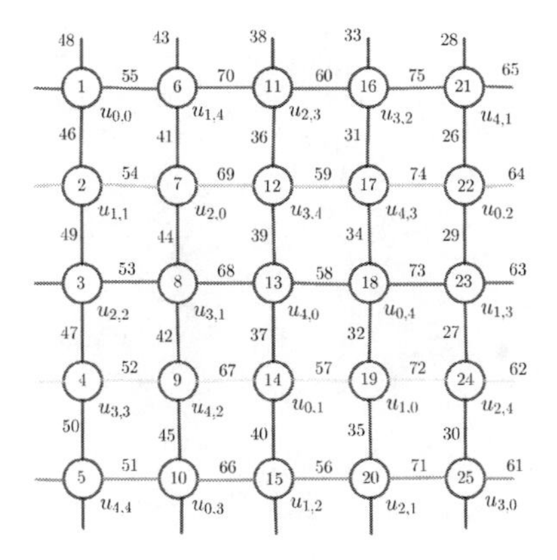

Figure 14.6: VMT labeling of $C_5 \otimes C_5$ with magic constant $k = 215$

Similarly, one may verify that the sum of the label of the vertex and labels of edges incident on it is 215 for every vertex v in $C_5 \otimes C_5$ (See Figure 14.6). Therefore, $g \circ f$ admits a vertex magic total labeling of $C_5 \otimes C_5$. Hence, it is VMTL of graph $C_5 \otimes C_5$.

With the range of vertex-magic constant obtained from Equation (14.1), we can have precisely three possible integer values of constant $\frac{13n^2+5}{2}$, $\frac{15n^2+5}{2}$ and $\frac{17n^2+5}{2}$. As we have a different choice of VMTL of $C_m \times C_n$ described in [1] by Dalibor Fronček, Peter Kovář and Tereza Kovářová based on Theorem 14.1. For these three different integer values of the constant the labels of vertices and edges of tensor product $C_n \otimes C_n$ are shown in the following table:

	$g(v_{i,j})$	$g(v_{i,j}v_{i+1,j})$	$g(v_{i,j}v_{i,j+1})$	k
1	$1,2,\ldots,n^2$	$n^2+1,n^2+2,\ldots,2n^2$	$2n^2+1,n^2+2,\ldots,3n^2$	$\frac{17n^2+5}{2}$
2	$1,2,\ldots,n^2$	$2n^2+1,n^2+2,\ldots,3n^2$	$n^2+1,n^2+2,\ldots,2n^2$	$\frac{17n^2+5}{2}$
3	$n^2+1,n^2+2,\ldots,2n^2$	$1,2,\ldots,n^2$	$2n^2+1,n^2+2,\ldots,3n^2$	$\frac{15n^2+5}{2}$
4	$n^2+1,n^2+2,\ldots,2n^2$	$2n^2+1,n^2+2,\ldots,3n^2$	$1,2,\ldots,n^2$	$\frac{15n^2+5}{2}$
5	$2n^2+1,n^2+2,\ldots,3n^2$	$1,2,\ldots,n^2$	$n^2+1,n^2+2,\ldots,2n^2$	$\frac{13n^2+1}{2}$
6	$2n^2+1,n^2+2,\ldots,3n^2$	$n^2+1,n^2+2,\ldots,2n^2$	$1,2,\ldots,n^2$	$\frac{13n^2+1}{2}$

In the above table, the labelings 1 and 6, 3 and 5, 2 and 3 are dual labelings of each other.

REFERENCES

1. Dalibor Fronček, Peter Kovář and Tereza Kovářová . Vertex magic total labeling of products of cycles. *Australasian Journal of Combinatorics*, 33, 169-181, 2005.
2. J. A. MacDougall, M. Miller, Slamin and W. D. Wallis. Vertex magic total labeling of graphs. *Utilitas Mathematica- combinatorial Mathematics*, 61, 3-21, 1990.
3. M. Bača, F. Bertault, J. A. MacDougall, Miller, Simanjuntak and R. Slamin. Vertex-antimagic total labelings of graphs. *Discussiones Mathematicae Graph Theory*, 23, 67-83, 2003.
4. W. D. Wallis. *Magic Graphs*. Birkhäuser, 2001.
5. J. A. Gallian. A dynamic survey of graph labeling. *The Electronic Journal of Combinatorics*, #DS6, 2018.

15 Antimagic Labeling of Some Star and Bistar related Graphs

Tarunkumar Chhaya
L. E. Polytechnic
Ahmedabad, Gujarat (INDIA)
E-mail: tarunchhaya@yahoo.com

Kailas K. Kanani
Department of Mathematics,
Government Engineering College,
Rajkot, Gujarat (INDIA)
E-mail: kananikkk@yahoo.co.in

In this chapter, we investigate antimagic labeling in the context of some standard graph operations. We prove that splitting graph of star graph $K_{1,n}$ and splitting graph of bistar $B_{m,n}$ are antimagic. We also prove that degree splitting graph of $B_{n,n}$ and restricted square graph $B^2_{m,n}$ are antimagic.

15.1 INTRODUCTION

Graph labeling is an assignment of numbers to the vertices or edges or both subject to certain condition(s). If the domain of function is a set of vertices (edges), then such labeling is called a vertex (edge) labeling. In this chapter we discuss antimagic labeling in the context of some standard graph operations on star and bistar.

Here, we consider simple, finite and connected graphs. Gallian[1] (2019) contains a dynamic survey of various graph labeling problems with an extensive bibliography.

In this chapter, we consider the following standard definitions. For any undefined term one can refer to Gross and Yellen[2].

Definition (Bistar)**.** The bistar $B_{m,n}$ is the graph obtained by joining the center(apex) vertices of $K_{1,m}$ and $K_{1,n}$ by an edge.

Definition (Splitting graph)**.** The splitting graph $S'(G)$ of a graph G is obtained by adding a new vertex v' corresponding to each vertex v of G such that $N(v) = N(v')$.

Definition (Degree splitting graph). Let $G=(V(G),E(G))$ be a graph with $V=S_1\cup S_2\cup\ldots\cup S_t\cup T$ where each S_i is a set of all vertices of the same degree with at least two elements and $T=V-\left(\bigcup_{i=1}^{t} S_i\right)$. The degree splitting graph of G denoted by $DS(G)$ is obtained from G by adding vertices $w_1,w_2,w_3,\ldots,w_t$ and joining to each vertex of S_i for $1\leq i\leq t$.

Definition (Square of graph). Let $G=(V(G),E(G))$ be a simple connected graph. The square of the graph of G is denoted by G^2 and defined as the graph with the same vertex set as that of G and two vertices are adjacent in G^2 if they are at a distance 1 or 2 apart in G.

It is noted that the ***restricted*** $B^2_{m,n}$ is the graph obtained from $B_{m,n}$ by joining all the pendent vertices of the $K_{1,m}$ with the apex vertex of $K_{1,n}$ and all the pendent vertices of the $K_{1,n}$ with the apex vertex of $K_{1,m}$.

15.2 MAGIC LABELING

Inspired by the magic square of positive integers, Sedlacek[4] indtroduced the concept of magic labeling during 1963.

Definition (Magic). A graph $G=(V(G),E(G))$ is said to be magic if it has real valued edge labeling such that,

1. distinct edges have distinct non negative labels;
2. the sum of the labels of the edges incident to a particular vertex is same for all the vertices.

In other words, for a graph $G=(V(G),E(G))$, the bijection $f:E(G)\rightarrow\{1,2,3,\ldots,q\}$ is called **magic labeling**, if sums of edge labels around any vertex in G is same.

Illustration 15.2.1. The magic labeling is shown in Figure 15.1.

15.3 ANTIMAGIC LABELING

Hartsfield and Ringel[3] introduced the concept of antimagic labeling in 1990.

Definition (Antimagic). A graph $G=(V(G),E(G))$ with q edges is said to be antimagic if its edges can be labelled with distinct positive integers $\{1,2,\ldots,q\}$ such that sums of the labels of the edges incident to each vertex are distinct.

Illustration 15.3.1. Antimagic labeling of cycle C_5 is shown in Figure 15.2.

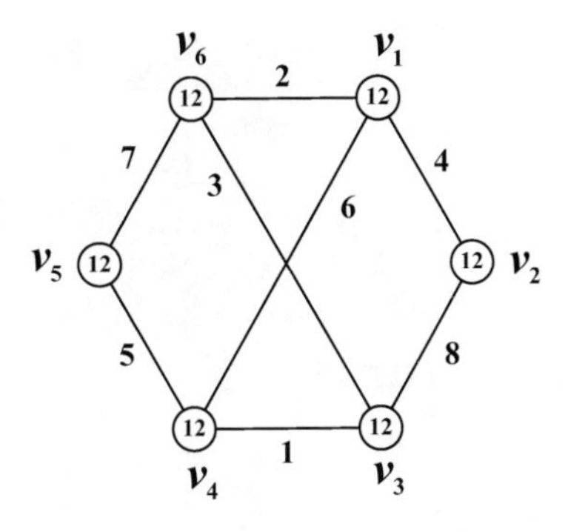

Figure 15.1: Magic labeling of a 6-vertices, 8-edges graph

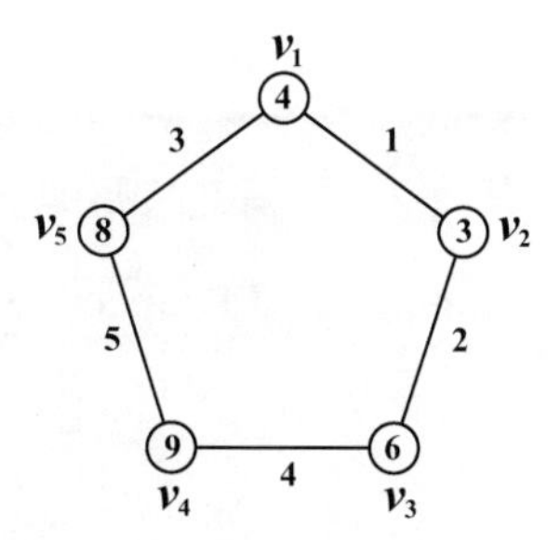

Figure 15.2: Antimagic labeling of cycle C_5

15.4 SOME KNOWN RESULTS OF ANTIMAGIC LABELING

Hartsfield and Ringel[3] proved the following results:

1. Every cycle C_n is antimagic.
2. Every wheel W_n is antimagic.
3. The complete graph K_n is antimagic, if and only if, $n \geq 3$.

Hartsfield and Ringel[3] conjectured the following :

1. All trees except K_2 are antimagic.
2. All connected graphs except K_2 are antimagic.

Vaidya and Vyas[5] have discussed antimagic labeling of some path and cycle related graphs and proved the following results:

1. The middle graph of path P_n is antimagic.
2. The middle graph of cycle C_n is antimagic.

3. The total graph of path P_n is antimagic.
4. The total graph of cycle C_n is antimagic.
5. The splitting graph of path P_n is antimagic.
6. The splitting graph of cycle C_n is antimagic.
7. The shadow graph of path P_n is antimagic.
8. The shadow graph of cycle C_n is antimagic.

15.5 OUR INVESTIGATIONS

Theorem 15.1

The splitting graph $S'(K_{1,n})$ of star $K_{1,n}$ is antimagic. ■

Proof. Let $K_{1,n}$ be the star with the vertex set $\{v_0, v_i : 1 \leq i \leq n\}$; where $v_1, v_2, \ldots, v_n$ are pendent vertices. In order to obtain $G = S'(K_{1,n})$ add the vertices $\{v'_0, v'_i : 1 \leq i \leq n\}$. It is noted that $|V(G)| = 2n+2$ and $|E(G)| = 3n$.

The edge labeling $f : E(G) \to \{1, 2, \ldots, 3n\}$ is defined as follows:
$f(v_0 v'_i) = i; \quad 1 \leq i \leq n;$
$f(v_0 v_i) = i+n; \quad 1 \leq i \leq n;$
$f(v_i v'_0) = i+2n; \quad 1 \leq i \leq n.$

The labeling pattern defined above satisfies the vertex conditions and edge conditions of antimagic labeling. That is, the splitting graph $S'(K_{1,n})$ of star $K_{1,n}$ is antimagic. □

Illustration 15.5.1. The splitting graph $S'(K_{1,4})$ of star $K_{1,4}$ and its antimagic labeling are shown in the Figure 15.3.

Theorem 15.2

The splitting graph $S'(B_{m,n})$ of bistar $B_{m,n}$ is antimagic. ■

Proof. Let $B_{m,n}$ be the bistar with vertex set $\{u_0, v_0, u_i, v_j : 1 \leq i \leq m; 1 \leq j \leq n\}$ where u_i, v_j are pendent vertices. In order to obtain $S'(B_{m,n})$ add u'_0, v'_0, u'_i, v'_j vertices corresponding to u_0, v_0, u_i, v_j; where $1 \leq i \leq m$ and $1 \leq j \leq n$. Let $G = S'(B_{m,n})$. It is noted that, $|V(G)| = 2m+2n+4$ and $|E(G)| = 3m+3n+3$.

The edge labeling $f : E(G) \to \{1, 2, \ldots, 3m+3n+3\}$ is defined as follows:
$f(u'_i u_0) = i; \quad 1 \leq i \leq m;$

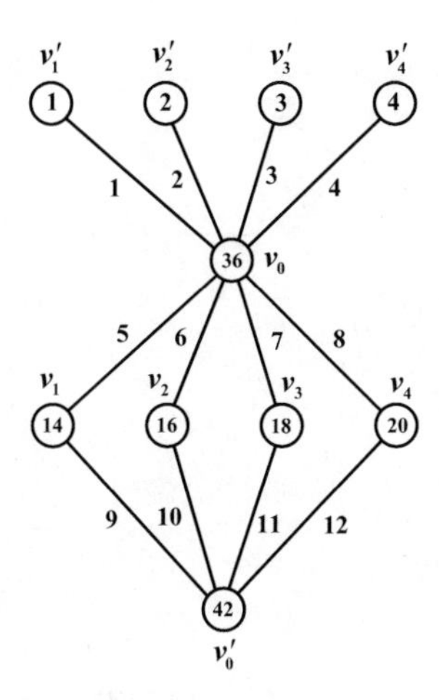

Figure 15.3: Antimagic labeling of the splitting graph $S'(K_{1,4})$ of star $K_{1,4}$

$f(u_iu_0) = i+m+n; \quad 1 \le i \le m;$
$f(u_0'u_i) = i+2m+2n; \quad 1 \le i \le m.$

$f(v_i'v_0) = i+m; \quad 1 \le i \le n;$
$f(v_iv_0) = i+2m+n; \quad 1 \le i \le n;$
$f(v_0'v_i) = i+3m+2n; \quad 1 \le i \le n.$

$f(u_0v_0') = 3m+3n+1;$
$f(u_0'v_0) = 3m+3n+2;$
$f(u_0v_0) = 3m+3n+3.$
The labeling pattern defined above satisfies the vertex conditions and edge conditions of antimagic labeling. That is, the splitting graph $S'(B_{m,n})$ of bistar $B_{m,n}$ is antimagic. □

Illustration 15.5.2. The $S'(B_{5,3})$ and its antimagic labeling are shown in Figure 15.4.

Theorem 15.3

The degree splitting graph $DS(B_{n,n})$ of bistar $B_{n,n}$ is antimagic. ■

Proof. Let $B_{n,n}$ be the bistar with vertex set $\{u_0, v_0, u_i, v_i : 1 \le i \le n\}$; where u_i, v_i are pendent vertices. Here $V(B_{n,n}) = V_1 \cup V_2$; where $V_1 = \{u_i, v_i, 1 \le i \le n\}$ and $V_2 = \{u_0, v_0\}$. Now in order to obtain $DS(B_{n,n})$ from $B_{n,n}$ we add vertices w_1 and w_2 corresponding to V_1 and V_2. Let $G = DS(B_{n,n})$. Here edge

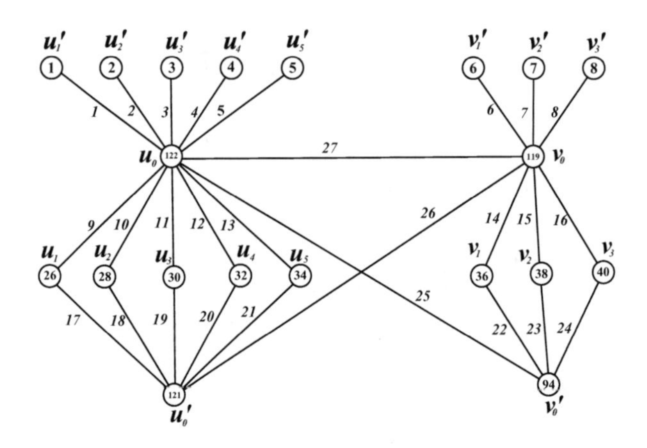

Figure 15.4: Antimagic labeling of the splitting graph $S'(B_{5,3})$ of bistar $B_{5,3}$

set $E(G) = \{u_0v_0, u_0w_2,$
$v_0w_2\} \cup \{u_0u_i, v_0v_i, w_1u_i, w_1v_i : 1 \leq i \leq n\}$. It is noted that, $|V(G)| = 2n+4$ and $|E(G)| = 4n+3$.

The edge labeling $f : E(G) \to \{1, 2, \ldots, 4n+3\}$ is defined as follows:
$f(w_1u_i) = i; \quad 1 \leq i \leq n;$
$f(w_1v_i) = n+i; \quad 1 \leq i \leq n;$
$f(u_iu_0) = 2n+i; \quad 1 \leq i \leq n;$
$f(v_iv_0) = 3n+i; \quad 1 \leq i \leq n.$

$f(u_0v_0) = 4n+1;$
$f(u_0w_2) = 4n+2;$
$f(v_0w_2) = 4n+3.$

The labeling pattern defined above satisfies the vertex conditions and edge conditions of antimagic labeling. That is, the degree splitting graph $DS(B_{n,n})$ of bistar $B_{n,n}$ is antimagic. □

Illustration 15.5.3. The degree splitting graph $DS(B_{3,3})$ of bistar $B_{3,3}$ and its antimagic labeling are shown in Figure 15.5.

Theorem 15.4

The restricted square graph $B^2_{m,n}$ of bistar $B_{m,n}$ is antimagic. ■

Proof. Let $B_{m,n}$ be the bistar with vertex set $\{u_0, v_0, u_i, v_j, 1 \leq i \leq m, 1 \leq j \leq n\}$, where u_i, v_j are pendent vertices. Let $G = B^2_{m,n}$ be the restricted

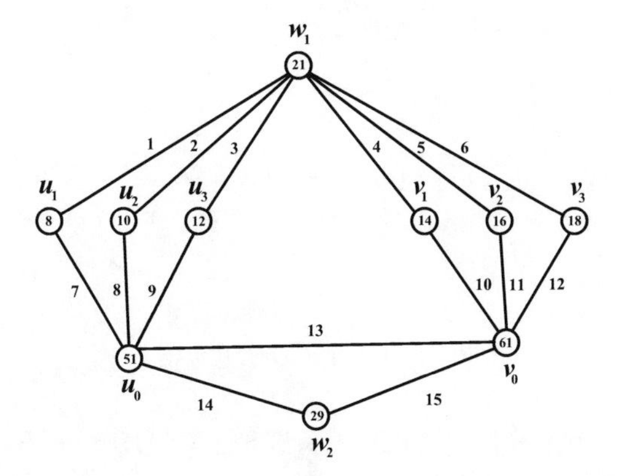

Figure 15.5: Antimagic labeling of the degree splitting graph $DS(B_{3,3})$ of bistar $B_{3,3}$

square graph with $V(G) = V(B_{m,n})$ and $E(G) = E(B_{m,n}) \cup \{u_0v_j, v_0u_i : 1 \leq i \leq m; 1 \leq j \leq n\}$.
It is noted that $|V(G)| = m+n+2$ and $|E(G)| = 2m+2n+1$.

The edge labeling $f : E(G) \to \{1,2,\ldots,2m+2n+1\}$ is defined as follows:
$f(u_0u_i) = i; \qquad 1 \leq i \leq m;$
$f(v_0u_i) = m+i; \qquad 1 \leq i \leq m.$

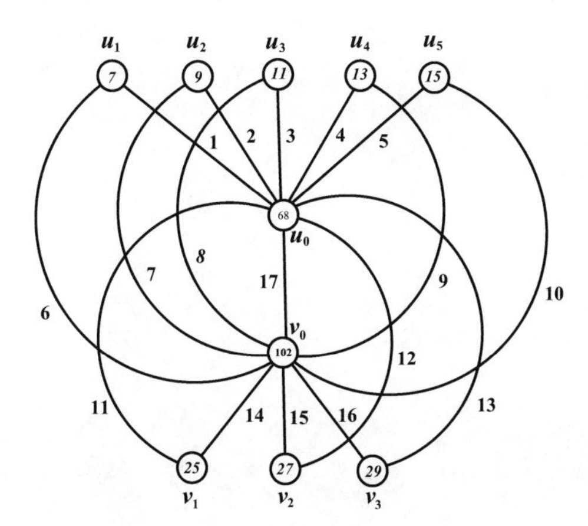

Figure 15.6: Antimagic labeling of the restricted square graph $B^2_{5,3}$ of bistar $B_{5,3}$

$f(u_0 v_j) = 2m + j; \qquad 1 \le j \le n;$
$f(v_0 v_j) = 2m + n + j; \qquad 1 \le j \le n;$
$f(u_0 v_0) = 2m + 2n + 1.$

The labeling pattern defined above satisfies the vertex conditions and edge conditions of antimagic labeling. That is, the restricted square graph $B^2_{m,n}$ of bistar $B_{m,n}$ is antimagic. □

Illustration 15.5.4. The restricted square graph $B^2_{5,3}$ of bistar $B_{5,3}$ and its antimagic labeling are shown in Figure 15.6.

15.6 CONCLUSION AND SCOPE FOR FURTHER RESEARCH

We have proved some results related to antimagic labeling using some graph operations on star and bistar. To find other graph families which are antimagic is an open area of research.

REFERENCES

1. J. A. Gallian. A dynamic survey of graph labeling. *The Electronics Journal of Combinatorics*, 2019.
2. J. Gross and J. Yellen. *Graph Theory and its Application.* CRC Press, Inc. Boca Raton, FL, USA, 2004.
3. N. Hartsfield and G. Ringel. *Pearls in Graph Theory: A Comprehensive Introduction.* Academic Press, Boston, 1990.
4. J. Sedlacek. Problem 27: In theory of graph and its application. In *Proceedings Symposium Smolenice*, 163-167, 1963.
5. S. K. Vaidya and N. B. Vyas. Antimagic labeling of some path and cycle related graphs. *Annals of Pure and Applied Mathematics*, 3(2): 119-128, 2013.

16 Distance Magic and Distance Antimagic Labeling of Some Product Graphs

N P Shrimali
Department of Mathematics,
Gujarat University,
Ahmedabad, Gujarat (INDIA)
E-mail: narenp05@gmail.com

Y M Parmar
Department of Mathematics,
Gujarat University,
Ahmedabad, Gujarat (INDIA)
E-mail: ymp.maths@gmail.com

Let $G=(V,E)$ be a graph of order n. Let $f:V(G)\to\{1,2,\dots,n\}$ be a bijection. For any vertex $q\in V(G)$, the sum $\sum_{p\in N(q)} f(p)$ is called the weight of the vertex q and is denoted by $w(q)$. If there exists a positive integer γ such that $w(q)=\gamma$, for every $q\in V(G)$, then f is called a distance magic labeling. The constant γ is called the magic constant for f. A graph which admits a distance magic labeling is called a distance magic graph. If $w(q)\neq w(r)$ for any two distinct vertices q and r, then f is called a distance antimagic labeling. A graph which admits a distance antimagic labeling is called a distance antimagic graph. In this chapter, we discuss the existence of distance magic labeling and distance antimagic labeling for $C_3^t \,\square\, C_4$, $C_3^t\times C_4$, $C_3^t\boxtimes C_4$ and $C_4\odot C_3^t$.

16.1 INTRODUCTION

Here, we consider that all graphs G with vertex set $V(G)$ and edge set $E(G)$ are finite and simple. We adopt Gross and Yellen [5] for various graphs and its theoretic notations and for number theoretical results, we follow Burton [3]. For acquiring the latest update, we follow a dynamic survey on graph labeling by Gallian [4].

A *distance magic labeling* of a graph G of order n is a bijection $f:V(G)\to\{1,2,\dots,n\}$ such that $\sum_{p\in N(q)} f(p)=\gamma$, for all $q\in V(G)$, where $N(q)$ is the set of all vertices of $V(G)$ which are adjacent to q. The constant γ is called the *magic constant* of the distance magic labeling for f. A graph which admits a distance magic labeling is called a distance magic graph. For any vertex $q \in V(G)$, the neighbor sum $\sum_{p \in N(q)} f(p)$ is called the weight of the vertex q and is denoted by $w(q)$.

The concept of distance magic labeling was introduced and studied by many researchers with different terminologies. For example, the term sigma labeling is used by Vilfred [10], the term 1-vertex magic labeling [1-VML] is used by Miller et al. [7], the term neighborhood magic labeling is used by B.D.Acharya et al. [1] and the term distance magic labeling is used by Sugeng et al. [9]. The following lemmas are proved by Miller et al. [7].

Lemma 16.1

[7] A necessary condition for the existance of a distance magic vertex labeling f of a graph G is

$$\gamma v = \sum_{p \in V(G)} d(p) f(p)$$

where $d(p)$ is the degree of vertex p and v is the number of vertices of G. ■

Lemma 16.2

[7] If G contains two vertices p and q such that $|N(p) \cap N(q)| = d(p) - 1 = d(q) - 1$, then G has no distance magic labeling. ■

A *distance antimagic labeling* of a graph G of order n is a bijection $f : V(G) \to \{1, 2, \ldots, n\}$ such that $\sum_{p \in N(q)} f(p) = w(q)$ for all $q \in V(G)$, where $N(q)$ is the set of all vertices of $V(G)$ which are adjacent to q and $w(p) \neq w(q)$ for every pair of vertices $p, q \in V(G)$. A graph which admits a distance antimagic labeling is called a distance antimagic graph.
A distance antimagic labeling was introduced by Simanjuntak and Wijaya [8] in 2013. They proved the following Lemma:

Lemma 16.3

[8] If a graph contains two vertices with the same neighborhood then it is not distance antimagic. ■

They also proved that cycles, suns, wheel W_n for $n \neq 5$, fan F_n for $n \neq 4$, friendship graphs f_n are distance antimagic graphs and multipartite graphs are not distance antimagic graphs.

Definition. [6] A Cartesian product, denoted by $G \square H$ is a graph with vertex set $V(G) \times V(H)$. Vertices (p,q) and (p',q') in $G \square H$ are adjacent if and only if $p = p'$ and q is adjacent to q' in H or $q = q'$ and p is adjacent to p' in G.

Definition. [6] A Direct product, denoted by $G \times H$ is a graph with vertex set $V(G) \times V(H)$. Vertices (p,q) and (p',q') in $G \times H$ are adjacent if and only if p is adjacent to p' in G and q is adjacent to q' in H.

Definition. [6] A strong product, denoted by $G \boxtimes H$ is a graph with vertex set $V(G) \times V(H)$. Vertices (p,q) and (p',q') in $G \boxtimes H$ are adjacent if and only if $p = p'$ and q is adjacent to q' in H or $q = q'$ and p is adjacent to p' in G or p is adjacent to p' in G and q is adjacent to q' in H.

Definition. A corona product $G \odot H$ of two graphs G and H is obtained by taking one copy of G and $|V(G)|$ copies of H; and by joining each vertex of the i-th copy of H to the i-th vertex of G, where $1 \leq i \leq |V(G)|$.

M. Anholcer and S. Cichacz [2] have already proved that $C_3^t \circ C_4$ is not distance magic. In this chapter, we apply some more products between C_3^t and C_4. We investigate the existence of distance magic labeling and distance antimagic labeling of $C_3^t \square C_4$, $C_3^t \times C_4$, $C_3^t \boxtimes C_4$ and $C_4 \odot C_3^t$ graphs.

Throughout this chapter, friendship graph C_3^t consists of $t-$ triangles with a common vertex r and vertices p_k, q_k for $k = 1, 2, \ldots, t$. i.e. p_k, q_k are the vertices of the kth copy of cycle C_3. Let $C_4 = c_0c_1c_2c_3c_0$ be a cycle. Here, $G = C_3^t \square C_4$, $C_3^t \boxtimes C_4$ and $C_3^t \times C_4$ are graphs with vertex set $V(G) = \{p_k^j, q_k^j, r^j \; / 1 \leq k \leq t, 0 \leq j \leq 3\}$, where $p_k^j = (p_k, c_j), q_k^j = (q_k, c_j), r^j = (r, c_j)$.

16.2 CARTESIAN PRODUCT OF C_3^t AND C_4

Theorem 16.1

The graph $C_3^t \square C_4$ is not distance magic. ■

Proof. Suppose that the graph $G = C_3^t \square C_4$ is a distance magic graph under distance magic labeling f. So weights of each vertex are equal.

Now,

$$w(r^0) = \sum_{k=1}^{t} (f(p_k^0) + f(q_k^0)) + f(r^1) + f(r^3)$$

$$\text{and} \quad w(r^2) = \sum_{k=1}^{t} (f(p_k^2) + f(q_k^2)) + f(r^1) + f(r^3)$$

Since, $w(r^0) = w(r^2)$,

$$\sum_{k=1}^{t} (f(p_k^0) + f(q_k^0)) = \sum_{k=1}^{t} (f(p_k^2) + f(q_k^2)) \tag{16.1}$$

Now, for $1 \leq k \leq t$

$$w(p_k^0) = f(q_k^0) + f(p_k^1) + f(p_k^3) + f(r^0)$$

$$\text{and} \quad w(p_k^2) = f(q_k^2) + f(p_k^1) + f(p_k^3) + f(r^2)$$

Since, $w(p_k^0) = w(p_k^2)$,

$$f(q_k^0) + f(r^0) = f(q_k^2) + f(r^2), \ \forall k \tag{16.2}$$

Analogously, we can derive the following equation,

$$f(p_k^0) + f(r^0) = f(p_k^2) + f(r^2), \forall k \tag{16.3}$$

Adding the above $2t$ equations, we get

$$\sum_{k=1}^{t} (f(p_k^0) + f(q_k^0)) + 2tf(r^0) = \sum_{k=1}^{t} (f(p_k^2) + f(q_k^2)) + 2tf(r^2) \tag{16.4}$$

From, (16.1) and (16.4), $f(r^0) = f(r^2)$, which is not possible.
Hence, the graph $C_3^t \,\square\, C_4$ is not a distance magic graph. □

Theorem 16.2

The graph ${C_3}^t \,\square\, C_4$ is distance antimagic. ■

Proof. We define a vertex labeling $f : V(G) \to \{1, 2, \ldots, |V(G)|\}$ as follows:
For $k = 1, 2, \ldots, t$,

$$f(p_k^0) = 8k - 7;$$

$$f(p_k^1) = 8k - 2;$$

$$f(p_k^2) = 8k-6;$$
$$f(p_k^3) = 8k-3;$$
$$f(q_k^0) = 8k-5;$$
$$f(q_k^1) = 8k;$$
$$f(q_k^2) = 8k-4;$$
$$f(r^0) = 8t+2;$$
$$f(r^1) = 8t+1;$$
$$f(r^2) = 8t+4;$$
$$f(r^3) = 8t+3$$

Under the above labeling, weights for each vertex are as follows:
For $1 \leq k \leq t$,

$$w(p_k^0) = 8t+24k-8;$$
$$w(p_k^1) = 8t+24k-12;$$
$$w(p_k^2) = 8t+24k-5;$$
$$w(p_k^3) = 8t+24k-11;$$
$$w(q_k^0) = 8t+24k-6;$$
$$w(q_k^1) = 8t+24k-10;$$
$$w(q_k^2) = 8t+24k-3;$$
$$w(q_k^3) = 8t+24k-9;$$
$$w(r^0) = 8t^2+12t+4;$$
$$w(r^1) = 8t^2+22t+6;$$
$$w(r^2) = 8t^2+14t+4;$$
$$w(r^3) = 8t^2+20t+6$$

Since weights of each vertex are distinct, ${C_3}^t \,\square\, C_4$ is a distance antimagic graph. □

Illustration 16.2.1. The distance antimagic labeling for $C_3^2 \,\square\, C_4$ is given in Figure 16.1.

16.3 DIRECT PRODUCT OF C_3^t AND C_4

Theorem 16.3

The graph ${C_3}^t \times C_4$ is not distance magic. ■

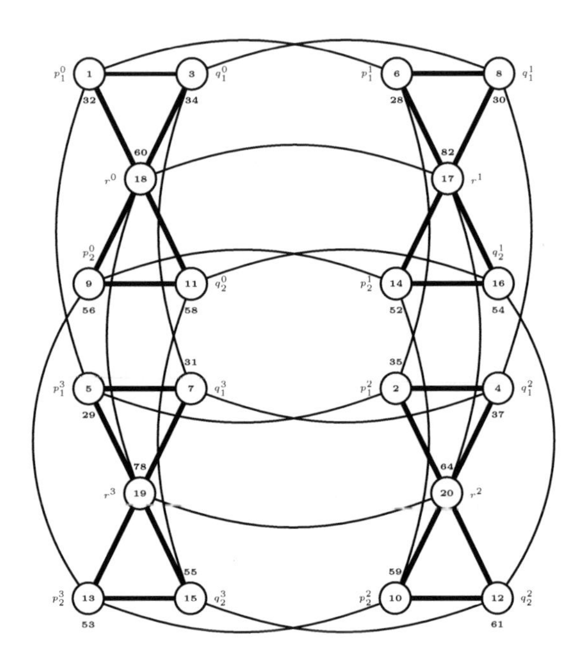

Figure 16.1: Distance antimagic labeling for $C_3^2 \,\square\, C_4$

Proof. Suppose that $G = {C_3}^t \times C_4$ is a distance magic graph under distance magic labeling f and that magic constant is γ. We have

$$\begin{aligned}\gamma = w(r^0) &= \sum_{k=1}^{t} \left(f(p_k^1) + f(q_k^1) + f(p_k^3) + f(q_k^3)\right) \\ = w(r^1) &= \sum_{k=1}^{t} \left(f(p_k^0) + f(q_k^0) + f(p_k^2) + f(q_k^2)\right)\end{aligned} \tag{16.5}$$

and

$$\begin{aligned} w(p_k^0) &= f(q_k^1) + f(q_k^3) + f(r^1) + f(r^3), \\ w(q_k^0) &= f(p_k^1) + f(p_k^3) + f(r^1) + f(r^3), \\ w(p_k^1) &= f(q_k^0) + f(q_k^2) + f(r^0) + f(r^2), \\ w(q_k^1) &= f(p_k^0) + f(p_k^2) + f(r^0) + f(r^2) \end{aligned} \tag{16.6}$$

which implies that

$$\begin{aligned} w(r^0) &= \sum_{k=1}^{t} (w(p_k^0)+w(q_k^0)) - 2t(f(r^1)+f(r^3)) \\ &= w(r^1) = \sum_{k=1}^{t} (w(p_k^1)+w(q_k^1)) - 2t(f(r^0)+f(r^2)) \end{aligned} \tag{16.7}$$

so that,

$$2t\gamma - 2t(f(r^1)+f(r^3)) = 2t\gamma - 2t(f(r^0)+f(r^2)) \tag{16.8}$$

Thus,

$$f(r^0)+f(r^2) = f(r^1)+f(r^3) = \delta(\text{say}) \tag{16.9}$$

Now, for all $k = 1,2,\ldots,t$ and $j \equiv 0 \ (mod\ 4)$

$$\begin{aligned} w(p_k^j) &= f(q_k^{j-1})+f(q_k^{j+1})+f(r^{j-1})+f(r^{j+1}), \\ w(q_k^j) &= f(p_k^{j-1})+f(p_k^{j+1})+f(r^{j-1})+f(r^{j+1}) \end{aligned} \tag{16.10}$$

But $w(p_k^j) = w(q_k^j)$. So,

$$f(p_k^{j-1})+f(p_k^{j+1}) = f(q_k^{j-1})+f(q_k^{j+1}) = \mu \ (\text{say}), \forall k,j \tag{16.11}$$

By Equations (16.5) and (16.11)

$$\gamma = 2t\mu$$

and by Equations, (16.9), (16.10) and (16.11)

$$\gamma = \delta + \mu$$

Now, by the above two equations, we get

$$\mu = \frac{\delta}{2t-1} \tag{16.12}$$

Let us find the sum of labels of all vertices except r^j vertices.

$$\sum_{\substack{k=1 \\ j=0,1,2,3}}^{t} (f(p_k^j)+f(q_k^j)) = \frac{(8t+4)(8t+5)}{2} - (f(r^0)+f(r^2)+f(r^1)+f(r^3)), \tag{16.13}$$

since, $|V(G)| = 8t+4$.
By (16.9) and (16.11), we get

$$4t\mu = \frac{(8t+4)(8t+5)}{2} - 2\delta \tag{16.14}$$

Thus, by Equations (16.12)

$$\delta = \frac{32t^3+20t^2-8t-5}{4t-1} \tag{16.15}$$

i.e.,

$$f(r^0)+f(r^2) = f(r^1)+f(r^3) = \frac{32t^3+20t^2-8t-5}{4t-1} \tag{16.16}$$

Now, $f(r^0), f(r^2) \in \{1,2,\ldots,8t+4\}$. If we take $f(r^0) = 8t+4$ and $f(r^2) = 8t+3$, then $f(r^0)+f(r^2) = \delta = 16t+7$, which is less than $\frac{32t^3+20t^2-8t-5}{4t-1}$. So, equality in (16.16) cannot possible.

Hence, ${C_3}^t \times C_4$ is not a distance magic graph. □

Here, the graph ${C_3}^t \times C_4$ is not a distance magic graph as well as by Lemma 16.3 it is not distance antimagic because neighborhoods of vertex p_k^0 and p_k^2, for $k = 1,2,\ldots,t$ are always the same.

16.4 STRONG PRODUCT OF C_3^t AND C_4

Theorem 16.4

The graph $C_3^t \boxtimes C_4$ is distance antimagic. ■

Proof. We define a vertex labeling $f : V(G) \to \{1,2,\ldots,|V(G)|\}$ as follows:

For $1 \leq k \leq t$ and $j \equiv 0 \ (mod\ 4)$

$$f(p_k^j) = 2jt+2t-k+1;$$
$$f(q_k^j) = 2jt+k;$$
$$f(r^j) = 8t-j+4$$

Under the above labeling, weights for each vertex are as follows:

For $1 \leq k \leq t$, $j \equiv 0 \ (mod\ 4)$,

$$w(p_k^j) = 28t+k+14+2jt+4(j+1)t+4(j-1)t-j-(j+1)-(j-1);$$
$$w(q_k^j) = 30t-k+15+2jt+4(j+1)t+4(j-1)t-j-(j+1)-(j-1);$$
$$w(r^j) = 6t^2+19t+8+4jt^2+4(j+1)t^2+4(j-1)t^2-(j+1)-(j-1)$$

Here, we can see that, each vertex has distinct weights.
Therefore, the graph $C_3^t \boxtimes C_4$ is a distance antimagic graph. □

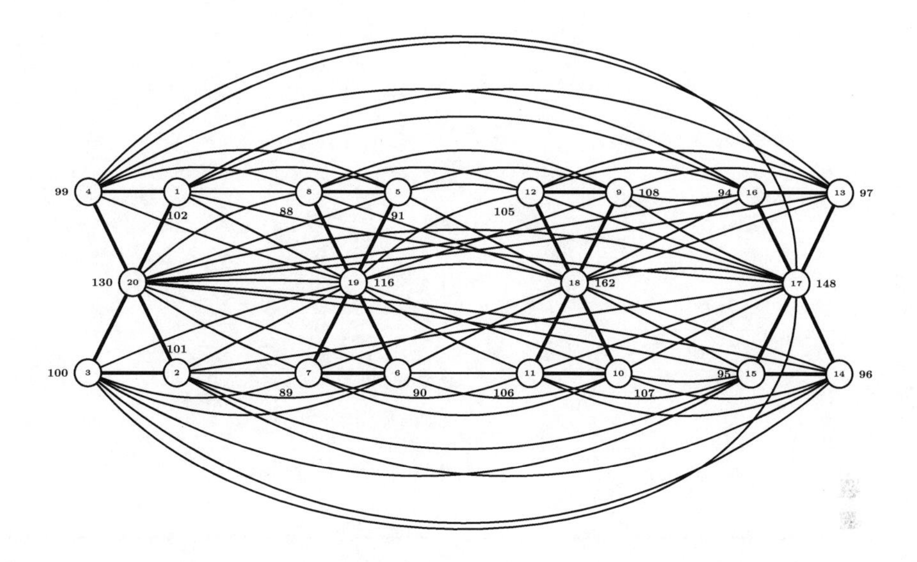

Figure 16.2: The distance antimagic labeling of $C_3^2 \boxtimes C_4$

Illustration 16.4.1. The graph $C_3^2 \boxtimes C_4$ with distance antimagic labeling is given in Figure 16.2.

Here, the graph $C_3^t \boxtimes C_4$ is distance antimagic but it is not distance magic. Since for this graph, $|N(p_1^0) \cap N(q_1^0)| = d(p_1^0) - 1 = d(q_1^0) - 1$, so by Lemma 16.2 it is not a distance magic graph.

16.5 CORONA PRODUCT OF C_4 AND C_3^t

Theorem 16.5

The graph $C_4 \odot C_3^t$ is distance antimagic. ■

Proof. Let $G = C_4 \odot {C_3}^t$ and vertex set $V(G) = \{p_k^j, q_k^j, r^j, c_j | 1 \leq k \leq t, 0 \leq j \leq 3\}$ where p_k^j, q_k^j, r^j $(1 \leq k \leq t)$ are the vertices of the $(j+1)^{th}$ copy of ${C_3}^t$ which are adjacent to vertex c_j of C_4 for $j = 0, 1, 2, 3$.
We define a vertex labeling $f : V(G) \to \{1, 2, \ldots, |V(G)|\}$ as follows:

For, $1 \leq k \leq t$ and $j \equiv 0 \ (mod\ 4)$,

$$f(p_k^j) = j + 8k - 3;$$
$$f(q_k^j) = j + 8k + 1;$$

$$f(r^j) = 8t - j + 8;$$
$$f(c_j) = 4 - j$$

Under the above labeling, weights for each vertex are as follows:

For $1 \leq k \leq t$ and $j \equiv 0 \ (mod\ 4)$,

$$w(p_k^j) = 8t + 8k - j + 13;$$
$$w(q_k^j) = 8t + 8k - j + 9;$$
$$w(r^j) = 8t^2 + 6t + 2tj - j + 4;$$
$$w(c_j) = 8t^2 + 14t + 2tj - (j+1) - j - (j-1) + 16$$

One can easily verify that, all weights are distinct.
Hence, the graph $C_4 \odot C_3^t$ is a distance antimagic graph. □

Illustration 16.5.1. The distance antimagic labeling for $C_4 \odot C_3^2$ is given in Figure 16.3.

The corona product of C_4 and C_3^t is a distance antimagic graph but it is not a distance magic graph by Lemma 16.2, since $|N(p_1^0) \cap N(q_1^0)| = d(p_1^0) - 1 = d(q_1^0) - 1$.

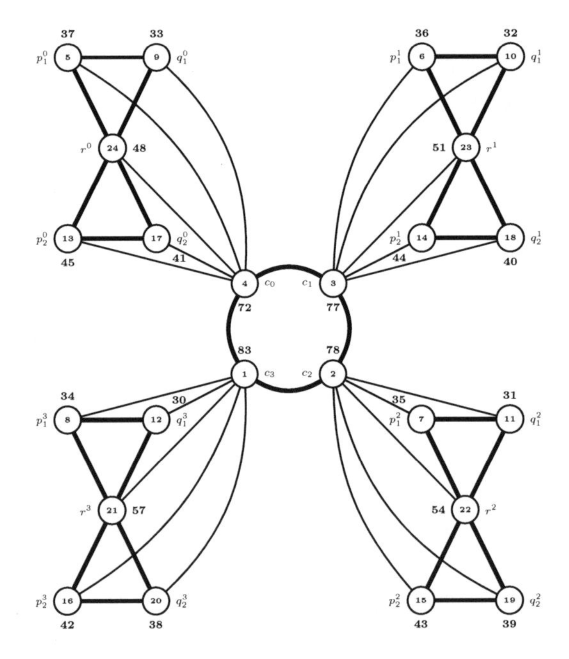

Figure 16.3: The distance antimagic labeling of $C_4 \odot C_3^2$

16.6 CONCLUDING REMARK

Here, we have investigated the existence of distance magic labeling and distance antimagic labeling of certain products of the friendship graph and cycle C_4. To explore some new distance magic and distance antimagic graphs is an open problem.

REFERENCES

1. B. Acharya, S. Rao, T. Singh and V. Parameswaran. Neighbourhood magic graphs. *National Conference on Graph Theory, Combinatorics and Algorithm*, 2004.
2. M. Anholcer and S. Cichacz. Note on distance magic products $G \circ C_4$. *Graphs and Combinatorics*, 31, 1117-1124, 2015.
3. D. M. Burton. *Elementary Number Theory.* Tata McGraw-Hill, 2007.
4. J. A. Gallian. A dynamic survey of graph labeling. *The Electronics Journal of Combinatorics*, # DS6, 2018.
5. J. Gross and J. Yellen. *Graph Theory and its Applications.* CRC Press, 2005.
6. R. Hammack, W. Imrich and S. Klavžar. *Handbook of Product Graphs*, CRC Press, Boca Raton, FL. 2011.
7. M. Miller, C. Rodger and R. Simanjuntak. Distance magic labeling of graphs. *Australas. J. Combin.*, 28, 305-315, 2003.
8. R. Simanjuntak and K. Wijaya. On distance antimagic graphs. arXiv:1312.7405v1[math.CO], 2013.
9. K. Sugeng, D. Fronček, M. Miller, J. Ryan and J. Walker. On distance magic labeling of graphs. *Journal of Combinatorial Mathematics and Combinatorial Computing*, 71, 39-48, 2009.
10. V. Vilfred. Σ-labelled Graphs and Circulant Graphs. Ph.D. Thesis, University of Kerala, Trivandrum, India 1994.

17 Graphs from Subgraphs

Joseph Varghese Kureethara
Department of Mathematics,
CHRIST (Deemed to be University),
Bengaluru (INDIA)
E-mail: frjoseph@christuniversity.in

Johan Kok
City of Tshwane,
South Africa
E-mail: jacotype@gmail.com

A graph is a pictorial representation of objects and their relations. Algebraically, it is represented in a minimum way as an ordered pair of two sets viz., set of vertices, representing objects and the set of edges, representing the relations. For a simple graph $G = (V, E)$, V denotes the set of vertices and E denotes the set of edges (two element sets of V). Once we have a graph G, we can generate other graphs from it. One method of creating new graphs is by the removal of some vertices or by the removal of some edges or by both. The graph thus obtained is known as a subgraph of G.

If we remove some vertices from the existing graph, the resulting subgraph is known as an induced subgraph. If we randomly remove some edges, the resultant subgraph need not be an induced subgraph. Hence, a subgraph need not be an induced subgraph but all induced subgraphs are subgraphs. Induced subgraphs are studied extensively in the area of structural studies of graphs and networks. An excellent study on induced subgraphs is done with respect to the study of perfect graphs. A perfect graph is a graph such that for every induced subgraph of it, the clique number[1] equals the chromatic number[2].

In the case of removal of a vertex and reinstating the same vertices, the original graph might not be obtained. i. e., $(G - v) + v$ need not be isomorphic to G. However, if the given graph is a regular graph, it is likely that $(G - v) + v = G$. A general question that can be asked is: for a subgraph $S \subset G$, is it possible to have $(G - S) + S = G$? It is not easy to answer affirmatively. However, if we know some nature of the original graph, and if we know some

[1]A clique of a graph G is a set X of vertices of G with the property that every pair of distinct vertices in X are adjacent in G. A maximal clique of a graph G is a clique X of vertices of G, such that there is no clique Y of vertices of G that contains all of X and at least one other vertex. Clique number of a graph is the maximum number of vertices in the maximal cliques of the graph.

[2]Chormatic number is the least number of colours that can be assigned to the vertices of a graph such that no two adjacent vertices receive the same colour.

hint on what is removed, we can reinstate the original graph by undoing the removals. Regular graphs form a large family of graphs where the above kind of revertibility is possible. Among them are complete graphs, generalized Petersen graphs, fullerine graphs, cycles, etc.

17.1 UNARY OPERATIONS ON GRAPHS

Retaining all the vertices of a graph but removing all the edges and then joining all the pairs of vertices that are not adjacent by edges generates the complement of a graph. If G is a graph, its complement is denoted as $\overline{G}$. It is not difficult to see that $\overline{(\overline{G})} = G$. There are some graphs that are isomorphic to their complement, called self-complementary graphs.

Power graphs are graphs that are obtained from a graph. Let G be a graph. The graph G^n is the nth power of the graph G with the following rule. The vertex set of G^n is that of G itself. Every pair of vertices that are at a distance 2 to n in G are made adjacent in G^n. The diameter of a graph, d, is the maximum of the set of distances between every pair of vertices in the graph. Hence, without much to sweat out, we can see that for a connected graph G, $G^d = K_{|V(G)|}$.

Theorem 17.1

For a finite connected graph G with p vertices, $G^k = K_p$. ■

17.2 BINARY OPERATIONS ON SUBGRAPHS

Given a graph G, through some binary operations of some of its subgraphs, we can get the original graph.

For example, $Q_n = Q_{n-1} \times K_2$, $K_{m,n} = \overline{K_m} + \overline{K_n}$, $K_{m+n} = K_m + K_n$.

Let us now see the consolidation of some general ideas regarding the construction of a graph from its subgraphs.

Definition. (Intersection Graph)[36] Let $\mathfrak{C}$ be a collection of subgraphs of G. Then, $G(\mathfrak{C})$ is a graph with vertex set $\mathfrak{C}$. The vertices S and T are adjacent if and only if $S \neq T$ and $S \cap T \neq$ ‰.

The following theorem is of great importance.

Theorem 17.2

[36] All graphs are intersection graphs. ■

Definition. (Intersection Adjacency) Let S and T be the subgraphs of a graph G. Then S and T are said to be adjacent if there exists a subgraph $H \subset G$ such that $S \neq T$ and $S \cap T = H$.

All possibilities of H can be explored. Let us see some of them.

1. $H \subset V(G)$ (Type 1)
2. $H \subset E(G)$ (Type 2)
3. $H \subset V(G)$ or $H \subset E(G)$ (Type 3)
4. $H \subset G$ (Type 4) where
    a. H is some subgraph other than K_1 or K_2
    b. H is an induced subgraph
5. $H = ‰$ (Type 5)

The conventional vertex-vertex adjacency using an edge also can be extended to sugraph-subgraph adjacency.

Definition. (Edge Adjacency (Type 6)) Let S and T be the subgraphs of a graph G. Then S and T are said to be adjacent if there exists $u \in V(S)$ and $v \in V(T)$ such that $uv \in E(G)$.

17.3 GRAPHS FROM SUBGRAPHS

Prisner brought out a monograph titled ***Graph Dynamics*** in 1995 and gave a detailed description of the concepts that came under the topics 'graph operators' and 'graph-valued functions'.[34] The monograph by McKee and McMorris in 1999 titled, ***Topics in Intersection Graph Theory*** is also an important work in the consolidation of the ideas of constructing new graphs from old graphs.[30]

The general idea of constructing a graph from the subgraph of a graph is as follows:

Definition. Let $\mathfrak{C}$ be a collection of subgraphs of G. Then, $G(\mathfrak{C})$ is a graph with vertex set $\mathfrak{C}$. The vertices S and T are adjacent in $G(\mathfrak{C})$ if and only if they are adjacent in G with respect to the six types of adjacencies specified in the definitions Type 1 and Type 6.

We can see that most of the existing derived graphs in the literature come under this definition. The rule of generating a graph from its subgraphs is known as a graph-valued function as it is a function defined on the collection of all graphs $\mathfrak{G}$.

17.3.1 GRAPHS FROM TYPE 1 ADJACENCY

Two famous graphs from the Type 1 adjacency are line graphs and total graphs.

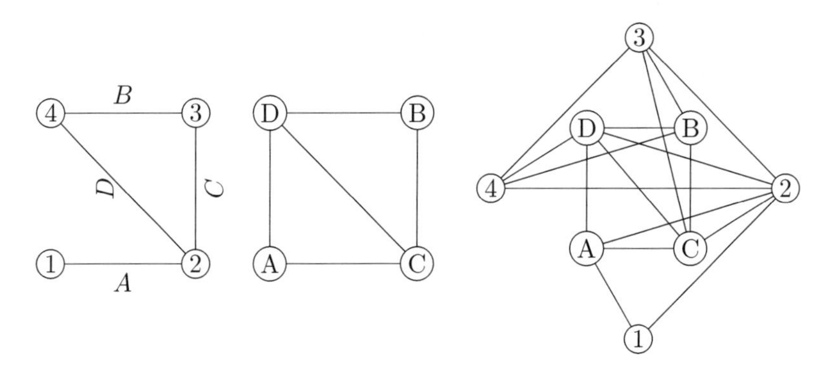

Figure 17.1: Paw, L(Paw) and T(Paw)

Example (Line Graph). [39] Let G be a graph and let $\mathfrak{C}$ be $E(G)$. i. e., the set of one subset of $E(G)$. Then $G(\mathfrak{C})$ is $L(G)$, the **line graph** of G (See Figure 17.1).

Example (Total Graph). [9] Let G be a graph and let $\mathfrak{C}$ be $V(G) \cup E(G)$. i. e., the set of one subset of $V(G)$ and $E(G)$. Then $G(\mathfrak{C})$ is $T(G)$, the **total graph** of G (See Figure 17.1).

Example (Entire Graph). Let G be a graph and let $\mathfrak{C}$ be $V(G) \cup E(G) \cup \mathbf{C}$ where $\mathbf{C}$ is the collection of faces in G. i. e., the set of one subset of $V(G)$ and $E(G)$ and the collection of faces in G. Then $G(\mathfrak{C})$ is $e(G)$, the **entire graph** of G.

The idea of entire graph is believed to be originated from the problem of entire colouring (vertices, edges, faces) of a planar graph.[20]

Example (k-overlap Clique Graph). [34] The k-overlap clique graph of a graph G has all maximal cliques of G as vertices. Two vertices are adjacent if their intersection contains at most k vertices.

Definition (Block Graph). [19] Let G be a graph and let $\mathfrak{B}$ be the collection of all blocks in G. Then the block graph $G(\mathfrak{B})$ has $\mathfrak{B}$ as the vertex set. Two vertices, i. e., the blocks B, $C \in \mathfrak{B}$ are adjacent if and only if B and C have a cut-vertex in common.

Definition (Full Graph). [23] The full graph $F(G)$ of a graph G is the graph whose vertex set is the union of the set of vertices, edges and blocks of G in which two vertices are adjacent if the corresponding members of G are adjacent or incident.

17.3.2 GRAPHS FROM TYPE 2 ADJACENCY

Example (Triangle Graph). [31] Let G be a graph and let $\mathfrak{C}$ be the set of C_3s of G. Then $G(\mathfrak{C})$ is the triangle graph of G with the vertex set $\mathfrak{C}$ and two vertices are adjacent of two triangles that share an edge.

Example (Cycle Graph). [34] Let G be a graph and let $\mathfrak{C}$ be the set of all induced cycles of G. Then $G(\mathfrak{C})$ is the cycle graph of G with vertex set $\mathfrak{C}$. Two vertices are adjacent, if the respective cycles share a common edge.

17.3.3 GRAPHS FROM TYPE 4 ADJACENCY

Example (k-Gallai Graph). [34] The k-Gallai graph of a graph G has the K_ks of G as their vertices. Two vertices are adjacent if the union of the respective two K_ks is a $K_{k+1} - e$.

Example (k-in-m Graph). [34] The k-in-m graph of a graph G has the K_ks of G as its vertices. Two vertices are adjacent if the respective K_ks lie in a K_r for some $r \leq m$.

Example (k-line Graph). [34] The k-line graph of a graph G has the K_ks of G as their vertices. Two vertices are adjacent if their intersection of the respective K_ks is a K_{k-1}.

Definition (Cut-vertex Graph). [19] Let G be a graph and let $\mathfrak{C}$ be the collection of all cut-vertices in G. Then the cut-vertex graph $G(\mathfrak{C})$ has $\mathfrak{C}$ as the vertex set. Two vertices, i. e., the cut-vertices B, $C \int \mathfrak{C}$ are adjacent if and only if B and C lie in a common block.

Non-adjacency also induces relation in the derived graphs. We require the defnition of null graph for defining jump graph.

Definition (Null Graph). Null graph, K_0 is the graph with $V(K_0) = ‰$ and $E(K_0) = ‰$. We consider it as a complete graph that is contained in every graph.

Example (Jump Graph). [13] Let G be a graph and let $\mathfrak{C}$ be $E(G)$. i. e., the set of one subset of $E(G)$. Then $G(\mathfrak{C})$ is $\overline{L(G)}$, the **jump graph** of G. i. e., two vertices in $G(\mathfrak{C})$ are adjacent if and only if the intersection of corresponding edges in G is the null graph, K_0.

17.3.4 GRAPHS FROM TYPE 6 ADJACENCY

The most trivial example for Type 6 adjacency is G itself.

Example. Let G be a graph and let $\mathfrak{C}$ be $V(G)$. i. e., the set of one subset of $V(G)$. Then $G(\mathfrak{C})$ is G itself.

Super line graph introduced by Bagga *et al.* is a generalised version of line graph.

Example. (Super Line Graph)[4] Let G be a graph and let $\mathfrak{C}_k$ be $E(G)$. i. e., the set of k subsets of $E(G)$. Then $G(\mathfrak{C}_k)$ is $\mathfrak{L}_k(G)$, the **super line graph** of G (See Figure 17.2).

When $k = 1$, we have the line graph. It is the Type 1 adjacency. When $k > 1$, Type 6 adjacency is applied.

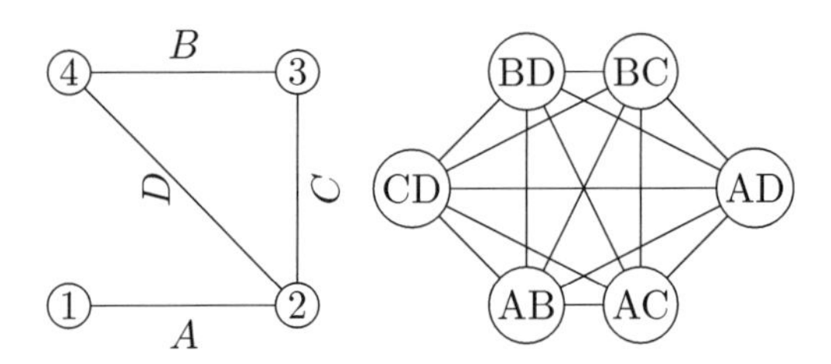

Figure 17.2: *Paw*, $\mathfrak{L}_2(Paw)$

17.3.5 SUBGRAPHS OF THE LINE GRAPH

Example (Gallai Graph). The Gallai graph of a graph G has the edges of G as their vertices. Two edges of G are adjacent in the Gallai graph of G if they are incident but do not span a triangle in G.

Example (Anti-Gallai Graph). The anti-Gallai graph of a graph G has the edges of G as its vertices. Two edges of G are adjacent in the anti-Gallai graph of G if they span a triangle in G.

Although not a subgraph of a line graph, medial graphs are from the family of line graphs.

Definition (Medial Graph). Let G be a connected planar graph and let $\mathfrak{C}$ be $E(G)$. i. e., the set of one subset of $E(G)$. Two vertices are adjacent if their corresponding edges intersect (for each face of G in which their corresponding edges occur consecutively).

17.4 VARIATIONS IN THE ADJACENCY DEFINITION

17.4.1 UNION ADJACENCY

Example (Path Graph). [10] Let G be a graph and let $\Pi_k(G)$ be the set of all paths of G on k vertices. Then the path graph $P_k(G)$ has the vertex set $\Pi_k(G)$. Two vertices in $P_k(G)$ are adjacent if the union of the respective paths forms either a path on $k+1$ vertices or a cycle on k vertices in G.

It is not difficult to see that $P_1(G) = G$ and $P_2(G) = L(G)$.

17.4.2 CHORD CROSSING ADJACENCY

Definition (Circle Graph). In graph theory, a circle graph is the intersection graph of a set of chords of a circle. That is, it is an undirected graph whose vertices can be associated with chords of a circle such that two vertices are adjacent if and only if the corresponding chords cross each other.

17.4.3 INCIDENCE ADJACENCY

Example (Incidence Graph). [11] Let G be a graph and let $\mathfrak{C} = \{ve \mid v \in V(G), e \in E(G), e = uv$, for some u$\in V(G)\}$ The incidence graph $I(G)$ is with vertex set $\mathfrak{C}_{12}$. Two vertices ve and uf are adjacent in $\mathfrak{C}$ if $v = u$ or $e = f$ or $vu = e$ or $vu = f$.

17.4.4 COUNTING BASED ADJACENCY

We see two derived graphs based on number of elements in the vertex set. At first we see the derived graph defined by Bandelt and Vel.

Definition (Simplex Graph). [5] Let G be a graph and let $\mathfrak{K}$ be the collection of all complete graphs in G including the null graph, K_0. Then the Simplex graph $G(\mathfrak{K})$ has $\mathfrak{K}$ as the vertex set. Two vertices, i. e., the complete graphs K, $L \subset \mathfrak{K}$ are adjacent if and only if $||V(K)| - |V(L)|| = 1$.

Definition (Matching Graph). [15] Let G be a graph and let $\mathfrak{M}$ be the collection of all **maximum matchings** in G. Then the matching graph $G(\mathfrak{M})$ has the vertex set $\mathfrak{M}$. Two vertices, i. e., the maximum matchings M, $N \subset \mathfrak{M}$ are adjacent if and only if $||M| - |N|| = 1$.

17.4.5 DISTANCE BASED ADJACENCY

Definition (Antipodal Graph). The antipodal graph of a graph G denoted by $A(G)$, is the graph on the same vertices as of G. Two vertices in $A(G)$ are adjacent if and only if the distance between them in G is equal to the diameter of G.

Definition (Antipodal Middle Graph). Antipodal middle graph of a graph G is the graph on the same vertices as of the middle graph $M(G)$. Two vertices are adjacent if and only if the distance between them in G is equal to the diameter of $M(G)$.

17.5 TRANSFORMATION GRAPHS

A recent development in the study of derived graphs is the introduction of transformation graphs. They are in a sense the generalizations of line graphs, total graphs, middle graphs, jump graphs etc.

Definition (Associativity). Let G be a graph and let $\mathfrak{C}$ be the collection of all vertices and edges in G. Let A, B $\in \mathfrak{C}$. The associativity of A and B is $+$ if they are adjacent or incident in G. The associativity of A and B is $-$ if they are neither adjacent or nor incident in G.

Table 17.1

Derived Graphs with respect to the Type of Adjacency

Graph	Vertex Set	1	2	3	4	5	6
G	$V(G)$						✓
Anti-Gallai Graph	$E(G)$	✓					
Block Graph		✓					
Circle Graph			✓				
Clique Graph		✓					
Cut-vertex Graph		✓					
Cycle Graph			✓				
Entire Graph	$V(G) \cup E(G) \cup C$			✓			
Full Graph		✓					
Gallai Graph	$E(G)$	✓					
Incidence Graph	$E(G)$	✓					
Jump Graph	$E(G)$					✓	
k-Gallai Graph					✓		
k-in-m Graph							✓
k-line Graph					✓		
k-overlap Clique Graph		✓					
Line Graph	$E(G)$	✓					
Medial Graph		✓					
Middle Graph		✓					
Path Graph					✓		
Super Line Graph	$E(G)$						✓
Total Graph	$V(G) \cup E(G)$	✓					✓
Transformation Graph		✓			✓		
Triangle Graph			✓				

Let xyz be a 3-permutation of the set $\{+, -\}$. We say that A and B correspond to the first term x (resp. the second term y or the third term z) of xyz if both A and B are in (resp. both A and B are in $E(G)$, or one of A and B is in $V(G)$ and the other is in $E(G)$).

Definition. (The Transformation Graph)[40] Let G be a graph and let $\mathfrak{C}$ be the collection of all vertices and edges in G. The transformation graph G^{xyz} of G has the vertex set $\mathfrak{C}$. Two vertices are adjacent if and only if their **associativity** in G is consistent with the corresponding term of xyz.

Figures 17.3-17.6 give all eight transformation graphs of Paw.

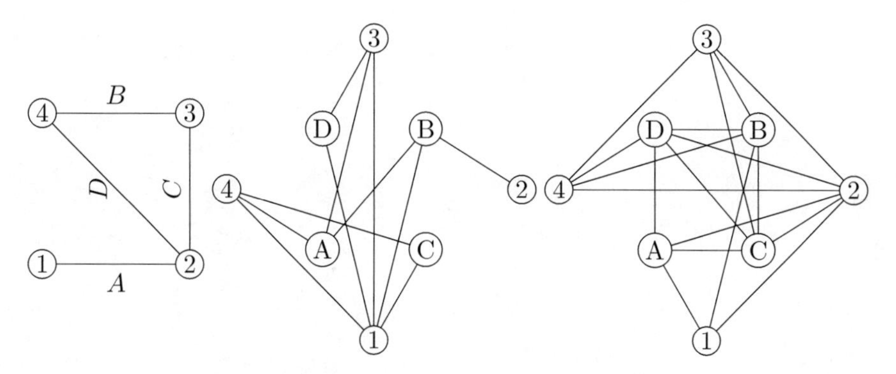

Figure 17.3: G=Paw, G^{---} and G^{+++}

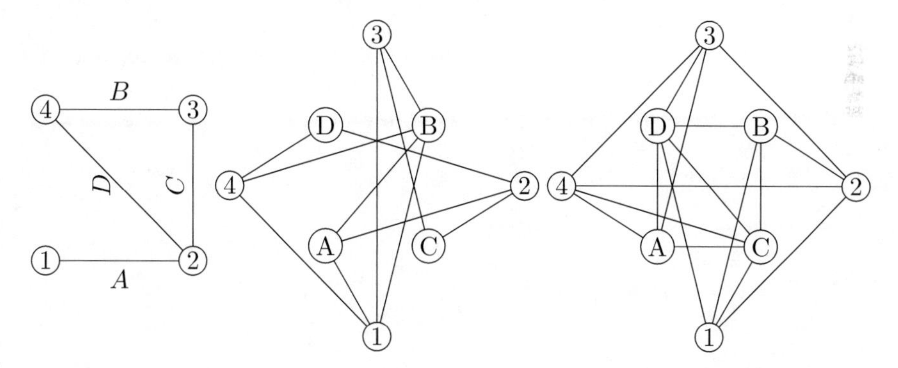

Figure 17.4: G=Paw, G^{+--} and G^{-++}

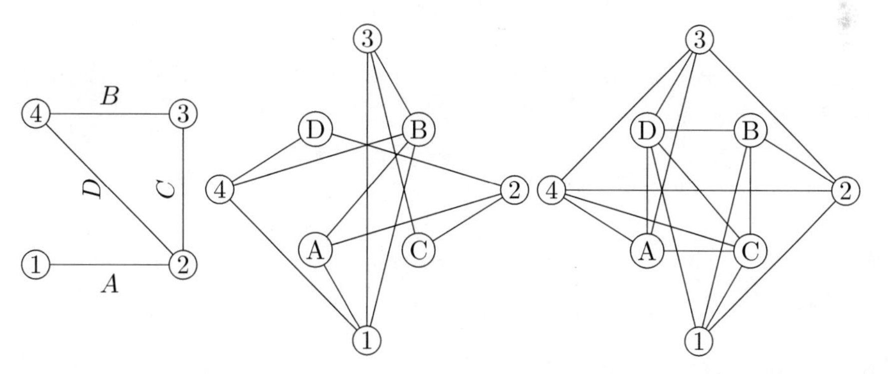

Figure 17.5: G=Paw, G^{--+} and G^{++-}

17.6 ITERATION, CONVERGENCE AND COMPLETION

Once we repeat the process of generating a new graph, we generate a sequence of iterated graphs.

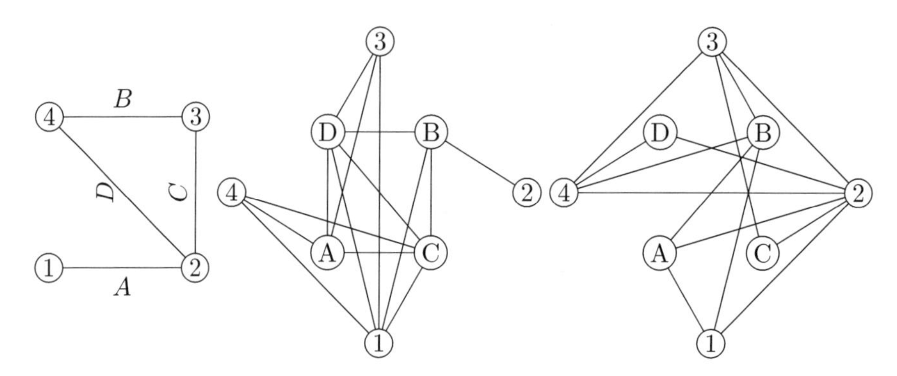

Figure 17.6: G=Paw, G^{-+-} and G^{+-+}

Definition. (Iterated graphs) Let G be a graph. Let H be the graph generated from the collection $\mathfrak{C}$ of subgraphs of G. i. e., $G(\mathfrak{C}) = H$. Then the iterated graphs with respect to $\mathfrak{C}$ are defined as follows:

$$H^i = \begin{cases} G, \; i = 0 \\ H(H^{i-1}), \; i > 0 \end{cases}$$

Question 1. What are the types of graphs that are formed at each level?

Question 2. Can we achieve a targeted graph?

Question 3. What are the forbidden graphs of each derived graph?

Question 4. When is the graph converging to itself?

Question 5. If G is a cycle graph then $L(G)$ and each subsequent graph in this sequence is isomorphic to G itself. Are these the only connected graphs for which $L(G)$ is isomorphic to G?

Question 6. If G is a claw $K_{1,3}$, then $L(G)$ and all subsequent graphs in the sequence are cycles. Are there similar situations?

Question 7. If G is a path graph, then each subsequent graph in the sequence is a shorter path until eventually the sequence terminates with a null graph or K_1. Are there similar situations?

17.7 RESEARCH PROBLEMS

We list here some open problems related to derived graphs.

Problem. Express a derived graph as another derived graph

Definition. Recall from [21] that, in a proper colouring of G all edges are good i. e., $uv \Leftrightarrow c(u) \neq c(v)$. For any proper colouring $\varphi(G)$ of a graph G the addition of all good edges, if any, is called the chromatic completion of

G in respect of $\varphi(G)$. The additional edges are called *chromatic completion edges.* The set of such chromatic completion edges is denoted by, $E_\varphi(G)$. The resultant graph G_φ is called a *chromatic completion graph* of G.

Definition. If a graph G is coloured with $\chi(G)$ colours, the colouring is called a *chromatic colouring* of G. The *chromatic completion number* of a graph G denoted by, $\zeta(G)$ is the maximum number of good edges that can be added to G over all chromatic colourings. Hence, $\zeta(G) = max\{|E_\chi(G)| : over\ all\ \varphi_\chi(G)\}$. This specific chromatic completion graph is denoted by, G_ζ.

Definition. For $G = (V(G), E(G))$ and $\varphi_\chi(G)$ a new derived graph can be defined i.e., $G_\zeta = (V(G_\zeta), E(G_\zeta))$, $V(G_\zeta) = \{v_{c_i} : v_{c_i}$ iff $c(v) = c_i, v \in V(G)\}$ and $E(G_\zeta) = \{u_{c_i}v_{c_j} : uv \in E(G)\} \cup \{u_{c_i}v_{c_j} : c(u) = c_i \neq c_j = c(v)\}$.

Problem. Characterize graphs G for which $\varepsilon(G) = \zeta(G)$.

Problem. Characterize graphs G, other than self-complementary graphs, for which $\varepsilon(G) = \zeta(\overline{G})$.

Problem. Characterize graphs G, other than self-complementary graphs, for which $\zeta(G) = \zeta(\overline{G})$.

REFERENCES

1. Alhulwah Khawlah Hamad. *Structures of Derived Graphs.* PhD Disertation, Western Miccingan University, 2017.
2. Alhulwah Khawlah Hamad, L. Garry Johns and Ping Zhang. On the connectedness of 3-line graphs. *Australasian J. Combinatorics*, 72, 113-127, 2018.
3. Jay Bagga. Old and new generalizations of line graphs. *International Journal of Mathematics and Mathematical Sciences*, 29, 1509-1521, 2004.
4. K. S. Bagga, L. W. Beineke and B. N. Varma. The line completion number of a graph. *Graph Theory, Combinatorics and Applications*, 2, 1197-1201, 1995.
5. H. J. Bandelt, M. van de Vel. Embedding Topological Median Algebras in Products of Dendrons. *Proceedings of the London Mathematical Society*, S3-58(3), 439-453, 1989, https://doi.org/10.1112/plms/s3-58.3.439.
6. B. Basavanagoud and V. R. Desai. On the total line-cut transformation graphs. *Journal of Computer and Mathematical Sciences*, 6(7): 371-387, 2015.
7. B. Basavanagoud and Veena Mathad. Graph equations for line graphs, jump graphs, middle graphs, splitting graphs and line splitting graphs.*Mapana-Journal of Sciences* 9(2): 53-61, 2010.
8. L. W. Beineke. Derived graphs and digraphs. *Beitriige zur Graphentheorie, Teubner, Leipzig*, 17-33, 1968.
9. Behzad and Mehdi. A criterion for the planarity of the total graph of a graph. *Proceedings of the Cambridge Philosophical Society*, 63(3), 679-681, 1967, 10.1017/S0305004100041657.
10. J. H. Broersma and Cornelis Hoede. Path graphs. *Journal of Graph Theory*, 13(4): 427-444, 1989.
11. A. Richard Brualdi and J. Jennifer Quinn Massey. Incidence and strong edge colorings of graphs. *Discrete Mathematics*, 122(1-3): 51-58, 1993.

12. Gary Chartrand. On hamiltonian line-graphs. *Transactions of the American Mathematical Society*, 134(3): 559-566, 1968.
13. Gary Chartrand, Hèctor Hevia, Elzbieta B. Jarrett and Michelle Schultz. Subgraph distances in graphs defined by edge transfers. *Discrete Mathematics*, 170(1-3): 63-79, 1997.
14. Cvetkoviĉ Dragoš, Michael Doob and Slobodan Simic. Generalized line graphs. *Journal of Graph Theory*, 5(4): 385-399, 1981.
15. Eroh Linda, and Michelle Schultz. Matching graphs. *Journal of Graph Theory*, 29(2): 73-86, 1998.
16. T. S. Evans and Lambiotte Renaud. Line graphs, link partitions, and overlapping communities, *Physical Review E*, 80(1): 016105, 2009.
17. Venkanagouda M. Goudar and G. Nagendrappa. Cutvertex jump graph.*Publications of Problems & Applications in Engineering Research*, 2(4): 129-133, 2011.
18. T. Hamada and I. Yoshimura. Traversability and connectivity of the middle graph of a graph. *Discrete Mathematics*, 14, 247-255, 1976.
19. F. Harary. A characterization of block-graphs. *Canadian Mathematical Bulletin*, 6(1), 1-6, 1963, doi:10.4153/CMB-1963-001-x.
20. H. Izbicki, *Verallgemeinerte Farbenzahlen, Beitràge zur Graphentheorie*, H. Sachs, H. Voss and H. Walther (Editors), Teubner, Leipzig, 81-84, 1968.
21. J. Kok, Chromatic Completion Number, *arXiv:1809.01136v2*, Communicated.
22. V. R. Kulli. The semitotal block graph and the total-block graph of a graph. *Indian J. Pure Appl. Math.*, 7, 625-630, 1976.
23. V. R. Kulli. On full graphs.*Journal of Computer and Mathematical Sciences*, 6(5): 261-267, 2015.
24. V. R. Kulli and M. S. Biradar. The middle blict graph of a graph. *International Research Journal of Pure Algebra*, 5(7): 111-117, 2015.
25. Van Bang Lê. Perfect k-line graphs and k-total graphs. *Journal of Graph Theory*, 17(1): 65-73, 1993.
26. Xue-Liang Li and Yan Liu. Path graphs versus line graphs-a survey. *Chinese Journal of Engineering Mathematics*, 24(5): 761-787, 2007.
27. Pralahad Mahagaonkar. A generalized review on line graphs in graph theory. *Asian Journal of Current Engineering and Maths*, 5-6, 2017.
28. Tabitha Agnes Mangam and Kureethara Joseph Varghese. Diametral paths in total graphs of complete graphs, complete bipartite graphs and wheels. *International Journal of Civil Engineering and Technology*, 8(5): 1212-1219, 2018.
29. Mathad, Veena and Narayankar, P. Kishori. On lict sigraphs. *Transactions on Combinatorics* 3(4): 11-18, 2014.
30. A. Terry McKee and F. R. McMorris. Topics in Intersection Graph Theory.*SIAM*, Philadelphia, 1999.
31. S. D. Monson, N. J. Pullman and R. Rees. A survey of clique and biclique coverings and factorizations of (0,1)-matrices.*Bull. Inst. Combin. Appl.*, 14: 17-86, 1995.
32. H. P. Patil and R. P. Raj. Solutions of graph equations involving line, middle and mycielski graphs.*Mapana-Journal of Sciences*, 12(3): 17-22, 2013.
33. Erich Prisner. A common generalization of line graphs and clique graphs. *Journal of Graph Theory*, 18(3): 301-313, 1994.
34. Erich Prisner. *Graph Dynamics*. Longman, England, 1995.

35. P. Reddy, Kota Siva, S. Kavita Permi and B. Prashanth. A note on line graphs. *Mathematical Combinatorics*, 1, 119-122, 2011.
36. E. Sur deux Szpilrajn-Marczewski. propriétés des classes d'ensembles, *Fund. Math.*, 33, 303-307, 1945. (Trans. Bronwyn Burlingham. "A Translation of Sur deux propriétés des classes d'ensembles," University of Alberta, 2009).
37. Jijo Thomas, and Joseph Varghese. On the decomposition of total graphs. *Advanced Modelling and Optimization*, 15(1): 81-84, 2013.
38. Zsolt Tuza. Perfect triangle families. *Bulletin of the London Mathematical Society*, 26(4): 321-324, 1994.
39. Hassler Whitney. Congruent graphs and the connectivity of graphs. *American Journal of Mathematics*, 54(1): 68-150, 1932, doi:10.2307/2371086.
40. Wu Baoyindureng, Zhang Li and Zhang Zhao. The transformation graph G^{xyz} when xyz=-++. *Discrete Mathematics*, 296(2): 263-270, 2005, https://doi.org/10.1016/j.disc.2005.04.002.
41. Tudor Zamfirescu. On the line-connectivity of line-graphs. *Mathematische Annalen*, 187(4): 305-309, 1970.

18 Unit Graphs having their Domination Number half their Order

Amit Kumar
Banasthali Vidyapith,
Banasthali, Rajasthan (INDIA)
E-mail: amitsu48@gmail.com

Pranjali
University of Rajasthan,
Jaipur, Rajasthan (INDIA)
E-mail: pranjali48@gmail.com

Mukti Acharya
Christ University,
Bengaluru (INDIA)
E-mail: pranjali48@gmail.com

Pooja Sharma
Banasthali vidyapith,
Banasthali, Rajasthan (INDIA)
E-mail: sapooja1984@gmail.com

In this chapter, we characterize the rings for which the extrema in the inequality $1 \leq \gamma(G(R)) \leq n$ holds, where γ is the domination number of unit graph $G(R)$. In the sequel, we determine the rings, whose unit graph has unique minimum dominating set. Further, we have considered the problem of characterizing the rings for which $G(R)$ has either domination number $\gamma(G(R)) = \frac{|R|}{2}$ or $\frac{|R|-1}{2}$. Furthermore, we determine the domination number of the line graph of unit graphs.

Oystein Ore, a well known lattice theorist, introduced the concept of domination in graphs in his famous book 'Theory of Graphs' published in 1962 and this concept lived almost in hybernation until 1975, when E. J. Cockayne and S. T. Hedetniemi unfolded its diverse aspects, by surveying all the available results, bringing to light new ideas and citing its application potential in a variety of scientific areas in their paper 'Towards a Theory of Domination in Graphs' which appeared later on in 'Networks' in 1977. Prior to their joint venture, it was only Claude Berge's pioneering book 'Graphs and Hypergraphs' in 1973 which included a bound on the domination number of a graph, Vizing's bound on the size of a graph with given order and domination number and very interesting applications of the idea of domination to surveillance networks and to Game Theory. Before it was shaped to become a theory in itself, it had acquired several disguised forms under the heading 'graph coverings' in a number of research papers.

Today, Ore's concept of domination in graphs has indeed become an independent theory. Our purpose in this chapter is to characterize the rings for which the extrema in the inequality $1 \leq \gamma(G(R)) \leq n$ holds, where γ is the domination number of unit graph $G(R)$. In the sequel, we determine the rings, whose unit graph has unique minimum dominating set. Further, we have considered the problem of characterizing the rings for which $G(R)$ has either domination number $\gamma(G(R)) = \frac{|R|}{2}$ or $\frac{|R|-1}{2}$.

18.1 UNIT GRAPH

Let $\mathbb{Z}_n$ be the ring of integers modulo n, where n is a positive integer. According to [4], the *unit graph* of $\mathbb{Z}_n$ to be the graph $G(\mathbb{Z}_n)$ with vertices $\mathbb{Z}_n$, and two distinct vertices x and y are adjacent if and only if $x+y$ is a unit of $\mathbb{Z}_n$. Thus for a positive integer n, it follows that $G(\mathbb{Z}_{2n})$ is a $\phi(2n)$-regular graph, where ϕ is the Euler phi function, and in case $n=2$, $G(\mathbb{Z}_{2n})$ can be expressed as the union of $\phi(2n)/2$ Hamiltonian cycles (see [1]). This investigation was then continued by Ashrafi et al. [1] and they were interested in generalizing the unit graph $G(\mathbb{Z}_n)$ to $G(R)$ for an arbitrary associative ring R. In the course of the investigation, they found that $G(R)$ is complete bipartite if and only if R is a local ring, and for finite commutative ring R, $G(R)$ is a cycle if and only if R is isomorphic to $\mathbb{Z}_4$ or $\mathbb{Z}_2[X]/\langle X^2\rangle$, or $\mathbb{Z}_6$ ([1, Theorem 3.4]. Some examples of unit graph over finite commutative rings are shown in Figures 18.1, 18.2 and 18.3. The following interesting result given in [1] provides some insight into the structure of unit graphs, and is useful in this chapter.

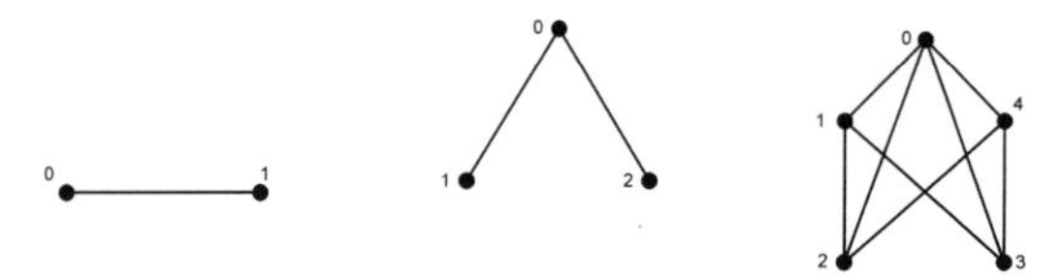

Figure 18.1: The unit graphs for the rings $\mathbb{Z}_2, \mathbb{Z}_3$ and $\mathbb{Z}_5$, respectively

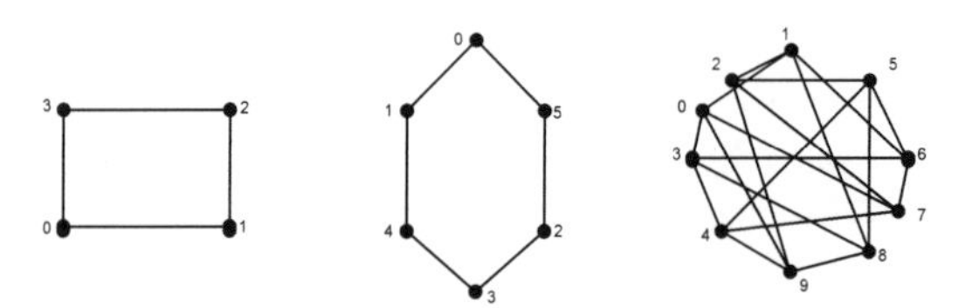

Figure 18.2: The unit graphs for the rings $\mathbb{Z}_4, \mathbb{Z}_6$ and $\mathbb{Z}_{10}$, respectively

Lemma 18.1

Let R be a finite commutative ring. Then for $G(R)$ the following statements hold:

(i) If $2 \notin U(R)$, then $G(R)$ is $|U(R)|$-regular;
(ii) If $2 \in U(R)$, then $G(R)$ is $(|U(R)|, |U(R)| - 1)$-semi regular.

■

Nowadays, there is an enormous amount of literature built upon several parameters of domination in graphs. It would be too unwieldy to cite all references, but a few pioneering papers like [1, 8] are worth comprehension. We shall also point out some common features of unit graphs whenever found prominent or pertinent as we proceed with this article. Our focus however will be on the notion of domination, which was introduced by a well-known lattice theorist, Ore [9]. Ore, introduced the notion of *domination* in graphs in his famous book '*Theory of Graphs*' [9], and this concept lived almost in hybernation until 1975, when Cockayne and Hedetniemi unfolded its diverse aspects, by surveying all the available results, bringing to light new ideas and citing its application potential in a variety of scientific areas in their paper [2]. At present, there are a plethora of different notions of domination, its motivations, generalizations and variations in graph theory. This is mainly due to its vast applications in other areas, especially in the study of *vulnerability* of a variety of important networks, especially communication, transmission and transportation networks. For more details on domination theory the reader is referred to Haynes et al. [6]. According to Ore; a nonempty subset D of $V := V(G)$ is called a *dominating set* if every vertex in $V - D$ is adjacent to at least one vertex in D. The *domination number* of a graph G, denoted by γ, is defined to be the minimum cardinality of a dominating set in G. Since the whole vertex set of $G(R)$ itself is a dominating set, the following inequality holds:

$$1 \leq \gamma(G(R)) \leq n \tag{18.1}$$

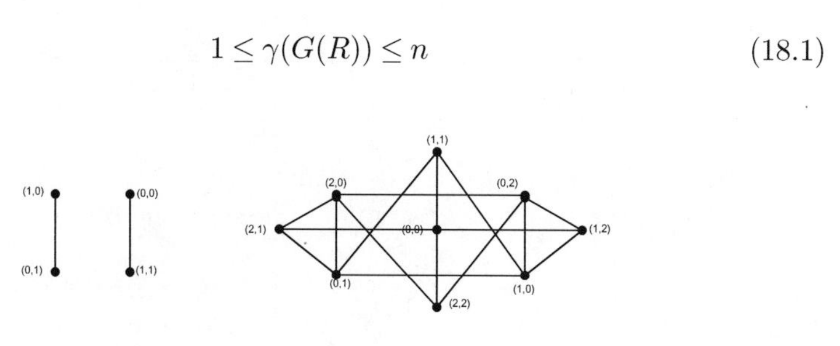

Figure 18.3: The unit graphs for the rings $\mathbb{Z}_2 \times \mathbb{Z}_2$ and $\mathbb{Z}_3 \times \mathbb{Z}_3$, respectively

and the bounds in the above inequality are rough. Due to the Inequality (18.1), the following problems of immediate interest would then arise :

Problem 1. Characterize the commutative rings for which $G(R)$ attain the bounds in the Inequality (18.1).

Problem 2. Determine the commutative rings for the which unit graphs have unique dominating set.

With this motivation, in this chapter, we give several basic results about domination in the unit graph to come closer to understanding the concept and characterize the rings whose $G(R)$ has domination number equal to one. We also show in Theorem 18.1 that if R is field of characteristic p, $p > 2$, then D is the unique minimum dominating set. In Theorem 18.5 and Corollary 18.1.1, we characterize the rings whose $G(R)$ has $\gamma(G(R)) = \frac{|R|}{2}$, and several new directions for further research are also indicated by raising questions.

We will consider only simple graphs, i.e., undirected graphs in which any two vertices are joined by at most one edge and without loops. Throughout, R will be a commutative ring with $1 \neq 0$. As usual, $\mathbb{Z}_n$, and $\mathbb{F}_q$ will denote the ring of integers modulo n, and a finite field with q elements, respectively; and $U(n)$ will be the set of units of $\mathbb{Z}_n$, and the vertices v_0, v_1 mean vertices of $G(R)$ are labeled with elements '0', '1' of R, respectively. To avoid trivialities, we implicitly assume when necessary that graphs are nonempty and rings are finite. Also, K_n will denote the complete graph on n vertices, and $H \circ K_1$ will denote the corona of any connected graph H with K_1. Because terminology and notation from ring theory and graph theory not defined in this chapter, we refer the reader to [7] and [5, 6], respectively.

18.1.1 RINGS R FOR WHICH $\gamma(G(R)) = \frac{|R|}{2}$

In this section we will answer Problem 1, and 2, and characterize the rings R for which $\gamma(G(R)) = \frac{|R|}{2}$. Towards solving Problem 1, first, we shall determine rough bounds for the Inequality (18.1), and then proceed for the sharp bounds by raising the problems.

Theorem 18.1

$\gamma(G(R)) = 1$ if and only if R is a field. Moreover, if R is a field of characteristic p, $p > 2$, then D is the unique minimum dominating set. ■

Proof. Necessity: Suppose $\gamma(G(R)) = 1$ (i.e., there exists at least one vertex in $G(R)$ whose degree is $|R| - 1$). We shall prove the result by contradiction. To do this, let R be not a field. Then there exists at least one element $0 \neq a \in R$ such that a^{-1} does not exist (i.e., $a \notin U(R)$). This gives that $|U(R)|$ is at least $|R| - 2$. But, in view of [1, Proposition 2.4], it is clear that the unit graph is either regular or semi-regular (according as n is even or odd, respectively). Therefore, due to the assumption that there exists a vertex v such that $deg(v) = |R| - 1$, this implies that $|U(R)|$ should be equal to $|R| - 1$, which is a contradiction to the assumption. Hence R must be a field.

Sufficiency: In order to prove $\gamma(G(R)) = 1$, it is enough to show that there exists at least one vertex $v \in V(G(R))$, which is adjacent to every vertex in $G(R)$. If R is a field, then each nonzero element is a unit (i.e., $|U(R)| = |R| - 1$). Clearly the vertex v_0 (with label '0') is adjacent to each unit of R in $G(R)$, this gives $deg(v_0) = |U(R)| = |R| - 1$. Therefore, $\gamma(G(R)) = 1$.

For the "moreover" statement, let R be a field with characteristic p, $p > 2$ and D is not the unique minimum dominating set. Then there exist at least two vertices of full degree (as $\gamma(G(R)) = 1$), but this is not possible as except the vertex v_0 other vertices are adjacent with at most $|U(R)| - 1$ vertices, a contradiction. Thus $D = \{v_0\}$ is the unique dominating set. Hence the proof. □

Remark. For $n > 1$, there does not exist a ring R with $|R| = n$, whose $G(R)$ have $\gamma(G(R)) = n$. Equivalently, no unit graph of order $n > 1$, can attain the upper bound in the Inequality (18.1). (cf. Lemma 18.1)

The above results lead us to consider the following two important fundamental problems for investigation.

Problem 3. Characterize the commutative rings for which

$$\gamma(G(R)) = \frac{|R|}{2}. \tag{18.2}$$

Problem 4. Characterize the commutative rings for which

$$\gamma(G(R)) = \frac{|R| - 1}{2}. \tag{18.3}$$

In order to answer Problem 3, first we establish the following results.

Lemma 18.2

Let $R \cong \underbrace{\mathbb{Z}_2 \times \mathbb{Z}_2 \times \mathbb{Z}_2 \times \cdots \times \mathbb{Z}_2}_{(t-times)} \times S$, $t \geq 0$. Then $G(R)$ has 2^t copies of $G(S)$, where S is isomorphic to any one of the rings $\mathbb{Z}_2$, $\mathbb{Z}_4$, $\mathbb{Z}_2[X]/\langle X^2 \rangle$, and $\mathbb{Z}_6$. ■

Proof. First, let $S \cong \mathbb{Z}_2$ and $t \geq 0$. If $t = 0$, then $G(\mathbb{Z}_2)$ is shown in Figure 18.1. If $t = 1$, then $R \cong \mathbb{Z}_2 \times S$ and $G(\mathbb{Z}_2 \times \mathbb{Z}_2) \cong 2K_2$ (cf., Figure 18.3). For $t = 2$, $R \cong \mathbb{Z}_2 \times \mathbb{Z}_2 \times \mathbb{Z}_2$, and then by [1, p. 20], it follows that $G(\mathbb{Z}_2 \times \mathbb{Z}_2 \times \mathbb{Z}_2) \cong 2(2G(\mathbb{Z}_2 \times \mathbb{Z}_2))$. Repeating the process t-times, for $t \geq 0$, we get $G(\underbrace{\mathbb{Z}_2 \times \mathbb{Z}_2 \times \mathbb{Z}_2 \times \cdots \times \mathbb{Z}_2}_{(t-times)} \times \mathbb{Z}_2)$ is isomorphic to 2^t copies of $G(\mathbb{Z}_2)$.

Next, consider $S \cong \mathbb{Z}_4$. By arguments analogous to those used in the previous case, one can deduce that if $t = 0$, then $G(\mathbb{Z}_4)$ is shown in Figure 18.2; and if $t = 1$, then $R \cong \mathbb{Z}_2 \times S$ and $G(\mathbb{Z}_2 \times \mathbb{Z}_4)$ have two disjoint four cycles, precisely $(0,0)-(1,1)-(0,2)-(1,3)-(0,0)$ and $(1,2)-(0,3)-(1,0)-(0,1)-(1,2)$, consequently, we get 2 copies of $G(\mathbb{Z}_4)$. Due to the result [1, p. 20], for $t = 2$, $G(\mathbb{Z}_2 \times \mathbb{Z}_2 \times \mathbb{Z}_4)$ has 2 copies of $G(\mathbb{Z}_2 \times \mathbb{Z}_4)$, and hence, $G(\mathbb{Z}_2 \times \mathbb{Z}_2 \times \mathbb{Z}_4)$ has 2^2 copies of $G(\mathbb{Z}_4)$. Continuing this process t-times, it is easily deduced that $G(R)$ have 2^t copies of $G(\mathbb{Z}_4)$ for their associated ring $R \cong \underbrace{\mathbb{Z}_2 \times \mathbb{Z}_2 \times \mathbb{Z}_2 \times \cdots \times \mathbb{Z}_2}_{(t-times)} \times \mathbb{Z}_4,\ t \geq 0$,

Let $S \cong \mathbb{Z}_2[X]/\langle X^2 \rangle$. The several possibilities to consider for $t \geq 0$ can also be proved analogously as it is noticed here that $G(\mathbb{Z}_4) \cong G(\mathbb{Z}_2[X]/\langle X^2 \rangle) \cong C_4$ (cycle graph with 4 vertices).

Finally, let $S \cong \mathbb{Z}_6$. If $t = 0$, then $G(\mathbb{Z}_6)$ is shown in Figure 18.3; and if $t = 1$, then $R \cong \mathbb{Z}_2 \times \mathbb{Z}_6$ and $G(R)$ consist of 2 copies of $G(\mathbb{Z}_6)$ which, in turn, implies the existence of C_6 (as $G(\mathbb{Z}_6) \cong C_6$) and two disjoint six cycles are precisely $(0,0)-(1,1)-(0,4)-(1,3)-(0,2)-(1,5)-(0,0)$ and $(1,2)-(0,3)-(1,4)-(0,1)-(1,0)-(0,5)-(1,2)$. If $t = 2$, then $R \cong \mathbb{Z}_2 \times \mathbb{Z}_2 \times \mathbb{Z}_6$, and $G(\mathbb{Z}_2 \times \mathbb{Z}_2 \times \mathbb{Z}_6) \cong 2G(\mathbb{Z}_2 \times \mathbb{Z}_6)$ (by [1, p. 20]). In a similar way by repeating the process t-times, we get the desired result. □

Theorem 18.2

If R and S are alike as in Lemma 18.2, then

$$\gamma(G(R)) = \begin{cases} 2^t, & if\ S \cong \mathbb{Z}_2; \\ 2^{t+1}, & otherwise. \end{cases}$$

■

Proof. The proof of Lemma 18.2 can be invoked to get the desired result. Consider $G(R)$, when $R \cong \underbrace{\mathbb{Z}_2 \times \mathbb{Z}_2 \times \mathbb{Z}_2 \times \cdots \times \mathbb{Z}_2}_{(t-times)} \times \mathbb{Z}_2,\quad t \geq 0$. Then by Lemma 18.2, $G(R)$ consists of 2^t copies of $G(\mathbb{Z}_2)$, and hence $\gamma(G(R)) = 2^t \cdot \gamma(G(\mathbb{Z}_2))$. Therefore $\gamma(G(R)) = 2^t$. Next, consider $G(R)$ when $R \cong \underbrace{\mathbb{Z}_2 \times \mathbb{Z}_2 \times \mathbb{Z}_2 \times \cdots \times \mathbb{Z}_2}_{(t-times)} \times S,\ t \geq 0$, where S is isomorphic to one of the rings $\mathbb{Z}_4$, $\mathbb{Z}_2[X]/\langle X^2 \rangle$ and $\mathbb{Z}_6$. Again in light of Lemma 18.2, $G(R)$ consists of 2^t copies of $G(S)$, and hence $\gamma(G(R)) = 2^t \cdot \gamma(G(S)$. Since $G(\mathbb{Z}_4)$, $G(\mathbb{Z}_2[X]/\langle X^2 \rangle)$ and $G(\mathbb{Z}_6)$, all have domination number equal to 2, it follows from the previous argument that $\gamma(G(R)) = 2^{t+1}$. □

Theorem 18.3

Let R be a finite commutative ring with unity and let H be a connected graph. Then $G(R)$, is a corona $H \circ K_1$ if and only if R is isomorphic to $\mathbb{Z}_2$. ■

Proof. Necessity: Let $G(R)$ be a corona $H \circ K_1$, for a connected graph H. It is obvious that if H is a finite connected graph of order n, then there are 'n' pendent vertices. We know by Lemma 18.1 that $G(R)$ is either $|U(R)|$-regular or $(|U(R)|, |U(R)|-1)$-semi regular. Therefore, if $G(R)$ is a corona $H \circ K_1$, then the value of $(|U(R)|-1)$ can be at most 1, that is $|U(R)| \leq 2$. Now we claim that $|U(R)| = 1$. We shall prove this claim by contradiction. Suppose $|U(R)| \neq 1$, this implies that $|U(R)| = 2$. If $|U(R)| = 2$, then $|R|$ may be even or odd.

First, let $|R|$ be even with $|U(R)| = 2$. Then in view of Lemma 18.1, $G(R)$ is $|U(R)|$-regular. Therefore, it cannot be a corona, a contradiction to the claim. Next, let $|R|$ be odd with $|U(R)| = 2$. Then the only such a ring R is $\mathbb{Z}_3$, and it is easy to see that $G(\mathbb{Z}_3)$ is not a corona, again a contradiction. Hence our claim is true, and precisely the ring for which $|U(R)| = 1$ is $\underbrace{\mathbb{Z}_2 \times \mathbb{Z}_2 \times \mathbb{Z}_2 \times \cdots \times \mathbb{Z}_2}_{(t-times)}$, $t \geq 1$, but then for all $t \geq 2$, $G(\underbrace{\mathbb{Z}_2 \times \mathbb{Z}_2 \times \mathbb{Z}_2 \times \cdots \times \mathbb{Z}_2}_{(t-times)})$ is disconnected, cannot be corona. Therefore, the only possibility for the ring is $\mathbb{Z}_2$.

Sufficiency: Let $R \cong \mathbb{Z}_2$. Then by Figure 18.1, $G(R) \cong K_2 \cong K_1 \circ K_1$. Consequently, it is a corona of K_2 with K_1. Thus the result follows. □

Payan and Xuong [10] and Fink et al. [3] established the following result:

Theorem 18.4

For a graph G with even order and no isolated vertices, $\gamma(G) = \frac{n}{2}$ if and only if the component of G are cycle C_4 or corona $H \circ K_1$ for any connected graph H. ■

At this stage we are ready to answer the Problem 3.

Theorem 18.5

$\gamma(G(R)) = \frac{|R|}{2}$ if and only if $R \cong \underbrace{\mathbb{Z}_2 \times \mathbb{Z}_2 \times \mathbb{Z}_2 \times \cdots \times \mathbb{Z}_2}_{(t-times)} \times S$, $t \geq 0$, where S is any one of the following rings $\mathbb{Z}_2$, $\mathbb{Z}_4$, and $\mathbb{Z}_2[X]/\langle X^2 \rangle$. ■

Proof. Necessity: Let $\gamma(G(R)) = \frac{|R|}{2}$. Then by Lemma 18.2, $G(R)$ is C_4 if and only if $R \cong \mathbb{Z}_4$ or $\mathbb{Z}_2[X]/\langle X^2\rangle$. Also, by Theorem 18.2 for $t \geq 0$, we know $G(\underbrace{\mathbb{Z}_2 \times \mathbb{Z}_2 \times \mathbb{Z}_2 \times \cdots \times \mathbb{Z}_2}_{(t-times)} \times \mathbb{Z}_4) \cong 2^t \cdot G(\mathbb{Z}_4)$, and $G(\underbrace{\mathbb{Z}_2 \times \mathbb{Z}_2 \times \mathbb{Z}_2 \times \cdots \times \mathbb{Z}_2}_{(t-times)} \times \mathbb{Z}_2[X]/\langle X^2\rangle) \cong 2^t \cdot G(\mathbb{Z}_2[X]/\langle X^2\rangle)$. Consequently, the components of $G(R)$ are C_4 if and only if $R \cong \underbrace{\mathbb{Z}_2 \times \mathbb{Z}_2 \times \mathbb{Z}_2 \times \cdots \times \mathbb{Z}_2}_{(t-times)} \times \mathbb{Z}_4$ or $R \cong \underbrace{\mathbb{Z}_2 \times \mathbb{Z}_2 \times \mathbb{Z}_2 \times \cdots \times \mathbb{Z}_2}_{(t-times)} \times \mathbb{Z}_2[X]/\langle X^2\rangle$.

On the other hand, $G(R)$ is a corona $H \circ K_1$ if and only if R is isomorphic to $\mathbb{Z}_2$ (by Theorem 18.3). Again by Lemma 18.2, $G(\underbrace{\mathbb{Z}_2 \times \mathbb{Z}_2 \times \mathbb{Z}_2 \times \cdots \times \mathbb{Z}_2}_{(t-times)} \times \mathbb{Z}_2) \cong 2^t \cdot G(\mathbb{Z}_2)$. Thus the component of $G(R)$ are corona $H \circ K_1$ if and only if $R \cong \underbrace{\mathbb{Z}_2 \times \mathbb{Z}_2 \times \mathbb{Z}_2 \times \cdots \times \mathbb{Z}_2}_{(t-times)} \times \mathbb{Z}_2$. Hence, $R \cong \underbrace{\mathbb{Z}_2 \times \mathbb{Z}_2 \times \mathbb{Z}_2 \times \cdots \times \mathbb{Z}_2}_{(t-times)} \times S$, $t \geq 0$, where S is any one of the following rings $\mathbb{Z}_2$, $\mathbb{Z}_4$, and $\mathbb{Z}_2[X]/\langle X^2\rangle$.

Sufficiency: Let $R \cong \underbrace{\mathbb{Z}_2 \times \mathbb{Z}_2 \times \mathbb{Z}_2 \times \cdots \times \mathbb{Z}_2}_{(t-times)} \times S$, $t \geq 0$, where S is any one of the following rings $\mathbb{Z}_2[X]/\langle X^2\rangle$, $\mathbb{Z}_2$ and $\mathbb{Z}_4$. Now our aim is to show that $\gamma(G(R)) = \frac{|R|}{2}$. To do so, let $t = 0$. Then $R \cong S$. To check, whether $\gamma(G(S)) = \frac{|S|}{2}$ or not, we have several possibilities to consider as follows:

(i) If $S \cong \mathbb{Z}_2$, then $\gamma(G(\mathbb{Z}_2)) = 1$. It implies that $\gamma(G(S)) = \frac{|S|}{2}$.

(ii) If $S \cong \mathbb{Z}_4$, then $\gamma(G(\mathbb{Z}_4)) = 2$ (as $G(\mathbb{Z}_4) \cong C_4$). This again implies that $\gamma(G(S)) = \frac{|S|}{2}$.

(iii) If $S \cong \mathbb{Z}_2[X]/\langle X^2\rangle$, then it is easy to verify that $\gamma(G(S)) = \frac{|S|}{2}$ as $G(\mathbb{Z}_4) \cong G(\mathbb{Z}_2[X]/\langle X^2\rangle)$.

The above analysis would lead one to conclude that $\gamma(G(S)) = \frac{|S|}{2}$.

Next, if $t \geq 1$, then clearly, $R \cong \underbrace{\mathbb{Z}_2 \times \mathbb{Z}_2 \times \mathbb{Z}_2 \times \cdots \times \mathbb{Z}_2}_{(t-times)} \times S$ and by Lemma 18.2, $G(R)$ is isomorphic to 2^t copies of $G(S)$. Therefore, $\gamma(G(R)) = 2^t \cdot \gamma(G(S))$. The remaining part of the proof follows immediately from Theorem 18.2. □

Corollary 18.1.1. *For a finite field $\mathbb{F}_m$, $\gamma(G(\mathbb{F}_m)) = \frac{|\mathbb{F}_m|}{2}$ if and only if $m = 2$.*

The following remarks are immediate consequences of Theorem 18.5.

Remark. It is worthwhile to note here that for the ring $\mathbb{Z}_n$, the unit graphs $G(\mathbb{Z}_n)$ and unitary addition Cayley graphs G_n are isomorphic (see [11, Remark 2.4]). Therefore, $\gamma(G_n) = \frac{n}{2}$ if and only if $n = 2$ or 4.

Remark. We know that the unitary addition Cayley graph G_n is isomorphic to the unitary Cayley graph X_n if and only if n is even (see [12, Theorem 5]). Therefore by Theorem 18.5, $\gamma(X_n) = \frac{n}{2}$ if and only if $n = 2$ or 4.

18.1.2 RINGS R FOR WHICH $\gamma(G(R)) = \frac{|R|-1}{2}$

The following Theorem provides the partial answer to Problem 4, when the ring $R \cong \mathbb{Z}_n$.

Theorem 18.6

$\gamma(G(\mathbb{Z}_n)) = \frac{n-1}{2}$ if and only if $n = 3$. ■

Proof. Necessity: Let $\gamma(G(\mathbb{Z}_n)) = \frac{n-1}{2}$. Then the following interesting cases arise for different values of n.
Case (i) If $n = p$, p is a prime, then $\gamma(G(\mathbb{Z}_n)) = 1$. Therefore $1 = \frac{p-1}{2}$; this implies that $p = 3$, and hence the precise ring is $\mathbb{Z}_3$.
Case (ii) If $n = p^k$, p is an odd prime, then by [1, Theorem 3.2], $G(\mathbb{Z}_n)$ is complete bipartite. Therefore $\gamma(G(\mathbb{Z}_{p^k})) = 2$ and $2 = \frac{p^k-1}{2}$ which implies that $p^k = 5$, $k > 1$; since it is not possible, it follows that in this case no ring exists.
Case (iii) If $n = p \cdot q$, $p, q > 2$, then again by [1, Theorem 3.2], $\gamma(G(\mathbb{Z}_n)) = 3$ which implies that $p \cdot q = 7$, which is not possible. Now, if $n = p_1^{a_1} p_2^{a_2} \cdots p_k^{a_k}$, then it may be argued again as above that the only possibility for the value of n is 3.

Sufficiency is easy to prove, for if we take $\mathbb{Z}_3$, then clearly, from Theorem 18.1, $\gamma(G(\mathbb{Z}_3)) = 1 = \frac{3-1}{2}$. Hence the result. □

The above discussion leads us to ask the following question:

Question For a given n, does there exist a ring R for which $\gamma(G(R)) = n$?

Towards validity of the above question, we found that if $n = 2^k$, $k \geq 1$, then there exists a ring R consisting of $2n$ elements having $\gamma(G(R)) = n$. However, the discussion about $\gamma(G(R))$ in Theorems 18.1 and 18.2, indicates that for every $n \in \{1, 2^k\}$, there is a ring R for which $\gamma(G(R)) = n$.

18.2 DOMINATION NUMBER OF LINE GRAPH OF UNIT GRAPH

The benefit of studying $L(G)$ of G is that one may recover the structure of any connected graph from its line graph, i.e., there is a one-to-one correspondence between the class of connected graphs and the class of connected line graphs. With the class of line graph of unit graphs at hand, it is natural to keep an eye on the properties of unit graphs and determine the properties of their corresponding line graphs.

18.2.1 LINE GRAPH OF UNIT GRAPH

In this section we study the basic properties of the line graph of unit graphs associated with finite commutative ring with unity.

Definition. The line graph $L(G)$ of a graph G, is a graph defined by $V(L(G)) := E(G)$ and $\{e_1, e_2\} \in E(L(G))$ if e_1 and e_2 are incident to a common vertex in G. If $\{x, y\} \in E(G)$ we will denote the corresponding vertex of $L(G)$ by $[x, y]$. Let R be a commutative ring with $1 \neq 0$ and $U(R)$ be the set of unit elements of R. In particular, the line graph $L(G(R))$ of unit graph $G(R)$ of R will have vertices of the form $[u, v]$ such that the sum of u and v is a unit of R, i.e., $u + v \in U(R)$.

It is easy to see that, for given rings R_1 and R_2, if $R_1 \cong R_2$, then $G(R_1) \cong G(R_2)$, and hence $L(G(R_1)) \cong LG((R_2))$ as for instance; for $\mathbb{Z}_2 \times \mathbb{Z}_3 \cong \mathbb{Z}_6$, clearly $G(\mathbb{Z}_2 \times \mathbb{Z}_3) \cong G(\mathbb{Z}_6) \cong C_6$ and $L(G(\mathbb{Z}_2 \times \mathbb{Z}_3)) \cong L(G(\mathbb{Z}_6)) \cong C_6$. Also one can notice that if $R_1 \ncong R_2$, but $G(R_1) \cong G(R_2)$, then $L(G(R_1)) \cong L(G(R_2))$, as for instance; $G(\frac{\mathbb{Z}_2[x]}{\langle x^2 \rangle}) \cong G(\mathbb{Z}_4) \cong C_4$ and $L(G(\frac{\mathbb{Z}_2[x]}{\langle x^2 \rangle})) \cong L(G(\mathbb{Z}_4)) \cong C_4$. However there do not exist rings R_1 and R_2 for which if $R_1 \ncong R_2$ and $G(R_1) \ncong G(R_2)$, yet $L(G(R_1)) \cong L(G(R_2))$.

Some examples of unit graph and its corresponding line graph over finite commutative rings are shown in Figure 18.4 and Figure 18.5, respectively. This point is also illustrated by means of the figures that, for given finite commutative ring R, if $R \cong \mathbb{Z}_4$, then $G(\mathbb{Z}_4) \cong L(G(\mathbb{Z}_4))$ and if $R \cong \mathbb{Z}_6$, then $G(\mathbb{Z}_6) \cong L(G(\mathbb{Z}_6))$

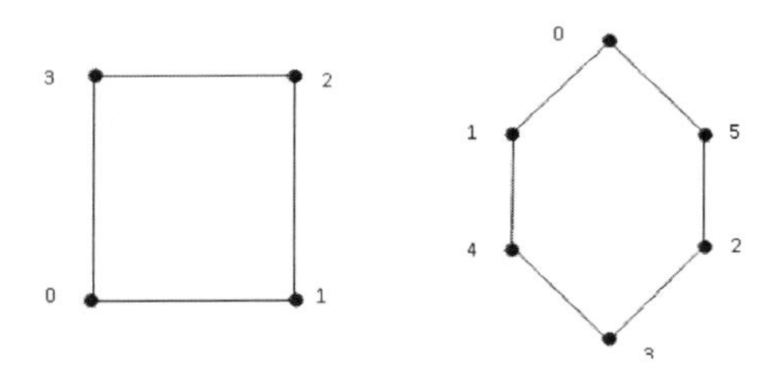

Figure 18.4: The unit graphs for the rings $\mathbb{Z}_4$ and $\mathbb{Z}_6$, respectively

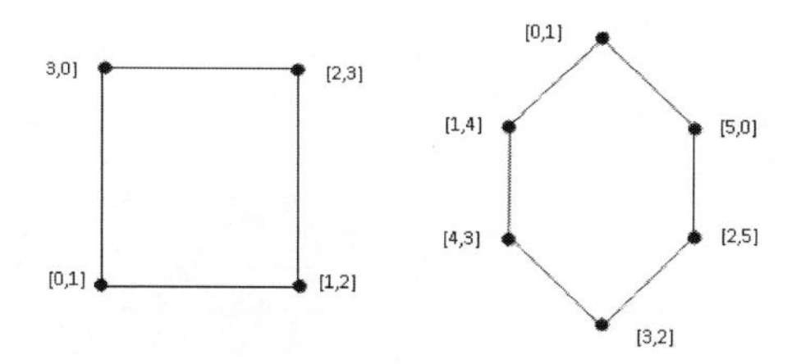

Figure 18.5: The line graph of unit graphs for the rings $\mathbb{Z}_4$ and $\mathbb{Z}_6$, respectively

Theorem 18.7

The following are the well known results:

i) If a graph G is connected, then $L(G)$ is connected and vice-versa.
ii) If a graph G contains a star graph with 'n' vertices, then its line graph will contain a clique of size '$n-1$'.
iii) If G is a star graph with cardinality 'n', then $L^n(G)$ is a complete graph with cardinality '$n-1$'.

■

Now we establish the following results:

Theorem 18.8

If $R \cong \underbrace{\mathbb{Z}_2 \times \mathbb{Z}_2 \cdots \times \mathbb{Z}_2}_{t-times} \times S$, where $t \geq 1$ and $S \cong \mathbb{Z}_2$, or $\mathbb{Z}_4$ or $\mathbb{Z}_6$ or $\frac{\mathbb{Z}_2[x]}{\langle x^2 \rangle}$, then

$$\gamma(L(G(R))) = \begin{cases} 2^t, & S \cong \mathbb{Z}_2; \\ 2^{t+1}, & \text{Otherwise.} \end{cases}$$

■

Proof. For the case when $R \cong \underbrace{\mathbb{Z}_2 \times \mathbb{Z}_2 \cdots \times \mathbb{Z}_2}_{t-times} \times S$, where $S \cong \mathbb{Z}_2$. Then clearly, $R \cong \underbrace{\mathbb{Z}_2 \times \mathbb{Z}_2 \cdots \times \mathbb{Z}_2}_{t-times} \times \mathbb{Z}_2$ and the unit graph $G(R) \cong \cup_{i=1}^{2^t} K_2$, and its corresponding line graph $L(G(R)) \cong \cup_{i=1}^{2^t} K_1$. Therefore $\gamma(L(G(R))) = 2^t$.

For the remaining cases when $S \cong \mathbb{Z}_4$ or $\mathbb{Z}_6$ or $\frac{\mathbb{Z}_2[x]}{\langle x^2 \rangle}$, we make use of the result "$L(G) \cong G$ if and only if G is a cycle graph". Now in view of

[1, Theorem 3.2.], $G(R)$ is cycle if and only if $R \cong \mathbb{Z}_4$ or $\mathbb{Z}_6$ or $\frac{\mathbb{Z}_2[x]}{\langle x^2 \rangle}$, and hence in all the cases $L(G(\mathbb{Z}_4)) \cong G(\mathbb{Z}_4)$ and $L(G(\mathbb{Z}_6)) \cong G(\mathbb{Z}_6)$. Therefore $\gamma(L(G(R))) = 2^{t+1}$. □

Theorem 18.9

Let R be a finite commutative ring with $|R| < 4$. Then $\gamma(L(G(R))) = 1$. ■

Proof. Let R be a finite commutative ring with $|R| < 4$. Then $|R|$ can be 2 or 3. If $|R| = 2$, then $R \cong \mathbb{Z}_2$ and $L(G(R)) \cong K_1$, and hence $\gamma(L(G(R))) = 1$. If $|R| = 3$, then $R \cong \mathbb{Z}_3$ and $G(R) \cong P_3$, and hence $L(G(R)) \cong K_2$. Therefore $\gamma(L(G(R))) = 1$. Hence the result. □

Theorem 18.10

Let R be a finite commutative ring with $|R| = 4$. Then $\gamma(L(G(R))) = 2$. ■

Proof. If $|R| = 4$, then R is precisely one of the following rings $\mathbb{Z}_4$, $\frac{\mathbb{Z}_2[x]}{\langle x^2 \rangle}$, $\mathbb{Z}_2 \times \mathbb{Z}_2$ and $\mathbb{F}_4$. For the rings $\mathbb{Z}_4$, $\frac{\mathbb{Z}_2[x]}{\langle x^2 \rangle}$, $\mathbb{Z}_2 \times \mathbb{Z}_2$ and $\mathbb{F}_4$, the corresponding unit graphs are C_4, C_4, $K_2 \cup K_2$ and K_4, respectively. However their corresponding line graph $L(G(R))$ are C_4, C_4, $K_1 \cup K_1$ and 4-regular graph on six vertices, respectively, and hence $\gamma(L(G(R))) = 2$. □

Remark. If $G(R) \cong L(G(R))$, then clearly $\gamma(G(R)) = \gamma(L(G(R))$; for example if we take $R \cong \mathbb{Z}_4$, then $G(R) \cong L(G(R)) \cong C_4$ and $\gamma(G(R)) = \gamma(L(G(R)) = 2$. However there exists a ring R for which $G(R) \ncong L(G(R))$, still $\gamma(G(R)) = \gamma(L(G(R))$; as for instance, if $R \cong \mathbb{Z}_3$, then $G(R) \cong P_3$ and $L(G(R)) \cong K_2$ both are non-isomorphic although $\gamma(G(R)) = \gamma(L(G(R)) = 1$

Remark. One can easily notice that $\gamma(L(G(R))) = 1$ if and only if $R \cong \mathbb{Z}_2$ or $\mathbb{Z}_3$.

REFERENCES

1. N. Ashrafi, H. R. Maimani, M. R. Pournaki and S. Yassemi. Unit graphs associated with rings. *Communications in Algebra*, 38, 2851-2871, 2010.
2. E. J. Cockayne and S.T. Hedetniemi. Towards a theory of domination in graphs. *Networks*, 7, 247-261, 1977.
3. J. F. Fink, M.S. Jacobson, L.F. Kinch and J. Roberts. On graph having their domination number half their order. *Period. Math. Hungar.*, 16, 287-293, 1985.

4. R. P. Grimaldi. Graphs from rings. *Proceedings of the Twentieth Southeastern Conference on Combinatorics, Graph Theory and Computing. Congr. Numer.*, 71, 95-103, 1990.
5. F. Harary. *Graph Theory.* Addison-Wesley Publ. Comp. Reading. MA., 1969.
6. T. Haynes, S. T. Hedetniemi and P. J. Slater. *Fundamentals of Domination in Graphs.* Marcel-Decker Inc. New York, 1998.
7. N. Jacobson. *Lectures in Abstract Algebra.* East-West Press Ltd., New Delhi, 1951.
8. H. R. Maimani, M. R. Pournaki and S. Yassemi. Necessary and sufficient conditions for unit graphs to be Hamiltonian. *Pacific Journal of Mathematics*, 249, 419-429, 2011.
9. O. Ore. Theory of Graphs. *American Mathematical Society Colloquium*, XXXVIII: 1962.
10. C. Payan and N. H. Xuong. Domination-balanced graph. *Journal of Graph Theory*, 6, 23-32, 1982.
11. Pranjali and M. Acharya. Energy and Wiener index of unit graphs. *Appl. Math. Inf. Sci.*, 9, 1339-1343, 2015.
12. D. Sinha, P. Garg and A. Singh. Some properties of unitary addition Cayley graphs. *Notes on Number Theory and Discrete Mathematics*, 17, 49-59, 2011.

19 The Pendant Number of Some Graph Products

Jomon K Sebastian
Savio HSS,
Devagiri, Kozhikode,
Kerala (INDIA)
E-mail: jomoncmi@gmail.com

Sudev Naduvath
Department of Mathematics,
CHRIST (Deemed to be University),
Bengaluru (INDIA)
E-mail: sudev.nk@christuniversity.in

Joseph Varghese Kureethara
Department of Mathematics,
CHRIST (Deemed to be University),
Bengaluru (INDIA)
E-mail: frjoseph@christuniversity.in

A path decomposition of a graph G is a collection $\mathbb{F}$ of its edge disjoint paths whose union is the given graph. The pendant number of the graph G is defined as the least number of end vertices of paths involved in the path decomposition of G. In this chapter, we discuss the pendant number of the Cartesian product as well as the direct product of graphs such as paths, cycles, stars and complete graphs.

Graphs bring out surprises through many counting problems. A good number of the combinatorial problems associated with the graphs are intriguing and challenging. Many combinatorial problems are optimization problems as well. Counting of paths, induced paths, stars, induced stars, cycles, induced cycles etc. gives ample challenges to researchers of all levels. In 2018, we introduced a counting problem associated with paths in graph. We also introduced a parameter called, the pendant number of a graph. In this work, we compute the pendant number of some graph products. For terms and definitions in Graph Theory, refer to [6]. The graphs under consideration in this chapter are undirected, simple, finite and connected graphs. We begin our discussions with some definitions that are required for better understanding of the content in this chapter.

A *decomposition* of a graph G is the collection of subgraphs whose union is G and edges in the subgraphs partition $E(G)$. If each subgraph is a path, then it is a *path-decomposition* of the graph G. The notion of pendant number of a graph is introduced in [1] and further studied in [3],[4] and in [5]. A similar study on the star number of graphs can be seen in [2].

We now see a particular type of path decomposition of a graph and the parameter *pendant number* associated with it.

Definition. [1] Let $V_P(G)$ be the set of all $u \in V(G)$ such that u is an end vertex of a path in $\mathbb{P}$-decomposition in G, then the *pendant number* of a graph G denoted as $\Pi_p(G) = \min\{|V_p(G)|\}$. That is, the pendant number of a graph is the least number of end vertices of paths in the path decomposition of the given graph.

Theorem 19.1

[1] Let G be a connected graph with n vertices. If G has l odd degree vertices, then $l \leq \Pi_p(G) \leq n$. ■

Theorem 19.2

[5] For an r-regular graph G, we have;

$$\Pi_p(G) = \begin{cases} 2 & \text{if } r \text{ is even;} \\ n & \text{if } r \text{ is odd.} \end{cases}$$

■

19.1 PENDANT NUMBER OF CARTESIAN PRODUCT OF GRAPHS

The Cartesian product of two graphs G and H, denoted as $G\Box H$, is as given below:

(i) $V(G\Box H) = \{(u,v)|u \in V(G) \text{ and } v \in V(H)\}$,
(ii) $E(G\Box H) = \{(u,v)(u',v')|u = u', vv' \in E(H) \text{ or } uu' \in E(G), v = v'\}$.

The graphs G and H are called factors of the product $G\Box H$. The Cartesian product of any two graph is commutative [7].

Let P_n and C_n denote the path and the cycle respectively on n vertices and let $K_{1,n}$ be a star on $n+1$ vertices.

The Cartesian products $P_2 \square P_n$, $P_m \square P_n; m,n > 2$, $P_m \square C_n$ and $C_m \square C_n$ are respectively known as *ladder graphs*, *grid graphs*, *prism graphs* and *torus graphs*. Let $u_a \in V(G)$ and $u'_b \in V(H)$, then for convenience, we use $u_{a,b}$ as the vertex in $G \square H$. Let $u_{a,b}, u_{c,d}$ be the vertices in $G \square H$. Then, $u_{a,b} - u_{c,d}$ represents an edge in $G \square H$.

Note that $P_m \square P_1$ is P_m itself and $P_2 \square P_2$ is the cycle graph C_4 and the pendant number of cycles has already been discussed in [1]. Therefore, among the ladder graphs, we need to consider $P_2 \square P_n$ with $n \geq 3$. Hence, we have

Proposition 19.1.1. *For $n \geq 3$, $\Pi_p(P_2 \square P_n) = 2(n-2)$.*

Proof. Consider the ladder graph $P_2 \square P_n$ with $n > 2$. It contains $2n$ vertices. Among these vertices, except four vertices in the extreme ends, all other vertices are of odd degree. By Theorem 19.1, the number of odd degree vertices remains as the lower bound for the pendant number. It is obvious that using these $2n-4 = 2(n-2)$ odd vertices as end vertices, we can make the path decomposition of the given graph.

□

Theorem 19.3

For the grid graph $P_m \square P_n; m,n > 1$, $\Pi_p(P_m \square P_n) = 2(m+n-4)$. ■

Proof. Let $\{u_1, u_2, \ldots, u_m\}$ be the vertex set of P_m and $\{u'_1, u'_2, \ldots, u'_n\}$ be the vertex set of P_n and $\{u_{1,1}, u_{1,2}, \ldots, u_{1,n}, u_{2,1}, u_{2,2}, \ldots, u_{2,n}, \ldots, u_{m,1}, u_{m,2}, \ldots, u_{m,n}\}$ be the vertex set of $P_m \square P_n$. Since, $P_m \square P_n$ is commutative, assume that m is a fixed number. Then, we use the method of induction on n to prove the result.

Let $n = 2$. Then, the pendant number of $P_m \square P_2$ is the same as $P_2 \square P_m$ which is $2(m-2) = 2(m+2-4); m > 2$ (by Proposition 19.1.1).

Assume the result is true for $n = k-1$. That is, $\Pi_p(P_m \square P_{k-1}) = 2(m+k-1-4) = 2(m+k-5)$. The vertices $\{u_{1,2}, u_{1,3}, \ldots, u_{1,k-2}, u_{m,2}, u_{m,3}, \ldots, u_{m,k-2}, u_{2,1}, u_{3,1}, \ldots, u_{m-1,1}\}$ and $\{u_{2,k-1}, u_{3,k-1}, \ldots, u_{m-1,k-1}\}$ are the $2(m+k-5)$ vertices of odd degree which are the end vertices of every path in the desired path decomposition.

Now, Consider $n = k$. Then, $(P_m \square P_k)$ contains m additional vertices. The vertices $u_{1,k-1}$ and $u_{m,k-1}$ become odd degree. Moreover, instead of $\{u_{2,k-1}, u_{3,k-1}, \ldots, u_{m-1,k-1}\}$, the corresponding vertices $\{u_{2,k}, u_{3,k}, \ldots, u_{m-1,k}\}$ are the odd degree vertices of $P_m \square P_k$. Thus, there are two more vertices of odd degree added to the graph when $n = k$. Then, the pendant number of $(P_m \square P_k)$ is $2(m+k-5)+2 = 2(m+k-4)$. Hence, the proof. □

Theorem 19.4

For the prism graph $P_m \square C_n$, $\Pi_p(P_m \square C_n) = 2n$. ■

Proof. Let $\{u_1, u_2, \ldots, u_m\}$ be the vertex set of P_m, $\{u'_1, u'_2, \ldots, u'_n\}$ be the vertex set of C_n and $\{u_{1,1}, u_{1,2}, \ldots, u_{1,n}, u_{2,1}, u_{2,2}, \ldots, u_{2,n} \ldots, u_{m,1}, u_{m,2}, \ldots, u_{m,n}\}$ be the vertex set of $P_m \square C_n$. Using the $2n$ odd degree vertices alone as end vertices of the paths, we can construct the desired path decomposition as follows: let P_a be a path starting from the edge $u_{1,1} - u_{2,1}$ that passes through all the even vertices and ends in the edge $u_{m-1,1} - u_{m,1}$ if m is even or ends in $u_{m-1,n} - u_{m,n}$ if m is odd. Let P_b be another path starting from the edge $u_{1,1} - u_{1,n}$ that passes through all the vertices in the form $u_{i,1}, u_{i,n}; 1 < i < m$ of $P_m \square C_n$ and ends in the edge $u_{m,n} - u_{m,1}$ if m is even and ends in $u_{m,1} - u_{m,n}$ if m is odd. Let P_c be the path on the vertices $u_{1,1}, u_{1,2}, \ldots, u_{1,n}$ and P_d be the path on the vertices $u_{m,1}, u_{m,2}, \ldots, u_{m,n}$. Let $P_i; 2 \leq i \leq n-1$ be the remaining paths starting from $u_{1,i}$ and ending in $u_{m,i}$. Hence, $\Pi_p(P_m \square C_n) = 2n$. □

Theorem 19.5

For the torus graph $C_m \square C_n$, $\Pi_p(C_m \square C_n) = 2$. ■

Proof. The torus graph $C_m \square C_n$ is always a 4-regular graph. Since the pendant number of any even regular graph is 2 (see 19.2), $\Pi_p(C_m \square C_n) = 2$. □

The Cartesian product $K_{1,n} \square P_2$ is called the *book graph* and $K_{1,n} \square P_m; m > 2$ is called *general book graph.*

Proposition 19.1.2. *For the book graph $K_{1,n} \square P_2$, $\Pi_p(K_{1,n} \square P_2) = 2$.*

Proof. Note that $K_{1,n} \square P_2$ will be either an even graph (if n is odd) or exact two vertices $u_{0,1}$ and $u_{0,2}$ are of odd degree (if n is even). In any case, the vertices $u_{0,1}$ and $u_{0,2}$ can be taken as the end vertices of every path in the desired path decomposition. Hence, $\Pi_p(K_{1,n} \square P_2) = 2$. □

Theorem 19.6

For the general book graph $K_{1,n} \square P_m$,

$$\Pi_p(K_{1,n} \square P_m; m > 2) = \begin{cases} (m-2)(n+1) & \text{if } n \text{ is odd;} \\ (m-2)n+2 & \text{if } n \text{ is even.} \end{cases}$$

■

Proof. Let u_0 be the root vertex and $u_i; 1 \leq i \leq n$ be the vertices of the star $K_{1,n}$ and $u'_j; 1 \leq j \leq m$ be the vertices of P_m. Let the vertices of $K_{1,n} \square P_m; m > 2$ be of the form $u_{i,j}; 0 \leq i \leq n; 1 \leq j \leq m$. There are two cases:

Case-1: Let n be odd. Then, $(m-2)(n+1)$ vertices of the form $u_{i,j}; 0 \leq i \leq n;\ 2 \leq j \leq m-1$ are of odd degree. The paths $u_{0,2} - u_{0,1} - u_{1,1} - u_{1,2} - \ldots, - u_{1,m} - u_{0,m} - u_{0,m-1}$, $u_{i,j} - u_{0,j} - u_{i+1,j}; i = 2,4,\ldots,n-1; 2 \leq j \leq m-1$, $u_{i,2} - u_{i,1} - u_{0,1} - u_{i+1,1} - u_{i+1,2}; i = 2,4,\ldots,n-1$, $u_{i,m-1} - u_{i,m} - u_{0,m} - u_{i+1,m} - u_{i+1,m-1}; i = 2,4,\ldots,n-1$, $u_{i,2} - u_{i,3} - \ldots, u_{i,m-1}; i = 0,2,3,4,\ldots,n$ and $u_{1,j} - u_{0,j}; j = 2,3,\ldots,m-1$ complete the path decomposition by taking $(m-2)(n+1)$ odd degree vertices.

Case-2: Let n be even. Then, only $u_{i,1}, u_{i,m}; 1 \leq i \leq n$ and $u_{0,j}; 2 \leq j \leq m-1$ of $K_{1,n} \square P_m$ are of even degree. Thus, there are $n(m-2)+2$ odd degree vertices in the given graph. By taking the paths $u_{0,1} - u_{i,1} - u_{i,2} - \ldots - u_{i,m} - u_{0,m}; 1 \leq i \leq n$, $u_{0,1} - u_{0,2} - \ldots - u_{0,m}$ and $u_{i,j} - u_{0,j} - u_{i+1,j}; i = 1,3,\ldots,n-1; j = 2,3,\ldots,m-1$ the path decomposition is complete which uses only the odd degree vertices in the graph as end vertices. Hence, the proof.

□

Theorem 19.7

$$\Pi_p(K_{1,n} \square C_m) = \begin{cases} mn & \text{if } n \text{ is even;} \\ m(n+1) & \text{if } n \text{ is odd.} \end{cases}$$

■

Proof. Let u_0 be the root vertex and $\{u_1, u_2, \ldots, u_n\}$ be the pendent vertices of the star graph. Let $\{u'_1, u'_2, \ldots, u'_m\}$ be the vertex set of C_m and $u_{i,j};\ 0 \leq i \leq n;\ 1 \leq j \leq m$ be a vertex in $K_{1,n} \square C_m$. In $K_{1,n} \square C_m$, we need to consider two cases:

Case-1: Let n be even. All the vertices except $u_{0,j}; 1 \leq j \leq m$ are of odd degree. The paths $u_{1,1} - u_{0,1} - u_{0,2} - \ldots - u_{0,m} - u_{1,m}$, $u_{2,1} - u_{0,1} - u_{0,m} - u_{2,m}$, $u_{1,j} - u_{0,j} - u_{2,j}; 2 \leq j \leq m-1$, $u_{i,j} - u_{0,j} u_{i+1,j}$ and $u_{i,j} - u_{0,j} - u_{i+1,j}; i = 3,5,\ldots,n-1; j = 1,2,\ldots,m$ decompose all the paths passing through the vertex $u_{0,j}; 1 \leq j \leq m$ in such a way that none of the paths ends in an even degree vertex. The remaining cycles in the form $u_{i,1} - u_{i,2} -, \ldots, u_{i,m} - u_{i,1}; 1 \leq i \leq n$ can be decomposed by taking any two vertices in the given cycle as an end vertex. Thus, all the vertices except $u_{0,j}; 1 \leq j \leq m$ are involved in the path decomposition. Hence, $\Pi_p(K_{1,n} \square C_m) = mn$.

Case-2: Let n be odd. Then, all the vertices of $K_{1,n}\Box C_m$ are of odd degree and by Theorem 19.1 $\Pi_p(K_{1,n}\Box C_m) = m(n+1)$.

□

Theorem 19.8

$$\Pi_p(K_{1,n}\Box K_{1,m}) = \begin{cases} 2 & \text{if both } m \text{ and } n \text{ are odd;} \\ n+m & \text{if both } m \text{ and } n \text{ are even;} \\ m+1 & \text{if } m \text{ is odd and } n \text{ is even;} \\ n+1 & \text{if } m \text{ is even and } n \text{ is odd.} \end{cases}$$

■

Proof. Let u_0 be the root vertex and $\{u_1, u_2, \ldots, u_n\}$ be the pendent vertices of the star graph $K_{1,n}$. Let u'_0 be the root vertex, $\{u'_1, u'_2, \ldots, u'_m\}$ be the pendent vertices of the star graph $K_{1,m}$ and $u_{i,j}$; $0 \leq i \leq n$; $0 \leq j \leq m$ be a vertex in $K_{1,n}\Box K_{1,m}$. Let $d(v)$ denote the degree of the vertex v of the given graph. In $K_{1,n}\Box K_{1,m}$, the following four cases are to be considered:

Case-1: Let both m and n be odd numbers. Then, there are vertices such that a vertex of degree $m+n$, n vertices of degree $m+1$, m vertices of degree $n+1$ and the remaining mn vertices of degree 2. Note that all vertices are having even degree and so the graph is Eulerian. Hence the graph can be decomposed into edge disjoint cycles. Let $\mathbb{P}$ denote the collection of paths in the desired path decomposition. Without loss of generality, let $m > n$. Let us start from an arbitrary vertex, say v_i, of degree $m+1$. Let $v_i, v_{i+1}, v_{i+2}, \ldots, v_j$ be the first path component in $\mathbb{P}$ in such a way that $d(v_{i+1}) = 2$ or $m+n$. If $d(v_{i+1}) = 2$, then $d(v_{i+2}) = n+1$. If $d(v_{i+1}) = m+n$, then $d(v_{i+2}) = m+1$ or $n+1$. Proceeding in this manner, the extension will be terminated at the vertex of degree $m+1$,say v_j, where the open neighbourhood of v_j, denoted as $n(v_j)$, has a neighbour which is already visited in the path concerned. Since $d(v_i), d(v_j) > 1$, again starting from v_i we can construct another path as mentioned above which terminates at v_j. Consider it as the second path component in $\mathbb{P}$. Since the degree of every vertex is even, this process can be repeated until every edge is covered exactly once by any one of the paths in $\mathbb{P}$. Since the degree of every vertex is even, the construction of path components can be performed in such a way that every edge of $K_{1,n}\Box K_{1,m}$ is contained in exactly one path component in $\mathbb{P}$ (See Figure 19.1 for illustration). Hence v_i and v_j are the end vertices of all path components in $\mathbb{P}$, which is a path decomposition of the given graph. Hence, its pendant number is 2.

Case-2: Let both m and n be even numbers. Then, there is a vertex of degree $m+n$, which is an even number. There are mn vertices of degree 2. The remaining $m+n$ vertices having degrees either $m+1$ or $n+1$ are of odd degree. By Theorem 19.1, all those $m+n$ vertices must be pendent vertices of

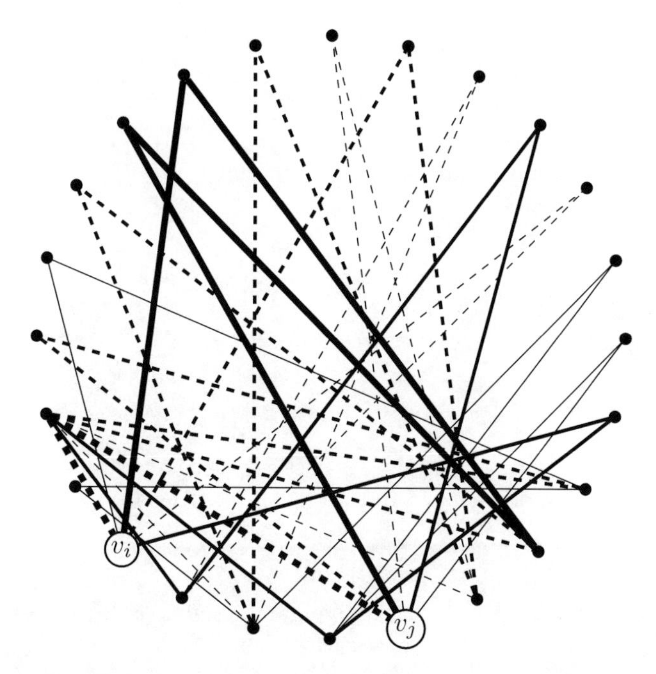

Figure 19.1: $K_{1,n}\Box K_{1,m}$ with $m = 5, n = 3$ (Different types of lines represent different paths from v_i to v_j)

some paths. Thus we have the freedom to choose any vertex of degree $m+1$ or degree $n+1$ as the starting vertex or ending vertex of the paths concerned. Without loss of generality, consider the paths starting from a vertex of degree $m+1$ passing through either a vertex of degree 2 or the vertex of degree $m+n$ and ending in a vertex of degree $n+1$. Note that all these paths are P_3 graphs. These paths make a path decomposition of the entire graph. Thus, those $m+n$ vertices alone can be the pendent vertices in the desired path decomposition. Hence, the pendant number is $m+n$.

Case-3: Let m be odd and n be even. Then, there is a vertex of degree $m+n$ and there are m vertices of degree $n+1$. These are of odd degree and hence must be counted for pendant number. Moreover, there are n vertices of degree $m+1$ and mn vertices of degree 2. Definitely, these $n(m+1)$ vertices are of even degree. Now, consider the paths of the form $v_i - v_{i+1} - v_{i+2} - v_{i+3}$ where v_i is the vertex of degree $m+n$, v_{i+1} is a vertex of degree $m+1$, v_{i+2} is a vertex of degree 2 and v_{i+3} is a vertex of degree $n+1$.There will be n such paths. Next n paths can be of the form $v_j - v_{j+1} - v_{j+2} - v_{j+3} - v_{j+4}$ where v_j is a vertex of degree $n+1$, v_{j+1} is a vertex of degree 2, v_{j+2} is a vertex of degree $m+1$, v_{j+3} is another vertex of degree 2 and v_{j+4} is another vertex of degree $n+1$. Continuing like this we can cover all the edges except those that are incident to the vertex of degree $m+n$ from the vertices $u_1, u_2, \ldots, u_m$, which are of degree $n+1$. These paths complete the

path decomposition of the given graph. Since every path starts and ends either in these m vertices of degree $n+1$ or in the single vertex of degree $m+n$, the pendant number of the given graph is $m+1$.
Case-4: Let m be even and n be odd. Since the box product is commutative, we can easily identify that the case is similar to *Case 3* and hence we get the result as $n+1$.

□

Definition. A connected graph G on n vertices is called a *lotus graph*, if the vertex set of G can be partitioned into two sets such that k vertices are of degree $2r$ and l vertices are of degree $2; k+l = n; r \in \mathbb{N}$.

Proposition 19.1.3. *The pendant number of a lotus graph is* 2.

Proof. Let G be a lotus graph on n vertices, of which k vertices are of degree $2r$ and l vertices are of degree 2 with $n = k+l$. If we take two vertices (preferably non-adjacent) of degree $2r$ as end vertices, then obviously all the paths have end points at these vertices. □

Examples of *lotus graphs*, whose pendant number is 2 (See Figures 19.2, 19.3, where the white nodes represent the end vertices of the paths in the given graphs).

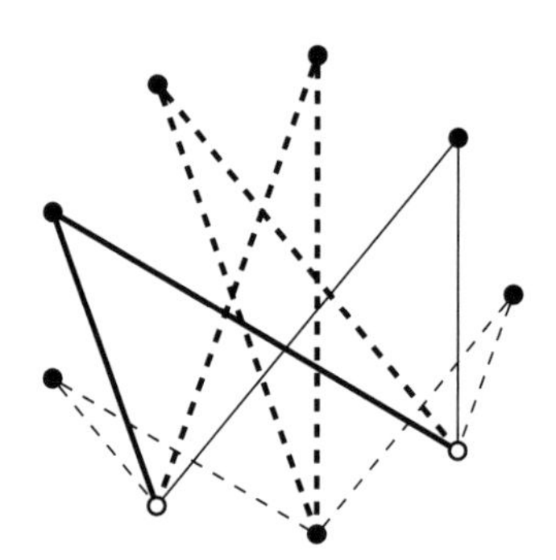

Figure 19.2: **(a)** A lotus graph G with $k = 9, l = 6, n = 15$

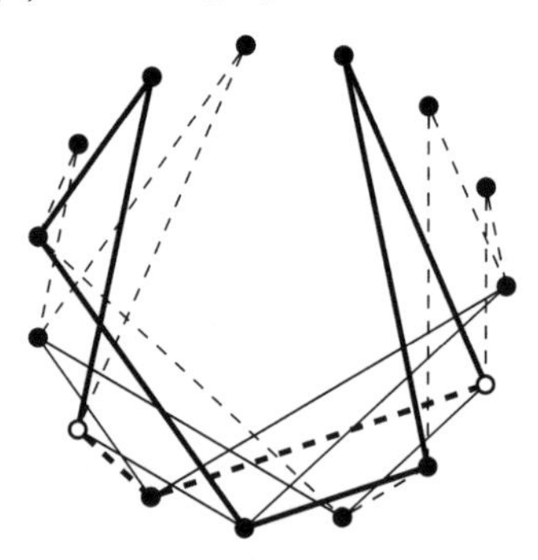

Figure 19.3: **(b)** A lotus graph G with $k = 3, l = 6, n = 9$

Theorem 19.9

Since $\Pi_p(K_n \Box P_m)$ for $n = 1,2,3$ are already proved and $m = 1$ is obvious, we consider $K_n; n > 3$ and $P_m; m > 1$.

$$\Pi_p(K_n \Box P_m) = \begin{cases} 2n & \text{if } n \text{ is odd;} \\ 2 & \text{if } m = 2 \& n \text{ is even;} \\ n(m-2) & \text{if } m > 2 \& n \text{ is even.} \end{cases}$$

■

Proof. Let $u_1, u_2, \ldots, u_n$ be the vertices of K_n, $u'_1, u'_2, \ldots, u'_m$ be the vertices of P_m and $u_{i,j}; 1 \leq i \leq n;\ 1 \leq j \leq m$ be the vertices of $K_n \Box P_m$. If n is odd, then the $2n$ vertices of the form $u_{1,1}, u_{2,1} \cdots, u_{n,1}$ and $u_{1,m}, u_{2,m} \cdots, u_{n,m}$ are of odd degree. The remaining vertices are even regular with degree $n+1$ and hence by a suitable path decomposition we can decompose the given graph into paths whose end points are any one of these $2n$ vertices.

If $m = 2$ and n is even, then $K_n \Box P_m$ becomes an n regular graph. Thus by Theorem 19.2, its pendant number is 2. If $m > 2$ and n is even, then except for the $2n$ vertices of the form $u_{1,1}, u_{2,1} \cdots, u_{n,1}$ and $u_{1,m}, u_{2,m} \cdots, u_{n,m}$, all other vertices are odd regular with degree $n+1$ and hence we can make a path decomposition of the given graph with these $n(m-2)$ odd degree vertices as end vertices of the paths. □

Theorem 19.10

$$\Pi_p(K_n \Box C_m) = \begin{cases} 2 & \text{if } n \text{ is odd;} \\ mn & \text{if } n \text{ is even.} \end{cases}$$

■

Proof. $K_n \Box C_m$ is an $n+1$ regular graph with mn vertices. If n is odd, then it will be an even regular graph. Hence by Theorem 19.2, its pendant number is 2. If n is even, then it is an odd regular graph and again by Theorem 19.2, the pendant number is mn. □

Theorem 19.11

$$\Pi_p(K_n \Box K_{1,m}) = \begin{cases} 2 & \text{if } n \text{ is even;} \\ n(m+1) & \text{if } n \text{ is odd.} \end{cases}$$

■

19.2 PENDANT NUMBER OF DIRECT PRODUCT OF GRAPHS

The direct product of two graphs G and H, denoted as $G \times H$, is defined as follows: The vertex set of $G \times H$ is $V(G) \times V(H)$ for which vertices (g,h) and (g',h') are adjacent precisely if $gg' \in E(G)$ and $hh' \in E(H)$ [7]. Thus,
$V(G \times H) = \{(g,h)|g \in V(G)\ and\ h \in V(H)\}$,
$E(G \times H) = \{(g,h)(g',h')|gg' \in E(G)\ and\ hh' \in E(H)\}$.
The direct product is synonymously known as tensor product, Kronecker product, cardinal product, relational product,conjunction, weak direct product or categorical product.

Let us recall the Weichsel's theorem on the product graphs. Let G and H be two connected and nontrivial graphs. If at least one of G or H contains an odd cycle, then $G \times H$ is connected. If both G and H are bipartite, then $G \times H$ contains exactly two components [8]. Since the direct products of paths, stars, even cycles, a path and an even cycle, a path and a star, a star and an even cycle are all bipartite and thereby disconnected, we do not find the pendant number of such products.

Theorem 19.12

The pendant number of the direct product of a path and an odd cycle is 2. ■

Proof. Let P_p be a path and C_q be an odd cycle. Then, $P_p \times C_q$ is a lotus graph with $k = q, r = 2$. Hence, by Proposition 19.1.3 $\Pi_p(P_m \times C_n) = 2$. □

Theorem 19.13

The pendant number of the direct product of any two odd cycles is 2. ■

Proof. Let G and H be two odd cycles. Then, the direct product $G \times H$ is always a 4– regular graph. By Theorem 19.2, its pendant number is 2. □

Theorem 19.14

For a star $K_{1,p}$ and an odd cycle C_q, $\Pi_p(K_{1,p} \times C_q) = 2$. ■

Proof. The direct product $K_{1,p} \times C_q$ is also a lotus graph with $k = q, r = 3$ and hence, by Proposition 19.1.3, $\Pi_p(K_{1,p} \times C_q) = 2$. □

19.3 CONCLUSION

Here, we determined the *pendant number* of the Cartesian product and the direct product of graphs among the graph classes such as paths, cycles, stars and complete graphs. We have identified the *pendant number* for their different combinations too. We introduced a new class of graphs namely, lotus graphs and identified that all lotus graphs have their *pendant number* as 2. Determining the *pendant number* of other graph classes and other graph products gives much scope for further research.

REFERENCES

1. J. K. Sebastian and J. V. Kureethara. Pendent number of graphs. *International Journal of Applied Mathematics*, 31(5), 679–689, 2018.
2. J. K. Sebastia, J. V. Kureethara, S. Naduvath and C. Dominic. On star decomposition and star number of some graph classes. *International Journal of Scientific Research in Mathematical and Statistical Sciences* 5(6): 81-85, 2018.
3. J. K. Sebastia, J. V. Kureethara, S. Naduvath and C. Dominic. On the pendent number of some new graph classes. *Research and Reviews: Discrete Mathematical Structures*, 6(1): 15-21, 2019.
4. J. K. Sebastia, J. V. Kureethara, S. Naduvath and C. Dominic. A study on the pendent number of graph products. *Acta Universitatis Sapientiae, Informatica*, 11(1): 24-40, 2019.
5. J. K. Sebastia, S. Naduvath and J. V. Kureethara. The pendent number of line graphs and total graphs. *communicated.*
6. G. Chartrand and P. Zhang. *Introduction to Graph Theory.* McGraw Hill Education (India) Pvt. Ltd, Chennai, 2017.
7. R. Hammack, W. Imrich and S. Klavzar. *Handbook of Product Graphs.* CRC Press, 2011.
8. P. M. Weichsel. The Kronecker product of graphs. *American Mathematical Society, Proceedings of the American Mathematical Society*, 13(1): 47-52, 1962.

20 Wiener Index of Tensor Product of Cycle Graph and Some Other Graphs

H. S. Mehta
Department of Mathematics,
Sardar Patel University,
Vallabh Vidhyanagar,
Anand, Gujarat (INDIA).
E-mail: hs_mehta@spuvvn.edu

J. George
Shri Alpesh N. Patel P. G. Institute of Science & Research,
Anand, Gujarat (INDIA).
E-mail: johnsongeorge157@gmail.com

The tensor product of graphs is a very well-known graph product and has been studied in detail. Also the Wiener index of a connected graph G is defined as, $W(G) = \sum_{\{x,x'\} \subset V(G)} d(x,x')$. In this chapter, we obtain $W(C_{2m+1} \otimes H)$ for any connected, bipartite graph H with diameter at most 4. Also, we obtain $W(C_{2m+1} \otimes H)$ for some non-bipartite graphs H as the wheel graph W_n, Helm graph H_n etc.

20.1 INTRODUCTION

Let $G = (V(G), E(G))$ and $H = (V(H), E(H))$ be finite, simple and connected graphs. Then tensor product $G \otimes H$ of G and H is defined as the graph with vertex set $V(G \otimes H) = V(G) \times V(H)$ and edge set

$$\{(x,y)(x',y') : xx' \in E(G) \,\&\, yy' \in E(H), x, x' \in V(G), y, y' \in V(H)\}.$$

The tensor product graph is a very well known graph product and has been studied in detail ([2],[3]).

For a graph G, $d(x,x')$ denotes the distance between two vertices x and x'. The diameter of the graph is defined as

$$diam(G) = max\{d(x,x') : x, x' \in V(G)\}.$$

One of the important concepts related to distance is the Wiener index. For a connected graph G, it is defined as

$$W(G) = \frac{1}{2} \sum_{x,x' \in V(G)} d(x,x') = \sum_{\{x,x'\} \subset V(G)} d(x,x').$$

The Wiener index has many applications in various fields, e.g., cryptography, chemistry etc. ([2],[3]). The Wiener index of $C_m \otimes C_{2n+1}, P_m \otimes C_{2n+1}$ & $K_m \otimes H$ are studied in [3],[4],[5].

In this chapter, we obtain the Wiener index of $C_{2m+1} \otimes H$ for any connected bipartite graph H with $diam(H) \leq 4$. Then we also obtain $W(C_{2m+1} \otimes H)$ for some non-bipartite graphs H as the wheel graph W_n, helm graph H_n etc.

We shall use the following definition and result from [3].

Definition. [3] Let G be a graph and $x, x' \in V(G)$. Define $d'_G(x,x')$ as follows:

1. If $d_G(x,x')$ is odd(even), then $d'_G(x,x')$ is defined as the length of the shortest even(odd) walk joining x and x' in G, and if there is no shortest even(odd) walk, then $d'_G(x,x') = +\infty$.
2. If $d_G(x,x') = +\infty$, then $d'_G(x,x') = +\infty$.

Note that $d(x,x') \leq d'(x,x')$, $\forall x, x' \in V(G)$ and if G is a connected bipartite graph, then $d'(x,x') = \infty$.

Theorem 20.1

Let G and H be connected graphs and $(x,y),(x',y') \in V(G \otimes H)$. Then

1. If $d_G(x,x')$ and $d_H(y,y')$ have same parity, then

$$d_{G\otimes H}((x,y),(x',y')) = \max\{d_G(x,x'), d_H(y,y')\}.$$

2. If $d_G(x,x')$ and $d_H(y,y')$ have different parity, then

$$d_{G\otimes H}((x,y),(x',y')) \\ = \min\{\max\{d_G(x,x'), d'_H(y,y')\}, \max\{d'_G(x,x'), d_H(y,y')\}\}.$$

■

Remark. We shall use the following notations:

$$W(G \otimes H) \\ = \sum_{\{(x,y),(x',y')\} \subset V(G\otimes H)} d((x,y),(x',y'))$$

$$= \sum_{\substack{\{(x,y),(x,y')\}\subset V(G\otimes H)\\ x=x'\,\&\,y,y'\in V(H)}} d((x,y),(x',y')) + \sum_{\substack{\{(x,y),(x',y')\}\subset V(G\otimes H)\\ y=y'\,\&\,x,x'\in V(G)}} d((x,y),(x',y'))$$

$$+2 \sum_{\substack{\{(x,y),(x',y')\}\subset V(G\otimes H)\\ \{x,x'\}\subset V(G)\,\&\,\{y,y'\}\subset V(H)}} d((x,y),(x',y'))$$

We fix the following notations:

Let $N(H) = \{\{y,y'\} \subset V(H) : d(y,y') \neq 0\}$. Then $|N(H)|$ is the number of distinct pairs of vertices in H. Let $|N_i(H)| =$ number of pairs of vertices in H at distance i.

$$\sum_{\substack{y,y'\in V(H)\\ d(y,y')=0}} 1 = P, \quad \sum_{\substack{\{y,y'\}\subset V(H)\\ d(y,y')\neq 0\\ \&\, even}} 1 = Q \;\& \quad \sum_{\substack{\{y,y'\}\subset V(H)\\ d(y,y')-odd}} 1 = R$$

Then note that $Q+R = |N(H)|$ and $P = |V(H)|$.

For the basic results and concepts, we refer to [1], [2].

20.2 WIENER INDEX OF TENSOR PRODUCT OF CYCLE GRAPH AND BIPARTITE GRAPH

In this section, we obtain the Wiener index of the tensor product of a cycle graph C_{2m+1} and any bipartite graph H with $diam(H) \leq 4$. First we assume that $m \geq 4$.

Theorem 20.2

Let H be a connected bipartite graph with $diam(H) \leq 4$. Then
$W(C_{2m+1} \otimes H) = |V(C_{2m+1})|W(H) + 2|V(C_{2m+1})|(|N_3(H)| + 2|N_4(H)|)$
$+ 2|V(H)|^2 W(C_{2m+1})$ ■

Proof. We will divide the summation in the Wiener index into six parts based on $d(x,x') = 0,1,2,3,4\,\&\,i\,(i \geq 5)$ for $x,x' \in V(C_{2m+1})$. Note that $|N_i(C_{2m+1})| = 2m+1, \forall i$. We first compute $\sum_{y,y'\in V(H)} d((x,y),(x',y'))$ for one pair $\{x,x'\}$ with $d(x,x') = i$. Note that for any $y,y' \in V(H)$, $d(y,y') \leq 4$ and, $d'(y,y') = \infty$, as H is bipartite.

Part:A. We fix $x,x' \in V(C_{2m+1})$ with $d(x,x') = 0$. Then $d'(x,x') = 2m+1$. Therefore

$$d = d((x,y),(x',y')) = \begin{cases} d(y,y') & \text{if } d(y,y') \text{ is even} \\ 2m+1 & \text{if } d(y,y') \text{ is odd} \end{cases}$$

Thus

$$\sum_{y,y' \in V(H)} d = \sum_{\substack{\{y,y'\} \subset V(H) \\ \& \, even}} d(y,y') + \sum_{\substack{\{y,y'\} \subset V(H) \\ d(y,y') - odd}} (2m+1)$$

$$= \sum_{\substack{\{y,y'\} \subset V(H) \\ d(y,y') \neq 0 \\ \& \, even}} d(y,y') + (2m+1)R$$

Part:B. Fix $\{x,x'\} \subset V(C_{2m+1})$ with $d(x,x') = 1$. Then $d'(x,x') = 2m$. Therefore

$$d = d((x,y),(x',y')) = \begin{cases} 2m & \text{if } d(y,y') \text{ is even} \\ d(y,y') & \text{if } d(y,y') \text{ is odd} \end{cases}$$

Thus $$\sum_{y,y' \in V(H)} d = \sum_{\substack{y,y' \in V(H) \\ d(y,y') = 0}} (2m) + 2 \sum_{\substack{\{y,y'\} \subset V(H) \\ d(y,y') \neq 0 \\ \& \, even}} (2m) + 2 \sum_{\substack{\{y,y'\} \subset V(H) \\ d(y,y') - odd}} d(y,y')$$

$$= 2mP + 4mQ + 2 \sum_{\substack{\{y,y'\} \subset V(H) \\ d(y,y') - odd}} d(y,y').$$

Part:C. Fix $\{x,x'\} \subset V(C_{2m+1})$ with $d(x,x') = 2$. Then $d'(x,x') = 2m-1$. Therefore

$$d = d((x,y),(x',y')) = \begin{cases} d(y,y') & \text{if } d(y,y') \neq 0 \, \& \, even \\ 2 & \text{if } d(y,y') = 0 \\ 2m-1 & \text{if } d(y,y') \text{ is odd} \end{cases}$$

Thus $$\sum_{y,y' \in V(H)} d = 2P + 2 \sum_{\substack{\{y,y'\} \subset V(H) \\ d(y,y') \neq 0 \\ \& \, even}} d(y,y') + 2(2m-1)R.$$

Part:D. Fix $\{x,x'\} \subset V(C_{2m+1})$ with $d(x,x') = 3$. Then $d'(x,x') = 2m-2$. Therefore

$$d = d((x,y),(x',y')) = \begin{cases} 2m-2 & \text{if } d(y,y') \text{ is even} \\ 3 & \text{if } d(y,y') \text{ is odd} \end{cases}$$

Thus $$\sum_{y,y' \in V(H)} d = (2m-2)P + 2(2m-2)Q + 6R.$$

Part:E. We fix $\{x,x'\} \subset V(C_{2m+1})$ with $d(x,x') = 4$. Then $d'(x,x') = 2m-3$. Therefore

$$d = d((x,y),(x',y')) = \begin{cases} 4 & \text{if } d(y,y') \text{ is even} \\ 2m-3 & \text{if } d(y,y') \text{ is odd} \end{cases}$$

Thus,

$$\sum_{y,y' \in V(H)} d = 4P + 8Q + 2(2m-3)R$$

.

Thus from all parts A to E, we have

$$\sum_{i=0}^{4} \sum_{\substack{y,y' \in V(H) \\ d(x,x')=i}} d$$

$$= 2W(H) + \sum_{\substack{\{y,y'\} \subset V(H) \\ d(y,y') \neq 0 \\ \& even}} d(y,y') + 4(m+1)P + 4(2m+1)Q + (10m-1)R$$

$$= 2W(H) + 2|N_2(H)| + 4|N_4(H)| + 4(m+1)P + 4(2m+1)Q + (10m-1)R$$

Part:F. Finally we fix $\{x, x'\} \subset V(C_{2m+1})$ with $d(x,x') = i (i \geq 5)$. Then $d'(x,x') = 2m+1-i$. Therefore

$$d = d((x,y),(x',y')) = \begin{cases} i & \text{if } i \text{ and } d(y,y') \text{ have same parity} \\ 2m+1-i & \text{if } i \text{ and } d(y,y') \text{ have different parity} \end{cases}$$

Thus $\sum_{y,y' \in V(H)} d = \begin{cases} iP + 2iQ + 2(2m+1-i)R & \text{if } i \text{ is even} \\ (2m+1-i)P + 2(2m+1-i)Q + 2iR & \text{if } i \text{ is odd} \end{cases}$

Then

$$\sum_{i=5}^{m} \sum_{\substack{y,y' \in V(H) \\ d(x,x')=i}} d = \sum_{\substack{i=5 \\ i-even}}^{m} (iP + 2iQ + 2(2m+1-i)R)$$

$$+ \sum_{\substack{i=5 \\ i-odd}}^{m} ((2m+1-i)P + 2(2m+1-i)Q + 2iR)$$

$$= (m+1)(m-4)P + 2(m+1)(m-4)Q + 2m(m-4)R$$

Thus using $Q = |N_2(H)| + |N_4(H)|$, $R = |N_1(H)| + |N_3(H)|$ and $W(H) = |N_1(H)| + 2|N_2(H)| + 3|N_3(H)| + 4|N_4(H)|$, we get

$$\sum_{i=0}^{m} \sum_{\substack{y,y' \in V(H) \\ d(x,x')=i}} d$$

$$= \left(2W(H) + \sum_{\substack{\{y,y'\} \subset V(H) \\ d(y,y') \neq 0 \\ \& even}} d(y,y') + 4(m+1)P + 4(2m+1)Q + (10m-1)R \right)$$

$$+((m+1)(m-4)P+2(m+1)(m-4)Q+2m(m-4)R)$$
$$=W(H)+2m(m+1)|N(H)|+m(m+1)V(H)+2|N_3(H)|+4|N_4(H)|$$

Thus the Wiener index is given by

$$W(C_{2m+1}\otimes H)=|V(C_{2m+1})|W(H)+2|V(C_{2m+1})|(|N_3(H)|+2|N_4(H)|)$$
$$+2|V(H)|^2W(C_{2m+1})$$

□

It can be verified that for $H=P_n(n=4\,\text{or}\,5)$ and $C_{2n}(n=2,3\,\text{or}\,4)$, the above formulas coincide with formulas obtained in [4],[5].
Next we consider C_{2m+1} with $m=2\,\&\,3$.

Proposition 20.2.1. *Let H be any connected bipartite graph with $diam(H)\leq 4$. Then*

1. $W(C_5\otimes H)=5(3W(H)+8|N(H)|+6|N_0(H)|+2|N_1(H)|)$.
2. $W(C_7\otimes H)=7(3W(H)+20|N(H)|+12|N_0(H)|+2|N_1(H)|)$.

Proof. First we fix $x,x'\in V(C_{2m+1})\,(m=2,3)$ with $d(x,x')=i$.
For $i=0,1\,\&\,2$, we get the same terms as in Theorem 20.2. So we get

$$\sum_{i=0}^{2}\sum_{\substack{y,y'\in V(H)\\ d(x,x')=i}} d$$
$$=\left(\sum_{\substack{\{y,y'\}\subset V(H)\\ d(y,y')\neq 0\\ \&\,even}} d(y,y')+(2m+1)R\right)+\left(2mP+4mQ+2\sum_{\substack{\{y,y'\}\subset V(H)\\ d(y,y')-odd}} d(y,y')\right)$$
$$+\left(2P+2\sum_{\substack{\{y,y'\}\subset V(H)\\ d(y,y')\neq 0\\ \&\,even}} d(y,y')+2(2m-1)R\right)$$
$$=2W(H)+2|N_2(H)|+4|N_4(H)|+(2m+2)P+2(2m)Q+(6m-1)R$$

Thus, for $m=2$ we get,

$W(C_5\otimes H)=5(3W(H)+8|N(H)|+6|N_0(H)|+2|N_1(H)|)$
Note that, if $i=3$, then $d'(x,x')=4$. So,

$$d=d((x,y)(x',y'))=\begin{cases}4 & \text{if } d(y,y') \text{ is even}\\ 3 & \text{if } d(y,y') \text{ is odd}\end{cases}$$

$$\text{Thus}\qquad \sum_{\substack{y,y'\in V(H)\\ d(x,x')=3}} d=4P+8Q+6R$$

Now adding $i = 0,1,2\&3$, we get

$W(C_7 \otimes H) = 7(3W(H) + 20|N(H)| + 12|N_0(H)| + 2|N_1(H)|)$. □

20.3 WIENER INDEX OF TENSOR PRODUCT OF CYCLE GRAPH WITH SOME NON-BIPARTITE GRAPHS

In this section, we obtain $W(C_{2m+1} \otimes H)$ with H as some non-bipartite graphs with $diam(H) \leq 4$, e.g., W_n, Fl_n, H_n & CH_n.

Definition. A wheel graph W_n is a graph with $n+1$ vertices that contains C_n and one other vertex which is adjacent to every vertex of C_n. The vertex which is adjacent to every vertex is called the center vertex. The helm graph H_n is obtained from W_n by adding a pendent vertex to each vertex of C_n in W_n. The closed helm graph CH_n is obtained from H_n by adding edges between pendent vertices and a flower graph Fl_n is obtained from H_n by joining each pendent vertex to the center vertex.

Throughout this section we consider W_n and Fl_n with $n \geq 3$, H_n with $n \geq 4$ and CH_n with $n \geq 8$.

Theorem 20.3

For $m \geq 4$,

$$W(C_{2m+1} \otimes H_n) = (2m+1)\left(\frac{m(m+1)(2n+1)^2}{2} + 25n^2 - 25n + 2\right)$$
$$= (2n+1)^2 W(C_{2m+1}) + \frac{25}{6}(2m+1)W(H_n) + 2(2m+1).$$ ■

Proof. Let $V(H_n) = \{a_0, a_1, a_2, \ldots, a_n, b_1, b_2, \ldots, b_n\}$ be the vertex set of H_n, where a_0 is the center vertex, $a_i's$ are the vertices on the cycle of H_n and $b_i's$ are the pendent vertices. Let
$X_n = \{a_0, a_1, a_2, \ldots, a_n\}$, $X_n^c = V(H_n) \setminus X_n$,

$Y_{hn} = \{\{y,y'\} \subset V(H_n) : d(y,y') = 1 \,\&\, y = a_i \,\&\, y' = b_i, i \in \{1,2,\ldots n\}\}$ and
$Y_{hn}^c = \{\{y,y'\} \subset V(H_n) : d(y,y') = 1 \,\&\, \{y,y'\} \notin Y_{hn}\}$.

Note that for $y, y' \in V(H_n)$, if $d(y,y') = 0$, then $d'(y,y') = 3 \text{ or } 5$. If $d'(y,y') = 3$, then $y \in X_n$ & if $d'(y,y') = 5$, then $y \in X_n^c$. If $d(y,y') = 1$, then $d'(y,y') = 2 \text{ or } 4$. If $d'(y,y') = 2$, then $\{y,y'\} \in Y_{hn}^c$ & if $d'(y,y') = 4$, then $\{y,y'\} \in Y_{hn}$.

We fix a pair $\{x,x'\}$ with $d(x,x') = i$ for $x, x' \in V(C_{2m+1})$.

Part: A. Suppose $i = 0$. Then

$$d = d((x,y),(x',y')) = \begin{cases} d(y,y') & \text{if } d(y,y') = 2 \text{ or } 4 \\ d'(y,y') = 2 & \text{if } \{y,y'\} \in Y_{hn}^c \\ d'(y,y') = 4 & \text{if } \{y,y'\} \in Y_{hn} \text{ or } d(y,y') = 3 \end{cases}$$

Thus $\sum_{y,y' \in V(H_n)} d$

$$= \sum_{\substack{\{y,y'\} \subset V(H_n) \\ d(y,y')-\text{even}}} d(y,y') + \sum_{\{y,y'\} \in Y_{hn}^c} 2 + \sum_{\{y,y'\} \in Y_{hn}} 4 + \sum_{\substack{\{y,y'\} \subset V(H_n) \\ d(y,y')=3}} (d(y,y')+1)$$

$$= \sum_{\substack{\{y,y'\} \subset V(H_n) \\ d(y,y')-even}} d(y,y') + \sum_{\substack{\{y,y'\} \subset V(H_n) \\ d(y,y')=1}} 2 + 2 \sum_{\{y,y'\} \in Y_{hn}} 1 + \sum_{\substack{\{y,y'\} \subset V(H_n) \\ d(y,y')=3}} (d(y,y')+1)$$

$$= \sum_{\substack{\{y,y'\} \subset V(H_n) \\ d(y,y')-\text{even}}} d(y,y') + \sum_{\substack{\{y,y'\} \subset V(H_n) \\ d(y,y')-\text{odd}}} (d(y,y')+1) + 2|Y_{hn}|$$

$$= W(H_n) + R + 2|Y_{hn}|$$

Part:B. Suppose $i = 1$. Then

$$d = d((x,y),(x',y')) = \begin{cases} d'(y,y') & \text{if}\, d(y,y')\, \text{is even} \\ d(y,y') & \text{if}\, d(y,y')\, \text{is odd} \end{cases}$$

Thus $\sum_{y,y' \in V(H_n)} d$

$$= \sum_{y,y' \in X_n^c} 5 + \sum_{y,y' \in X_n} 3 + 2 \sum_{\substack{\{y,y'\} \subset V(H_n) \\ d(y,y')-even}} (d(y,y')+1) + 2 \sum_{\substack{\{y,y'\} \subset V(H_n) \\ d(y,y')-odd}} d(y,y')$$

$$= 5|X_n^c| + 3|X_n| + 2W(H_n) + 2Q = 2W(H_n) + 3P + 2Q + 2|X_n^c|$$

Part: C. Suppose $i = 2$. Then

$$d = d((x,y),(x',y')) = \begin{cases} 2 & \text{if}\, d(y,y') = 0\, \text{or}\, \{y,y'\} \in Y_{hn}^c \\ 4 & \text{if}\, d(y,y') = 3\, \text{or}\, \{y,y'\} \in Y_{hn} \\ d(y,y') & \text{if}\, d(y,y') = 2\, \text{or}\, 4 \end{cases}$$

Thus $\sum_{y,y' \in V(H_n)} d$

$$= \sum_{\substack{y,y' \in V(H_n) \\ d(y,y')=0}} 2 + 2 \sum_{\{y,y'\} \in Y_{hn}^c} 2 + 2 \sum_{\substack{\{y,y'\} \subset V(H_n) \\ d(y,y')=3}} (d(y,y')+1)$$
$$+ 2 \sum_{\{y,y'\} \in Y_{hn}} 4 + 2 \sum_{\substack{\{y,y'\} \subset V(H_n) \\ d(y,y')-even}} d(y,y')$$

$$= 2P + 2 \sum_{\substack{\{y,y'\} \subset V(H_n) \\ d(y,y')=1}} (d(y,y')+1) + 2 \sum_{\{y,y'\} \in Y_{hn}} 2$$
$$+ 2 \sum_{\substack{\{y,y'\} \subset V(H_n) \\ d(y,y')=3}} (d(y,y')+1) + 2 \sum_{\substack{\{y,y'\} \subset V(H_n) \\ d(y,y')-even}} d(y,y')$$

$$= 2P + 2 \sum_{\substack{\{y,y'\} \subset V(H_n) \\ d(y,y')-odd}} (d(y,y')+1) + 4|Y_{hn}| + 2 \sum_{\substack{\{y,y'\} \subset V(H_n) \\ d(y,y')-even}} d(y,y')$$

$$= 2W(H_n) + 2P + 2R + 4|Y_{hn}|$$

Part: D. Suppose $i = 3$. Then

$$d = d((x,y),(x',y')) = \begin{cases} 3 & \text{if } y,y' \in X_n \text{ or } d(y,y') = 1,2, \text{or } 3 \\ 5 & \text{if } y,y' \in X_n^c \text{ or } d(y,y') = 4 \end{cases}$$

Thus

$$\sum_{y,y' \in V(H_n)} d = \sum_{y,y' \in X_n} 3 + 2 \sum_{\substack{\{y,y'\} \subset V(H_n) \\ d(y,y')=1}} 3 + 2 \sum_{\substack{\{y,y'\} \subset V(H_n) \\ d(y,y')=2}} (d(y,y')+1)$$

$$+ 2 \sum_{\substack{\{y,y'\} \subset V(H_n) \\ d(y,y')=3}} 3 + \sum_{y,y' \in X_n^c} 5 + 2 \sum_{\substack{\{y,y'\} \subset V(H_n) \\ d(y,y')=4}} (d(y,y')+1)$$

$$= 3|X_n| + 6 \sum_{\substack{\{y,y'\} \subset V(H_n) \\ d(y,y')-odd}} 1 + 2 \sum_{\substack{\{y,y'\} \subset V(H_n) \\ d(y,y')-even}} (d(y,y')+1) + 5|X_n^c|$$

$$= 3|X_n| + 6R + 2 \sum_{\substack{\{y,y'\} \subset V(H_n) \\ d(y,y')-even}} d(y,y') + 2Q + 5|X_n^c|$$

Thus $$\sum_{y,y' \in V(H_n)} d = 2 \sum_{\substack{\{y,y'\} \subset V(H_n) \\ d(y,y')-even}} d(y,y') + 3P + 2Q + 6R + 2|X_n^c|.$$

Part: E. Suppose $i \geq 4$. Then $d = d((x,y),(x',y')) = i$ if $d(y,y') = 0,1,2,3$ & 4.

Thus $$\sum_{y,y' \in V(H_n)} d = \sum_{\substack{y,y' \in V(H_n) \\ d(y,y')=0}} i + 2 \sum_{\substack{y,y' \in V(H_n) \\ d(y,y') \neq 0}} i = 2i|N(H_n)| + iP$$

Thus from all parts A to E, we have

$$W(C_{2m+1} \otimes H_n) = (2m+1)\left(\frac{m(m+1)(2n+1)^2}{2} + 25n^2 - 25n + 2\right) \qquad \square$$

Corollary 20.3.1. *For* $m \geq 4$

$$W(C_{2m+1} \otimes CH_n) = (2m+1)\left(\frac{m(m+1)(2n+1)^2}{2} + 25n^2 - 43n + 2\right)$$

$$= (2n+1)^2 W(C_{2m+1}) + \frac{(2m+1)}{2}\left(25n^2 - 43n + 2\right).$$

Proof. Consider the notations of CH_n as in the previous theorem. Also let $N_2'(CH_m) = \{\{x,x'\} \subset V(CH_m) : d(x,x') = 2, x = b_i \,\&\, x' = b_{i+2}, 1 \le i \le m\}$. Note that here we change the second graph H_n of Theorem 20.3 to CH_n. Thus $d(y,y') = 2$ with $d'(y,y') = 3$ in H_n will be divided into two parts $d(y,y') = 2$ with $d'(y,y') = 3 \text{ or } 5$. But the term $d(y,y') = 2$ with $d'(y,y') = 3$ appears only in part B and part D of Theorem 20.3. All other terms will be similar.
In part B, $d((x,y),(x',y')) = d = d'(y,y') = 3$ if $d(y,y') = 2$ will be replaced by

$$d = \begin{cases} d'(y,y') = 3 & \text{if } d(y,y') = 2 \,\&\, d'(y,y') = 3 \\ d'(y,y') = 5 & \text{if } d(y,y') = 2 \,\&\, d'(y,y') = 5 \end{cases}$$

Thus the term $2 \sum\limits_{\substack{\{y,y'\} \subset V(H_n) \\ d(y,y')=2}} 3$ will be replaced by

$2 \sum\limits_{\{y,y'\} \in N_2(CH_n) \setminus N_2'(CH_n)} 3 + 2 \sum\limits_{\{y,y'\} \in N_2'(CH_n)} 5$. Similarly for part D, the term

$2 \sum\limits_{\substack{\{y,y'\} \subset V(H_n) \\ d(y,y')=2}} 3$ will be replaced by

$2 \sum\limits_{\substack{\{y,y'\} \in N_2(CH_n) \setminus N_2'(CH_n) \\ d(y,y')=2}} 3 + 2 \sum\limits_{\{y,y'\} \in N_2'(CH_n)} 5$. Therefore the Wiener index is given by

$$W(C_{2m+1} \otimes CH_n) = (2n+1)^2 W(C_{2m+1}) + \frac{(2m+1)}{2}\left(25n^2 - 43n + 2\right).$$

□

Finally, we consider C_{2m+1} for $m = 2 \,\&\, 3$.

Theorem 20.4

For $n \ge 4$,

1. $W(C_7 \otimes H_n) = 7(48n^2 + n + 8)$.
2. $W(C_5 \otimes H_n) = 5(32n^2 - 5n + 5)$.

■

Proof. 1. Note that for the pair $\{x,x'\}$ with $d(x,x') = i\,(i = 0,1,2)$, the proof will be the same as in Theorem 20.3. Now for $i = 3$ (in Part D), the case of $d(y,y') = 4$ or $y,y' \in X_n^c$, $d = d((x,y),(x',y')) = 5$ is now replaced by $d = 4$.

Thus

$$\sum_{\substack{y,y'\in V(H_n)\\ d(x,x')=3}} d = 3|X_n|+6R+6|N_2(H_n)|+4|X_n^c|+8|N_4(H_n)|$$

Therefore

$$\begin{aligned}W(C_7\otimes H_n) &= (W(H_n)+R+2|Y_{hn}|)+(2W(H_n)+3P+2Q+2|X_n^c|)\\ &\quad +(2W(H_n)+2P+2R+4|Y_{hn}|)\\ &\quad +(3|X_n|+6R+6|N_2(H_n)|+4|X_n^c|+8|N_4(H_n)|)\\ &= 7(48n^2+n+8)\end{aligned}$$

2. For the pair $\{x,x'\}$ with $d(x,x')=0$, the proof is the same as in Theorem 20.3. If $i=1$ (Part B), then as earlier, if $d(y,y')=4$ or $y,y'\in X_n^c$, then $d=5$ is replaced by $d=4$. If $i=2$, then if $d(y,y')=3$ or $\{y,y'\}\in Y_{hn}$, then $d=4$ is replaced by $d=3$. Thus adding $i=0,1\,\&\,2$, we get

$$\begin{aligned}W(C_5\otimes H_n) &= (W(H_n)+R+2|Y_{hn}|)+(2W(H_n)+3P+2|N_2(H_n)|+|X_n^c|)\\ &\quad +(2W(H_n)+2P+2|Y_{hn}|+2|N_1(H_n)|)\\ &= 5(32n^2-5n+5)\end{aligned}$$

□

Corollary 20.3.2.

1. $W(C_5\otimes CH_n)=5(32n^2-19n+5)$.
2. $W(C_7\otimes CH_n)=7(48n^2-15n+8)$.

Proof. Here the second graph H_n is replaced by CH_n. For $d(y,y')=2$, the term $d'(y,y')=3$ in Corollary 20.4 is divided here into two in parts $d'(y,y')=3$ or $d'(y,y')=5$. Note that these terms appear in part B and part D only. Thus for $C_7\otimes CH_n$, the changes will be as follows:
For $i=1$,

$$d=\begin{cases} d'(y,y')=3 & \text{if } d(y,y')=2\,\&\,d'(y,y')=3\\ d'(y,y')=5 & \text{if } d(y,y')=2\,\&\,d'(y,y')=5\end{cases}$$

For $i=3$,

$$d=\begin{cases} d'(y,y')=3 & \text{if } d(y,y')=2\,\&\,d'(y,y')=3\\ d'(x,x')=4 & \text{if } d(y,y')=2\,\&\,d'(y,y')=5\end{cases}$$

Also, note that for the graph $C_5\otimes CH_n$, in case $i=1$, the changes are exactly the same as for $i=3$. Thus we get

$$W(C_5\otimes CH_n)=5(32n^2-19n+5)$$

$$W(C_7\otimes CH_n)=7(48n^2-15n+8)$$

□

Theorem 20.5

For $m \geq 2$,

$$W(C_{2m+1} \otimes W_n) = |V(W_n)|^2 W(C_{2m+1}) + 3|V(C_{2m+1})|W(W_n) + 2(2m+1).$$

■

Proof. Note that for $x, x' \in V(C_{2m+1})$, if $d(x,x') = i$, then $d'(x,x') = 2m+1-i$ and for every $y, y' \in V(W_n)$, $d(y,y') = 0, 1 \text{ or } 2$. We fix one pair $\{x, x'\}$ with $d(x,x') = i$. Thus if $i = 0$, then we get $d = d((x,y),(x',y')) = 1 \text{ or } 2$. If $i = 1$, then

$$d = d((x,y),(x',y')) = \begin{cases} 1 & \text{if } d(y,y') = 1 \\ 3 & \text{otherwise} \end{cases}$$

If $i \geq 2$, then $d = d((x,y),(x',y')) = i, \forall y, y' \in V(W_n)$.

Thus, we get

$$W(C_{2m+1} \otimes W_n) = (2m+1)\left(2|N(W_n)| + 3P + 6|N_2(W_n)| + 2|N_1(W_n)| + \sum_{i=2}^{m}(iP + 2i|N(W_n)|)\right)$$

Since, $W(W_n) = n(n-1)$, $|N_1(W_n)| = 2n$, $|N_2(W_n)| = \frac{n(n-3)}{2}$ & $|N(W_n)| = \frac{n(n+1)}{2}$, we get

$$W(C_{2m+1} \otimes W_n) = |V(W_n)|^2 W(C_{2m+1}) + 3|V(C_{2m+1})|W(W_n) + 2(2m+1)$$

□

Corollary 20.3.3. *For* $m \geq 2$,

$$W(C_{2m+1} \otimes Fl_n) = |V(Fl_n)|^2 W(C_{2m+1}) + 3|V(C_{2m+1})|W(Fl_n) + 2(2m+1).$$

Proof. Note that the diameter of Fl_n is 2. So as in the case of W_n, we get the result.

□

REFERENCES

1. R. Balakrishnan and K. Ranganathan. *A Textbook of Graph Theory*. Springer, 2000.
2. R. Hammack, W. Imrich and S. Klavzar. *Handbook of Product Graphs*. CRC Press, 2011.

3. S. Moradi. A note on tensor product of graphs. *Iran. J. of Math. Sc. and Info.*, 7, 73-81, 2012.
4. K. Pattabiraman. Wiener index of the tensor product cycles. *J. of Prime Research in Math.*, 10, 1-18, 2015.
5. K. Pattabiraman and P. Paulraja. Wiener index of the tensor product of path and a cycle. *Dicuss. Math. Graph Theory*, 31, 737-751, 2011.
6. P. M. Weichsel. The kronecker product of graphs. *Proc. Amer. Math. Soc.*, 13, 47-52, 1962.

21 Wiener Index of Some Zero-Divisor Graphs

S. K. Vaidya
Saurashtra University
Rajkot, Gujarat (INDIA)
E-mail: samirkvaidya@yahoo.co.in

M. R. Jadeja
Shree Manibhai Virani and Smt. Navalben Virani Science College(Autonomous)
Rajkot, Gujarat (INDIA)
E-mail: jadejamanoharsinh111@gmail.com

The Wiener index of a graph is defined as the sum of the distance between all pairs of vertices in the graph. The zero-divisor graph $\Gamma(R)$ of a commutative ring R is a graph whose vertices are non-zero zero-divisors of R and two vertices are adjacent if their product is zero. Here we investigate the Wiener Index for some zero divisor graphs.

21.1 INTRODUCTION

The concept of the zero-divisor graph of commutative ring R was first introduced by I. Beck [5] in 1988. According to him all elements of R were vertices of the zero-divisor graph. Anderson and Naseer [2] continued with the same definition. Then in 1999 Anderson and Livingston [3] redefined the concept in which the vertices of the zero-divisor graph are non-zero zero-divisors. We continue with the definition and notation introduced in [3].
We consider that the ring R means commutative ring with unity. If R is a ring then $Z(R)$ and $Z^*(R)$ denote the set of zero-divisors and set of non-zero zero-divisors of the ring R respectively.

Definition. A *zero-divisor* in a ring R is an element x for which $\exists y \neq 0$ in R such that $xy = 0$.

Definition. The *zero divisor graph* of a ring R, denoted as $\Gamma(R)$, is a graph with $V(\Gamma(R)) = Z^*(R)$ and $x, y \in V(\Gamma(R))$ are adjacent if $xy = 0$.

Illustration 21.1.1. Let $\mathbb{Z}_{12}$ be a ring; then $Z^*(\mathbb{Z}_{12}) = \{2,3,4,6,8,9,10\}$ and $\Gamma(\mathbb{Z}_{12})$ is shown in Figure 21.1

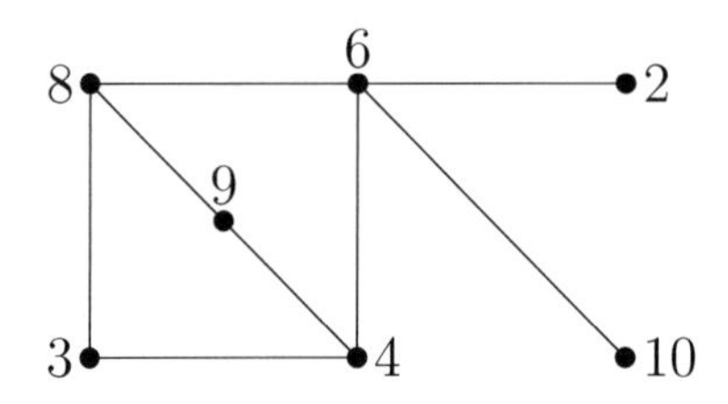

Figure 21.1: $\Gamma(\mathbb{Z}_{12})$

The concept of the Wiener index was introduced by H. Wiener [10] in 1947. The Wiener index was used for predicting the boiling points of paraffins but later a strong correlation between the Wiener index and the chemical properties of a compound was identified. Nowadays this index is a tool used for preliminary screening of drug molecules. The Wiener index also predicts the binding energy of a protein-ligand complex at a preliminary stage.
The concept of the Wiener index of a zero-divisor graph was introduced by M. Ahmadi and R. Nezhad [1] for the ring $\mathbb{Z}_n$ for $n = p^2$ and $n = pq$, where p and q are distinct primes. Then later on S. Reddy *et al.* [9] discussed the Wiener of the zero-divisor graph $\Gamma(\mathbb{Z}_n)$ for $n = p^3$ and $n = p^2q$. For the graph theoretic terminology we refer to Balakrishnan and Ranganathan [8] while any undefined term and notation related to algebra we rely upon Atiyah and Macdonald [4] .

Definition. Let $G = (V, E)$ be a graph; then the *Wiener index* of G is denoted as $W(G)$ and defined as

$$W(G) = \frac{1}{2} \sum_{x,y \in V(G)} d(x, y)$$

where $d(x, y)$ is the distance from x to y.

Illustration 21.1.2. Let us consider the graph $\Gamma(\mathbb{Z}_{27})$ as shown in Figure 21.2.
Divide the vertex set of $\Gamma(\mathbb{Z}_{27})$ into two sets such that $A = \{3, 6, 12, 15, 21, 24\}$ and $B = \{9, 18\}$.

Now as shown in Figure 21.2 every vertex of A is adjacent to every vertex of B and vice-versa.
Therefore $d(x, y) = 1\ \forall x \in A\,, \forall y \in B$

Also every vertex of B is adjacent to all vertices of B and vertices of A are mutually non-adjacent.

Therefore $d(x, y) = 1\ \forall x, y \in B$ and $d(x, y) = 2\ \forall x, y \in A$

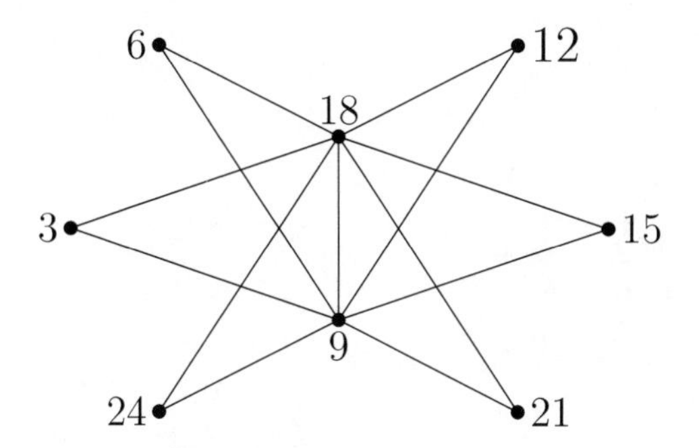

Figure 21.2: $\Gamma(\mathbb{Z}_{27})$

Hence,

$$W(\Gamma(\mathbb{Z}_{27})) = \sum_{x\in A, y\in B} d(x,y) + \sum_{x,y\in A} d(x,y) + \sum_{x,y\in B} d(x,y)$$

$$W(\Gamma(\mathbb{Z}_{27})) = 12+30+1$$

$$W(\Gamma(\mathbb{Z}_{27})) = 43$$

In the following sections we investigate the Wiener index of the graphs $\Gamma(\mathbb{Z}_n)$, $\Gamma(\mathbb{Z}_n[i])$ and $\Gamma(R_1 \times R_2)$.

21.2 WIENER INDEX OF THE ZERO-DIVISOR GRAPH $\Gamma(\mathbb{Z}_n)$

Theorem 21.1

Let p be any prime.
Then $W(\Gamma(\mathbb{Z}_{p^n})) = \frac{1}{2}((n-6)p^{n-1} - (n-1)p^n + 2p^{2n-2} + p^{n-\lfloor \frac{n-1}{2} \rfloor - 1} + 2)$. ■

Proof. Let $n = p^n$. Then the set of non-zero zero-divisors of $\mathbb{Z}_n$ is $Z^*(\mathbb{Z}_n) = \{p, 2p, 3p, \ldots, (p^{n-1}-1)p\}$. We partition the set $Z^*(\mathbb{Z}_n)$ as $Z^*(\mathbb{Z}_n) = \bigcup_{i=1}^{n-1} A_i$, where $A_i = \{k_i p^i \mid k_i = 1,2,3,\ldots,p^{n-i}-1 \; and \; p \nmid k_i\}$, then $|A_i| = p^{n-i} - p^{n-i-1}$.
It is obvious that every element of A_1 is adjacent with every element of A_{n-1}, every element of A_2 is adjacent with every element of A_{n-1} and A_{n-2} and in general, every element of A_i is adjacent to every element of $A_{n-1}, A_{n-2}, \ldots, A_{n-i}$.

For every $x \in A_1$:

$$W(x) = \sum_{y \in A_1, x \neq y} d(x,y) + \sum_{y \in A_2} d(x,y) + \ldots + \sum_{y \in A_{n-1}} d(x,y)$$

$$= \sum_{y \in A_1, x \neq y} 2 + \sum_{y \in A_2} 2 + \ldots + \sum_{y \in A_{n-1}} 1$$

$$= 2(p^{n-1} - p^{n-2} - 1) + 2(p^{n-2} - p^{n-3}) + \ldots + 1(p-1)$$

$$= 2p^{n-1} - p - 3$$

Now for every $x \in A_2$:

$$W(x) = \sum_{y \in A_1} d(x,y) + \sum_{y \in A_2, x \neq y} d(x,y) + \ldots + \sum_{y \in A_{n-2}} d(x,y) + \sum_{y \in A_{n-1}} d(x,y)$$

$$= \sum_{y \in A_1, x \neq y} 2 + \sum_{y \in A_2} 2 + \ldots + \sum_{y \in A_{n-2}} 1 + \sum_{y \in A_{n-1}} 1$$

$$= 2(p^{n-1} - p^{n-2}) + 2(p^{n-2} - p^{n-3} - 1) + \ldots + 1(p^2 - p) + 1(p-1)$$

$$= 2p^{n-1} - p^2 - 3$$

In general, for any $x \in A_i$

$$W(x) = \begin{cases} 2p^{n-1} - p^i - 3 & \text{if i} < \frac{\text{n}}{2} \\ 2p^{n-1} - p^i - 2 & \text{if i} \geq \frac{\text{n}}{2} \end{cases}$$

Therefore,

$$W(\Gamma(\mathbb{Z}_{p^n})) = \frac{1}{2} \left[\sum_{x \in A_1} d(x,y) + \sum_{x \in A_2} d(x,y) + \ldots + \sum_{x \in A_{n-1}} d(x,y) \right]$$

$$= \frac{1}{2} \left[\sum_{i < \frac{n}{2}} (p^{n-i} - p^{n-i-1})(2p^{n-1} - p^i - 3) \right.$$

$$\left. + \sum_{i \geq \frac{n}{2}} (p^{n-i} - p^{n-i-1})(2p^{n-1} - p^i - 2) \right]$$

$$= \frac{1}{2}((n-6)p^{n-1} - (n-1)p^n + 2p^{2n-2} + p^{n - \lfloor \frac{n-1}{2} \rfloor - 1} + 2)$$

□

In 2008 Osba *et al.* [7] explored the concept of zero-divisor graph of Gaussian integers modulo n. In the following theorem we proved a result for zero divisor graph of $\mathbb{Z}_{q^2}[i]$.

Theorem 21.2

Let q be any prime such that $q \equiv 3\ (mod\ 4)$.
Then $W(\Gamma(\mathbb{Z}_{q^2}[i])) = \frac{1}{2}(q^2-1)(q^2-2)$. ∎

Proof. The graph $\Gamma(\mathbb{Z}_{q^2}[i])$ is complete and $|V(\Gamma(\mathbb{Z}_{q^n}[i]))| = q^{2n-2}-1$ as proved by E Osba *et al.* [7]. Therefore $|V(\Gamma(\mathbb{Z}_{q^2}[i]))| = q^2-1$.
Hence $W(\Gamma(\mathbb{Z}_{q^2}[i])) = \frac{1}{2}(q^2-1)(q^2-2)$. □

Now we investigate the Wiener Index of some zero-divisor graphs of the direct product of rings.

21.3 WIENER INDEX OF ZERO-DIVISOR GRAPHS FROM DIRECT PRODUCT OF RINGS

Definition. Let R_1 and R_2 be rings; then the *direct product* of rings R_1 and R_2 is defined as $R_1 \times R_2 = \{(x,y) \mid x \in R_1\ and\ y \in R_2\}$

Theorem 21.3

Let R_1 and R_2 be any integral domains with n_1 and n_2 elements respectively. Then $W(\Gamma(R_1 \times R_2)) = n_1^2 + n_2^2 + n_1 n_2 - 4n_1 - 4n_2 + 5$. ∎

Proof. Let R_1 and R_2 be any integral domains with n_1 and n_2 elements respectively. So $Z^*(R_1 \times R_2) = A \cup B$, where $A = \{(0,x) \mid x \in R_2^*\}$ and $B = \{(y,0) \mid y \in R_1^*\}$, moreover $|Z^*(R_1 \times R_2)| = n_1 + n_2 - 2$, $|A| = n_2 - 1$ and $|B| = n_1 - 1$. For every $x \in A$:

$$\begin{aligned} W(x) &= \sum_{y\in B} d(x,y) + \sum_{y\in A, x\neq y} d(x,y) \\ &= \sum_{y\in B} 1 + \sum_{y\in A, x\neq y} 2 \\ &= (n_1-1) + 2(n_2-2) \\ &= 2n_2 + n_1 - 5 \end{aligned}$$

and for every $x \in B$:

$$\begin{aligned} W(x) &= \sum_{y \in A} d(x,y) + \sum_{y \in B, x \neq y} d(x,y) \\ &= \sum_{y \in A} 1 + \sum_{y \in B, x \neq y} 2 \\ &= (n_2 - 1) + 2(n_1 - 2) \\ &= 2n_1 + n_2 - 5 \end{aligned}$$

Therefore,

$$\begin{aligned} W(\Gamma(R_1 \times R_2)) &= \frac{1}{2}\left[\sum_{x \in A} d(x,y) + \sum_{x \in B} d(x,y)\right] \\ &= \frac{1}{2}[(n_2 - 1)(2n_2 + n_1 - 5) + (n_1 - 1)(2n_1 + n_2 - 5)] \\ &= n_1^2 + n_2^2 + n_1 n_2 - 4n_1 - 4n_2 + 5 \end{aligned}$$

□

Corollary 21.3.1. *Let $\mathbb{Z}_p \times \mathbb{Z}_p$ be a ring with p being any prime. Then $W(\Gamma(\mathbb{Z}_p \times \mathbb{Z}_p)) = 3p^2 - 8p + 5$.*

Proof. Let $\mathbb{Z}_p \times \mathbb{Z}_p$ be a ring. Now as we know that $\mathbb{Z}_p$ has no non-zero-divisor this implies $\mathbb{Z}_p$ is an integral domain with $p-1$ elements. Hence by Theorem 21.3 we have $W(\Gamma(\mathbb{Z}_p \times \mathbb{Z}_p)) = 3p^2 - 8p + 5$.

□

Corollary 21.3.2. *Let p be any prime such that $p \equiv 1(mod 4)$. Then $W(\Gamma(\mathbb{Z}_p[i])) = 3p^2 - 8p + 5$.*

Proof. In 2008 E. Osba *et al.* [7] proved that $\Gamma(\mathbb{Z}_p[i])$ is isomorphic to $K_{p-1,p-1}$ if p is prime and $p \equiv 1(mod 4)$. Now proof follows from the Theorem 21.3. □

Corollary 21.3.3. *Let q_1 and q_2 be distinct prime such that q_1 & $q_2 \equiv 3\ (mod\ 4)$. Then $W(\Gamma(\mathbb{Z}_{q_1 q_2}[i])) = q_1^4 + q_2^4 - 7q_1^2 - 7q_2^2 + q_1^2 q_2^2 + 15$.*

Proof. As we know that $\mathbb{Z}_q$ is a field if q is prime such that $q \equiv 3(mod 4)$ $\mathbb{Z}_{q_1 q_2} \cong \mathbb{Z}_{q_1} \times \mathbb{Z}_{q_2}$ as rings. The result follows from the Theorem 21.3. □

Theorem 21.4

Let $\mathbb{Z}_p \times \mathbb{Z}_{p^2}$ be a ring with p being any prime.
Then $W(\Gamma(\mathbb{Z}_p \times \mathbb{Z}_{p^2})) = \frac{1}{2}\left[10p^4 - 18p^3 - 11p^2 + 3p + 2\right]$. ■

Proof. Let $\mathbb{Z}_p \times \mathbb{Z}_{p^2}$ be a ring. Note that $\mathbb{Z}_p$ has no non-zero zero-divisor and in $\mathbb{Z}_{p^2}$ the non-zero zero-divisors are multiples of p. Now we partition the set $Z^*\left(\mathbb{Z}_p \times \mathbb{Z}_{p^2}\right)$ as follows
$Z^*\left(\mathbb{Z}_p \times \mathbb{Z}_{p^2}\right) = A_1 \cup A_2 \cup A_3 \cup A_4$ where

$$\begin{aligned}
A_1 &= \{(x_1, kp)\,|x_1 \in \mathbb{Z}_p^* \,\&\, k = 1,2,3,\ldots,p-1\}\\
A_2 &= \{(0, x_2)\,|x_2 \in \mathbb{Z}_{p^2}^*\}\\
A_3 &= \{(x_1, 0)\,|n_1 \in \mathbb{Z}_p^*\}\\
A_4 &= \{(0, kp)\,|k = 1,2,3,\ldots,p-1\}
\end{aligned}$$

Then $|A_1| = (p-1)^2$, $|A_2| = (p^2 - p)$, $|A_3| = (p-1)$, $|A_4| = (p-1)$.
For every $x \in A_1$:

$$\begin{aligned}
W(x) &= \sum_{y\in A_2} d(x,y) + \sum_{y\in A_3} d(x,y) + \sum_{y\in A_4} d(x,y) + \sum_{y\in A_1, x\neq y} d(x,y)\\
&= \sum_{y\in A_2} 3 + \sum_{y\in A_3} 2 + \sum_{y\in A_4} 1 + \sum_{y\in A_1, x\neq y} 2\\
&= 3(p^2 - p) + 2(p-1) + (p-1) + 2((p-1)^2 - 1)\\
&= 5p^2 - 4p - 3
\end{aligned}$$

For every $x \in A_2$:

$$\begin{aligned}
W(x) &= \sum_{y\in A_1} d(x,y) + \sum_{y\in A_3} d(x,y) + \sum_{y\in A_4} d(x,y) + \sum_{y\in A_2, x\neq y} d(x,y)\\
&= \sum_{y\in A_1} 3 + \sum_{y\in A_3} 1 + \sum_{y\in A_4} 2 + \sum_{y\in A_2, x\neq y} 2\\
&= 3(p-1)^2 + (p-1) + 2(p-1) + 2(p^2 - p - 1)\\
&= 5p^2 - 5p - 2
\end{aligned}$$

For every $x \in A_3$:

$$W(x) = \sum_{y\in A_1} d(x,y) + \sum_{y\in A_2} d(x,y) + \sum_{y\in A_4} d(x,y) + \sum_{y\in A_3, x\neq y} d(x,y)$$

$$= \sum_{y\in A_1} 2 + \sum_{y\in A_2} 1 + \sum_{y\in A_4} 1 + \sum_{y\in A_3, x\neq y} 2$$

$$= 2(p-1)^2 + (p^2-p) + (p-1) + 2(p-2)$$

$$= 3p^2 - 2p - 3$$

For every $x \in A_4$:

$$W(x) = \sum_{y\in A_1} d(x,y) + \sum_{y\in A_2} d(x,y) + \sum_{y\in A_3} d(x,y) + \sum_{y\in A_4, x\neq y} d(x,y)$$

$$= \sum_{y\in A_1} 1 + \sum_{y\in A_2} 2 + \sum_{y\in A_3} 1 + \sum_{y\in A_4, x\neq y} 1$$

$$= (p-1)^2 + 2(p^2-p) + (p-1) + (p-2)$$

$$= 3p^2 - 2p - 2$$

Therefore

$$W(\Gamma(\mathbb{Z}_p \times \mathbb{Z}_{p^2})) = \frac{1}{2}\left[\sum_{x\in A_1} d(x,y) + \sum_{x\in A_2} d(x,y) + \sum_{x\in A_3} d(x,y) + \sum_{x\in A_4} d(x,y)\right]$$

$$= \frac{1}{2}[(5p^2-4p-3)(p-1)^2 + 5p^2 - 5p - 2(p^2-p)$$

$$+ (3p^2-2p-3)(p-1) + (3p^2-2p-2)(p-1)]$$

$$= \frac{1}{2}\left[10p^4 - 18p^3 - 11p^2 + 3p + 2\right]$$

□

Theorem 21.5

Let $\mathbb{Z}_p \times \mathbb{Z}_{pq}$ be a ring with p, q being distinct primes.
Then $W(\Gamma(\mathbb{Z}_p \times \mathbb{Z}_{pq})) = \frac{1}{2}[2p^4 + 12p^3q + 10p^2q^2 - 13p^3 - 34p^2q - 13pq^2 + 27p^2 + 4q^2 + 13pq + 9q - 13]$. ■

Proof. Let $\mathbb{Z}_p \times \mathbb{Z}_{pq}$ be a ring with p, q being distinct primes. Note that the set of zero-divisors of $\mathbb{Z}_p \times \mathbb{Z}_{pq}$ is given by $Z^*(\mathbb{Z}_p \times \mathbb{Z}_{pq}) = A_1 \cup A_2 \cup A_3 \cup A_4 \cup A_5 \cup A_6$, where

$$A_1 = \{(x_1, k_1p)\,|x_1 \in \mathbb{Z}_p^* \;\&\; k_1 = 1,2,3,\ldots,q-1\}$$
$$A_2 = \{(0, x_2)\,|x_2 \in \mathbb{Z}_{pq}^* \; and \; x_2 \; is \; non \; zero \; divisor\;\}$$

$$A_3 = \{(x_1, k_2 q)\,|x_1 \in \mathbb{Z}_p^* \;\&\; k_2 = 1,2,3,\ldots,p-1\}$$
$$A_4 = \{(0, k_2 q)\,|k_2 = 1,2,3,\ldots,p-1\}$$
$$A_5 = \{(x_1, 0)\,|n_1 \in \mathbb{Z}_p^*\}$$
$$A_6 = \{(0, k_1 p)\,|k_1 = 1,2,3,\ldots,q-1\}$$

Then $|A_1| = (p-1)(q-1)$, $|A_2| = (p-1)(q-1)$, $|A_3| = (p-1)^2$, $|A_4| = (p-1)$, $|A_5| = (p-1)$, & $|A_6| = (q-1)$.
For every $x \in A_1$:

$$\begin{aligned}
W(x) =& \sum_{y\in A_2} d(x,y) + \sum_{y\in A_3} d(x,y) + \sum_{y\in A_4} d(x,y) + \sum_{y\in A_5} d(x,y) + \sum_{y\in A_6} d(x,y) \\
&+ \sum_{y\in A_1, x\neq y} d(x,y) \\
=& \sum_{y\in A_2} 3 + \sum_{y\in A_3} 3 + \sum_{y\in A_4} 1 + \sum_{y\in A_5} 2 + \sum_{y\in A_6} 2 + \sum_{y\in A_1, x\neq y} 2 \\
=& 3(p-1)(q-1) + 3(p-1)^2 + (p-1) + 2(p-1) + 2(q-1) \\
&+ 2((p-1)(q-1)-1) \\
=& 3p^2 + 5pq - 9p - 3q - 9
\end{aligned}$$

for every $x \in A_2$:

$$\begin{aligned}
W(x) =& \sum_{y\in A_1} d(x,y) + \sum_{y\in A_3} d(x,y) + \sum_{y\in A_4} d(x,y) + \sum_{y\in A_5} d(x,y) + \sum_{y\in A_6} d(x,y) \\
&+ \sum_{y\in A_2, x\neq y} d(x,y) \\
=& \sum_{y\in A_1} 3 + \sum_{y\in A_3} 3 + \sum_{y\in A_4} 2 + \sum_{y\in A_5} 1 + \sum_{y\in A_6} 2 + \sum_{y\in A_2, x\neq y} 2 \\
=& 3(p-1)(q-1) + 3(p-1)^2 + 2(p-1) + (p-1) + 2(q-1) \\
&+ 2((p-1)(q-1)-1) \\
=& 3p^2 + 5pq - 9p - 3q - 9
\end{aligned}$$

for every $x \in A_3$:

$$\begin{aligned}
W(x) =& \sum_{y\in A_1} d(x,y) + \sum_{y\in A_2} d(x,y) + \sum_{y\in A_4} d(x,y) + \sum_{y\in A_5} d(x,y) + \sum_{y\in A_6} d(x,y) \\
&+ \sum_{y\in A_3, x\neq y} d(x,y)
\end{aligned}$$

$$= \sum_{y\in A_1} 3 + \sum_{y\in A_2} 3 + \sum_{y\in A_4} 2 + \sum_{y\in A_5} 2 + \sum_{y\in A_6} 1 + \sum_{y\in A_3, x\neq y} 2$$

$$= 3(p-1)(q-1) + 3(p-1)(q-1) + 2(p-1) + 2(p-1) + (q-1)$$

$$+ 2((p-1)^2 - 1)$$

$$= 2p^2 + 6pq - 6p - 5q + 1$$

for every $x \in A_4$:

$$W(x) = \sum_{y\in A_1} d(x,y) + \sum_{y\in A_2} d(x,y) + \sum_{y\in A_3} d(x,y) + \sum_{y\in A_5} d(x,y) + \sum_{y\in A_6} d(x,y)$$

$$+ \sum_{y\in A_4, x\neq y} d(x,y)$$

$$= \sum_{y\in A_1} 1 + \sum_{y\in A_2} 2 + \sum_{y\in A_3} 2 + \sum_{y\in A_5} 1 + \sum_{y\in A_6} 1 + \sum_{y\in A_4, x\neq y} 2$$

$$= (p-1)(q-1) + 2(p-1)(q-1) + 2(p-1)^2 + (p-1) + (q-1) + 2(p-2)$$

$$= 2p^2 + 3pq - 4p - 2q - 1$$

for every $x \in A_5$:

$$W(x) = \sum_{y\in A_1} d(x,y) + \sum_{y\in A_2} d(x,y) + \sum_{y\in A_3} d(x,y) + \sum_{y\in A_4} d(x,y) + \sum_{y\in A_6} d(x,y)$$

$$+ \sum_{y\in A_5, x\neq y} d(x,y)$$

$$= \sum_{y\in A_1} 2 + \sum_{y\in A_2} 1 + \sum_{y\in A_3} 2 + \sum_{y\in A_4} 1 + \sum_{y\in A_6} 1 + \sum_{y\in A_5, x\neq y} 2$$

$$= 2(p-1)(q-1) + (p-1)(q-1) + 2(p-1)^2 + (p-1) + (q-1) + 2(p-2)$$

$$= 2p^2 + 3pq - 4p - 2q - 1$$

for every $x \in A_6$:

$$W(x) = \sum_{y\in A_1} d(x,y) + \sum_{y\in A_2} d(x,y) + \sum_{y\in A_3} d(x,y) + \sum_{y\in A_4} d(x,y) + \sum_{y\in A_5} d(x,y)$$

$$+ \sum_{y\in A_6, x\neq y} d(x,y)$$

$$= \sum_{y\in A_1} 2 + \sum_{y\in A_2} 2 + \sum_{y\in A_3} 1 + \sum_{y\in A_4} 1 + \sum_{y\in A_5} 1 + \sum_{y\in A_6, x\neq y} 2$$

$$=2(p-1)(q-1)+2(p-1)(q-1)+(p-1)^2+(p-1)+(q-1)+2(p-2)$$

$$=p^2+4pq-3p-3q-1$$

Therefore

$$W(\Gamma(\mathbb{Z}_p\times\mathbb{Z}_{pq}))=\frac{1}{2}\left[\sum_{x\in A_1}d(x,y)+\sum_{x\in A_2}d(x,y)+\sum_{x\in A_3}d(x,y)+\sum_{x\in A_4}d(x,y)\right.$$

$$\left.+\sum_{x\in A_5}d(x,y)+\sum_{x\in A_6}d(x,y)\right]$$

$$=\frac{1}{2}[2p^4+12p^3q+10p^2q^2-13p^3-34p^2q-13pq^2+27p^2$$

$$+4q^2+13pq+9q-13]$$

□

Theorem 21.6

Let $\mathbb{Z}_p\times\mathbb{Z}_{p^3}$ be a ring with p being any prime.
Then $W(\Gamma(\mathbb{Z}_p\times\mathbb{Z}_{p^3}))=\frac{1}{2}[10p^6-10p^5+4p^3-p^2-3p+2]$. ■

Proof. Let $\mathbb{Z}_p\times\mathbb{Z}_{p^3}$ be a ring. Note that $\mathbb{Z}_p$ has no non-zero zero-divisor and in $\mathbb{Z}_{p^3}$ the non-zero zero-divisors are multiples of p and p^2.We partition the set
$Z^*(\mathbb{Z}_p\times\mathbb{Z}_{p^3})$ as follows:
$Z^*(\mathbb{Z}_p\times\mathbb{Z}_{p^3})=A_1\cup A_2\cup A_3\cup A_4\cup A_5\cup A_6$, where

$$A_1=\{(0,x_2)\,|x_2\in\mathbb{Z}^*_{p^3}\ and\ x_2\ is\ non\ zero\ divisor\ \}$$
$$A_2=\{(x_1,k_2p^2)\,|x_1\in\mathbb{Z}^*_p\ \&\ k_1=1,2,3,\ldots,p-1\}$$
$$A_3=\{(x_1,k_1p)\,|x_1\in\mathbb{Z}^*_p\ \&\ k_2=1,2,3,\ldots,p^2-1\ and\ p\nmid k_1\}$$
$$A_4=\{(x_1,0)\,|x_1\in\mathbb{Z}^*_p\}$$
$$A_5=\{(0,k_1p)\,|k_2=1,2,3,\ldots,p^2-1\ and\ p\nmid k_1\}$$
$$A_6=\{(0,k_2p^2)\,|k_1=1,2,3,\ldots,p-1\}$$

Then $|A_1|=p^2(p-1)$, $|A_2|=(p-1)^2$, $|A_3|=p(p-1)^2$, $|A_4|=(p-1)$,
$|A_5|=p(p-1)$, & $|A_6|=(p-1)$.
For every $x\in A_1$:

$$W(x)=\sum_{y\in A_2}d(x,y)+\sum_{y\in A_3}d(x,y)+\sum_{y\in A_4}d(x,y)+\sum_{y\in A_5}d(x,y)+\sum_{y\in A_6}d(x,y)$$

$$+\sum_{y\in A_1,x\neq y} d(x,y)$$

$$=\sum_{y\in A_2} 3+\sum_{y\in A_3} 3+\sum_{y\in A_4} 1+\sum_{y\in A_5} 2+\sum_{y\in A_6} 2+\sum_{y\in A_1,x\neq y} 2$$

$$=3(p-1)^2+3p(p-1)^2+(p-1)+2p(p-1)+2(p-1)+2(p^2(p-1)-1)$$

$$=5p^3-3p^2-2p-2$$

for every $x\in A_2$:

$$W(x)=\sum_{y\in A_1} d(x,y)+\sum_{y\in A_3} d(x,y)+\sum_{y\in A_4} d(x,y)+\sum_{y\in A_5} d(x,y)+\sum_{y\in A_6} d(x,y)$$

$$+\sum_{y\in A_2,x\neq y} d(x,y)$$

$$=\sum_{y\in A_1} 3+\sum_{y\in A_3} 3+\sum_{y\in A_4} 2+\sum_{y\in A_5} 1+\sum_{y\in A_6} 1+\sum_{y\in A_2,x\neq y} 2$$

$$=3p^2(p-1)+3p(p-1)^2+2(p-1)+p(p-1)+(p-1)+2((p-1)^2-1)$$

$$=6p^3-6p^2+p-3$$

for every $x\in A_3$:

$$W(x)=\sum_{y\in A_1} d(x,y)+\sum_{y\in A_2} d(x,y)+\sum_{y\in A_4} d(x,y)+\sum_{y\in A_5} d(x,y)+\sum_{y\in A_6} d(x,y)$$

$$+\sum_{y\in A_3,x\neq y} d(x,y)$$

$$=\sum_{y\in A_1} 3+\sum_{y\in A_2} 3+\sum_{y\in A_4} 2+\sum_{y\in A_5} 2+\sum_{y\in A_6} 1+\sum_{y\in A_3,x\neq y} 2$$

$$=3p^2(p-1)+3(p-1)^2+2(p-1)+2p(p-1)+(p-1)+2(p(p-1)^2-1)$$

$$=5p^3-2p^2-3p-2$$

for every $x\in A_4$:

$$W(x)=\sum_{y\in A_1} d(x,y)+\sum_{y\in A_2} d(x,y)+\sum_{y\in A_3} d(x,y)+\sum_{y\in A_5} d(x,y)+\sum_{y\in A_6} d(x,y)$$

$$+\sum_{y\in A_4,x\neq y} d(x,y)$$

$$=\sum_{y\in A_1} 1+\sum_{y\in A_2} 2+\sum_{y\in A_3} 2+\sum_{y\in A_5} 1+\sum_{y\in A_6} 1+\sum_{y\in A_4, x\neq y} 2$$

$$=p^2(p-1)+2(p-1)^2+2p(p-1)^2+p(p-1)+(p-1)+2(p-2)$$

$$=3p^2-2p^2-3$$

for every $x \in A_5$:

$$W(x)=\sum_{y\in A_1} d(x,y)+\sum_{y\in A_2} d(x,y)+\sum_{y\in A_3} d(x,y)+\sum_{y\in A_4} d(x,y)+\sum_{y\in A_6} d(x,y)$$

$$+\sum_{y\in A_5, x\neq y} d(x,y)$$

$$=\sum_{y\in A_1} 2+\sum_{y\in A_2} 1+\sum_{y\in A_3} 2+\sum_{y\in A_4} 1+\sum_{y\in A_6} 1+\sum_{y\in A_5, x\neq y} 2$$

$$=2p^2(p-1)+(p-1)^2+2p(p-1)^2+(p-1)+(p-1)+2(p(p-1)-1)$$

$$=4p^3-3p^2-3$$

for every $x \in A_6$:

$$W(x)=\sum_{y\in A_1} d(x,y)+\sum_{y\in A_2} d(x,y)+\sum_{y\in A_3} d(x,y)+\sum_{y\in A_4} d(x,y)+\sum_{y\in A_5} d(x,y)$$

$$+\sum_{y\in A_6, x\neq y} d(x,y)$$

$$=\sum_{y\in A_1} 2+\sum_{y\in A_2} 1+\sum_{y\in A_3} 1+\sum_{y\in A_4} 1+\sum_{y\in A_5} 1+\sum_{y\in A_6, x\neq y} 1$$

$$=2p^2(p-1)+(p-1)^2+p(p-1)^2+(p-1)+p(p-1)+(p-2)$$

$$=3p^3-2p^2-2$$

Therefore

$$W(\Gamma(\mathbb{Z}_p\times\mathbb{Z}_{p^3}))=\frac{1}{2}[\sum_{x\in A_1} d(x,y)+\sum_{x\in A_2} d(x,y)+\sum_{x\in A_3} d(x,y)+\sum_{x\in A_4} d(x,y)$$

$$+\sum_{x\in A_5} d(x,y)+\sum_{x\in A_6} d(x,y)]$$

$$=\frac{1}{2}[p^2(p-1)(5p^3-3p^2-2p-2)+(p-1)^2(6p^3-6p^2+p-3)$$

$$+p(p-1)^2(5p^3-2p^2-3p-2)+(p-1)(3p^2-2p^2-3)$$

$$+p(p-1)(4p^3-3p^2-3)+(p-1)(3p^3-2p^2-2)]$$

$$=\frac{1}{2}[10p^6-10p^5+4p^3-p^2-3p+2]$$

□

21.4 CONCLUDING REMARK

The concept of the Wiener index is a frontier between algebraic graph theory and chemistry. We have investigated the Wiener index of various graphs like $\Gamma(\mathbb{Z}_n)$, $\Gamma(\mathbb{Z}_n[i])$ and $\Gamma(R_1 \times R_2)$ obtained from commutative rings.

REFERENCES

1. Mohammas Reza Ahmadi and Reza Jahani-Nezhad. Energy and Wiener index of zero-divisor graphs. *Iranian Journal of Mathematical Chemistry*, 2(1): 45-51, 2011.
2. D. D. Anderson and M. Naseer. Beck's coloring of a commutative ring. *J. Algebra*, 159, 500-514, 1993.
3. D. F. Anderson and P. S. Livingston. The zero-divisor graph of a commutative ring. *J. Algebra*, 217, 434-447, 1999.
4. M. F. Atiyah and I. G. Macdonald. *Introduction to Commutative Algebra.* Addison-Wesley Publishing Company, 1969.
5. I. Beck. Coloring of commutative rings. *J. Algebra*, 116, 208-226, 1988.
6. I. Gutman, S. Klavzar and B. Mohar. Fifty years of the Wiener index. *MATCH Commun. Math. Comput. Chem.*, 35, 1-259, 1997.
7. Emad Abu Osba, Salah Al-Addasi and Nafiz Abu Jaradeh. Zero-divisor graph for the ring of Gaussian integers modulo n. *Communications in Algebra*, 36, 3865-3877, 2008.
8. R. Balakrishnan and K. Ranganathan. *A Textbook of Graph Theory.* Second edition, Springer, 2012.
9. Surendranath Reddy, B. Jain, S. Rupali and N. Laxmikanth. Eigenvalues and Wiener index of the zero-divisor graph $\Gamma[\mathbb{Z}_n]$. 2017, eprint arXiv:1707.05083.
10. H. Wiener. Structural determination of the paraffin boiling points. *J. Am. Chem. Soc.*, 69, 17-20, 1947.

22 Algebraic Signed Graphs: A Review

Pranjali
University of Rajasthan,
Jaipur, Rajasthan (INDIA)
E-mail: pranjali48@gmail.com

Amit Kumar
Banasthali vidyapith,
Banasthali, Rajasthan (INDIA)
E-mail: amitsu48@gmail.com

In this chapter we survey the research conducted on algebraic graphs, with a focus on the odyssey of algebraic signed graphs associated with a finite commutative ring R. In particular, we focus on the problems which are based on balanceness and consistency of algebraic signed graphs. To demonstrate the concept suitable examples are also given.

The field of graph theory has undergone incredible development during the past century and during the past quarter-century, the revolution of the subject has continued, with individual areas (such as algebraic combinatorics, algorithmic graph theory) escalating to the point of having essential sub-branches themselves. Algebraic combinatorics is an area of mathematics that employs methods of abstract algebra in diverse combinatorial contexts and vice versa. Associating a graph with an algebraic structure is a research subject in this area and has attracted huge attention. In fact, research in this subject aims at exploring the relationship between algebra and graph theory and their applications.

The area of algebraic graph theory is the study of algebra and graph theory. In this chapter, our focus has been placed on the problems that are related to the study of algebraic graphs and it gives an interplay between ring-theoretic and graph theoretic properties. The focus is also placed on very interesting and widely studied variants of graphs, namely, the problems based on balanced signed total graphs of commutative rings, and we have also conducted a study on various properties and applications of these concepts. We hope that the notes would be useful for anyone interested to work further on current problems in the theory of signed graphs. This is a predominantly interesting subject for some graph theorists and algebraists, since it relates two different areas of mathematics.

22.1 ALGEBRAIC GRAPH

In this section we shall discuss the structure of some algebraic graphs. The purpose of studying these graphs is that one may find some results about the algebraic structures and vice-versa. There are three major problems in this area: (1) characterization of the resulting algebraic graphs, (2) characterization of the algebraic structures with isomorphic graphs and (3) realization of the connections between the structures and the corresponding graphs. The term algebraic graphs using the binary operation was coined by Pranjali [13], the formal definition is as follows:

Definition. Let $(A,*)$ be an algebraic structure. A graph $G := (V,E)$ is called an *algebraic graph*, if $V \subseteq A$ and the adjacency rule is due to *binary operation* '$*$' of algebraic structure.

The Zero-divisor graph, Cayley graph, Total graph, Zero ring graph, Co-maximal graph are the examples of an algebraic graph.

Remark. In spite of choosing algebraic structure group and ring, the zero-divisor graphs have been associated with lattice, semi group and other algebraic graphs.

22.1.1 EXISTING ALGEBRAIC GRAPHS

In this section we shall demonstrate the structure of algebraic graphs.

In 1878, Cayley [9] introduced the first algebraic graph called the Cayley graph associated with an algebraic structure 'group', which is an important concept relating two different branches of mathematics group theory and graph theory. The Cayley graphs are frequently used to render the abstract structure of a group easily visible by way of representing this structure in graph form. The following definition was given by Cayley [9]:

Definition. Let H be a finite group and let $S \subseteq H$ be a subset. The Cayley graph, denoted by $C(H,S)$ has vertex set H and two distinct vertices $x, y \in H$ are joined by a directed edge from x to y if and only if there exists $s \in S$ such that $x = sy$. Each edge is labeled to denote that it corresponds to $s \in S$. If Γ is a graph such that there exists a group H and generating set $S \subset H$ with $G \cong C(H,S)$, then Γ is said to be Cayley.

Example. If $H = \mathbb{Z}_4$ and $S = \{1,3\}$, then $C(H,S)$ is the cycle on 'n' vertices.

Other than the Cayley graph there are many algebraic graphs which are associated with an algebraic structure 'group', but here we are interested in those algebraic graphs which are associated with an algebraic structure 'ring'.

In 1988, the concept of zero-divisor graph $\Gamma(R)$ of a commutative ring R was introduced by Beck in [7]. According to Beck, let zero-divisor graph $\Gamma(R)$ of R have $Z(R)$ as the vertex set and two distinct vertices x and y be defined to be adjacent if and only if $x \cdot y = 0$, because the element $0 \in R$ would create a vertex of *full degree* in $\Gamma(R)$.

In 1995, Sharma and Bhatwadekar [18], introduced another graphical structure on R, which later came to be known as Comaximal graphs.

Definition. [18] Let R be a commutative ring with identity. The comaximal graph of R is a graph whose vertices as elements of R and two distinct vertices x and y are adjacent if and only if $Rx + Ry = R$.

In 1999, Anderson and Livingston [2] modified the definition of zero-divisor graph as follows:

Definition. [2] Let R be a commutative ring with unity and $Z^0(R)$ be its set of nonzero zero-divisors. The zero-divisor graph $\Gamma(R)$ of R is a graph whose vertices are the set of nonzero zero-divisors of R and two distinct vertices $x, y \in Z^0(R)$ are adjacent if and only if $x \cdot y = 0$.

Example. The zero-divisor graph of ring $\mathbb{Z}_6$ and $\mathbb{Z}_{25}$ is the path P_3 and K_4, respectively.

In 2008, Boggess et al. [5] introduced the notion of *unitary Cayley graph* as follows:

Definition. [5] Let $\mathbb{Z}_n$ be the ring of integers modulo n and $U(n)$ denotes the set of all units of the ring $\mathbb{Z}_n$. The unitary Cayley graph, X_n is the graph whose vertices are the elements of $\mathbb{Z}_n$ and two distinct vertices x and y are adjacent if and only if $x - y \in U(n)$.

Example. The unitary Cayley graph of ring $\mathbb{Z}_6$ is the cycle graph C_6.

At the same time, in 2008, the concept of *total graph* was initiated by Anderson and Badawi [3] and the formal definition of total graph is as follows:

Definition. [3] The total graph of a commutative ring R, denoted by $T(\Gamma(R))$ is a graph, whose vertices are the elements of R and two distinct vertices x and y are said to be adjacent if and only if $(x+y) \in Z(R)$, where $Z(R)$ denotes the set of zero-divisors of R.

Example. The total graph of ring $\mathbb{Z}_4$ is disconnected and is isomorphic to two copies of K_2.

In 1999, the concept of *unit graph*, denoted as $G(\mathbb{Z}_n)$, was defined by Grimaldi [10], based on the elements and the units of $\mathbb{Z}_n$, which was further studied by Ashrafi et al. [4] for any arbitrary associative ring R.

Definition. [4] Let R be a ring with nonzero identity. The unit graph of R, denoted by $G(R)$, is a graph whose vertices are the elements of R, and two distinct vertices x and y are adjacent if and only if $x + y$ is a unit of R.

However, it is found that from the literature in 2011, with the same definition the graph was named *unitary addition Cayley graph.*

Example. The unit graph of rings $\mathbb{Z}_3$ and $\mathbb{Z}_4$ is the path P_3 and C_4 respectively.

In order to generalize the concept of unitary Cayley graph form $\mathbb{Z}_n$ to arbitrary ring R, the concept of counit graph [11] and total graph was initiated.

Definition. Let R be a ring and $U(R)$ be the set of unit elements of R. The counit graph of R, denoted $\tilde{G}(R)$, is the graph obtained by setting all the elements of R to be the vertices and defining distinct vertices x and y to be adjacent if and only if $x-y \in U(R)$.

Example. The counit graph of $\mathbb{Z}_4$ is C_4.

The counit graph is alike to the *unit graph.* For this graph, one may deduce akin properties such as those of unit graphs. In some cases, dealing with the counit graph is better than with the unit graph.

Keeping this in mind, Maimani et al. [12] defined a graph similar to the *total graph* known as *cototal* graph. The formal definition is as follows:

Definition. Let R be a ring and $Z(R)$ be the set of zero-divisors of R. The cototal graph of R, denoted $\tilde{T}(\Gamma(R))$, is the graph obtained by setting all the elements of R to be the vertices and distinct vertices x and y to be adjacent if and only if $x-y \in Z(R)$

Example. The cototal graph of $\mathbb{Z}_4$ is $2K_2$.

In 2014, Pranjali and Acharya [15] introduce the notion of *zero ring graph* (the graph whose vertices are the elements of zero ring) and investigated the basic properties of the zero ring graph.
The following is a formal definition of the new notion given in [15].

Definition. Let $(R^0,+,\cdot)$ be a finite zero ring. Then the zero ring graph, denoted as $\Gamma(R^0)$, is a graph whose vertices are the elements of zero ring R^0 and two distinct vertices x and y are adjacent if and only if $x+y \neq 0$, where '0' is the additive identity of R^0.

Example. The zero ring graphs $\Gamma(M_2^0(\mathbb{Z}_4))$ and $\Gamma(M_2^0(\mathbb{Z}_5))$ are isomorphic to $K_4-\{e\}$ and C_4+K_1, respectively.

22.2 SIGNED GRAPH AND ITS PARAMETERS

The signed graphs are the natural generalization of graphs and they arose while building a prototype model for dealing with highly intricate problems in social psychology, especially dealing with structural models of cognitive interpersonal relationships in a social group. They are much studied in the literature because of their imperative use in the analysis of models of a variety of cognition-based social processes.

On the other hand, they are found to have strong connection with many classical mathematical systems too. For example, they enable one to describe and analyze the geometry of subsets of the classical root systems. They appear in topological graph theory and group theory. They are a natural context for investigating questions about odd and even cycles in graphs.

Let G be an arbitrary graph. Then a *signature* σ on G is a rule that labels each edge of G with either a positive or a negative sign. The graph G equipped with a signature σ is called a *signed graph*, denoted by $\Sigma := (G, \sigma)$, where $G = (V, E)$ is an *underlying graph* and $\sigma : E \to \{+, -\}$ is the signature that makes each edge of G either '+' or '−'. The edge which receives the positive (negative) sign is called a positive (negative) edge. A signed graph is said to be an *all-positive* if all its edges are positive and an *all-negative* signed graph is defined similarly. A signed graph is called *homogeneous* if it is either all-positive or all-negative and *heterogeneous* otherwise.

By $d^-(v)$ $(d^+(v))$, we mean the negative (positive) degree of a vertex v, i.e., the number of negative (positive) edges incident at v in Σ. For a detailed study of signed graphs, the reader is referred to the bibliography on signed graph by Zaslavsky [24].

22.2.1 CRITERIA FOR BALANCE

One of the essential concepts in the theory of signed graph is that of *balance*. Harary [23] set up the enthralled concept of *balanced signed graphs* for the study of social networks, in which a positive edge stands for a positive relation and a negative edge stands for a negative relation. A signed graph is *balanced* if every cycle has an even number of negative edges and a signed graph which is not balanced is called an *unbalanced* signed graph. The following is the well-known result given by Harary [23].

Lemma 22.1

A signed graph Σ is balanced if and only if V can be partitioned into two disjoint subsets V_1 and V_2, one of them possibly may be empty such that all negative edges lie across V_1 and V_2. ■

22.2.2 CRITERIA FOR CONSISTENCY

The study on consistent marked graphs was established by Beineke and Harary [8]. The concept was inspired by communication networks.

A *marked signed graph* is an ordered pair $\Sigma_\mu = (\Sigma, \mu)$, where $\Sigma = (G, \sigma)$ is a signed graph and $\mu : V(\Sigma) \to \{+, -\}$ is a function from the vertex set $V(\Sigma)$ into the set $\{+, -\}$, called the *marking* of Σ.

Definition. [19] A cycle in Σ_μ is said to be *consistent* if it contains an even number of negative vertices; a vertex is negative (positive) if it receives '−' ('+') sign under μ. A given marked signed graph Σ is said to be consistent if every cycle in it is consistent. In particular, σ induces a unique marking μ_σ defined by

$$\mu_\sigma(v) = \prod_{e \in E_v} \sigma(e),$$

where E_v is the set of edges incident at v in Σ, called a *canonical marking* of Σ. If every vertex of a given signed graph Σ is canonically marked, then a cycle Z in Σ is said to be *canonically consistent* (C-consistent) and the signed graph Σ is said to be C-consistent if every cycle in it is C-consistent.

22.2.3 CRITERIA FOR SIGN-COMPATIBILITY

The concept of sign-compatibility for a graph was discussed in [19]. A signed graph Σ is *sign-compatible* if there exists a marking μ to its vertices such that the end vertices of each negative edge receive '−' marking and no positive edge should receive '−' marking to both its ends; otherwise Σ is called *sign-incompatible.* The following characterization of sign-compatible signed graphs has been established, which is somewhat antithetic to the Harary partition criterion for balance in signed graphs.

Theorem 22.1

A signed graph Σ is sign-compatible if and only if $V(\Sigma)$ can be partitioned into subsets V_1 and V_2 such that the all-negative subgraph of Σ is precisely the subgraph induced by exactly one of the subset V_1 and V_2. ■

22.3 ALGEBRAIC SIGNED GRAPH

This section is concerned with a natural extension of the notion of algebraic graph in the realm of signed graph for a finite commutative ring R. Apart from these concepts, an attempt has also been made to investigate certain parameters of signed graphs derived from algebraic structure.

Definition. A signed graph $\sum := (G, \sigma)$ is called an *algebraic signed graph,* if its underlying graph $G := (V, E)$ is an algebraic graph and $\sigma : E \to \{+, -\}$ is the signature that labels each edge of G either by '+' or '−'.

Since σ is not unique, there are $2^{|E|}$ choices for σ out of which $2^{|E|} - 2$ will produce the heterogeneous signed graph.

Some examples of algebraic signed graph that are associated with algebraic structure(group/ring) are Signed Total graph, Signed Zero-Divisor Graph, Co-maximal Meet Signed Graph, etc.

Problem. For a given algebraic structure one can construct an algebraic graph and algebraic signed graph and then further can investigate its parameter. But the reverse is quite tedious.

22.3.1 EXISTING ALGEBRAIC SIGNED GRAPHS

This section depicts some well-known algebraic signed graphs.

In 2011, Sinha and Garg [22] introduced the *unitary Cayley signed graph*, denoted by S_n on the group $\mathbb{Z}_n$. In a similar vein in 2013, Sinha and Dhama [21] introduced the concept of *unitary addition Cayley signed graph* denoted by Σ_n on the group $\mathbb{Z}_n$. They have derived certain results for various parameters by posing some condition for value of n.

Notice that the studies on both unitary Cayley signed graph and unitary addition Cayley signed graph were associated to the particular group $\mathbb{Z}_n$. To generalize the study on signed graph from group $\mathbb{Z}_n$ to arbitrary ring R, in 2017, Pranjali and Acharya [14] introduced the new algebraic signed graph, often called 'signed total graph' for an arbitrary ring R. In 2018, Sinha and Rao [20] introduced the concept of *co-maximal signed graphs* under three possible signatures function σ and obtained some results.

22.4 SIGNED TOTAL GRAPHS

In this section, we shall give a glimpse of the results for *signed total graph.*

From a survey of the literature, it is known that there does not exist any structure theorem for the total graph, when $Z(R)$ is not an ideal. Another motivation of the work reported here is to give a characterization of balanced signed total graph, when $Z(R)$ is not an ideal (without knowing the structure of the total graph). The formal definition of the new notion established in [14] is as follows:

Definition. [14] A signed total graph is an ordered pair $T_\Sigma(\Gamma(R)) := (T(\Gamma(R)), \sigma)$, where $T(\Gamma(R)$ is the total graph of a commutative ring R and for an edge (a,b) of $T_\Sigma(\Gamma(R))$, σ is defined as

$$\sigma(ab) = \begin{cases} + & \text{if } a \in Z(R) \text{ or } b \in Z(R), \\ - & \text{otherwise.} \end{cases}$$

The signed total graphs $T_\Sigma(\Gamma(\mathbb{Z}_4))$ and $T_\Sigma(\Gamma(\mathbb{Z}_6))$ are shown in Figure 22.1, in which positive edges are drawn as a solid line segment and negative edges as a dotted line segment.

22.4.1 BALANCED SIGNED TOTAL GRAPHS

Now we shall give a characterization of balanced signed total graphs in both the cases when $Z(R)$ is an ideal of R and when $Z(R)$ is not an ideal of R,

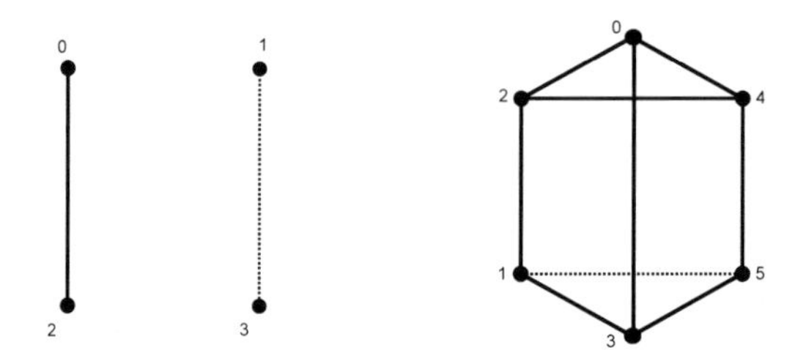

Figure 22.1: Signed total graph $T_\Sigma(\Gamma(\mathbb{Z}_4))$ and $T_\Sigma(\Gamma(\mathbb{Z}_6))$

which have been appeared in [14]. Here we have characterized all the rings R, whose signed total graphs are balanced.

In the case when $Z(R)$ is a non-trivial ideal of R, we have the following theorem.

Theorem 22.2

Let R be a finite commutative ring with unity such that $Z(R)$ is a non-trivial ideal of R with $|Z(R)| = \alpha > 2$. Then the signed total graph $T_\Sigma(\Gamma(R))$ is balanced if and only if $2 \notin Z(R)$. ■

Theorem 22.3

Let R be a finite commutative ring with unity such that $Z(R)$ is not an ideal of R. Then $T_\Sigma(\Gamma(R))$ is balanced if and only if

$$R \cong \underbrace{\mathbb{Z}_2 \times \mathbb{Z}_2 \times \mathbb{Z}_2 \times \cdots \times \mathbb{Z}_2}_{t-times}, \quad t > 1 \quad \text{or} \quad \mathbb{Z}_3 \times \mathbb{F}_{2^k}$$

■

Theorem 22.4

Let R be a finite commutative ring with unity such that $|Z(R)| = \alpha > 1$. Then $T(\Gamma(R)) \cong T_\Sigma(\Gamma(R))$ if and only if R is either a field of characteristic 2 or R is isomorphic to $\underbrace{\mathbb{Z}_2 \times \mathbb{Z}_2 \times \mathbb{Z}_2 \times \cdots \times \mathbb{Z}_2,}_{t-times}$ $t \geq 1$. ■

Now, we shall study C-consistency of $T_\Sigma(\Gamma(R))$ for both the cases when $Z(R)$ is an ideal or not an ideal of R.

Theorem 22.5

Let R be a commutative ring with unity such that $Z(R)$ is an ideal of R. Then $T_\Sigma(\Gamma(R))$ is C-consistent if and only if $2 \notin Z(R)$. ■

Theorem 22.6

Let R be commutative ring with unity such that $Z(R)$ is not an ideal of R. Then $T_\Sigma(\Gamma(R))$ is C-consistent if and only if $R \cong \prod_{i=1}^{t} \mathbb{F}_i$, where each $\mathbb{F}_i$ is a field of characteristic 2. ■

22.5 SIGNED UNIT GRAPHS

In this section, we intend to discuss the notion of a *signed unit graph*, introduced by Pranjali et al. [16]. The formal definition of the new notion is as follows:

Definition. A signed unit graph is an ordered pair $G_\Sigma(R) := (G(R),\ \sigma)$, where $G(R)$ is the unit graph of a commutative ring R and for an edge (a,b) of $G_\Sigma(R)$, σ is defined as

$$\sigma(a,b) = \begin{cases} + \,, & \text{if } \ a \in U(R) \text{ or } b \in U(R); \\ - \,, & \text{otherwise.} \end{cases}$$

The signed unit graphs $G_\Sigma(\mathbb{Z}_2 \times \mathbb{Z}_2)$ and $G_\Sigma(\mathbb{Z}_6)$ are shown in Figure 22.2, in which positive edges are drawn as a solid line segment and negative edges as a dotted line segment. It is clear from Figure 22.2 that $G_\Sigma(\mathbb{Z}_2 \times \mathbb{Z}_2)$ and $G_\Sigma(\mathbb{Z}_6)$ are both balanced.

The following result appeared in [16], in which authors characterize the finite commutative rings R with $1 \neq 0$ for which $G_\Sigma(R)$ is balanced.

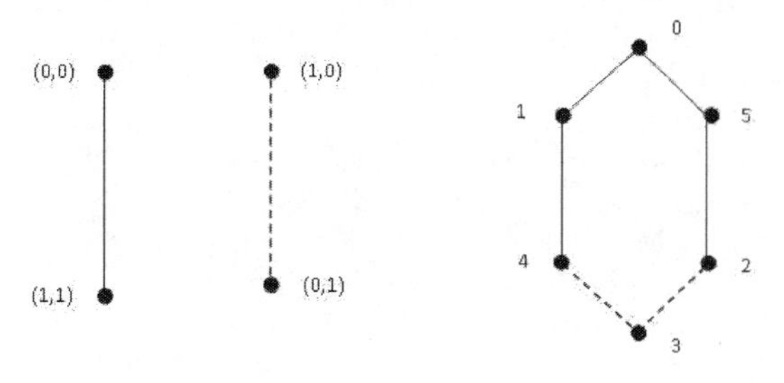

Figure 22.2: The signed unit graphs $G_\Sigma(\mathbb{Z}_2 \times \mathbb{Z}_2)$ and $G_\Sigma(\mathbb{Z}_6)$

Theorem 22.7

Let R be a finite commutative ring $1 \neq 0$. The signed unit graph $G_\Sigma(R)$ is homogeneous (all-positive) if and only if R is a local ring. ■

Corollary 22.5.1. *If R is a local ring, then $G_\Sigma(R)$ is balanced.*

In the contrast of Corollary 22.5.1, one can naturally have the following question: For a non-local ring R, is $G_\Sigma(R)$ still balanced?

The answer is not affirmative, as for instance, the signed unit graph $G_\Sigma(R)$ for the ring $R \cong \mathbb{Z}_3 \times \mathbb{F}_4$ consists of a negative triangle, namely, $(1,0)-(1,1)-(0,a)-(1,0)$ with exactly one negative edge in $G_\Sigma(R)$. Consequently, $G_\Sigma(R)$ is not balanced.

Yet, this still does not give a negative answer to the original question; it only says the answer is not affirmative for a few rings. Thus one has the following problem:

Problem. Determine all those non-local rings R for which $G_\Sigma(R)$ is balanced.

This motivates us to ask what conditions should be imposed on a non-local ring, such that $G_\Sigma(R)$ is balanced, and to observe that this has been settled in the following theorem.

Theorem 22.8

Let R be a finite commutative ring with $1 \neq 0$. The signed unit graph $G_\Sigma(\Gamma(R))$ is balanced if and only if R is either a local ring or R is the direct product of local rings with among them at least one having $\mathbb{Z}_2$ as a quotient. ■

REFERENCES

1. N. Ashrafi, H. R. Maimani, M. R. Pournaki and S. Yassemi. Unit graphs associated with rings. *Communications in Algebra*, 38, 2851-2871, 2010.
2. D. F. Anderson and P.S. Livingston. The zero-divisor graph of a commutative ring. *J. Algebra*, 217, 434-447, 1999.
3. D. F. Anderson and A. Badawi. The total graph of a commutative ring. *Journal of Algebra*, 320, 2706-2719, 2008.
4. N. Ashrafi, H. R. Maimani, M. R. Pournaki and S. Yassemi. Unit graphs associated with rings. *Communications in Algebra*, 38, 2851-2871, 2010.
5. M. Boggess, T. Jackson-Henderson, I. Jiménez and R. Karpman. The structure of unitary Cayley graphs. *The Journal of the Summer Undergraduate Mathematical Science Research Institute*, 1-23, 2008.

6. M. Behboodi, and R. Beyranvand. On the structure of commutative rings with $p_1^{k_1} \ldots p_n^{k_n} (1 \leq k_i \leq 7)$ zero-divisors. *European Journal of Pure and Applied Mathematics*, 3, 303-316, 2010.
7. I. Beck. Coloring of commutating rings. *J. Algebra*, 116, 208-226, 1998.
8. L. W. Beineke and F. Harary. Consistency in marked graph. *Journal of Mathematical Psychology.* 18(3): 260-269, 1978.
9. A. Cayley. Desiderata and suggestions: No. 2. The theory of groups: graphical representation. *American Journal of Mathematics*, 1(2): 174-176, 1878.
10. R. P. Grimaldi. Graphs from rings. In Proceedings of the *Twentieth Southeastern Conference on Combinatorics, Graph Theory and Computing.* Congressus Numerantium. 71, 95-103, 1990.
11. A. Lucchini and A. Maróti, Some results and questions related to the generating graph of a finite group. *Ischia Group Theory* 2008, World Scientific Publishing Company, 183-208, 2009.
12. H. R. Maimani, M. R. Pournaki, A. Tehranian and S. Yassemi. Graphs attached to rings revisited. *Arabian Journal for Science and Engineering*, 36(6): 997-2011, 2011.
13. Pranjali. Odyssey of Algebraic Graphs, unpublished.
14. Pranjali and M. Acharya. Balanced signed total graphs of commutative rings. *Graphs and Combinatorics*, 32(4): 1585-1597, 2016.
15. Pranjali and M. Acharya. Graphs associated with zero rings. *General Mathematics Notes.* 24(2): 53-69, 2014.
16. Pranjali, A. Kumar and P. Sharma. Balanced Signed Unit Graphs. *submitted* 2019.
17. Pranjali, A. Gaur and M. Acharya. *C*-consistency in signed total graph of commutative rings, *Discrete Mathematics, Algorithms and Applications*, 8(3): 1650041, 2016.
18. P. K. Sharma and S. M. Bhatwadekar. A note on graphical representation of rings. *Journal of Algebra*, 176, 124-127, 1995.
19. D. Sinha. New Frontiers in the Theory of Signed Graph. Ph.D. Thesis, University of Delhi, (Faculty of Technology), 2005.
20. D. Sinha and A. Kumari Rao. Co-maximal signed graphs of commutative rings. *Turkish Journal of Mathematics*, 42, 1203-1220, 2018.
21. D. Sinha, P. Garg and A. Singh. Some properties of unitary addition Cayley graphs. *Notes on Number Theory and Discrete Mathematics*, 17(3): 49-59, 2011.
22. D. Sinha and Pravin Garg. On the unitary Cayley signed graphs. *Electronic Journal of Combinatorics*, 18(1): #P229, 2011.
23. F. Harary. On the notion of balance of a signed graph. *Michigan Mathematical Journal*, 2, 143-146, 1953.
24. T. Zaslavsky. A mathematical bibliography of signed and gain graphs and allied areas. *Electronic Journal of Combinatorics*, Dynamic Surveys in Combinatorics, #DS8, 1-346, 2012.

23 Nullity and Energy of Complete Tripartite Graphs

Pranjali
University of Rajasthan,
Jaipur, Rajasthan (INDIA)
E-mail: pranjali48@gmail.com

Renu Naresh
Department of Mathematics and Statistics
Banasthali Vidyapith,
Banasthali, Rajasthan (INDIA)
E-mail: renunaresh1@gmail.com

We consider a special class of complete tripartite graphs such as the $(r,2)$ fan graph, $(r,3)$ fan graph and $(r,4)$ cone graph, which are denoted by $T_{1,1,r}, T_{1,2,r}$ and $T_{2,2,r}$ respectively. Let $A(T)$ be the adjacency matrix of a complete tripartite graph T_{n_1,n_2,n_3}, where n_1 is the number of vertices in the first partition of vertex set, n_2 is the second partition of the vertex set and n_3 is the third partition of the vertex set of $T_{1,1,r}, T_{1,2,r}$ and $T_{2,2,r}$, if for some non zero vector V such that $AV = \lambda V$, where λ is an eigenvalue corresponding to vector V namely an eigen vector. The set of eigenvalues of the $A(T)$ form the spectrum of the graph T_{n_1,n_2,n_3}. The number of occurence zero eigenvalues in the spectrum of a graph T_{n_1,n_2,n_3} is known as nullity of the graph T_{n_1,n_2,n_3}, denoted as $\eta(T_{n_1,n_2,n_3})$ and the absolute sum of eigenvalues of the spectrum is called energy of complete tripartite graphs T_{n_1,n_2,n_3}. In this chapter, we have evaluated the nullity and the energy of T_{n_1,n_2,n_3} in terms of n_3 (one of the partitions) respectively. Moreover, we have discussed the chemical nature of isomorphic organic compounds.

23.1 INTRODUCTION AND PRELIMINARIES

Fan and Qian [3] evaluated the nullity set of bipartite graphs and they have also characterized the bipartite graphs with nullity $(n-4)$ and $(n-6)$ respectively. Farooq and et al. [4] have obtained the nullity set of $n-$vertices tripartite graphs and also characterized some tripartite graphs with nullity $(n-4)$ and $(n-6)$. Throughout the chapter we have considered special classes of complete tripartite graphs such as $(r,2)$ fan graph, $(r,3)$ fan graph and $(r,4)$ cone graph, which are denoted by $T_{1,1,r}$, $T_{1,2,r}$ and $T_{2,2,r}$ respectively. In complete tripartite graphs vertex set can be partitioned into three subsets X_1, X_2 and X_3 such that $G[X_1]$, $G[X_2]$ and $G[X_3]$ are non-empty induced graphs: such a

partition (X_1, X_2, X_3) and every vertex of every set is adjoining to vertices of other two vertex sets, where n_1, n_2 and n_3 are the number of vertices in the vertex set (X_1, X_2, X_3) respectively .
Now, we have explored a complete tripartite graph T_{n_1,n_2,n_3} with three partitions $(X_{n_1}, X_{n_2}, X_{n_3})$ in such a manner that there are $n_1+n_2+n_3 = n$ vertices and m edges, where n_1, $n_2 \in \{1,2\}$ and n_3 are fixed positive integer r (say). Then the adjacency matrix $A(T)$ of graph T_{n_1,n_2,n_3} is defined by

$$A(T) = \begin{bmatrix} O & P & Q \\ P^t & O & R \\ Q^t & R^t & O \end{bmatrix}$$

Where P, Q, R and O denote the matrices with all entries 1 and 0 respectively. Specifically, matrices P, Q, R are defined with order $(n_1 \times n_2)$, $(n_1 \times n_3)$ and $(n_2 \times n_3)$ and O is a square matrix with order $(n_1 \times n_1)$, $(n_2 \times n_2)$ and $(n_3 \times n_3)$. Let $A(T)$ be the adjacency matrix of a graph T_{n_1,n_2,n_3}, where n_1 is the number of vertices in the first partition of vertex set, n_2 is the second partition of vertex set and n_3 the third partition of vertex set of $T_{1,1,r}$, $T_{1,2,r}$ and $T_{2,2,r}$, if for some non zero vector V $AV = \lambda V$, where λ is an eigenvalue corresponding to vector V namely eigen vector. The set of eigenvalues of the $A(T)$ form the spectrum of the graph T_{n_1,n_2,n_3}. Energy and nullity are more useful concepts and a lot of work has been done on these concepts see [5, 7].

23.2 NULLITY OF COMPLETE TRIPARTITE GRAPHS

The number of occurences of zero eigenvalue in the spectrum of a graph T_{n_1,n_2,n_3} is known as nullity of the graph T_{n_1,n_2,n_3}, denoted as $\eta(T_{n_1,n_2,n_3})$. The nullity of graphs is very applicable in chemistry. It is responsible for the behavior of unsaturated conjugate hydrocarbons. By Huckel Molecular Orbital Theory [2],

1. If the nullity of a molecular graph is zero, then the isomorphic unsaturated conjugate hydrocarbons have stable, closed shell, electron configuration, low chemical reactivity and exist in nature.
2. If the nullity of a molecular graph is greater than zero, then the respective unsaturated conjugate hydrocarbons have unstable, open shell, electron configuration, high reactivity and are non-existent.

Techniques to determine the nullity have appeared in [6]

23.2.1 TECHNIQUES FOR NULLITY OF GRAPH

Now we discuss some techniques for finding the nullity of a graph.

23.2.1.1 Zero-sum Weighting Technique

A function f defined as $f: V(G) \to R($ *real numbers*$)$ is a vertex weighting of G, which allocates $R - \{0\}$ weight to each vertex $v_i \in V(G)$. A vertex $v \neq 0$ weighting of a graph G is called a zero-sum weighting provided that for each $v_i \in V(G), \sum f(v_j) = 0$, where the summation is taken over all $v_j \in N(G(v))$.

Proposition 23.2.1. In any graph, the maximum number of independent non-zero variables in a high zero-sum weighting equals the nullity of graph.

23.2.1.2 Co-neighbor Technique

If v_1 and v_2 are two vertices, which are not adjacent but have the same set of neighbors, then v_1 and v_2 are said to be co-neighbor vertices. Let v_i and v_j be co-neighbor vertices of a connected graph G, then $\eta(G) = \eta(G - v_i) + 1 = \eta(G - v_j) + 1$. Here, adopting the zero-sum weighting technique, we have obtained the nullity set of complete tripartite graphs T_{n_1,n_2,n_3}.

Theorem 23.1

Let T_{n_1,n_2,n_3} be complete tripartite graphs, where $n_1, n_2 \in \{1,2\}$ and n_3 are fixed positive integer r. Then

$$\eta(T_{n_1,n_2,n_3}) = \begin{cases} r-1, & \text{if } n_1 = n_2 = 1; \\ r, & \text{if } n_1 = 1 \& n_2 = 2 \text{ and vice versa}; \\ r+1, & \text{if } n_1 = n_2 = 2. \end{cases}$$

■

Proof. Case(i) When $n_1 = n_2 = 1$ and $n_3 = r \in \mathbb{I}_+$:
T_{n_1,n_2,n_3} is nothing but $T_{1,1,r}$. Firstly, we construct the graph of $T_{1,1,r}$ so that the number of vertices in X_1 is u_1, in X_2 is v_1 and in X_3 is $w_1, w_2, w_3, \ldots, w_r$. On applying zero-sum weighting technique, assign real weights on $u_1, v_1, w_1, w_2, w_3, \ldots, w_r$ and find that we have used zero weights in u_1, v_1 and $a_1, a_2, a_3, \ldots, a_{r-1}$ non-zero $(r-1)$ weights on $w_1, w_2, w_3, \ldots, w_{r-1}$ and $-(a_1 + a_2 + a_3 + \cdots + a_{r-1})$ weight on w_r. It means that we found $(r-1)$ weights in $T_{1,1,r}$. Therefore, $\eta(T_{n_1,n_2,n_3}) = (r-1)$.
Case(ii) When $n_1 = 1, \& n_2 = 2$ and n_3 is fixed positive integer r(say):
We consider that T_{n_1,n_2,n_3} is $T_{1,2,r}$. Then the vertex set $V = \{u_1, v_1, v_2, w_1, w_2, w_3, \ldots, w_r\}$. In zero-sum weighting technique, we have assigned real weights zero weight on u_1, a_1, $-a_1$ on v_1, v_2, $a_2, a_3, a_4, \ldots, a_r$ on $w_1, w_2, w_3, \ldots, w_{r-1}$ and $-(a_2 + a_3 + a_4 + \cdots + a_r)$ on w_r. In this way, we have assigned r non-zero weights in $T_{1,2,r}$. Therefore, $\eta(T_{1,2,r}) = r$.
Case(iii) When $n_1 = n_2 = 2$ and $n_3 = (r \geq 1); n_3 \in \mathbb{N}$:
Let us assume that $T_{2,2,r}$ are the complete tripartite graphs with vertices $u_1, u_2, v_1, v_2, w_1, w_2, w_3, \ldots, w_r$ in the usual manner. Now, apply zero-sum

weighting technique, we have assigned a_1, $-a_1$ on u_1, u_2; a_2, $-a_2$ on v_1, v_2; $a_3, a_4, a_5, \ldots, a_{r+1}$ on $w_1, w_2, w_3, \ldots, w_{r-1}$ and $-(a_3+a_4+a_5+\cdots+a_{r+1})$ on w_r. This gives that $(r+1)$ non-zero weights have been used in $T_{2,2,r}$. Therefore, $\eta(T_{2,2,r}) = (r+1)$. □

23.3 ENERGY OF COMPLETE TRIPARTITE GRAPHS

The energy of a graph is the most useful in quantum chemistry. By HMO [1], in the molecular graph the vertex is the atom, the edge is the bond and the adjacency matrix is the Huckel matrix. The general solution of the HMO model is as given:

$$H = \alpha I + \beta A$$

where, H is the Hamiltonian matrix, I is the identity matrix and α, β are parameters known as the coulomb and resonance respectively. Then, the corresponding energy levels are

$$E_\pi = \alpha + \beta\lambda$$

and

$$E_\pi = \sum_{i=1}^{n} h_i \lambda_i$$

In 1970, Gutman determine the total $\pi-$ electron energy by the HMO, in graph theory is defined as: the energy of graph is the absolute sum of eigenvalues of the spectrum of graph G.
$T_{1,1,r}$ and $T_{2,2,r}$ are those complete tripartite graphs, which attained the bounds in the inequality 23.2

Theorem 23.2

Let $E(T_{n_1,n_2,n_3})$ be the energy of complete tripartite graphs, where $n_1, n_2 \in \{1,2\}$ and $n_3 \in \mathbb{N}$. Then

$$(1+\sqrt{(8r+1)}) \leq E(T_{n_1,n_2,n_3}) \leq 2(1+\sqrt{(4r+1)}). \tag{23.1}$$

■

Proof. Consider complete tripartite graph T_{n_1,n_2,n_3} , where $n_1 \& n_2 \in \{1,2\}$ and $n_3 = r \geq 1$. In order to calculate the energy of complete tripartite graph T_{n_1,n_2,n_3}, we shall tackle the following cases:
Case(i) When $n_1, \& n_2 = 1$ and $n_3 = r \in \mathbb{N}$:
Firstly consider $T_{1,1,r}$ complete tripartite graph and let $A(T)$ be the adjacency matrix, whose spectrum is $\{\frac{1}{2}(1 \pm \sqrt{(8r+1)}), -1, 0, 0, \ldots 0\ (r-1)\ times\}$. Then

$$E(T_{1,1,r}) = \sum |\lambda_i|; i = 1,2,3,\ldots,r+2$$

$$\rightarrow E(T_{1,1,1}) = 4 = (1+\sqrt{9}) = (1+\sqrt{(8r+1)})$$

and

$$E(T_{1,1,2}) = 5.1231 = (1+\sqrt{17}) = (1+\sqrt{(8r+1)})$$

......

$$E(T_{1,1,5}) = 7.40312 = (1+\sqrt{41}) = (1+\sqrt{(8r+1)})$$

and so on. Therefore, we can see that $E(G) = (1+\sqrt{(8r+1)})$
Case(ii) When $n_1 = 1, \& n_2 = 2$ or vice- versa and $n_3 \in \mathbb{I}_+$ (say r):
Let us take complete tripartite graph is as $T_{1,2,r}$. Then

$$E(T_{1,2,3}) = 7.53287$$

$$\Rightarrow 6 < 7.53287 < 9.2111$$

$$\rightarrow (1+\sqrt{25}) < 7.53287 < 2(1+\sqrt{13})$$

$$\rightarrow E(T_{1,1,3}) < E(T_{1,2,3}) < E(T_{2,2,3})$$

and so on

$$\rightarrow E(T_{1,1,r}) < E(T_{1,2,r}) < E(T_{2,2,r})$$

Therefore,

$$(1+\sqrt{(8r+1)}) \leq E(T_{1,2,r}) \leq 2(1+\sqrt{(4r+1)}).$$

In particular cases $r = 1$ and $r = 2$, $T_{1,2,r}$ attains the left and right bounds in the inequality 23.2 respectively.
Case(iii) When $n_1 = n_2 = 2$ and $n_3 = r \in \mathbb{N}$
Let us consider $T_{2,2,r}$ to be the complete tripartite graph whose spectrum is $\{(1 \pm \sqrt{(4r+1)}), -2, 0, 0, \ldots 0\ (r+1)\ times\}$. Then

$$E(T_{2,2,2}) = 8 = 2(1+\sqrt{9}) = 2(1+\sqrt{(4r+1)}); r = 2$$

and

$$E(T_{2,2,3}) = 9.2111 = 2(1+\sqrt{13}) = 2(1+\sqrt{(4r+1)}); r = 3,$$

and so on. Therefore,

$$E(G) = 2(1+\sqrt{(4r+1)}).$$

Energy of complete bipartite graph is shown in Figure 23.1. □

23.4 CONCLUSION

From the above theorems and 23.1, we conclude some results on the isomorphic chemical organic compound:

1. By HMO, all respective unsaturated conjugated hydrocarbons are highly reactive, unstable and non-existent in nature except $T_{1,1,1}$.
2. All isomorphic chemical organic compounds are non-hyperenergetic graphs.
3. Only $T_{1,1,1}$ is a borderenergetic graph.
4. $T_{1,1,6}$ and $T_{2,2,2}$ are non- cospectral equienergetic graphs.

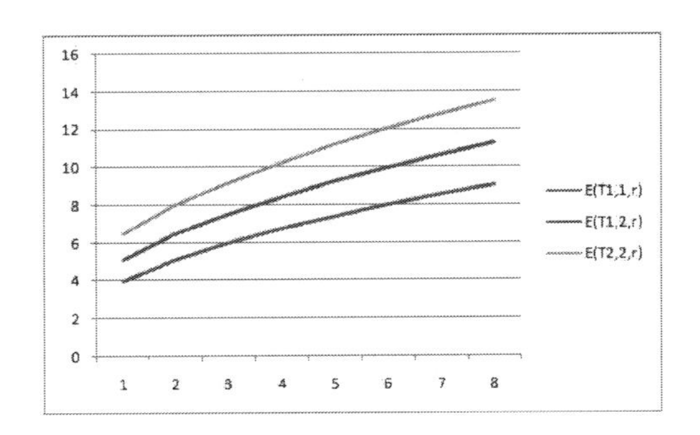

Figure 23.1: Energy of complete tripartite graphs

REFERENCES

1. N. A. Alawn, N. M. Al-Saidi and R. T. Rasheed. The energy of a tripartite graph. In AIP Conference Proceedings, 2086(1): 030006). *AIP Publishing*, 2019.
2. B. Borovicanin and I. Gutman. Nullity of graphs. *Applications of Graph Spectra, Math. Inst., Belgrade*, 107-122, 2009.
3. Y. Z. Fan and K. S. Qian. On the nullity of bipartite graphs. *Linear Algebra and its Applications*, 430(11-12): 2943-2949, 2009.
4. R. Farooq, M. A. Malik, Q. Naureen and S. Pirzada. On the nullity of a family of tripartite graphs. *Acta Universitatis Sapientiae, Informatica*, 8(1): 96-107, 2016.
5. R. Naresh and U. Sharma. Nullity of corona of a path with Smith graphs. *European Journal of Pure and Applied Mathematics*, 10(5): 1050-1057, 2017.
6. K. R. Sharaf and K. B. Rasul. On the nullity of expanded graphs. *Gen*, 21(1), 97-117, 2014.
7. U. Sharma and R. Naresh. Nullity of expanded Smith graphs. *International Journal of Scientific and Engineering Research*, 7(11): 1815-1824, 2016.

24 Some New Results on Restrained Edge Domination Number of Graphs

S. K. Vaidya
Saurashtra University
Rajkot, Gujarat (INDIA)
E-mail: samirkvaidya@yahoo.co.in

P. D. Ajani
Atmiya University
Rajkot, Gujarat (INDIA)
E-mail: paragajani@gmail.com

For a graph $G = (V, E)$, a subset D of E is a restrained edge dominating set of G if every edge not in D is adjacent to an edge in D as well as an edge in $E - D$. The restrained edge domination number of G, denoted by $\gamma_{re}(G)$ is the minimum cardinality of a restrained edge dominating set of G. In this chapter, we characterize a restrained edge dominating set and also investigate a restrained edge domination number of book graph B_n, crown Cr_n, armed crown ACr_n and friendship graph F_n.

24.1 INTRODUCTION

The concept of domination in a graph is one of the fastest growing areas within and outside of graph theory. It has received considerable attention due to its diversified applications and its potential to handle real life situations. We begin with the simple, finite, connected and undirected graph $G = (V, E)$ of order n, where V is the set of vertices and E is the set of edges of G. The open neighbourhood $N(v)$ of $v \in V$ is the set of vertices adjacent to v and the closed neighbourhood of v is the set $N[v] = N(v) \cup \{v\}$. The minimum degree among the vertices of graph G is denoted by $\delta(G)$ while the maximum degree among the vertices of graph G is denoted by $\Delta(G)$ and the maximum degree among the edges of graph G is denoted by $\Delta'(G)$. An edge e of a graph G is said to be incident with vertex v if v is an end vertex of e. Two vertices u and v of G are said to be adjacent vertices, if there is an edge between u and v. Two edges e and f of G having a vertex v in common are called adjacent edges. In a graph G, a vertex of degree one is called a pendent vertex and an

edge incident with a pendent vertex is called a pendent edge. The cardinality of vertex set V of G is the number of vertices in V, denoted by $|V|$ while the cardinality of edge set E of G is the number of edges in E denoted by $|E|$. For Graph theoretic terms and notation we refer Harary [6].
A set $S \subseteq V$ of vertices in a graph G is called a dominating set if every vertex $v \in V$ is either an element of S or is adjacent to an element of S. A dominating set S is a minimal dominating set if no proper subset $S' \subset S$ is a dominating set. The minimum cardinality of a dominating set of G is called domination number, denoted by $\gamma(G)$ and the corresponding dominating set is called a γ-set of G. A brief account of dominating set and its related concepts can be found in Haynes *et al.* [7].
There are many variants of dominating sets available in the existing literature. Some of them, total domination [3], equitable domination [13], global domination [11], independent domination [2, 10], are worth mentioning. One such variant is restrained domination; a set $S \subseteq V$ is a restrained dominating set if every vertex not in S is adjacent to a vertex in S as well as to a vertex in $V - S$. The minimum cardinality of a restrained dominating set S is called the restrained domination number of G, denoted by $\gamma_r(G)$. The concept of restrained domination was introduced by Telle and Proskurowski [14] as a vertex partitioning problem. Restrained domination of a complete graph, multipartite graphs and the graphs with minimum degree two is well studied by Domke *et al* [4, 5] while restrained domination in the context of path, cycle and wheel is discussed by Vaidya and Ajani [15, 16, 17]. These variants are introduced by identifying one or more characteristics of the elements of vertex subset or edge subset.
An edge analogue of a dominating set is also available. A subset $F \subseteq E$ is an edge dominating set if each edge in E is either in F or is adjacent to an edge in F. An edge dominating set F is called a minimal edge dominating set if no proper subset F' of F is an edge dominating set. The minimum cardinality among all minimal edge dominating sets is called the edge domination number, denoted by $\gamma_{re}(G)$. Mitchell and Hedetniemi [9] introduced the concept of edge domination. Yannakakis and Gavril [19] have explored edge dominating sets in graphs while the complementary edge domination in graphs is well studied by Kulli and Soner [8]. Arumugam and Velammal [1] have discussed the edge domination in graphs and edge domination in some path and cycle related graphs is investigated by Vaidya and Pandit [18].
We continue to study an edge analogue of restrained domination in graphs. For a graph $G = (V, E)$, a set $F \subseteq E$ is a restrained edge dominating set if every edge not in F is adjacent to an edge in F and also adjacent to an edge in $E - F$. The minimum cardinality of the restrained edge dominating set of G is called the restrained edge domination number, denoted as $\gamma_{re}(G)$. This concept was conceived by Soner and Ghobadi [12].
The present work is aimed to investigate some new results and characterize the concept of restrained edge domination in graphs.

24.2 PRELIMINARIES

Definition. The *degree of an edge* $e = uv$ of G is defined by $deg(e) = deg(u) + deg(v) - 2$ and it is equal to the number of edges adjacent to it. The maximum degree of an edge in G is denoted by $\Delta'(G)$.

Definition. The graph $K_{1,n}$ is called a star. We recognize the vertex with degree n as the apex of $K_{1,n}$.

Definition. The *bistar* $B_{n,n}$ is a graph obtained by joining the apex vertices of two copies of $K_{1,n}$ by an edge.

Definition. The *book graph* B_m is defined as $S_m \times P_2$, where S_m is a star graph.

Definition. The *crown* Cr_n is obtained by joining a pendent edge to each vertex of cycle C_n.

Definition. The *armed crown* is a graph in which path P_2 is attached at each vertex of cycle C_n by an edge. It is denoted by ACr_n where n is the number of vertices in cycle C_n.

Definition. The *friendship graph* F_n is a one point union of n-copies of cycles C_3.

Definition. The *switching of a vertex* v of G means removing all the edges incident to v and adding edges joining v to every vertex which is not adjacent to v in G. We denote the resultant graph by $\widetilde{G}$.

24.3 CHARACTERIZATION OF RESTRAINED EDGE DOMINATING SET

Theorem 24.1

For a graph G with order $n \geq 4$, $\gamma_{re}(G) = 1$ if and only if G is a star or bistar. ■

Proof. Let G be any graph and $\gamma_{re}(G) = 1$. It follows that $F \subseteq E(G)$ is a restrained edge dominating set with $|F| = 1 = |\{e\}|$, which implies that an edge e is adjacent to all the remaining edges of G. Also each edge in $E - F$ is adjacent to an edge in F and also to an edge in $E - F$. It is very clear to see that G is star or bistar.

Conversely, let G be a star $K_{1,n}$ with $|V(G)| = n + 1$ and $|E(G)| = n$. Let $e \in E(G)$ be any arbitrary edge, clearly an edge e will dominate all the remaining $(n-1)$ edges of G. If $e \in F \subseteq E(G)$ then each edge in $E-F$ is adjacent to an edge in F as well as an another edge in $E-F$. In this case, F is a restrained edge dominating set of G with minimum cardinality. Hence $\gamma_{re}(G) = 1$. If G is a bistar then it follows the same argument as above. □

24.4 RESTRAINED EDGE DOMINATION NUMBER OF STAR RELATED GRAPHS

Theorem 24.2

For book graph, $\gamma_{re}(B_n) = n$. ■

Proof. Let $\{e'', e_1, e_2, \cdots, e_n, e'_1, e'_2, \cdots, e'_n, e''_1, e''_2, \cdots, e''_n\}$ be the edge set of book graph B_n with $|V(B_n)| = 2(n+1)$ and $|E(B_n)| = 3n+1$.

We construct an edge set $F \subseteq E(B_n)$ as follows.
$F = \{e_1\} \cup \{e'_2\} \cup \{e''_i\}$, for $3 \leq i \leq n$ with $|F| = n$. As $F \subseteq E(B_n)$ and each edge in $E-F$ is adjacent to an edge in F and also adjacent to another edge in $E-F$, which implies that the above set F is a restrained edge dominating set of B_n. Note that for any edge $f \in F$, the set $F - \{f\}$ does not dominate the edges in $N(f)$ of B_n, so the set F is a minimal restrained edge dominating set of B_n.

Now $\Delta'(B_n) = 2n = deg(e'')$, $deg(e_i) = 4 = deg(e'_i)$, for $i = 1, 2, \cdots, n$ and $deg(e''_i) = 2$, for $i = 1, 2, \cdots, n$. So the edges in the above set F will dominate the maximum number of distinct edges of B_n. Therefore any set containing the edges less than the number of edges in set F will not dominate all the edges of (B_n). It follows that the above set F is the restrained edge dominating set with minimum cardinality.

Hence set F is a minimal restrained edge dominating set with minimum cardinality among all the minimal restrained edge dominating sets of (B_n). Thus $\gamma_{re}(B_n) = n$. □

Illustration 24.4.1. The book graph B_3 is shown in Figure 24.1 where the set of edges $\{e_1, e'_2, e''_3\}$ is its restrained edge dominating set of minimum cardinality.

Theorem 24.3

For a star, $\gamma_{re}(\widetilde{K_{1,n}}) = 2$. ■

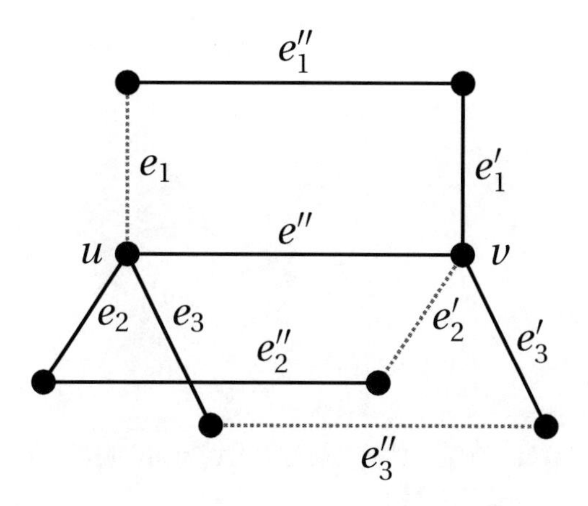

Figure 24.1: $\gamma_{re}(B_3) = 3$

Proof. Let $K_{1,n}$ be the star with edge set $E(K_{1,n}) = \{e_1, e_2, \cdots, e_n\}$ and vertex set $V(K_{1,n}) = \{v, v_1, v_2, \cdots, v_n\}$, where v is the apex and $v_1, v_2, \cdots, v_n$ are pendent vertices. Let $(\widetilde{K_{1,n}})$ be the graph obtained by switching of an arbitrary vertex of $K_{1,n}$. Without loss of generality we switch the vertex v_1. Then $E(\widetilde{K_{1,n}}) = \{e_1, e_2, \cdots, e_{n-1}, e_1', e_2', \cdots, e_{n-1}'\}$ with $|V(\widetilde{K_{1,n}})| = n+1$ and $|E(\widetilde{K_{1,n}})| = 2(n-1)$.

Consider the edge set $F \subseteq E(\widetilde{K_{1,n}})$ as $F = \{e_1, e_1'\}$ with $|F| = 2$. Since each edge of $E - F$ is adjacent to an edge in F and also adjacent to an another edge in $E - F$, the above set F is a restrained edge dominating set of $(\widetilde{K_{1,n}})$.

To prove F is a restrained edge dominating set with minimum cardinality, if possible, suppose F' is a restrained edge dominating set such that $|F'| = 1 < 2 = |F|$. Now $\Delta'(\widetilde{K_{1,n}}) = n-1 = deg(e_i) = deg(e_i')$, for $i = 1, 2, 3, \cdots, n-1$. Clearly $\Delta'(\widetilde{K_{1,n}}).1 = (n-1).1 = (n-1) < 2(n-1) = |E(\widetilde{K_{1,n}})|$. Therefore F' cannot be a restrained edge dominating set of $(\widetilde{K_{1,n}})$, which implies that F is a restrained edge dominating set of $(\widetilde{K_{1,n}})$ with minimum cardinality. Hence $\gamma_{re}(\widetilde{K_{1,n}}) = 2$. □

Illustration 24.4.2. Switching of star $(\widetilde{K_{1,5}})$ is shown in Figure 24.2 where the set of edges $\{e_1, e_1'\}$ is its restrained edge dominating set of minimum cardinality.

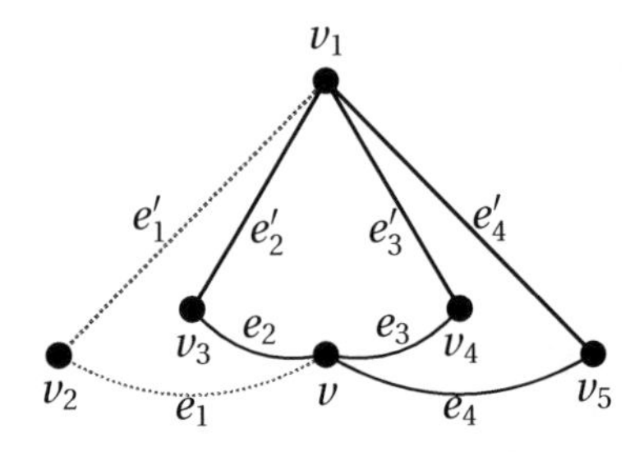

Figure 24.2: $\gamma_{re}(\widetilde{K_{1,5}}) = 2$

24.5 RESTRAINED EDGE DOMINATION NUMBER OF CYCLE RELATED GRAPHS

Theorem 24.4

For crown, $\gamma_{re}(Cr_n) = \left\lceil \frac{n}{2} \right\rceil$. ■

Proof. Let $e_1, e_2, \cdots, e_n$ be the edges of C_n. In order to obtain crown Cr_n, we add pendent edges $e'_1, e'_2, \cdots, e'_n$ to each vertex of cycle C_n. So $E(Cr_n) = \{e_1, e_2, \cdots, e_n, e'_1, e'_2, \cdots, e'_n\}$ with $|V(Cr_n)| = 2n = |E(Cr_n)|$. We construct an edge set $F \subseteq E(Cr_n)$ as follows.
$F = \{e_1, e_3, e_5, \cdots, e_{2i+1}\}$, where $0 \leq i \leq \left\lceil \frac{n}{2} \right\rceil - 1$ with $|F| = \left\lceil \frac{n}{2} \right\rceil$. Note that the above set $F \subseteq E(Cr_n)$ is a restrained edge dominating set of Cr_n, as each edge in $E - F$ is adjacent to an edge in F as well as an another edge in $E - F$.

We claim that F is a restrained edge dominating set with minimum cardinality. If possible, suppose F' is a restrained edge dominating set such that $|F'| = \left\lceil \frac{n}{2} \right\rceil - 1 < \left\lceil \frac{n}{2} \right\rceil = |F|$. Now $\Delta'(Cr_n) = 4 = deg(e_i)$, for $i = 1, 2, 3, \cdots, n$. In order to attain minimum cardinality, F' does not contain the edges where each edge among them can dominate distinct $2n$ edges of Cr_n. Moreover $\Delta'(Cr_n). \left\lceil \frac{n}{2} \right\rceil - 1 = 4. \left\lceil \frac{n}{2} \right\rceil - 1 = 2n - 1 < 2n = |E(Cr_n)|$. Therefore F' cannot be a restrained edge dominating set of Cr_n. This implies that F is a restrained edge dominating set with minimum cardinality for Cr_n. Thus $\gamma_{re}(Cr_n) = \left\lceil \frac{n}{2} \right\rceil$. □

Illustration 24.5.1. The crown Cr_4 is shown in Figure 24.3 where the set of edges $\{e_1, e_3\}$ is its restrained edge dominating set of minimum cardinality.

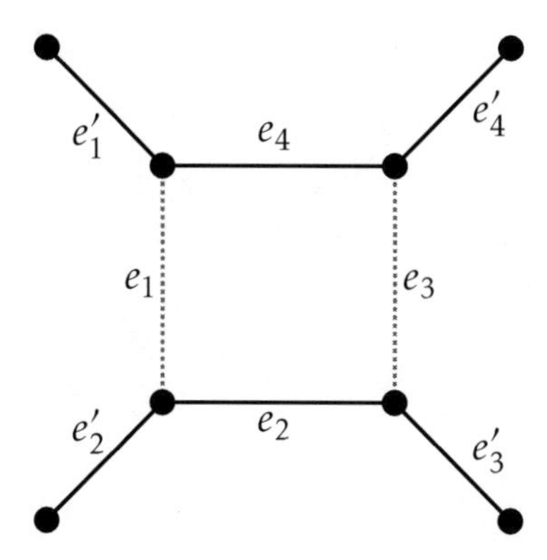

Figure 24.3: $\gamma_{re}(Cr_4) = 2$

Theorem 24.5

For armed crown, $\gamma_{re}(ACr_n) = \left\lceil \frac{n}{3} \right\rceil + n$. ■

Proof. Let $e_1, e_2, \cdots, e_n$ be the edges of C_n. To construct armed crown ACr_n, we attach pendent edges $e_1'', e_2'', \cdots, e_n''$ at each vertex of cycle C_n by edges $e_1', e_2', \cdots, e_n'$ respectively. So $E(ACr_n) = \{e_1, e_2, \cdots, e_n, e_1', e_2', \cdots, e_n', e_1'', e_2'', \cdots, e_n''\}$ with $|V(ACr_n)| = 3n = |E(ACr_n)|$.

Now $\Delta'(ACr_n) = 4 = deg(e_i)$, for $i = 1, 2, 3, \cdots, n$ and $deg(e_i'') = 1$, for $i = 1, 2, 3, \cdots, n$. Note that the pendent edges $e_1'', e_2'', \cdots, e_n''$ are mutually non-adjacent. Moreover pendent edges dominate themselves as well as their adjacent edges $e_1', e_2', \cdots, e_n'$ respectively. So every restrained edge dominating set of ACr_n must contain the pendent edges $e_1'', e_2'', \cdots, e_n''$. Now, by known result, at least $\left\lceil \frac{n}{3} \right\rceil$ edges are required to dominate all the remaining edges of cycle C_n, which implies that, at least $\left\lceil \frac{n}{3} \right\rceil + n$ edges are required to dominate all the edges of armed crown ACr_n.

If $F \subseteq E(ACr_n)$ is a restrained edge dominating set then $|F| = \left\lceil \frac{n}{3} \right\rceil + n$; the set F is of minimum cardinality because removal of any of the edge from set F will not dominate all the edges of ACr_n. Moreover every edge of $E - F$ is adjacent to an edge in F as well as an another edge in $E - F$. Hence, $\gamma_{re}(ACr_n) = \left\lceil \frac{n}{3} \right\rceil + n$. □

Illustration 24.5.2. Armed crown ACr_4 is shown in Figure 24.4 where the set of edges $\{e_1, e_3, e_1'', e_2'', e_3'', e_4''\}$ is its restrained edge dominating set of minimum cardinality.

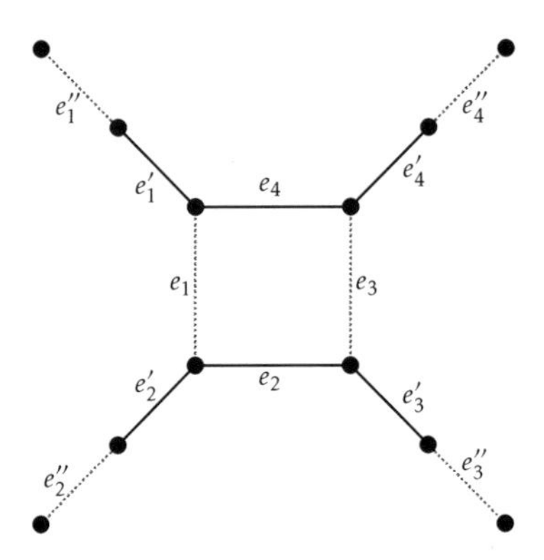

Figure 24.4: $\gamma_{re}(ACr_4) = 6$

Theorem 24.6

For friendship graph, $\gamma_{re}(F_n) = n$. ■

Proof. We take one point union of n copies of cycle C_3 to construct friendship graph F_n. Let $E(F_n) = \{e_1, e_2, \cdots, e_n, e'_1, e'_2, \cdots, e'_n, e''_1, e''_2, \cdots, e''_n\}$ be the edge set of F_n with $|V(F_n)| = 2n+1$ and $|E(F_n)| = 3n$.

Consider an edge set $F \subseteq E(F_n)$ as follows. $F = \{e''_1, e''_2, \cdots, e''_i\}$, for $i = 1, 2, 3, \cdots, n$ with $|F| = n$. Note that $F \subseteq E(F_n)$ and each edge of $E - F$ is adjacent to an edge in F as well as an another edge in $E - F$. It follows that the above set F is a restrained edge dominating set of F_n.

Now $\Delta'(F_n) = 2n = deg(e_i) = deg(e'_i)$, for $i = 1, 2, 3, \cdots, n$ and $deg(e''_i) = 2$, for $i = 1, 2, 3, \cdots, n$. Due to the adjacency nature of edges, for any edge $f \in F$, the set $F - \{f\}$ does not dominate the edges in $N(f)$ of F_n. So the set F is a minimal restrained edge dominating set of F_n. Moreover the edges in the above set F will dominate the maximum number of distinct edges of F_n. Therefore any set containing the edges less than the number of edges in set F will not dominate all the edges of F_n, which implies that the above set F is a restrained edge dominating set with minimum cardinality.

Hence set F is a minimal restrained edge dominating set with minimum cardinality among all minimal restrained edge dominating sets of F_n. Thus $\gamma_{re}(F_n) = n$. □

Illustration 24.5.3. Friendship graph F_3 is shown in Figure 24.5 where the set of edges $\{e''_1, e''_2, e''_3\}$ is its restrained edge dominating set of minimum cardinality.

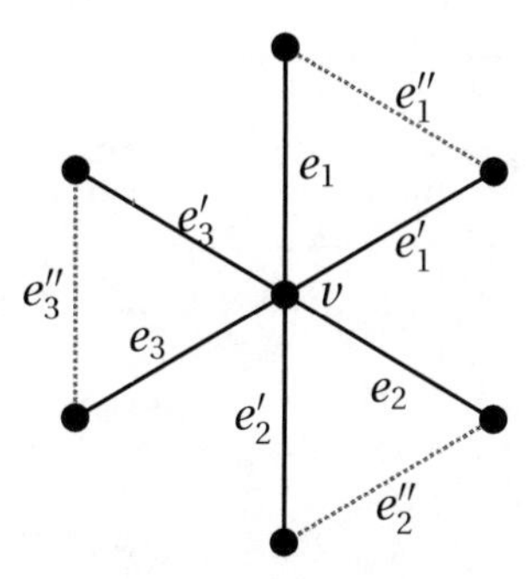

Figure 24.5: $\gamma_{re}(F_3) = 3$

24.6 CONCLUDING REMARKS

The concept of restrained edge domination is a variant of restrained domination and edge domination. The restrained edge dominating set relates the edge set and its complement. The restrained edge domination of some standard graphs is already known while we have characterized the restrained edge dominating set and also obtained the exact value of the restrained edge domination number of book graph B_n, friendship graph F_n, crown Cr_n and armed crown ACr_n. Similar results can be proved in the context of various types of dominating sets and different graph operations.

REFERENCES

1. S. Arumugam and S. Velammal. Edge domination in graphs. *Taiwanese Journal of Mathematics*, 2, 173-179, 1998.
2. C. Berge. *Theory of Graphs and its Applications.* Methuen, London, 1962.
3. E. J. Cockayne, R. M. Dawes and S. T. Hedetniemi. Total domination in graphs. *Networks*, 10, 211-219, 1980.
4. G. S. Domke, J. H. Hattingh, S. T. Hedetiniemi, R. C. Laskar and L. R. Markus. Restrained domination in graphs. *Discrete Mathematics*, 203, 61-69, 1999.
5. G. S. Domke, J. H. Hattingh, M. A. Henning and L. R. Markus. Restrained domination in graphs with minimum degree two. *J. Combin. Math.Combin. Comput.*, 35, 239-254, 2000.
6. F. Harary. *Graph Theory*, Addison-Wesley, Reading, Mass, 1969.
7. T. W. Haynes, S. T. Hedetniemi and P. J. Slater. Fundamentals of Domination in Graphs. Marcel Dekker, New York, 1998.
8. V. R. Kulli and N. D. Soner. Complementary edge domination in graphs. *Indian Journal of Pure and Applied Mathematics*, 28, 917-920, 1997.
9. S. Mitchell and S. Hedetniemi. Edge domination in trees. *Congr. Numer*, 19, 489-509, 1977.
10. O. Ore. Theory of Graphs. *Amer. Math. Soc. Colloq. Publ.*, 38 (Amer. Math. Soc., Providence, RI, 1662.

11. E. Sampathkumar. The global domination number of a graph. *Journal of Mathematical and Physical Sciences*, 23(5): 377-385, 1989.
12. N. D. Soner and S. Ghobadi. Restrained edge domination number in graphs. *Adv. Stud. Contemp. Math.*, 19, 143-149, 2009.
13. V. Swaminathan and K. M. Dharmalingam. Degree equitable domination on graphs. *Kragujevac Journal of Mathematics*, 35(1): 191-197, 2011.
14. J. A. Telle and A. Proskurowski. Algorithms for vertex partitioning problems on partial k-trees. *SIAM J. Discrete Mathematics*, 10, 529-550, 1997.
15. S. K. Vaidya and P. D. Ajani. Restrained domination number of some path related graphs. *Journal of Computational Mathematica*, 1(1): 114-121, 2017.
16. S. K. Vaidya and P. D. Ajani. On restrained domination number of graphs. *International Journal of Mathematics and Soft Computing*, 8(1): 17-23, 2018.
17. S. K. Vaidya and P. D. Ajani. On restrained domination number of some wheel related graphs. *Malaya Journal of Matematik*, 7(1): 104-107, 2019.
18. S. K. Vaidya and R. M. Pandit. Edge domination in some path and cycle related graphs. *ISRN Discrete Mathematics*, 2014, 1-5, 2014.
19. M. Yannakakis and F. Gavril. Edge dominating sets in graphs. *SIAM Journal on Applied Mathematics*, 38(3): 364-372, 1980.

25 Some New Graph Coloring Problems

Sudev Naduvath
Department of Mathematics,
CHRIST (Deemed to be University),
Bengaluru, Karnataka (INDIA)
E-mail: sudevnk@gmail.com

Johan Kok
City of Tshwane,
South Africa
E-mail: jacotype@gmail.com

Graph coloring is an assignment of colors, labels, or weights to the elements of a given graph G. A color class of G is a set of its vertices having the same color. A rainbow neighbourhood in a graph G is the closed neighbourhood of a vertex which consists of at least a vertex from every color class. In this chapter, we discuss some new types of graph coloring based on the rainbow neighbourhoods in the graph concerned and related results.

25.1 INTRODUCTION

In this chapter, we discuss some new graph coloring protocols, which have many practical applications, especially in connectivity and optimisation problems. For all terms and definitions in graph theory, we refer to [3, 7, 19]. For the terminology of graph coloring, refer to [2, 11]. Unless mentioned otherwise, all graphs we consider in this are simple, finite and connected.

Graph coloring can be considered as an assignment of colors to the elements of graphs under consideration. Graph coloring has emerged as a fertile and fruitful research area since its inception in the second half of the nineteenth century. Based on different real life problems, several types of graph coloring protocols and related parameters have been introduced and studied extensively.

A *vertex coloring* of a graph G (called a k-coloring) can formally be defined as a function $c: V(G) \to \mathcal{C} = \{c_i : 1 \leq i \leq \ell\}$, where each c_i is a color or a label or a weight. A *proper coloring* of a graph G is a vertex coloring where no two adjacent vertices have the same color. The minimum number of colors required for a proper coloring of G is called the *chromatic number* of G, denoted by $\chi(G)$. A proper coloring of G consisting of exactly $\chi(G)$ colors is called a *chromatic coloring* of G. The set of vertices of G having the same color, say

c_i, is called a *color class* of G and is denoted by C_i. Let $\theta(c_i)$ be the number of vertices of G having the color $c_i \in C$, the given color set of G. Then the *color sum* of G with respect to the coloring c concerned is defined as $\sum_{i=1}^{\chi(G)} i\theta(c_i)$ (see [10]).

25.2 RAINBOW NEIGHBOURHOODS IN GRAPHS

A *rainbow neighbourhood* of a graph G is the closed neighbourhood of a vertex v which consists of at least one vertex from every color class of G (see [12]). Here, the vertex v is called a *rainbow neighbourhood vertex* of G. That is, a vertex v is a rainbow neighbourhood vertex of G if $N[v] \cap C_i \neq \emptyset$.

The number of rainbow neighbourhoods in a graph is particularly interesting because of many aspects such as adjacency, reachability and domination related constraints. It is also observed that the number of rainbow neighbourhoods in a given graph depends on the protocol we follow for coloring the vertices of G. In view of this fact, some new coloring protocols are defined in [12] as follows:

Definition (Rainbow Neighbourhood Coloring). [12] Let G be a finite connected graph and let I_r be a maximal independent set in $G - \bigcup_{j=0}^{r-1} I_j$, where $1 \leq r \leq \chi(G)$, in the sense that $I_0 = \emptyset$. Let $c : V(G) \to C = \{c_1, c_2, c_3, \ldots, c_{\chi(G)}\}$ be a proper chromatic coloring of G. Then,

(i) the coloring c is said to be the *rainbow neighbourhood coloring of the first kind* of G if $c^{-1}(c_r) = I_r$.
(ii) the coloring c is said to be the *rainbow neighbourhood coloring of the second kind* of G if $c^{-1}(c_r) = I_{(\chi(G)-r+1)}$.

Figure 25.1 provides illustrations of the two types of rainbow neighbourhood coloring of the Petersen graph.

Note that the first type of coloring mentioned above yields the minimum chromatic sum and the second type of coloring mentioned above yields the minimum chromatic sum. Another interesting fact about the above-mentioned colorings is that both colorings yield the same number of rainbow neighbourhoods in the graphs concerned. In view of this, by rainbow neighbourhood coloring, we mean the rainbow neighbourhood coloring of the first kind.

Thus, in the context of the elevated interest in the number of rainbow neighbourhoods in different graphs, we have the following definition:

Definition (Rainbow Neighbourhood Number). [12] The *rainbow neighbourhood number* of a graph G, denoted by $r_\chi(G)$, is the number of rainbow neighbourhood vertices in G with respect to a given rainbow neighbourhood coloring of G.

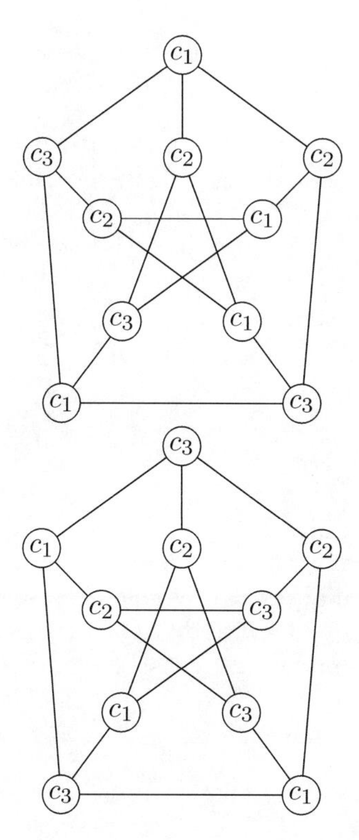

Figure 25.1: Rainbow neighbourhood coloring of a graph

The rainbow neighbourhood number of many fundamental and derived graph classes has been determined in [12, 15, 16]. Major results discussed in these articles are discussed in the following section.

25.2.1 RAINBOW NEIGHBOURHOOD NUMBER OF SOME BASIC GRAPH CLASSES

One of the most basic graph classes is the path graphs P_n. The chromatic number of any path $P_n; n \geq 2$ is 2. Therefore, for $n \geq 2$, we have $\chi(P_n) = 2$. Hence, if we color the vertices of a path alternatively by two colors, say c_1 and c_2, all the vertices of P_n will become rainbow neighbourhood vertices. Thus, $r_\chi(P_n) = n$. The same result can be established for any even cycle also. That is, for even integers n, $r_\chi(C_n) = n$. Hence, the result can be extended to bipartite graphs as follows:

Proposition 25.2.1. [12] *If G is a bipartite graph, then $r_\chi(G) = n$.*

Proof. Note that all bipartite graphs are 2-colorable. If we color all vertices in one partition with color c_1 and all vertices in the other partition by color c_2, it is clear that the closed neighbourhood of every vertex consists of vertices from both color classes. Thus, the proof is straightforward. □

If C_n is an odd cycle, it is 3-colorable and its vertex set can be partitioned into three independence sets I_1, I_2 and I_3, where the cardinality of I_1 and I_2 are $\frac{n-1}{2}$ each and I_3 is a singleton set. Therefore, the first $n-1$ vertices can be colored alternatively by c_1 and c_2 and the last vertex can be colored by c_3. Then, it can be observed that only three vertices (v_1, v_{n-1} and v_n) are rainbow neighbourhood vertices. Thus,

Proposition 25.2.2. [12] *For a cycle C_n,* $r_\chi(C_n) = \begin{cases} 3; & n \text{ is odd,} \\ n; & n \text{ is even.} \end{cases}$

Proof. Case-1: Let n be even. Then $\chi(C_n) = 2$. Then, the result for cycle C_n is immediate from Proposition 25.2.2.

Case-2: Let n be odd. Then, $\chi(C_n) - 3$ and only one vertex in C_n has the color c_3. All other vertices are alternatively colored using the colors c_1 and c_2. Let the vertices be labeled clockwise and consecutively $v_1, v_2, v_3, \ldots, v_n$. Without loss of generality, assume that v_j, $j \in \{2, 3, 4, \ldots, \ell - 1\}$ is colored c_3. Clearly, only the vertices v_{j-1}, v_j, v_{j+1} have closed neighbourhoods containing all colors c_1, c_2, c_3. Hence the result follows. □

Now, we consider the class of complete graphs. It can be observed that every vertex of a complete graph K_n becomes a rainbow vertex with respect to any proper coloring of K_n (as K_n is n-colorable and each color class in K_n is a singleton). Hence, we have

Proposition 25.2.3. [12] *For any $n \geq 2$, $r_\chi(K_n) = n$.*

Proof. The proof is straightforward from the fact that every vertex of the complete graph is adjacent to all other vertices in that graph. □

If G is a complete k-partite graph, the rainbow neighbourhood number of G is provided in the following theorem (see [12]):

Theorem 25.1

For a complete k-partite graph, $K_{r_1, r_2, \ldots, r_k}$, where $k \geq 2$ and each $r_i \geq 1$, we have $r_\chi(K_{r_1, r_2, \ldots, r_k}) = \sum_{i=1}^{k} r_i$. ■

Proof. Since the vertices in the same partition of the graph $K_{r_1, r_2, \ldots, r_k}$ are not adjacent to each other, the vertices in the same partition will have the same color. Since every vertex of every partition is adjacent to all vertices of all other partitions in the graph we have $r_\chi(K_{r_1,r_2,\ldots,r_k}) = \sum_{i=1}^{k} r_i$. Hence, the result. □

The bounds for the rainbow neighbourhood number of graphs have been found in [12], as stated below:

Theorem 25.2

Any graph G of order n has $\chi(G) \leq r_\chi(G) \leq n$. ■

Proof. For $n=1$ the graph K_1 has $\chi(K_1)=1$ and $r_\chi(K_1)=1$. For $n=2$, the path P_2 has $\chi(P_2)=2$ and $r_\chi(P_2)=2$. For $n=3$, the path P_3 has $\chi(P_3)=2$ and $r_\chi(P_3)=3$. Also the cycle C_3 has $\chi(C_3)=3$ and $r_\chi(C_3)=3$. Therefore, $r_\chi(G) \geq \chi(G)$ for all graphs of order $1 \leq n \leq 3$.

Assume the result holds for all graphs of order $1 \leq n \leq k$. Consider a graph of order k. Hence $r_\chi(G) \geq \chi(G)$. Now attach a new vertex u to a number say, t, $1 \leq t \leq k$ vertices of G to obtain a new graph G'. If possible, identify a color in the chromatic coloring of G with which vertex u can be colored such that G' with or without recoloring of vertices, has a chromatic coloring. It implies that $\chi(G') = \chi(G)$ and also $r_\chi(G') \geq r_\chi(G)$. Hence the result holds for the graph G'. Alternatively, an additional color is indeed required to allow a chromatic coloring for G' hence, $\chi(G') = \chi(G)+1$. Then $N[u]$ yields a rainbow neighbourhood containing $\chi(G)+1$ colors. Further to that, then $N[v]$ yields a rainbow neighbourhood $\forall\ v \in N(u)$ because the induced subgraph $\langle N[u]\rangle$ is necessary complete (a clique). Therefore, $r_\chi(G') = \chi(G)+1 = \chi(G')$. So the result $r_\chi(G') \geq \chi(G')$ holds. Through induction, the result then holds for all graphs of order $n \in \mathbb{N}$.

Also, $r_\chi(G) \leq n$ is obvious and hence the result holds. □

The above bounds can be verified very easily. An odd cycle is an example for a graph with $\chi(G) = r_\chi(G)$, where a complete graph K_n has $\chi(G) = r_\chi(G) = n$.

Another interesting characterisation of the graphs all of whose vertices are rainbow neighbourhood vertices has been established in [12], as explained in the following theorem:

Theorem 25.3

For a connected graph G of order $n \geq 3$, $r_\chi(G) = \chi(G)$ if and only if G is an odd cycle or complete. ■

Proof. It is clear that if G is a cycle C_n, n is odd or a complete graph, K_n, then $r_\chi(C_n) = 3 = \chi(C_n)$ and $r_\chi(K_n) = n = \chi(K_n)$.

Conversely, let $n = 3$; then G is either the path P_3, or the cycle C_3 or put alternatively, the complete graph K_3. Since $r_\chi(P_3) = 3 \neq 2 = \chi(P_3)$ the result, $G = C_3$ (or K_3)$\Rightarrow r_\chi(G) = \chi(G)$ holds. Similarly for a graph on $n = 4$ vertices the result cannot hold for P_4, C_4, star $S_{1,3}$ or the 1-chord cycle. It only holds for that $G = K_4 \Rightarrow r_\chi(K_4) = \chi(K_4) = 4$. Assume the result holds for all graphs G of order $3 \leq \ell \leq t$ if G is either C_ℓ, ℓ is odd, or K_ℓ, $\forall \ell \in \mathbb{N}$.

Consider any graph G of order $t+1$. Clearly it holds that $G = C_{t+1}$, $t+1$ is odd $\Rightarrow r_\chi(G) = \chi(G) = 3$. Also, $G = K_{t+1} \Rightarrow r_\chi(G) = \chi(G) = t+1$. So let G be any other graph of order $t+1$ and assume $r_\chi(G) = \chi(G)$. Now we have $\chi(G) \leq r_\chi(G) < t+1$. Hence, it is possible to find and remove one vertex which does not yield a rainbow neighbourhood and the chromatic number remains the same. Therefore, we obtain a new graph G' of order t such that $G' \Rightarrow r_\chi(G') = \chi(G')$. The aforesaid is a contradiction, hence no such G exists. It implies that for any G of order $t+1$, only $G = C_{t+1}$, $t+1$ is odd $\Rightarrow r_\chi(G) = \chi(G)$ or $G = K_{t+1} \Rightarrow r_\chi(G) = \chi(G)$. □

Corollary 25.2.4. *Any connected graph of order n that has $r_\chi(G) = \chi(G)$ is either 3-chromatic or n-chromatic.*

Proof. The result follows directly from Theorem 25.3. □

In view of the above-mentioned facts, we can observe that any connected graph of order n that has $r_\chi(G) = \chi(G)$ is either 3-chromatic or n-chromatic.

Another interesting fact in this context is a construction of rainbow neighbourhood colorable graphs. The following theorem established such a construction (see [12]).

Theorem 25.4

For $m, n \in \mathbb{N}$, $m \geq n \geq 2$, it is possible to construct a graph G such that $r_\chi(G) = m$, and $\chi(G) = n$. ■

Proof. The above-mentioned construction is pretty straightforward. Consider the complete graph K_m. Then, obviously $\chi(K_n) = n$. Let G be the graph obtained by adding $m - n$ isolated vertices to this graph and then joining each of these vertices to all the vertices of K_m except one, say v_1. Thus, each of these $m - n$ vertices, being independent vertices, can assume the same color of v_1. Hence, $\chi(G) = \chi(K_n) = n$ and $r_\chi(G) = m$ (see Figure 25.2 for illustration).

□

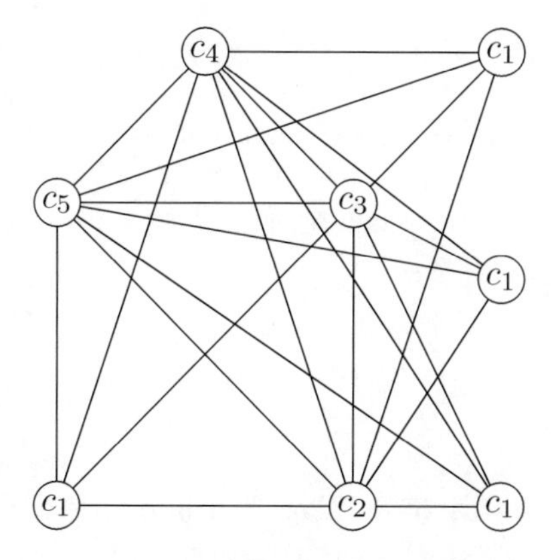

Figure 25.2: A graph with $\chi = 5$ and $r_\chi = 8$

25.2.2 RAINBOW NEIGHBOURHOOD NUMBER OF GRAPH OPERATIONS

In the following discussion, we discuss the rainbow neighbourhood number of different operations and products of some familiar graph classes. For the terminology of graph operations, we refer to [2] and for graph products, refer to [9].

The union of two graphs $G(V,E)$ and $H(U,F)$, denoted by $G \cup H$, is the graph with vertex set $V \cup U$ and the edge set $E \cup F$. The rainbow neighbourhood number of the union of two graphs has been discussed in the following theorem (see [12]).

Theorem 25.5

For any two graphs G and H, $r_\chi(G \cup H) \leq r_\chi(G) + r_\chi(H)$. ■

Proof. Since G and H have no common elements (both vertices and edges), it is obvious that the total number of rainbow neighbourhoods in $G \cup H$ will be the sum of the rainbow neighbourhoods in both graphs. That is, $r_\chi(G \cup H) = r_\chi(G) + r_\chi(H)$.

If G and H have some common elements, it can be noticed that $G \cup H = (G - G \cap H) \cup (H - G \cap H) \cup (G \cap H)$, where each of the three graphs are pairwise edge disjoint. Hence, we have $r_\chi(G \cup H) \leq r_\chi(G) + r_\chi(H) - r_\chi(G \cap H)$. Therefore, $r_\chi(G \cup H) \leq r_\chi(G) + r_\chi(H)$. This completes the proof. □

If $\chi(G) = \chi(H)$, then both G and H permit the same chromatic coloring with the same set of colors and thus the disjoint union of G and H cannot increase or decrease the respective values $r_\chi(G)$ and $r_\chi(H)$. Hence, when

$\chi(G) = \chi(H)$, then $r_\chi(G \cup H) = r_\chi(G) + r_\chi(H)$. Therefore, the bound mentioned in 25.5 is sharp.

The *join* of two graphs G and H, denoted by $G + H$ is the graph obtained by drawing edges from every vertex of G to all vertices of H. The number of rainbow neighbourhoods in the joins of different graphs have been discussed in [12] as follows:

Theorem 25.6

For any graph G, the rainbow neighbourhood number of the graph $G' = K_1 + G$ is $r_\chi(G') = 1 + r_\chi(G)$. Furthermore, $r_\chi(tK_1 + G) = t + r_\chi(G)$, where tK_1 is the disjoint union of t number of K_1 (that is, the empty graph of order t). ■

Proof. The proof is immediate from the fact that each of the vertices K_1 in $tK_1 + G$ will be adjacent to all vertices of G and hence will be a rainbow neighbourhood vertex. □

The above result is a consequence of the facts that the (central) vertex k_1 is adjacent to every vertex of the graph G and $\chi(G + K_1) = \chi(G) + 1$ and hence will also be a rainbow neighbourhood vertex of $G + K_1$. Therefore, the following result is immediate.

Theorem 25.7

For any two (connected) graphs G and H, $r_\chi(G + H) = r_\chi(G) + r_\chi(H)$. ■

Proof. Let the coloring $(c_1, c_2, c_3, \ldots, c_{\ell_1})$ be a chromatic coloring of G and $(c_{\ell_1+1}, c_{\ell_1+2}, c_{\ell_1+3}, \ldots, c_{\ell_1+\ell_2})$ be a chromatic coloring of H. Then, we note that the coloring $(c_1, c_2, c_3, \ldots, c_{\ell_1}, c_{\ell_1+1}, c_{\ell_1+2}, c_{\ell_1+3}, \ldots, c_{\ell_1+\ell_2})$ is a chromatic coloring of the graph $G + H$. Clearly, any vertex $v \in V(G)$ that yields a rainbow neighbourhood in G also yields a rainbow neighbourhood in $G + H$ and vice versa. Also, any vertex $u \in V(G)$ that does not yield a rainbow neighbourhood in G, cannot yield a rainbow neighbourhood in $G + H$ and vice versa. Therefore, the result follows. □

25.2.3 IMPORTANT OBSERVATIONS

Following from Theorem 25.2 we have that $r_\chi(G) \geq \chi(G) \Rightarrow r_\chi(L(G)) \geq \chi(L(G))$ and $r_\chi(\overline{G}) \geq \chi(\overline{G})$. Hence, all known Nordhauss-Gaddum lower bounds apply in respect to the sum and the product of rainbow neighbourhood numbers for G, $L(G)$ and $\overline{G}$. So do other lower bounds for $\chi(G)$ and correspondingly for $L(G)$ and $\overline{G}$, apply. Note we adopt the convention that $\overline{K_1} = K_1$ and $L(K_1) = K_1$. For a graph of order $n \geq 1$ and size $q \geq 0$, we list a few of these bounds below.

(i) $2\sqrt{n} \leq r_\chi(G) + r_\chi(\overline{G}) \leq 2n$, and $n \leq r_\chi(G) \cdot r_\chi(\overline{G}) \leq n^2$.
(ii) $2 \leq r_\chi(G) + r_\chi(L(G)) \leq n + \ell \cdot \Delta(G)$, and $1 \leq r_\chi(G) \cdot r_\chi(L(G)) \leq n\ell \cdot \Delta(G)$.
(iii) It is known that if graph G is t-regular then $\chi(G) \geq \frac{n}{n-t}$ and therefore, $\chi(\overline{G}) \geq \frac{n}{t+1}$. Hence, $r_\chi(G) + r_\chi(\overline{G}) \geq \frac{n(n+1)}{(n-t)(t+1)}$ and $r_\chi(G) \cdot r_\chi(\overline{G}) \geq \frac{n^2}{(n-t)(t+1)}$.
(iv) Let $\gamma(G)$ be the domination number of G. Since $\gamma(G) \leq n - \Delta(G)$ it follows that $r_\chi(G) + \gamma(G) \leq 2n - \Delta(G)$ and $r_\chi(G) \cdot \gamma(G) \leq n(n - \Delta(G))$. Also since we have the upper bound $\gamma(G) \leq \lceil \frac{n+1-\delta(G)}{2} \rceil$, similar inequalities are found in terms of $\delta(G)$.

25.2.4 RAINBOW NEIGHBOURHOOD NUMBER OF SOME CYCLE RELATED GRAPHS

A *wheel graph*, denoted by W_n, is a graph obtained by drawing edges between every vertex of a cycle C_n to an external vertex. That is, $W_n = C_n + K_1$. Here C_n is called the *rim* of W_n. In view of Theorem 25.5, the rainbow neighbourhood number of W_n can be determined [12] as follows:

Theorem 25.8

For $n \geq 3$, the rainbow neighbourhood number of a wheel graph $W_{n+1} = C_n + K_1$ is given by $r_\chi(W_{n+1}) = \begin{cases} 4, & \text{if } n \text{ is odd}, \\ n+1, & \text{if } n \text{ is even}. \end{cases}$ ■

Proof. Since the central vertex of the wheel graph is adjacent to all rim vertices, it will always be a rainbow neighbourhood vertex and hence the proof follows from the fact that $r_\chi(W_{n+1}) = 1 + r_\chi(C_n)$. □

A *double-wheel graph*, denoted by DW_n, is the graph $2C_n + K_1$, that is, it consists of two cycles of size n, where the vertices of the two cycles are all connected to a common central vertex. Note that the central vertex is adjacent to all other vertices in DW_n and hence the central vertex is always a rainbow neighbourhood vertex. If n is even, all vertices of the two rims will also be rainbow neighbourhood vertices of DW_n. If n is odd, exactly three vertices in each rim will be rainbow neighbourhood vertices. Invoking these facts, the following theorem establishes the rainbow neighbourhood number of double wheel graphs (see [15]).

Theorem 25.9

$$r_{\chi}(DW_n) = \begin{cases} 7; & \text{if } n \text{ is odd} \\ 2n+1; & \text{if } n \text{ is even.} \end{cases}$$ ■

Proof. Since the central vertex of the wheel graph is adjacent to all vertices of both of its rims, it will always be a rainbow neighbourhood vertex and hence the proof follows from the fact that $r_{\chi}(DW_n) = 1 + 2 \cdot r_{\chi}(C_n)$. □

A *helm graph* is a graph obtained from a wheel by attaching one pendent edge to each vertex of the cycle. The rainbow neighbourhood number of a helm graph is established in [15] as explained in the following result:

Theorem 25.10

For a helm graph $H_n, n \geq 3$,

$$r_{\chi}(H_n) = \begin{cases} 4; & \text{if } n \text{ is odd,} \\ n+1; & \text{if } n \text{ is even.} \end{cases}$$

■

Proof. Since the pendent vertices of the helm graph cannot be rainbow neighbourhood vertices, it is clear that $r_{\chi}(H_n) = r_{\chi}(W_{n+1})$ and hence the proof is straightforward. □

A *closed helm*, denoted by CH_n is a graph obtained from a helm by joining each pendent vertex to form a cycle and a flower as the graph obtained from a helm by joining each pendent vertex to the central vertex of the helm (see [6]). Observe that the corresponding vertices of the inner cycle and the outer cycle are adjacent in CH_n; they must have different colors in CH_n. But we can use the same colors to color the vertices (but in reverse orders) of these cycles such that no two adjacent vertices receive the same color. Invoking these facts, we have the following result as a consequence of Theorem 25.10 (see [15]).

Theorem 25.11

$$r_{\chi}(CH_n) = \begin{cases} 4; & \text{if } n \text{ is odd,} \\ n+1; & \text{if } n \text{ is even.} \end{cases}$$ ■

A *flower graph*, denoted by F_n, is the graph obtained from a helm graph H_n by joining each of its pendent vertices to its central vertex. Note that the chromatic coloring of F_n is similar to that of the corresponding helm graph H_n, except for the fact that the central vertex of F_n must have a color, different from the colors of all other vertices. Hence, the following result is immediate from Theorem 25.10 (see [15]).

Theorem 25.12

$$r_\chi(F_n) = \begin{cases} 4; & \text{if } n \text{ is odd,} \\ n+1; & \text{if } n \text{ is even.} \end{cases}$$ ■

A web graph, denoted by Wb_n, is the graph obtained by attaching one pendent edge to each vertex of the outer cycle of the double wheel graph DW_n. Note that the coloring of a web graph is similar to that of a closed helm graph and the result is also similar (see [15]) as stated below:

Theorem 25.13

$$r_\chi(Wb_n) = \begin{cases} 7; & \text{if } n \text{ is odd} \\ 2n+1; & \text{if } n \text{ is even.} \end{cases}$$ ■

Numerous graph classes have been introduced and studied in detail in the literature for many decades. Hence, the studies of the rainbow neighbourhood number of such graph classes will be a suitable area for further investigations.

25.2.5 RAINBOW NEIGHBOURHOOD NUMBER OF SOME GRAPH TRANSFORMATIONS

Some interesting investigations of the rainbow neighbourhood number of some derived graphs of certain fundamental graph classes have been done in recent years. Some of the major findings in this area are discussed in the following section.

Graph complements are widely studied for many interesting properties. The following discussion establishes some interesting results in this area. The rainbow neighbourhood number of complements of paths has been determined in [16] as per the following theorem.

Theorem 25.14

$$r_\chi(\bar{P_n}) = \begin{cases} n, & \text{if } n \text{ is even} \\ n-1, & \text{if } n \text{ is odd.} \end{cases}$$ ■

Proof. Let $P_n = (v_1, v_2, \dots, v_n)$ and $G = \bar{P_n}$. Let $k = \lceil \frac{n}{2} \rceil$. Since the clique number $\omega(G) = k$, it follows that $\chi(G) \geq k$. Now, if n is even, let $\mathcal{C}_i = \{v_{2i-1}, v_{2i}\}$, where $1 \leq i \leq k$. If n is odd, let

$$\mathcal{C}_i = \begin{cases} \{v_{2i-1}, v_{2i}\}; & \text{if } 1 \leq i \leq k-1, \\ \{v_n\}; & \text{if } i = k. \end{cases}$$

Then $\mathcal{C} = \{\mathcal{C}_1, \mathcal{C}_2; \dots, \mathcal{C}_k\}$ is a vertex coloring of G and hence $\chi(G) \leq k$ Thus $\chi(G) = k$. Now if n is even, then $N[v] \cap \mathcal{C}_i \neq \emptyset$ for all $v \in V$ and for all i. If n is odd, $N[v] \cap \mathcal{C}_i \neq \emptyset$ for all $v \in V - \{v_{n-1}\}$ and for all i and $N[v_{n-1}] \cap \mathcal{C}_k = \emptyset$. Hence it follows that

$$r_\chi(G) = \begin{cases} n; & \text{if } n \text{ is even}, \\ n-1; & \text{if } n \text{ is odd}. \end{cases}$$

□

The rainbow neighbourhood number of complements of cycles has been determined in [16] as per the following theorem.

Theorem 25.15

$$r_\chi(\bar{C_n}) = \begin{cases} n; & \text{if } n \text{ is even} \\ n-2; & \text{if } n \text{ is odd}. \end{cases}$$

■

Proof. Let $C_n = (v_1, v_2, v_3, \dots, v_n, v_1)$ and let $G = \bar{C_n}$. Let $k = \lceil \frac{n}{2} \rceil$. As in the previous theorem, the clique number $\omega(G) = k$, we have $\chi(G) \geq k$. Now, if n is even, let $\mathcal{C}_i = \{v_{2i-1}, v_{2i}\}$, where $1 \leq i \leq k$. If n is odd, let

$$\mathcal{C}_i = \begin{cases} \{v_{2i-1}, v_{2i}\}; & \text{if } 1 \leq i \leq k-1, \\ \{v_n\}; & \text{if } i = k. \end{cases}$$

Then $\mathcal{C} = \{\mathcal{C}_1, \mathcal{C}_2; \dots, \mathcal{C}_k\}$ is a vertex coloring of G and hence $\chi(G) \leq k$ Thus $\chi(G) = k$. Now if n is even, then $N[v] \cap \mathcal{C}_i \neq \emptyset$ for all $v \in V$ and for all i. If n is odd, $N[v] \cap \mathcal{C}_i \neq \emptyset$ for all $v \in V - \{v_1, v_{n-1}\}$ and for all i and $N[v_1] \cap \mathcal{C}_1 = \emptyset$ and $N[v_{n-1}] \cap \mathcal{C}_k = \emptyset$. Hence, it follows that

$$r_\chi(G) = \begin{cases} n; & \text{if } n \text{ is even}, \\ n-2; & \text{if } n \text{ is odd}. \end{cases}$$

□

The rainbow neighbourhood number of complements of complete bipartite graphs has been determined in [16] as per the following theorem.

Theorem 25.16

$$r_\chi(\bar{K}_{r,s}) = \begin{cases} 2r; & \text{if } r = s, \\ \max\{r,s\}; & \text{if } r \neq s. \end{cases}$$ ■

Proof. Note that $\bar{K}_{r,s} = K_r \cup K_s$, the disjoint union of K_r and K_s. Since K_r and K_s are disjoint, the same set of colors can be used for coloring their vertices.

If $r = s$, then note that the same set of r colors is required for coloring the vertices of both components K_r and K_s. Then, all vertices in K_r and K_s belong to some rainbow neighbourhood in $\bar{K}_{r,s}$. Therefore, $r_\chi(\bar{K}_{r,s}) = r + s = 2r$.

Next, assume that $r \neq s$. Without loss of generality, let $r > s$. Then, r colors are required for coloring the vertices of the component K_r, but a fewer number of colors can be used for coloring the vertices in K_s. More precisely, $r - s$ colors will not be used in the coloring of K_s. Therefore, no vertex of K_s belongs to a rainbow neighbourhood in $\bar{K}_{r,s}$. Hence, $r_\chi(\bar{K}_{r,s}) = r = \max\{r,s\}$. □

The *line graph* of a graph G (see [8]), denoted by $L(G)$ is the graph obtained by taking each edge of G as a vertex and two vertices in $L(G)$ are adjacent if and only if their corresponding edges are adjacent in G.

An upper bound for the rainbow neighbourhood number of the line graphs of given graph classes has been determined in [12] as mentioned in the following result.

Theorem 25.17

If a graph G has ℓ vertices of degree $\Delta(G)$, then $r_\chi(L(G)) \leq \ell \cdot \Delta(G)$. ■

Proof. Assume graph G has m vertices of maximum degree $\Delta(G) \neq \delta(G)$. Hence, in the line graph, m maximum cliques $K_{\Delta(G)}$ exist, and only the vertices of these maximum cliques can yield a rainbow neighbourhood on the colors $(c_1, c_2, c_3, \ldots, c_{\Delta(G)})$ in $L^{\cdot\cdot}(G)$. Clearly after contracting the broken line edges these vertices remain the same and $\chi(L(G)) = \Delta(G)$ remains because pairwise, two distinct cliques share at most one common vertex. Hence the result is straightforward. □

In view of the above theorem, the rainbow neighbourhood number of regular graphs can be determined as given below (see [16]):

Theorem 25.18

For $r \geq 3$, if G is an r-regular graph, then $r_\chi(L(G)) = \varepsilon(G)$, the size of G. ■

The Mycielski graph of a given graph G with vertex set $V(G) = \{v_1, v_2, v_3, \ldots, v_n\}$ is the graph obtained by applying the following steps.

1. Corresponding to each vertex v_i in $V(G)$, introduce a new vertex u_i and let $U = \{u_i : 1 \leq i \leq n\}$. Add edges from each vertex u_i of U to the vertex v_j if $v_i v_j \in E(G)$.
2. Take another vertex u and add edges from u to all vertices in U.

The new graph thus obtained is called the *Mycielski graph* of G and is denoted by $\mu(G)$ (see [13]).

The rainbow neighbourhood number of the Mycielskian of an arbitrary graph G is determined in [18] as in the following result.

Theorem 25.19

For any graph G, $r_\chi(\mu(G)) = r_\chi(G) + 1$. ■

Proof. Let $V = \{v_1, v_2, \ldots, v_n\}$ be the vertex set of G. Let $U = \{u_1, u_2, \ldots, u_n\}$ be the set of newly introduced vertices such that u_i corresponds to the vertex v_i such that u_i is adjacent to v_j and v_k provided v_j and v_k are adjacent to v_i. Take another vertex $u \neq u_i, u \neq v_i$ for $1 \leq i \leq k$, and add edges from u to u_i for all i. Let $\mathcal{C} = \{c_1, c_2, \ldots, c_k\}$ be the chromatic coloring of the graph G. Since u_i is not adjacent to v_i, u_i and v_i can have the same color. Since u is adjacent to all u_i, no color in $\mathcal{C}$ can be assigned to the vertex u. Therefore, we have to consider a new color, say c_{k+1} for the vertex u. Since no vertices in V are adjacent to the vertex u in $\mu(G)$, it is clear that the vertices in V will not be in the rainbow neighbourhood of $\mu(G)$. If v_i is a vertex in the rainbow neighbourhood of G, it can be observed that the corresponding vertex u_i will be in a rainbow neighbourhood of $\mu(G)$, since u_i is adjacent to the vertices in all color classes in $\mathcal{C}$ and is adjacent to the vertex u having color c_{k+1}. Therefore, every vertex in U belongs to some rainbow neighbourhoods of $\mu(G)$. Moreover the vertex u is adjacent to all vertices of U, which belong to different color classes in $\mathcal{C}$ and hence belong to some rainbow neighbourhoods of $\mu(G)$. Therefore, $r_\chi(\mu(G)) = r_\chi(G) + 1$. □

Further studies in this area of graph transformations are highly promising as only a few studies have been reported in this area so far.

25.2.6 RAINBOW NEIGHBOURHOOD NUMBER OF GRAPH PRODUCTS

The product of two graphs is defined in many different ways. According to the adjacency conditions, different graph products are defined in the literature.

For the definitions of different types of graph products, we refer to [9]. The first among these products is the Cartesian product of two graphs.

The *Cartesian product* of two graphs G and H, denoted by $G\square H$ is a graph such that $V(G\square H)$ is the Cartesian product $V(G)\times V(H)$ (and hence the vertices of $G\square H$ are defined as ordered pairs) and two vertices (u_i, v_r) and (u_j, v_s) are adjacent in $G\square H$ if and only if either $u_i = u_j$ and v_r is adjacent to v_s in H, or $v_r = v_s$ and u_i is adjacent to u_j in G. Note that $\chi(G\square H) = \max\{\chi(G), \chi(H)\}$. Finding the rainbow neighbourhood number of two arbitrary graphs is very complex. Hence, designing an efficient algorithm to determine this parameter will be a promising research problem. In the following results, we study the rainbow neighbourhood number of the Cartesian product of some fundamental graph classes.

The Cartesian product of two paths P_m and P_n is called a *grid graph.* Note that the grid graphs are bipartite. Hence, the rainbow neighbourhood number of the grid graph $G = P_m\square P_n$ is determined in [16] as explained in the following result.

Theorem 25.20

For $m, n \geq 2$, $r_\chi(P_m\square P_n) = mn$. ■

Proof. Since the Cartesian product of two bipartite graphs is also a bipartite graph, the grid graph $G = P_m\square P_n$ is bipartite and any minimal proper coloring of G contains 2 colors, say c_1 and c_2. Since for every vertex v in G, $2 \leq d(v) \leq 4$ and hence every vertex in one color class of G will be adjacent to at least two (and at most four) vertices of the other color class. Therefore, all vertices in $G = P_m\square P_n$ belong to some rainbow neighbourhood in G. Hence, $r_\chi(P_m\square P_n) = mn$. □

The Cartesian product of paths P_m and a cycle C_n is called a *prism graph.* If n is even, it is clear that $P_m\square C_n$ is a bipartite graph. Hence, we have the following theorem.

Theorem 25.21

[16] $r_\chi(P_m\square C_n) = \begin{cases} 3m; & \text{if } n \text{ is odd,} \\ mn; & \text{if } n \text{ is even.} \end{cases}$ ■

Proof. The result can be proved exactly as in the proof of Theorem 25.20. □

Consider two graphs G and H of order n_1, n_2 respectively. Then, a *torus grid* is the Cartesian product of two cycles C_m and C_n. The following theorem describes the rainbow neighbourhood number of torus grid graphs.

Theorem 25.22

[16] For $m,n \geq 3$, $r_\chi(C_m \Box C_n) = \begin{cases} mn; & \text{if } m,n \text{ are even} \\ 3n; & \text{if } m \text{ is odd and } n \text{ is even} \\ 3\max\{m,n\}; & \text{if } m \text{ and } n \text{ are odd.} \end{cases}$ ■

Proof. The result can be proved exactly as in the proof of Theorem 25.20. □

Given two graphs G and H, the corona $G \odot H$ is defined as the graph obtained by taking $|V(G)|$ copies of H and inserting edges between a vertex of G and each vertex of the corresponding copy of H (see [1]). The following result in [12] discussed the rainbow neighbourhood number of the corona of two connected graphs.

Theorem 25.23

$$r_\chi(G \odot H) = \begin{cases} n_1(1 + r_\chi(H)); & \text{if } \chi(H) \geq \chi(G) - 1; \\ r_\chi(G); & \text{otherwise.} \end{cases}$$ ■

Proof. For any vertex $v \in V(G)$ with $c(v) = c_i$ recolor all vertices $u \in V(H)$, which have $c(u) = c_i$ to the color $c_{\chi(H)+1}$. Then, $\chi(H) \geq \chi(G) - 1$, is a direct consequence of the fact that the graph $G' = K_1 + G$ has $r_\chi(G') = 1 + r_\chi(G)$.

Otherwise, it is clear that all vertices $v \in V(G)$ that yield a rainbow neighbourhood will yield a rainbow neighbourhood in $G \odot H$. Therefore, $r_\chi(G \odot H) \geq r_\chi(G)$. But, no vertex $w \in V(H)$ can yield a rainbow neighbourhood in $G \odot H$. This can be verified as follows. Assume that a vertex $w \in V(H)$ of the t-th copy of H joined to $v \in V(G)$ is a vertex yielding a rainbow neighbourhood in $G \odot H$. It means that vertex w has at least one neighbour for each color c_i, $1 \leq i \leq \ell_2 < \ell_1 - 1$ as well as the neighbour v with the color $c(v) = c_{\ell_2+1}$. Since, c_{ℓ_2+1} can at best be the color c_{ℓ_1-1}, the color $c_{\ell_1} \notin N[w]$ in $r_\chi(G \odot H)$ which is a contradiction. Therefore, $r_\chi(G \odot H) = r_\chi(G)$. □

Problems on the rainbow neighbourhood number of all other graph products are still open for further studies.

25.3 J-COLORING OF GRAPHS

Motivated by the studies mentioned above, a new graph coloring, namely *Johan coloring* or the *J-coloring* of a graph is introduced in [17] as follows.

Definition. [14] A proper k-coloring $\mathbb{C}$ of a graph G is called the *Johan coloring* or the *J-coloring* of G if $\mathbb{C}$ is the maximal coloring such that every vertex of G belongs to a rainbow neighbourhood of G. A graph G is *J-colorable* if it admits a J-coloring.

Definition. [14] The *J-coloring number* of a graph G, denoted by $\mathfrak{J}(G)$, is the maximum number of colors in a J-coloring of G.

A J-coloring of a graph is illustrated in Figure 25.3.

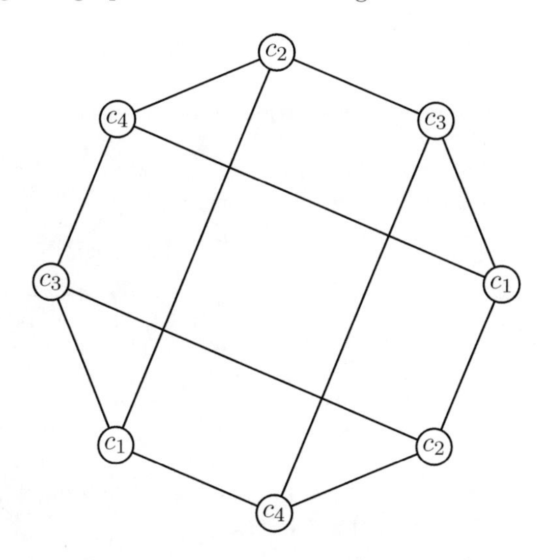

Figure 25.3: J-coloring of a graph

Definition. [14] A proper k-coloring $\mathbb{C}$ of a graph G is called the *modified Johan coloring* or the *J^*-coloring* of G if $\mathbb{C}$ is the maximal coloring such that every internal vertex of G belongs to a rainbow neighbourhood of G. A graph G is *J^*-colorable* if it admits a J^*-coloring.

Definition. [14] The *J^*-coloring number* of a graph G, denoted by $\mathfrak{J}^*(G)$, is the maximum number of colors in a J^*-coloring of G.

Figure 25.4 illustrates a J^*-coloring of a graph.

The following results provided some bounds for the J-coloring number and/or J^*-coloring number of a graph G (see [14]).

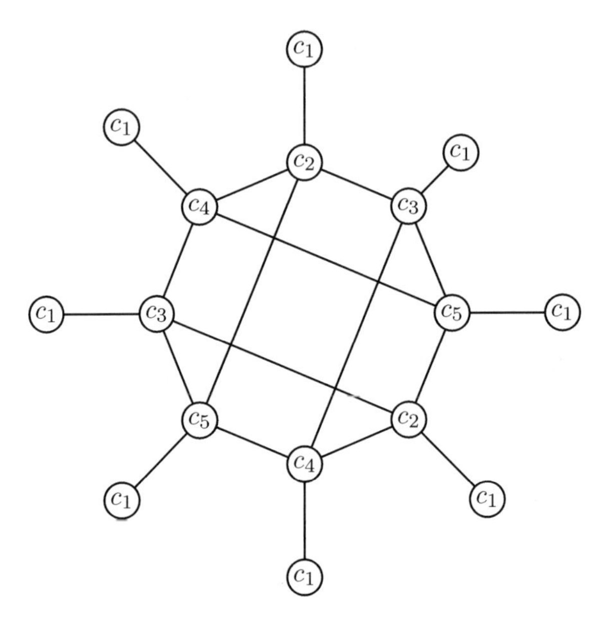

Figure 25.4: A J^*-coloring of a graph

Theorem 25.24

Let G be a connected graph. Then, $\mathcal{J}(G) \leq \delta(G)+1$. ■

Proof. Let c be the coloring function defined on G which defines a J-coloring $\mathcal{C}$ on G. Then, all vertices of G are included in a rainbow neighbourhood of G and hence $c(N[v]) = \mathcal{C}$ for all $v \in V(G)$. □

A relation between the J-chromatic number and the J^*-chromatic number of graphs has been established in [14], as mentioned below:

Proposition 25.3.1. *For a graph G admitting a J-coloring as well as a J^*-coloring, $\mathcal{J}(G) \leq \mathcal{J}^*(G) \leq \delta(G)+1$.*

Proof. If G is a graph without pendent vertices, then $\mathcal{J}(G) = \mathcal{J}^*(G)$. If G has pendent vertices, then by Proposition 2.4, $\mathcal{J}(G) \leq 2$.

If G is a star graph $K_{1,n-1}$, then $2 \leq \mathcal{J}^*(G) \leq n$. Otherwise, let G' be the connected subgraph of G obtained by removing the pendent vertices of G. Since, $d_{G'}(v) \geq 2$, $\forall v \in V(G')$, we have $\mathcal{J}^*(G) \geq 2$. Therefore, we have $\mathcal{J}(G) \leq \mathcal{J}^*(G)$. □

The J-coloring number of certain basic graph classes are determined in the following results (see [14]).

Proposition 25.3.2. *For $n \geq 3$, $\mathfrak{J}(P_n) = 2$ and $\mathfrak{J}^*(P_n) = 3$.*

Proof. Since any proper coloring of a path P_n consists of 2 colors, we have $\mathfrak{J}(P_n) \geq 2$. Also, since P_n has two pendent vertices, $\delta(P_n) = 1$. Then, we have $\mathfrak{J}(P_n) \leq 2$. Therefore, $\mathfrak{J}(P_n) = 2$.

Since all internal vertices of P_n have degree 2, by Proposition 25.3.1, we have $\mathfrak{J}^*(P_n) \leq 3$. Now, assign color c_1 to the vertices v_i of P_n if $i \equiv 1(\text{mod } 3)$, assign color c_2 to the vertices v_i of P_n if $i \equiv 2(\text{mod } 3)$ and assign color c_3 to the vertices v_i of P_n if $i \equiv 0(\text{mod } 3)$. Clearly, every internal vertex will be adjacent to one vertex of color c_1 and one vertex of c_2. Then, this coloring is a $\mathfrak{J}^*$-coloring of P_n. Therefore, $\mathfrak{J}^*(P_n) = 3$. □

Theorem 25.25

[14] A cycle C_n is J-colorable if and only if $n \equiv 0\,(\text{mod } 2)$ or $n \equiv 0\,(\text{mod } 3)$. ■

Proof. Let C_n be a cycle on n vertices. Since $d(v) = 2$ for all $v \in V(C_n)$, by Proposition 25.24, we have $\mathfrak{J}(C_n) \leq 3$. Then, we consider the following cases:

Case-1a: First assume that $n \equiv 0\,(\text{mod } 3)$. Then, assign color c_1 to the vertices v_i of C_n, where $i \equiv 1(\text{mod } 3)$, assign color c_2 to the vertices v_i of C_n, when $i \equiv 2(\text{mod } 3)$ and assign color c_3 to the vertices v_i of C_n if $i \equiv 0(\text{mod } 3)$ (See Figure 25.3). Clearly, this coloring is a J-coloring of the cycle C_n.

Case-1b: Next, assume that $n \equiv 0\,(\text{mod } 2)$ and $n \not\equiv 0\,(\text{mod } 3)$. It is to be noted that no 3-coloring exists such that every vertex of C_n is in a rainbow neighbourhood. Then, the coloring in which the vertices are colored alternately by c_1 and c_2 will itself be the J-coloring of C_n.

Now, assume that $n \not\equiv 0\,(\text{mod } 2)$ as well as $n \not\equiv 0\,(\text{mod } 3)$. Since $n \not\equiv 0\,(\text{mod } 2)$, C_n is an odd cycle and hence any proper coloring of C_n consists of at least 3 colors. By the assumptions, we have $n \equiv \pm 1\,(\text{mod } 6)$. Then,

Case-2a: Let $n \equiv 1\,(\text{mod } 6)$. It is also equivalent to say that $n \equiv 1\,(\text{mod } 3)$. Then, for a positive integer k, the cycle C_n has $3k+1$ vertices. Now, for all $i \equiv 1\,(\text{mod } 3)$, color the vertices v_i, v_{i+1}, v_{i+2} respectively by c_1, c_2 and c_3. But, the vertex v_n must be colored by c_2, as we cannot assume colors c_1 or c_3 (since $v_nv_1, v_nv_{n-1} \in E(C_n)$ and v_1 has color c_1 and v_{n-1} has color c_3). Hence, the vertex v_1 is not in a rainbow neighbourhood, since it is adjacent to two vertices having the color c_2 and not adjacent to any vertex having color c_3. Therefore, C_n has no J-coloring in this context.

Case-2b: Let $n \equiv -1\,(\text{mod } 6)$. It is also equivalent to say that $n \equiv 5\,(\text{mod } 6)$ or $n \equiv 2\,(\text{mod } 3)$. Then, for a positive integer k, the cycle C_n has $3k+2$ vertices. As mentioned above, for all $i \equiv 1(\text{mod } 3)$, color the vertices v_i, v_{i+1}, v_{i+2} respectively by c_1, c_2 and c_3. But, the vertex v_n, irrespective of the possible colors it can take, is adjacent to vertices v_1 and v_{n-1} having the color c_1 and is not adjacent to a vertex of color c_3. Therefore, in this case also, C_n has no J-coloring.

Therefore, the cycle C_n does not have a J-coloring, unless either $n \equiv 0 \pmod 2$ or $n \equiv 0 \pmod 3$. This completes the proof. □

Theorem 25.26

For any graph G which admits a J^*-coloring, we have $\mathfrak{J}^*(G) \leq \Delta(G)+1$. ■

Note that any proper coloring of a complete graph K_n is also a J-coloring of K_n and hence we have the following result straight forward.

Proposition 25.3.3. [14] *For a complete graph K_n, we have $\mathfrak{J}(K_n) = n$.*

The following result discussed the J-coloring number of complete l-partite graphs.

Proposition 25.3.4. [14] *For a complete l-partite graph $K_{n_1,n_2,\ldots,n_l}$, we have the J-coloring number $\mathfrak{J}(K_{n_1,n_2,\ldots,n_l}) = l$.*

The studies on the J-colorability of individual graphs/graph classes, graph transformations etc. are yet to be settled.

25.3.1 J-COLORABILITY OF GRAPH OPERATIONS

In view of the above mentioned concepts and facts, we have the following theorems (see [17]).

Theorem 25.27

If G is a tree of order $n \geq 2$, then $\mathfrak{J}(G) < \mathfrak{J}^*(G)$. ■

Proof. A tree G of order $n \geq 2$ has at least two pendent vertices, say u and v. Therefore, the maximum number of colors which will allow both vertices u and v to yield rainbow neighbourhoods is $\chi(G) = 2$. Therefore, G admits a J-coloring and $\mathfrak{J}(G) = 2$.

Any internal vertex w of G has $d(w) \geq 2$. Therefore, $\mathfrak{J}^*(G) \leq 3$. Consider any diameter path of G say $P_{diam(G)}$. Beginning at a pendent vertex of the diameter path, label the vertices consecutively $v_1, v_2, v_3, \ldots, v_{diam(G)}$. Color the vertices consecutively $c(v_1) = c_1$, $c(v_2) = c_2$, $c(v_3) = c_3$, $c(v_4) = c_1$, $c(v_5) = c_2$, $c(v_6) = c_3$ and so on such that

$$c(v_{diam(G)}) = 1; \quad \text{if } diam(G) \equiv 1 \pmod 3 \tag{25.1}$$

$$c(v_{diam(G)}) = 2; \quad \text{if } diam(G) \equiv 2(\text{mod } 3) \tag{25.2}$$
$$c(v_{diam(G)}) = 3; \quad \text{if } diam(G) \equiv 0(\text{mod } 3). \tag{25.3}$$

Clearly, in respect to path $P_{diam(G)}$, it is a proper coloring and all internal vertices yield a rainbow neighbourhood on 3 colors. Consider any maximal path starting from, say, $v \in V(P_{diam(G)})$. Hence, v is a pendent vertex to that maximal path. Color the vertices consecutively from v as follows:

(a) If $c(v) = c_1$ in $P_{diam(G)}$, color as $c_1, c_2, c_3, c_1, c_2, c_3, \ldots, \underbrace{c_1 \text{ or } c_2 \text{ or } c_3}$

(b) If $c(v) = c_2$ in $P_{diam(G)}$, color as $c_2, c_3, c_1, c_2, c_3, c_1, \cdots, \underbrace{c_2 \text{ or } c_3 \text{ or } c_1}$.

(c) If $c(v) = c_3$ in $P_{diam(G)}$, color as $c_3, c_1, c_2, c_3, c_1, c_2, \cdots, \underbrace{c_3 \text{ or } c_1 \text{ or } c_2}$.

It follows from mathematical induction that all maximal branching can receive such coloring which remains a proper coloring with all internal vertices $v \in V(G)$ having $|c(N[v])| = 3$. Furthermore, all nested branching can be colored in a similar way until all vertices of G are colored. Therefore, $\mathfrak{J}^*(G) \geq 3$. Hence, $\mathfrak{J}(G) < \mathfrak{J}^*(G)$. □

An upper bound for the J^*-coloring of a graph has been determined in [17] as follows:

Theorem 25.28

For any graph G which admits a J^*-coloring, we have $\mathfrak{J}^*(G) \leq \Delta(G) + 1$. If $\mathfrak{J}^*(G) > \mathfrak{J}(G)$ for a graph G, then G has at least one pendent vertex. ■

Another interesting result on the relation between $\mathfrak{J}^*(G)$ and $\mathfrak{J}(G)$ can be seen in [17] as follows:

Theorem 25.29

If $\mathfrak{J}^*(G) > \mathfrak{J}(G)$ for a graph G, then G has at least one pendent vertex. ■

Proof. Since all $v \in V(G)$ are internal vertices and any vertex u for which $d(u) = \delta(G)$ must yield a rainbow neighbourhood, it follows that any maximal proper coloring $\mathcal{C}$ are bound to $|\mathcal{C}| = |N[u]| = \delta(G) + 1$. Therefore, if $\mathfrak{J}^*(G) > \mathfrak{J}(G)$, then G has at least one pendent vertex. □

The following theorem characterised those graphs which admit a J-coloring in [17].

Theorem 25.30

A graph G of order n admits a J-coloring if and only if $r_\chi(G) = n$. ■

Proof. If $r_\chi(G) = n$, then every vertex of G belongs to a rainbow neighbourhood. Hence, either the chromatic coloring $\varphi : V(G) \mapsto \mathcal{C}$ is maximal or a maximal coloring $\varphi' : V(G) \mapsto \mathcal{C}'$ exists.

An immediate consequence of the definition of J-coloring is that if graph G admits a J-coloring then each vertex $v \in V(G)$ yields a rainbow neighbourhood. This consequence also follows from the the result that for any connected graph G, $\mathcal{J}(G) \leq \delta(G) + 1$. Hence, it follows that either the J-coloring is minimal or a minimal coloring $\varphi' : V(G) \mapsto \mathcal{C}'$ exists such that $r_\chi(G) = n$. □

The following theorem established another necessary and sufficient condition for a graph G to have a J-coloring with respect to a minimal coloring of G (see [17]).

Theorem 25.31

A graph G admits a J-coloring if and only if each $v \in V(G)$ yields a rainbow neighbourhood with respect to a minimal-coloring of G. ■

Proof. If in a χ^--coloring of G, each $v \in V(G)$ yields a rainbow neighbourhood it follows that the corresponding proper coloring can be maximised to obtain a J-coloring.

Conversely, assume that a graph G admits a J-coloring. Then, it follows that the corresponding proper coloring can be minimised to obtain a minimal proper coloring for which each $v \in V(G)$ yields a rainbow neighbourhood. Let the aforesaid set of colors be $\mathcal{C}'$. Assume that a minimum set of colors $\mathcal{C}$ exists which is a χ^--coloring of G and $|\mathcal{C}| < |\mathcal{C}'|$. It implies that there exists at least one vertex $v \in V(G)$ for which at least one distinct pair of vertices, say $u, w \in N(v)$ exists such that u and v are non-adjacent. Furthermore, $c(u) = c(w)$ under the coloring $\varphi : V(G) \mapsto \mathcal{C}$.

Assume that there is exactly one such v and exactly one such vertex pair $u, w \in N(v)$. But then both u and w yield rainbow neighbourhoods in G under the proper coloring $\varphi : V(G) \mapsto \mathcal{C}$, which is a contradiction to the minimality of $\mathcal{C}'$. By mathematical induction, similar contradictions arise for all vertices similar to v. This completes the proof. □

The following result discusses the condition for the join of two graphs to admit a J-coloring.

Theorem 25.32

The join $G+H$ admits a J-coloring if and only if both graphs G and H admit a J-coloring. ■

Proof. Assume that both G and H admit a J-coloring. Without loss of generality, let $\mathfrak{J}(G) \leq \mathfrak{J}(H)$. Assume that $\varphi : V(G) \mapsto \mathcal{C}$, $\mathcal{C} = \{c_1, c_2, c_3, \ldots, c_\ell\}$ and $\varphi' : V(H) \mapsto \mathcal{C}'$, $\mathcal{C}' = \{c_1, c_2, c_3, \ldots, c_{\ell'}\}$ is a J-coloring of G and H, respectively. For each $v \in V(G)$, $c(v) = c_i$ recolor $c(v) \mapsto c_{i+\ell'}$. Denote the new color set by $\mathcal{C}_{i+\ell'}$. Clearly, each vertex $v \in V(G)$ is adjacent to at least one of each color in $G+H$, hence each such vertex yields a rainbow neighbourhood in $G+H$. Similarly, each vertex $u \in V(H)$ is adjacent to at least one of each color in $G+H$ and hence each such vertex yields a rainbow neighbourhood in $G+H$. Furthermore, since both $|\mathcal{C}|$, $|\mathcal{C}'|$ are maximal color sets, the set $|\mathcal{C}_{i+\ell'} \cup \mathcal{C}'|$ is maximal. Therefore, $G+H$ admits a J-coloring.

The converse follows trivially from the fact that the additional edges between G and H as defined for join form an edge cut in $G+H$. □

The above theorem is a consequence of the fact that every vertex of the first graph G will be adjacent to all vertices of the second graph, say H in their join $G+H$. In a similar manner, the J-colorability of the corona of two J-colorable graphs has been discussed in [17] as stated in the following theorem.

Theorem 25.33

If graphs G and H admit J-colorings, then $G \circ H$ admits a J-coloring if and only if either $G = K_1$ or $\mathfrak{J}(G) = \mathfrak{J}(H) + 1$. ■

The admissibility of J-coloring by the join of two graphs G and H has been established in following theorem (see [17]).

Theorem 25.34

If graphs G and H of order n and m respectively admit a J-coloring, then

(i) $G \square H$ admits a J-coloring.
(ii) $\mathfrak{J}(G \square H) = \max\{\mathfrak{J}(G), \mathfrak{J}(H)\}$.

■

Proof. (i) Without loss of generality assume $\mathcal{J}(H) \geq \mathcal{J}(G)$. Also, assume that $V(G) = \{v_i : 1 \leq i \leq n\}$ and $V(H) = \{u_i : 1 \leq i \leq m\}$. From the definition of $G\Box H$ it follows that $V(G\Box H) = \{(v_i, u_j) : 1 \leq i \leq n, 1 \leq j \leq m\}$. For $i = 1$, if $u_j \sim u_k$ in H, where $\sim$ denotes the adjacency, then $(v_1, u_j) \sim (v_1, u_k)$ and hence we obtain an isomorphic copy of H. Such a copy admits a J-coloring identical to that of H in respect to the vertex elements $u_1, u_2, u_3, \ldots, u_m$. Now obtain the disjoint union with the copies of H corresponding to $i = 2, 3, 4, \ldots, n$. Apply the definition of $G\Box H$ for u_1 and if $v_i \sim v_j$ in G, then $(v_i, u_1) \sim (v_j, u_1)$. An interconnecting copy of G is obtained which results in the first iteration connected graph. Similarly, this copy of G admits a J-coloring identical to that of G in respect to the vertex elements $v_1, v_2, v_3, \ldots, v_n$. Proceeding iteratively to add all copies of G for $i = 2, 3, 4, \ldots, n$ in terms of the definition of $G\Box H$, clearly shows that a J-coloring is admitted.

(ii) The second part of the result follows from the similar reasoning used to prove and hence, $\chi(G\Box H) = \max\{\chi(G), \chi(H)\}$. □

The J-coloring of edge deleted subgraphs of a J-colorable graph has been discussed in [17] as follows:

Theorem 25.35

For any connected graph G which does not admit a J-coloring, a minimal set of edges, E', which need not necessarily be unique, can be removed such that $G - E'$ admits a J-coloring. ■

Proof. Since any connected graph G of order n and size $\varepsilon(G) = p$ has a spanning subtree and any tree admits a J-coloring, at most $p - (n-1)$ edges must be removed from G. Therefore, if $p - (n-1)$ is not a minimal number of edges to be removed then a minimal set of edges E', $|E'| < p - (n-1)$ must exist whose removal results in a spanning subgraph G' which allows a J-coloring. □

25.3.2 THE PAUCITY NUMBER OF GRAPHS

It is to be noted that all graphs do not admit J-coloring, in general. As stated in the previous section, for a graph G which is not J-colorable, by removing or adding some edges of G, we can make the corresponding derived graph J-colorable. In this context, a new parameter was defined called the J-paucity number in [5] as follows:

Definition (J-paucity number). [5]Let G be a graph which does not admit a J-coloring. Then, the *J-paucity number* of G, denoted by $\varrho(G)$, is defined as the minimum number of edges to be added to G so that the modified graph G' becomes J-colorable with respect to a $(\delta(G)+1)$-coloring of G.

For example, consider a cycle C_n. We can notice that the J-paucity number is r if $n \equiv r(\text{mod } 3)$. Further investigation on the paucity number of graphs promises much for intense research in this area.

25.4 CONCLUSION

In this chapter, we have discussed a couple of new coloring protocols of some fundamental graph classes and studied some parameters related to those colorings. The study on the parameters related to those coloring protocols demands strong background and support of efficient algorithms. Finding out an efficient algorithm to examine the existence of the property in a given graph will be a valuable contribution in this area of research.

We have considered three types of parameters with respect to a single concept of rainbow neighbourhoods in graphs. The studies can be extended further to different other types of colorings, different graph classes, different graph transformations etc. The same can be extended to edge coloring of graphs also. All these facts highlight a wide scope for further research in this area.

REFERENCES

1. S. Barik, S. Pati and B. K. Sarma. The spectrum of the corona of two graphs. *SIAM J. Discrete Math.*, 21(1): 47-56, 2007.
2. J. A. Bondy and U. S. R. Murty. *Graph Theory with Applications.* Macmillan, London, 1976.
3. J. A. Bondy and U. S. R Murty. *Graph Theory.* Springer, 2008.
4. G. Chartrand and P. Zhang. *Chromatic Graph Theory,* Chapman and Hall/CRC, 2008.
5. F. Fornasiero and S. Naduvath. On J-colorability of certain derived graph classes. *Acta Univ. Sapientiae Inform.*, 11(2): 159-173, 2019.
6. J. A. Gallian. A dynamic survey of graph labeling. *Electron. J. Combin.*, #DS6, 2018.
7. F. Harary. *Graph Theory.* Narosa, New Delhi, 2001.
8. R. Hemminger and L. Beineke. Line graphs and line digraphs, in *Selected Topics in Graph Theory.* Academic Press, 271-305, 1978.
9. W. Imrich and S. Klavzar. *Product Graphs: Structure and Recognition.* Wiley, 2000.
10. J. Kok, N. K. Sudev and K. P. Chithra. Generalised coloring sums of graphs. *Cogent Math.*, 3(1): 1140002: 1-11, 2016.
11. T.R. Jensen and B. Toft. *Graph Coloring Problems.* John Wiley & Sons, 39, 2011.
12. J. Kok, S. Naduvath, and M. K. Jamil. Rainbow neighbourhood number of graphs. *Proyecciones J. Math.*, 38(3): 469-484, 2019.

13. W. Lin, J. Wu, P.C.B Lam, and G. Gu. Several parameters of generalized Mycielskians. *Discrete Appl. Math.*, 154(8): 1173-1182, 2006.
14. S. Naduvath. On certain J-coloring parameters of graphs. *Nat. Acad. Sci. Lett.*, 43(1): 53-57, 2020.
15. S. Naduvath, S. Chandoor, S. J. Kalayathankal and J. Kok. A note on the rainbow neighbourhood number of certain graph classes. *Nat. Acad. Sci. Lett.*, 42(2): 135-138, 2019.
16. S. Naduvath, S. Chandoor, S. J. Kalayathankal and J. Kok. Some new results on the rainbow neighborhood number of graphs. *Nat. Acad. Sci. Lett.*, 42(3): 249–252, 2019.
17. S. Naduvath and J. Kok. J-coloring of graph operations. *Acta Univ. Sapientiae. Inform.*, 11(1): 95-108, 2019.
18. N. K. Sudev, C. Susanth and S. J. Kalayathankal. On the rainbow neighbourhood number of mycielski type graphs. *Int. J. Appl. Math.*, 31(6): 797-803, 2018.
19. D.B. West. *Introduction to Graph Theory.* Prentice Hall, Upper Saddle River, NJ, 1996.

26 Total Global Dominator Coloring of Graphs

K.P. Chithra
Department of Mathematics,
CHRIST (Deemed to be University),
Bengaluru, Karnataka (INDIA)
E-mail: chithra.kp@res.christuniversity.in

Mayamma Joseph
Department of Mathematics,
CHRIST (Deemed to be University),
Bengaluru, Karnataka (INDIA)

Let $\{V_1, V_2, V_3 \ldots, V_k\}$ be the collection of color classes corresponding to a proper coloring c of a graph $G = (V, E)$, where V_i represents the set of vertices with color c_i. A color class V_i is called a proper dom-color class of a vertex $v \in V$ if v dominates every vertex in V_i and $V_i \neq \{v\}$. A color class V_j is called an anti dom-color class of v if v does not dominate any of the vertices in V_j. The coloring c is called a total global dominator coloring of G if every vertex in V has a proper dom-color class and an anti dom-color class. The minimum number of colors required for the total global dominator coloring of G is called the total global dominator chromatic number and is denoted by $\chi_{tgd}(G)$. Here, we determine the exact values of $\chi_{tgd}(G)$ for some color classes and study their properties.

26.1 INTRODUCTION

For the terminology and results of graph theory, we refer to [6], for more about domination in graphs refer to [7] and for the terminology of graph coloring, we rely upon [2]. Unless mentioned otherwise, all graphs considered in this chapter are simple, connected, undirected and finite.

For a graph $G = (V, E)$, a set $D \subset V$ is called a *dominating set* if every vertex $v \in V$ is either an element of D or adjacent to some vertex in D. The minimum cardinality of a minimal dominating set in G is called the *domination number* of G and is denoted by $\gamma(G)$. A vertex v is said to dominate a vertex, say u, if either $v = u$ or v is adjacent to u. In a similar manner, the vertex v is said to dominate a set $S \subseteq V(G)$ if v is adjacent to every vertex in S.

If the graph induced by D does not contain an isolated vertex, then D is called a *total dominating set* and the cardinality of the minimum total dominating set is called the *total domination number* and is denoted by γ_t.

D is called a *global dominating set* if it is a dominating set of both G and $\overline{G}$ and the minimum cardinality of a minimal global dominating set is called the *global domination number* and is denoted by γ_g. The global domination number of derived graphs and related results can be found in [11, 12]. If D is a total dominating set of both G and $\overline{G}$, then D is called a *total global dominating set* (tgd-set) and the minimum cardinality of a minimal tgd-set is called the *total global domination number* and is denoted by γ_{tg} [10].

A graph coloring is an assignment of colors to the vertices or edges or both of a graph G. The minimum number of colors required to color the vertices of a graph G such that the adjacent vertices possess different colors is known as the *chromatic number* of G and is denoted by $\chi(G)$. A coloring of a graph G with $\chi(G)$ colors is called the *chromatic coloring* or $\chi-$*coloring* of G. Let $\{V_1, V_2, V_3 \ldots, V_k\}$ be a partition of vertex set V with respect to a coloring c such that a vertex $v \in V_i$ if $c(v) = c_i$ then V_i is called the color class of v with respect to c.

Since domination and coloring have been two areas of study for researchers for several decades, efforts have been made to link these two concepts. The first significant attempt in this area was done in [4] and the coloring concerned is called a *dominator coloring*. A coloring c of a graph G is called the *dominator coloring* of G if every vertex dominates at least one color class. The *dominator chromatic number* of a graph G, denoted by $\chi_d(G)$, is the minimum number of colors required for a dominator coloring of G. The concept was further studied by Gera [3], Vijayalakshmi [13], Henning [8], Arumugam, Bagga and Chandrasekar [1]. Later, a variation of dominator coloring, namely the *total dominator coloring* was introduced and studied by Kazemi [9]. A coloring c is called the *total dominator coloring* if every vertex in V is adjacent to every vertex of some color class. The *total dominator chromatic number* of a graph G, denoted by $\chi_d^t(G)$, is the minimum number of colors required for a total dominator coloring of G.

Motivated by the studies on dominator coloring and global domination in graphs, the notion of *global dominator coloring* was introduced by Sahul Hamid and Rajeswari [5] as a coloring in which every vertex of G has a dom-color class and an anti dom-color class. A color class V_i is called a *dom-color class* of a vertex v if v dominates every vertex of V_i and v is called a *dominator* of V_i. A color class V_j is called an *anti dom-color class* of v if v does not dominate any of the vertices in V_j. The minimum number of colors required for the global dominator coloring of G is called a *global dominator chromatic number* of G and is denoted by $\chi_{gd}(G)$. The global dominator chromatic number of some fundamental graph classes such as paths, cycles and the Petersen graph have been discussed in [5]. Furthermore, the bounds for the global dominator chromatic number and the relationships between global dominator chromatic number and other parameters such as chromatic number, dominator chromatic number and global domination number have also been examined in that article.

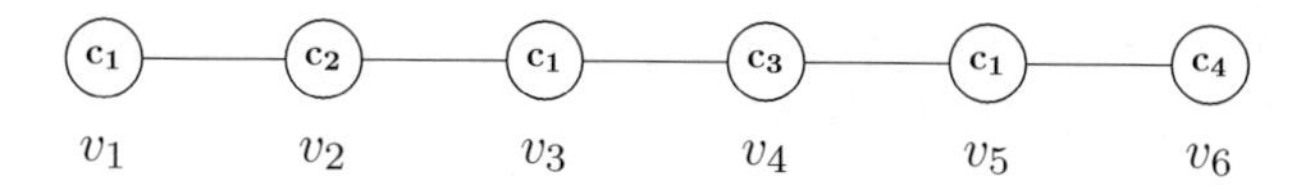

Figure 26.1: χ_{gd} colorings

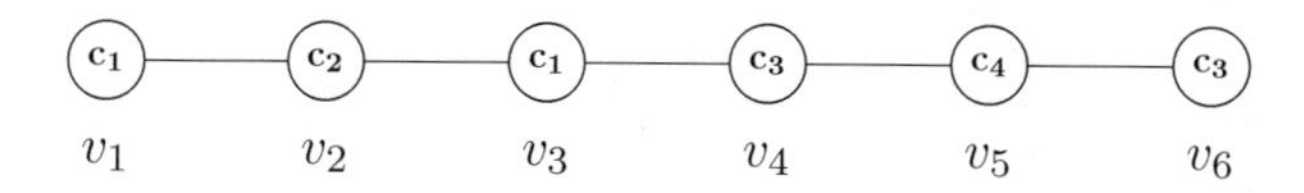

Figure 26.2: χ_{gd} colorings

Figure 26.1 and Figure 26.2 illustrate the global dominator coloring of P_6 in two different ways. In Figure 26.1 the dom-color class of v_1, v_2 and v_3 is $V_2 = \{v_2\}$, dom-color class of v_4 and v_5 is $V_3 = \{v_4\}$ and dom-color class of v_6 is $V_4 = \{v_6\}$. Here the dom-color classes of vertices v_2, v_4 and v_6 are singleton sets consisting of the corresponding vertices only. The coloring scheme in Figure 26.2 has the additional property that the dom-color class of none of the vertices is itself.

Motivated by this observation, we define the concept of a proper dom-color class as follows: A *proper dom-color class* V_i of a vertex v is a dom-color class such that $v \notin V_i$ (that is, if $V_i \neq \{v\}$). This idea of the proper dom-color class leads to the introduction of another variation of global dominator coloring, which we call the total global dominator coloring.

A coloring c is called a *total global dominator coloring* (tgd-coloring) of G if every vertex in V has a proper dom-color class and an anti dom-color class. The minimum number of colors required for the total global dominator coloring of G is called the *total global dominator chromatic number* and is denoted by $\chi_{tgd}(G)$.

A total global dominator coloring of P_6 is shown in Figure 26.2, where the proper dom-color classes for the vertices $v_1, v_2, v_3, v_4, v_5, v_6$ are $V_2, V_1, V_2, V_4, V_3, V_4$ respectively and $V_4 = \{v_5\}$ is the anti dom-color class for the vertices v_1, v_2, v_3 and the anti dom-color class for the other three vertices is $V_2 = \{v_2\}$. Since each of the vertices has a proper dom-color class and an anti dom-color class the coloring followed in Figure 26.2 is the total global dominator coloring.

In the light of the above notions, a total dominator coloring of a graph G can be viewed as a proper coloring in which every vertex of G has a proper dom-color class and hence $\chi_{tgd}(G) \geq \chi_d^t(G)$. It is clear that all graphs which admit a total global dominator coloring also admit a global dominator coloring and every total global dominator coloring is also a global dominator coloring. Hence $\chi_{gd}(G) \leq \chi_{tgd}(G)$.

26.2 BOUNDS FOR TOTAL GLOBAL DOMINATOR CHROMATIC NUMBER

In this section, we determine the conditions for a graph G of order n to admit tgd-coloring and the bounds for $\chi_{tgd}(G)$. Here we also find the relationships between the parameters $\chi, \chi_{gd}, \chi_d^t$ and χ_{tgd}.

Theorem 26.1

If a graph G admits a tgd-coloring, then $4 \leq \chi_{tgd}(G) \leq n$. ■

Proof. Let G admit tgd-coloring and without loss of generality let $v \in V(G)$ belongs to the color class V_1. Let V_2 and V_3 be its proper dom-color class and anti dom-color class respectively. Hence, there exist at least 3 color classes in G. Assume that G has exactly 3 color classes, say V_1, V_2, V_3. Since v is adjacent to every element in V_2, V_1 cannot be an anti-dom color class for any element in V_2. Hence, V_3 should be an anti dom-color class for every element in V_2 also. Therefore, V_1 and V_2 become the anti dom-color class for every element $u \in V_3$. Hence, there is no proper dom-color class for u, which is a contradiction, hence G should have more than 3 color classes. Since the graph admits tgd-coloring, there is a proper dom-color class for u, say $V_4 \in G$. Then V_3 becomes a proper dom-color class for V_4 and V_1 or V_2 or both can be anti dom-color classes. Therefore $\chi_{tgd}(G) \geq 4$. n being the number of vertices of G $\chi_{tgd}(G) \leq n$. Therefore, $4 \leq \chi_{tgd}(G) \leq n$. This completes the proof. □

From Theorem 26.1 it is clear that the graphs which admit tgd-coloring have at least 4 vertices. Therefore for further discussion we consider only graphs of order greater than 3.

Theorem 26.2

A graph G of order $n \geq 4$ admits tgd-coloring if and only if $1 \leq d(v) \leq n-2$. ■

Proof. Let G be a graph of order at least 4. Assume that G admits tgd-coloring. Then $d(v) \geq 1$ as every vertex v of G has a proper dom-color class. Also, since every vertex v should have an anti dom-color class, there exists at least one vertex which is not adjacent to v. Therefore, $d(v) < n-1$. Hence, we have $1 \leq d(v) \leq n-2$.

Conversely, assume that $1 \leq d(v) \leq n-2$, $\forall v \in V(G)$. Now we have to show that G admits a tgd-coloring. To establish such a coloring, we consider the following coloring scheme.

Color the n vertices with n distinct colors. Let v be an arbitrary vertex in $V(G)$, since $d(v) \geq 1$, v is adjacent to at least one vertex say u in $V(G)$. Hence $\{u\}$ is a proper dom-color class of v. Since $d(v) \leq n-2$, there exists at least one vertex, say w, in $V(G)$ which is not adjacent to v. Hence, $\{w\}$ is an anti dom-color class of v. v being arbitrary, it follows that for every vertex $v \in V(G)$, there exists a proper dom-color class and an anti dom-color class. That is, G admits tgd-coloring. □

Remark. If a graph G admits a tgd-coloring, then $1 \leq \delta(G) \leq \Delta(G) < n-1$. Therefore, a graph without isolated vertices that admits global dominator coloring also admits a tgd-coloring.

Invoking Theorem 26.2 we can infer that complete graphs, star graphs, wheel graphs and double wheel graphs do not admit a tgd-coloring.

Theorem 26.3

Given any positive integer $k \geq 4$, there exists a connected graph G of order $n \geq k$ such that $\chi_{tgd}(G) = k$. ■

Proof. Construct a complete graph K_{k-2} with vertices $\{v_1, v_2, v_3, \ldots, v_{k-2}\}$. Color these vertices with the colors $c_1, c_2, c_3, \ldots, c_{k-2}$ respectively. Attach a pendent vertex v_{k-1} at vertex v_1 and color it with color c_{k-1} and attach the $n-k+1$ pendent vertices to the vertex v_2 and color them with color c_k. The resultant graph G is a graph with n vertices and $\chi_{tgd}(G) = k$. □

We know that $\chi(G)$ is the minimum number of colors used in a proper coloring of G and we also know that tgd-coloring is a proper coloring. In the following theorem we find a relation between χ and χ_{tgd} for any graph G.

Theorem 26.4

For any graph G, we have $\chi_{tgd}(G) \geq 1 + \chi(G)$. ■

Proof. The inequality $\chi_{tgd}(G) \geq \chi(G)$ is immediate as any tgd-coloring is a proper coloring. Hence it remains to show that $\chi_{tgd}(G) \neq \chi(G)$.

If possible let $\chi(G) = \chi_{tgd}(G) = k$. Since $\chi(G) = k$, there exists a vertex $v \in V$ such that for all i, $1 \leq i \leq k$, $N[v] \cap V_i \neq \emptyset$. Then the vertex v has no anti dom-color class, which means that G cannot have a tgd-coloring with k colors. Hence $\chi_{tgd}(G) \geq 1 + \chi(G)$. □

We can see that there are graphs which admit the equality $\chi_{tgd}(G) = 1 + \chi(G)$. The chromatic number of C_5 is 3 and $\chi_{tgd}(C_5)$ is 4, therefore $\chi_{tgd}(C_5) = 1 + \chi(C_5)$.

The following lemma gives a necessary condition for the total global dominator chromatic number of a graph to be equal to its total dominator chromatic number.

Lemma 26.1

If G is a connected graph with $\chi_d^t(G) \geq \Delta(G)+2$, then $\chi_{tgd}(G) = \chi_d^t(G)$. ■

Proof. Let G be a connected graph admitting total dominator coloring. Note that every total global dominator coloring is also a total dominator coloring. Hence $\chi_d^t \leq \chi_{tgd}$. Now consider a total dominator coloring c of G. Let $\chi_d^t(G) = k \geq \Delta+2$. Let $V_1, V_2, V_3, \ldots, V_k$ be the color classes induced by the coloring c. Also let v_i be a vertex in the ith partition. Then, the vertex v_i can have at most $\Delta(G)$ neighbours other than v_i. Since $\chi_d^t(G) \geq \Delta(G)+2$, there is a color class that contains none of the neighbours of v_i, say $V_j, j \neq i$. Hence, V_j is an anti dom-color class of v_i. v_i being an arbitrary vertex, all vertices in $V(G)$ have an anti dom-color class. Since G admits a total dominator coloring, all vertices in $V(G)$ have a proper dom-color class. Thus all vertices in $V(G)$ have a proper dom-color class and an anti dom-color class. Hence, the total dominator coloring of G is also a total global dominator coloring with $\chi_{tgd}(G) = k$. □

The following theorem establishes a relation between $\chi_{tgd}(G)$ and $\chi_d^t(G)$ for disconnected graphs.

Proposition 26.2.1. *If G is a disconnected graph admitting tgd-coloring, then $\chi_{tgd}(G) = \chi_d^t(G)$.*

Proof. Let G be a disconnected graph and also let G admit a total dominator coloring. Hence, each vertex v of G properly dominates a color class.This color class acts as a proper dom-color class for v. Since the graph is disconnected, the proper dom-color class of a vertex in one component will be an anti dom-color class for the vertex in another component. This completes the proof. □

The next theorem is a realization problem.

Theorem 26.5

Given three consecutive positive integers $k, k+1, k+2, k \geq 3$, there exists a connected graph G such that $\chi(G) = k, \chi_d^t(G) = \chi_{gd}(G) = k+1$ and $\chi_{tgd}(G) = k+2$. ■

Proof. Given an integer $k \geq 3$, consider the complete graph K_k with vertex set $\{v_1, v_2, v_3, \ldots, v_k\}$ and the graph K_2 with vertices u_1 and u_2 . Construct a graph G by joining the vertex v_k of K_k with the vertex u_1 of K_2 by an edge. Note that K_k being a clique of G, any proper coloring of G must have at least k colors.

Claim 1: $\chi(G) = k$.

Define a coloring c such that $c(v_i) = c_i, 1 \leq i \leq k$; $c(u_1) = c_1$ and $c(u_2) = c_2$. Then c is a proper coloring, proving that $\chi(G) = k$.

Claim 2: $\chi_d^t(G) = k+1$.

We observe that k colors are not sufficient to obtain a total dominator coloring of G. This is because, u_1 is the only vertex that can be in the proper dom-color class of u_2 and therefore the color assigned to u_1 should be different from the k colors given to the vertices of K_k. Hence $\chi_d^t(G) \geq k+1$. Further, the coloring c defined by $c(v_i) = c_i, 1 \leq i \leq k$; $c(u_1) = c_{k+1}$ and $c(u_2) = c_2$ is a total dominator coloring so that $\chi_d^t(G) = k+1$.

Claim 3: $\chi_{gd}(G) = k+1$.

In this case also note that k colors are not enough to obtain a global dominator coloring of G. This is because, u_2 is the only vertex that can be in the anti dom-color class of v_k and therefore the color assigned to u_2 should be different from the k colors given to the vertices of K_k. Hence $\chi_{gd}(G) \geq k+1$. Further, the coloring c defined by $c(v_i) = c_i, 1 \leq i \leq k$; $c(u_1) = c_1$ and $c(u_2) = c_{k+1}$ is a global dominator coloring so that $\chi_{gd}(G) = k+1$.

Claim 4: $\chi_{tgd}(G) = k+2$.

From claim 3, it follows that $\chi_{gd}(G) = k+1$ and we have $\chi_{tgd}(G) \geq \chi_{gd}(G) = k+1$. Since tgd-coloring is a total dominator coloring the color of the vertex u_1 should be different from the k colors of the vertices of K_k and being a global dominator coloring the color of the vertex u_2 should be different from the $k+1$ colors used and hence $\chi_{tgd}(G) = k+2$.

□

The following theorem establishes a relationship between $\chi_{tgd}(G)$ and $\gamma_{tg}(G)$.

Theorem 26.6

For any graph G, we have $\chi_{tgd}(G) \geq \gamma_{tg}(G)$. ■

Proof. Let G be a graph with total global dominator chromatic number χ_{tgd}. Let D be the set formed by taking exactly one vertex from all the χ_{tgd} color classes, then D will be a dominating set of both G and its complement $\overline{G}$.

First we prove that D is a tgd-set of G. Let $v \in V(G)$ be arbitrary. Then there exists a $u \neq v \in D$ that is adjacent to v as all vertices in G have a proper dom-color class. Since v is an arbitrary vertex of G, it follows that any vertex in $V(G)$ is adjacent to at least one vertex in D. Therefore D is a tgd-set of G.

Next we prove that D is a tgd-set of $\overline{G}$. For any arbitrary vertex $v \in V(\overline{G})$ there exists a $w \neq v \in D$ that is adjacent to v as all vertices in $V(G)$ have an anti dom-color class. Thus we have $v \in V(\overline{G})$ is adjacent to at least one vertex in D. As v is arbitrary, any vertex in $\overline{G}$ is adjacent to some vertex in D. Therefore D is a tgd-set of $\overline{G}$. Thus we have $\gamma_{tg} \leq |D| = \chi_{tgd}$. That is $\chi_{tgd}(G) \geq \gamma_{tg}(G)$.

□

There exist graphs for which $\chi_{tgd}(G) = \chi_{tgd}(G)$. For C_4 it is easy to verify that $\chi_{tgd}(G) = \gamma_{tg}(G) = 4$. From the above theorem we can see that $\gamma_{tg}(G)$ is a lower bound for $\chi_{tgd}(G)$. In the following theorem we find an upper bound for $\chi_{tgd}(G)$.

Theorem 26.7

If G admits tgd-coloring, then $\chi_{tgd}(G) \leq \gamma_{tg}(G) + \chi(G)$. ■

Proof. Consider a χ-coloring of G with color classes $\{V_1, V_2, V_3, \cdots, V_\chi\}$. Also let D be a minimum total global dominating set of G and let $D_i = D \cap V_i, \forall i, 1 \leq i \leq \chi$. Consider the following coloring c on G with color classes $\{\{v\}, v \in D\} \cup \{V_i \backslash D_i \forall i, 1 \leq i \leq \chi\}$ of G. Let u be an arbitrary vertex of G. D being a total global dominating set of G, there exist at least two vertices, say v_1, v_2, in D other than u such that u is adjacent to v_1 in G and is adjacent to v_2 in $\overline{G}$. Therefore, $\{v_1\}$ is a proper dom-color class and $\{v_2\}$ is an anti dom-color class of u in G. Since u is an arbitrary vertex of G, the coloring c is a total global dominator coloring of G. Therefore, $\chi_{tgd}(G) \leq \gamma_{tg}(G) + \chi(G)$. □

The above upper bound is sharp as we can see in the case of P_8. It can be verified that the value of χ_{tgd}, γ_{tg} and χ for P_8 are 6, 4 and 2 respectively.

26.3 TOTAL GLOBAL DOMINATOR CHROMATIC NUMBER FOR SOME CLASSES OF GRAPHS

The following results proved in [9] are necessary to investigate further about the total global dominator coloring of some graph classes.

Proposition 26.3.1. [9] *For* $n \geq 4$, $\chi_d^t(P_n) = \begin{cases} 4, & if\ n = 4, \\ 2\lceil \frac{n}{3} \rceil - 1, & if\ n \equiv 1(\mod 3), \\ 2\lceil \frac{n}{3} \rceil, & otherwise. \end{cases}$

Proposition 26.3.2. [9] *For* $n \geq 4$, $\chi_d^t(C_n) = \begin{cases} 4, & if\ n = 4, \\ 2\lceil \frac{n}{3} \rceil - 1, & if\ n \equiv 1(\mod 3), \\ 2\lceil \frac{n}{3} \rceil, & otherwise. \end{cases}$

First we determine the value of χ_{tgd} for paths and cycles and show that $\chi_{tgd}(P_n) = \chi_{tgd}(C_n)$.

Theorem 26.8

Let P_n be a path on $n \geq 4$ vertices and C_n be a cycle on $n \geq 4$ vertices, then

$$\chi_{tgd}(P_n) = \chi_{tgd}(C_n) = \begin{cases} 4, & if\ n = 4, \\ 2\lceil \frac{n}{3} \rceil - 1, & \text{if } n \equiv 1(\mod 3), \\ 2\lceil \frac{n}{3} \rceil, & otherwise. \end{cases}$$ ■

Proof. For $n = 4$ we can verify directly. For the remaining values of n, the result follows from Lemma 26.1, Proposition 30.1.1 and 26.3.2. □

Next we determine the value of $\chi_{tgd}(G)$ when G is a complete multipartite graph.

Theorem 26.9

The total global dominator chromatic number of complete l-partite graph G is $2l$. ■

Proof. Assume that G admits a tgd-coloring. Let v_i be a vertex in the ith partition. Then, the remaining $l-1$ partitions are proper dom-color classes for v_i. The elements of the anti dom-color class of v_i can be the elements from the ith partition other than v_i. Hence, the ith partition should have at least 2 vertices with different colors. There should be similar results for the other partitions. Clearly the color class in the jth partition, $j \neq i$, is a proper dom-color class of v_i. Hence the total number of minimum color classes will be $2l$. That is, $\chi_{tgd}(G) = 2l$. □

A multistar graph is formed by joining at least one pendent vertex to each vertex of a complete graph $K_l, l > 2$ and is denoted by $K_l(a_1, a_2, a_3, \ldots, a_l)$, where a_i denote the number of vertices at the vertex v_i [3].

Theorem 26.10

Let $G = K_l(a_1, a_2, a_3, \ldots, a_l)$. Then $\chi_{tgd}(G) = l + 2$. ■

Proof. Color the l vertices of the complete subgraph of G with l distinct colors. Let $v \in V(K_l)$ be the vertex with color c_1. Since every vertex of K_l has at least one pendent vertex, the used l colors cannot be repeated in tgd-coloring. Let c_{l+1} be the color of the pendent vertices adjacent to v. Then the used $l+1$ cannot be an anti dom-color class v, hence $\chi_{tgd}(G) \geq l+2$. If we assign color c_{l+2} to the pendent vertices not adjacent to v, then V_{l+2} act as an anti dom-color class for v and its pendent vertices and V_{l+1} act as an anti dom-color class for the remaining vertices in $V(G)$. From this coloring it is obvious that all vertices $V(G)$ have a proper dom-color class. Thus the defined coloring is a tgd-coloring and $\chi_{tgd}(G) = l+2$. □

The following two theorems gives the value of χ_{tgd} for the complement of paths and cycles.

Theorem 26.11

For $n \geq 4$, $\chi_{tgd}(\overline{P_n}) = \begin{cases} n - (\lfloor \frac{n-1}{4} \rfloor - 1), & \text{if } n \equiv 1(\mod 4), \\ n - \lfloor \frac{n-1}{4} \rfloor, & \text{otherwise.} \end{cases}$ ■

Proof. Let $V = \{v_1, v_2, v_3, \ldots, v_n\}$ be the vertex set of P_n (and hence of its complement $\overline{P_n}$), such that $v_i v_{i+1} \in E(P_n)$, $1 \leq i < n$. Hence, v_i and v_{i+1} are non-adjacent vertices in $\overline{P_n}$ for $1 \leq i < n$. To get a possible minimum coloring, it is sufficient if we assign the same colors to as many vertices as possible. Note that the independence number of $\overline{P_n}$ is 2 and hence the maximum cardinality of a color class of $\overline{P_n}$ is 2.

Without loss of generality, let $c(v_i) = c(v_{i+1}) = c_r$. Then, the vertex v_{i-1} should have a different color as it is adjacent to v_{i+1}. Hence, let $c(v_{i-1}) = c_s, s \neq r$. Since the vertex v_{i+1} has the same color of v_i, its only possible anti dom-color class is $\{v_{i-1}\}$ and hence the color of v_{i-1} cannot be assigned to any other vertex in $\overline{P_n}$. In a similar way, $\{v_{i+2}\}$ will be the anti dom-color class of v_{i+1} and hence its color, say $c_t \notin \{c_r, c_s\}$, cannot be assigned to any other vertex in $\overline{P_n}$. The color class of v_i cannot be the anti dom-color class of v_{i-1}, as v_{i+1} belongs to that color class and is adjacent to v_{i-1}. Therefore, there should be a vertex v_{i-2} preceding v_{i-1} with a color, say $c_l \notin \{c_r, c_s, c_t\}$. Similarly, there should be a vertex v_{i+3} succeeding v_{i+2} with a different color, say $c_h \notin \{c_r, c_s, c_t, c_l\}$, as $\{v_{i+1}\}$ cannot be the anti dom-color class of v_{i+2}. Hence, it clear that for each pair of vertices v_i, v_{i+1} assigned with the same colors, there should necessarily be two vertices preceding v_i and two vertices succeeding v_{i+1}, each having distinct colors. Thus, two vertices of $\overline{P_n}$ can have the same color only when $n \geq 6$. Furthermore, $\overline{P_n}$ consists of k distinct 2-element color classes if $n = 2+4k+d$, $0 \leq d < 4, k \geq 1$.

In the above-mentioned coloring scheme, it is evident that all vertices of $\overline{P_n}$ have dom-color classes, as for a vertex $v_i; 1 < i < n$, all color classes other than that of v_{i-1}, v_i and v_{i+1} are dom-color classes. Also, note that in a coloring scheme mentioned above, either $\{v_1\}$ will form a proper dom-color for all vertices other than v_1 and v_2 and $\{v_n\}$ will form a proper dom-color class for all vertices other than v_{n-1} and v_n. Therefore, the total global dominator chromatic number of of $\overline{P_n}$ is $n-2k+k=n-k$, where $k=\frac{n-d-2}{4}$. Therefore,

$$\chi_{tgd}(\overline{P_n}) = \begin{cases} \frac{3n+5}{4}, & \text{if } n \equiv 1(\bmod 4), \\ n-\lfloor\frac{n-1}{4}\rfloor, & \text{elsewhere.} \end{cases}$$

This completes the proof. □

Theorem 26.12

For $n \geq 4$, $\chi_{tgd}(\overline{C_n}) = n - \lfloor\frac{n}{4}\rfloor$. ■

Proof. Consider the vertex set of C_n as $V=\{v_1,v_2,v_3,\ldots,v_n\}$ such that $v_iv_{i+1} \in E(C_n)$, $1 \leq i \leq n$ in the sense that $v_{n+i}=v_i$. Hence, v_i and v_{i+1} are non-adjacent vertices in $\overline{C_n}$ for $1 \leq i < n$. To get a possible minimum coloring, it is sufficient if we assign the same colors to as many vertices as possible. Note that the independence number of $\overline{C_n}$ is 2 and hence the maximum cardinality of a color class of $\overline{C_n}$ is 2. $\overline{C_4}$ is the disjoint union of two K_2's. The total global coloring of this graph is assigning distinct colors to all the vertices, hence $\chi_{tgd}(\overline{C_4})=4$. Now we have to consider the case where $n \geq 5$. As explained in the proof of Theorem 26.11, two vertices v_i, v_{i+1} have the same color, if there exist two vertices preceding v_i and two vertices succeeding v_{i+1} with distinct colors. But note that the vertices v_1 and v_2 will act as the succeeding vertices of v_n. Thus, two vertices of $\overline{C_n}$ can have the same color only when $n \geq 5$. Furthermore, $\overline{C_n}$ consists of k distinct 2-element color classes if $n=4k+d$, $0 \leq d < 4, k \geq 1$.

Therefore, the total global dominator chromatic number of of $\overline{C_n}$ is $n-2k+k=n-k$, where $k=\frac{n-d}{4}$. Therefore, $\chi_{tgd}(\overline{C_n}) = n-\lfloor\frac{n}{4}\rfloor, n > 4$. □

Theorem 26.13

If G is an $(n-2)$-regular graph on n vertices, then $\chi_{tgd}(G)=n$. ■

Proof. Let G be an $n-2$-regular graph on n vertices. Therefore n will be even. Let $v_1,v_2,v_3,\ldots,v_n$ be the vertices of G. In this graph there will be exactly $\frac{n}{2}$ pair of non adjacent vertices will be there; ie, the anti dom-color class of each vertex will be a singleton set. Hence we have to color each vertex with distinct colors. It is clear that each vertex has an proper dom-color class. Hence this is a total global dominator coloring and $\chi_{tgd}(G)=n$. □

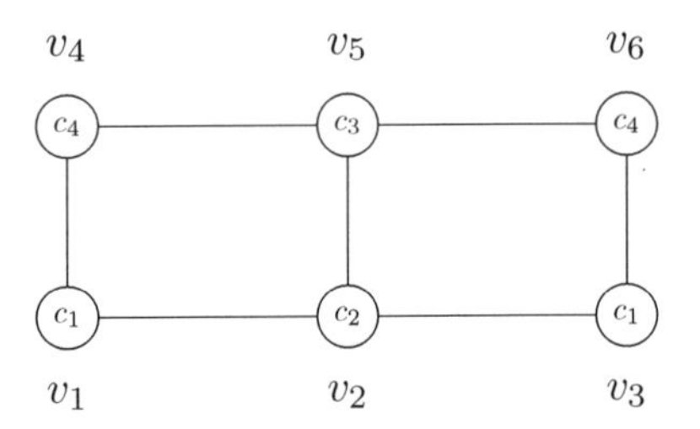

Figure 26.3: G

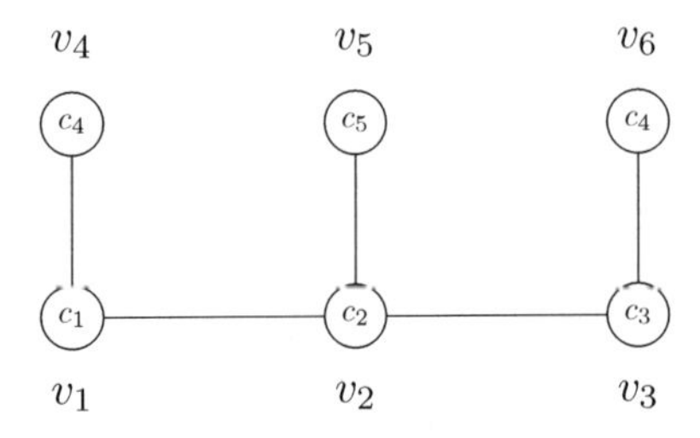

Figure 26.4: H

Among the graphs discussed above, we observe that $\chi_{tgd}(H) \leq \chi_{tgd}(G)$ where H is a subgraph of G. But it cannot be concluded that $\chi_{tgd}(H) \leq \chi_{tgd}(G)$, in general. The graph H in Figure 26.4 is a subgraph of graph G in Figure 26.3, but $\chi_{tgd}(H) \geq \chi_{tgd}(G)$.

26.4 CONCLUSION

Here we have initiated a study on a new coloring concept called tgd-coloring. The value of χ_{tgd} for some graphs and some of its bounds were determined. However, there is much scope for further research. Following are some of the open problems.

1. Characterize graphs for which
   a. $\chi_{tgd}(G) = 4$.
   b. $\chi_{tgd}(G) = \chi_d^t(G)$.
   c. $\chi_{tgd}(G) = \chi(G) + 1$.
   d. $\chi_{tgd}(G) = \gamma_{tg}(G)$.
   e. $\chi_{tgd}(G) = \chi(G) + \gamma_{tg}(G)$.
2. Detemine $\chi_{tgd}(G)$ when G is a trees.

ACKNOWLEDGEMENTS

The authors would like to thank Dr. I Sahul Hamid for giving us some useful insights on the concept of global dominator coloring and we also thank the anonymous referee for the valuable comments that led to the improvement of this article.

REFERENCES

1. S. Arumugam, J. Bagga and K. R. Chandrasekar. On dominator colorings in graphs. *Proceedings-Mathematical Sciences*, 122(4): 561-571, 2012.
2. G. Chartrand and P. Zhang. *Chromatic Graph Theory.* Chapman and Hall/CRC, 2008.
3. R. M. Gera. On dominator colorings in graphs. *Graph Theory Notes NY*, LII, 25-30, 2007.
4. R. M. Gera, C. W. Rasmussen and S. Horton. Dominator colorings and safe clique partitions. *Congr. Numer.*, 181, 19-32, 2006.
5. I. S. Hamid and R. Rajewari. Global dominator coloring of graphs. *Discuss. Math. Graph Theory*, 39, 325-339, 2019.
6. F. Harary. *Graph Theory.* Narosa Publishers, 2001.
7. T. W. Haynes, S. Hedetniemi and P. Slater. *Fundamentals of Domination in Graphs.* Marcel Dekker, New York, 1998.
8. A. Henning. Total dominator coloring and total domination in graphs. *Graphs. Combin.*, 31, 953-974, 2015.
9. A. P. Kazemi. Total dominator chromatic number of a graph. *Trans. Combin.*, 4(2): 57-68, 2015.
10. V. R. Kulli and B. Janakiram. The total global domination number of a graph. *Indian J. Pure Appl. Math.*, 27(6): 537-542, 1996.
11. S. K. Vaidya and R. M. Pandit. Some results on global dominating sets. *Proyecciones J. Math.*, 32(3): 235-244, 2013.
12. S. K. Vaidya and R. M. Pandit. Some new perspectives on global domination in Graphs. *ISRN Combin.*, (Article ID 201654): 4, 2013.
13. A. Vijayalakshmi. Total dominator coloring in Graphs. *Int. J. Adv. Res. Tech*, 1(4): 1-6, 2012.

27 Rainbow Vertex Connection Number of a Class of Triangular Snake Graph

Dharamvirsinh Parmar
Department of Mathematics,
C.U.Shah University,
Wadhwan, Gujarat (INDIA)
E-mail: dharamvir_21@yahoo.co.in

Bharat Suthar
Department of Mathematics,
C.U.Shah University,
Wadhwan, Gujarat (INDIA)
E-mail: bsuthar2011@gmail.com

Rainbow vertex colouring of a graph is a colouring the vertices of a graph, such that every pair of vertices is joined by at least one path in which all vertices except external vertices have different colours. In this chapter, we are going to discuss about the rainbow vertex connection number of triangular snake, double triangular snake, triple triangular snake, alternating triangular snake, double alternating triangular snake and quadrilateral snake graphs.

27.1 INTRODUCTION

Here, we consider all graphs are connected, finite and undirected graphs. A graph $G = (V(G), E(G))$ has a set of vertices $V(G)$ and a set of edges $E(G)$respectively. We refer Gross and Yellen [5] for all kinds of definitions and notations.

Definition. In connected graph G, the distance between two of its vertices v_i and v_j is the length of the shortest path between them. It is denoted by $d(v_i, v_j)$.

Definition. In a graph G, eccentricity of a vertex $v \in V(G)$, denoted by $E(v)$ is the distance from v to the vertex farthest from v in G. i.e. $E(v) = \underset{v_i \in V}{max}\ d(v, v_i)$.

Definition. In a graph G, vertex $v \in V(G)$ is called the center of G if $E(v)$ is minimum.

Definition. The diameter of a graph G is denoted by $diam(G)$, defined as $\max_{v_i \in V} E(v_i)$.

Definition. A path in an edge colored graph with no two edges having the same color is called rainbow path.

Definition. An edge colored graph G is called rainbow connected if any two vertices are connected by a rainbow path.
The rainbow connection number of a graph, denoted by $rc(G)$ is the smallest number of colors that are needed in order to make graph G rainbow connected. It is easy to see that for any connected graph G $diam(G) \leq rc(G)$.

Definition. A path in a vertex colored graph with no two internal vertices having the same color is called a rainbow vertex path.

Definition. A vertex colored graph G is called rainbow connected if any two vertices are connected by rainbow vertex path.

The rainbow connection number was introduced by Chartrand, Johns, McKeon, and Zhang in [4]. It has applications in transferring information of high security in multicomputer networks. We refer the reader to [7, 9] for details. The rainbow vertex connection number of a graph, denoted by $rvc(G)$ is the smallest number of colors that are needed in order to make graph G rainbow vertex connected. M. Krivelevich and R. Yuster in [7] gave the lower bound for $rvc(G)$, $diam(G) - 1 \leq rvc(G)$. Li. Hengzhe and Ma. Yingbin [6] discussed rainbow connection number and graph operation. Also Annammal and Mercy [2] derive the rainbow connection number of shadow graphs. Dian N. S. Simamora and A. N. M.Salman [3] discussed the rainbow (vertex) coloring of a pencil graph. In this chapter, we focus on the rainbow vertex connection number of the triangular snake graph, double triangular snake graph, triple triangular snake graph, alternating triangular snake, double alternating triangular snake and quadrilateral graph.

27.2 RAINBOW VERTEX CONNECTION NUMBER IN SELECTED GRAPHS

Definition. A triangular snake graph T_n is obtained from a path $u_1, u_2, \ldots, u_n$ by joining u_i and u_{i+1} to a new vertex v_i for $1 \leq i \leq n-1$. That is, every edge of a path is replaced by a triangle. See Figure 27.1.

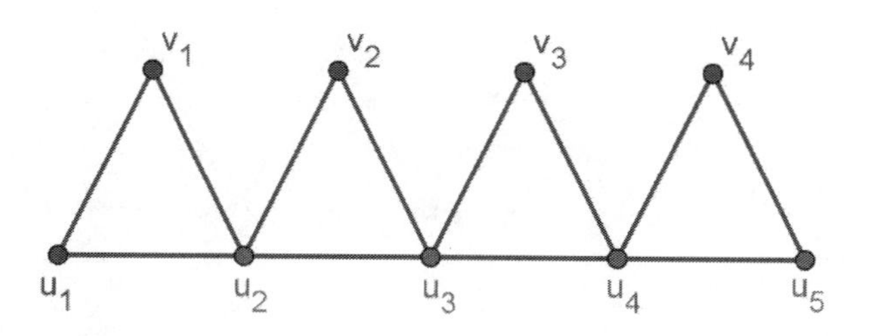

Figure 27.1: T_n

Theorem 27.1

If T_n is a triangular snake graph then $rvc(T_n) = n-2, n \geq 2$. ■

Proof. Let P_n be the path $u_1, u_2, \ldots, u_n$. Then it is clear that $diam(T_n) = n-1$. We can obtain a triangular snake graph from path $u_1, u_2, \ldots, u_n$ by joining u_i and u_{i+1} to a new vertex $v_i, 1 \leq i < n$. Hence we get new vertices $v_1, v_2, \ldots, v_{n-1}$.
Vertex coloring algorithm:
(1) Color the vertices u_1 and u_n by color c_1.
(2) Color the vertices u_i by color $c_{i-1}, 2 \leq i \leq n-1$.
(3) Color the vertices v_i by color $c_1, 1 \leq i \leq n-1$.
Consider any path in T_n.
Case-1 If any path with end vertices u_i and $u_j, 1 \leq i < j \leq n$ then the shortest path between u_i and $u_j, (P_1 : u_i, u_{i+1}, \ldots, u_j)$ is a rainbow path.
Case-2 If any path with end vertices v_i and $v_j, 1 \leq i < j < n$ then the shortest path between v_i and $v_j, (P_2 : v_i, u_{i+1}, \ldots, u_j, v_j)$ is a rainbow path.
Case-3 The shortest path between u_i and v_j,

$$P_3 = \begin{cases} u_i, u_{i+1}, \ldots, u_j, v_j, & i < j; \\ u_i, u_{i-1}, \ldots, u_{j+1}, v_j, & i > j; \end{cases} \text{ is rainbow path } (1 \leq i \leq n, 1 \leq j < n).$$

Thus, if we consider any path connecting the vertices there exists at least one path in T_n that forms a rainbow path. Hence $rvc(T_n) = n-2$. As per [1], the rainbow connection number for double triangular snake graph $D(T_n), n \geq 3$ is $n-2$. □

Illustration 27.2.1. Triangular snake graph T_5 and its rainbow vertex coloring are shown in Figure 27.2.

Definition. Triple triangular snake graph $T(T_n)$ consists of three triangular snakes that have a common path. See the Figure 27.3.

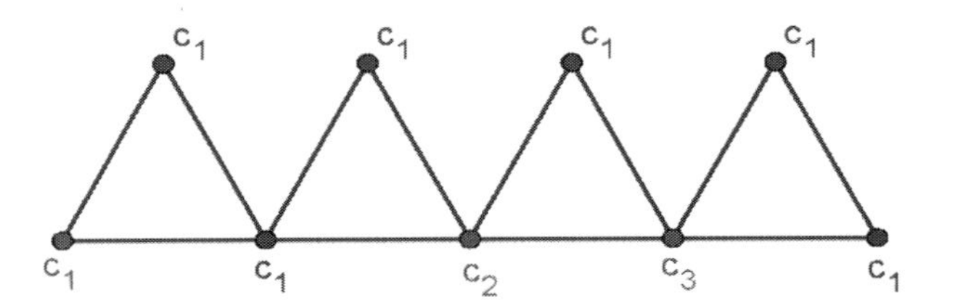

Figure 27.2: T_5

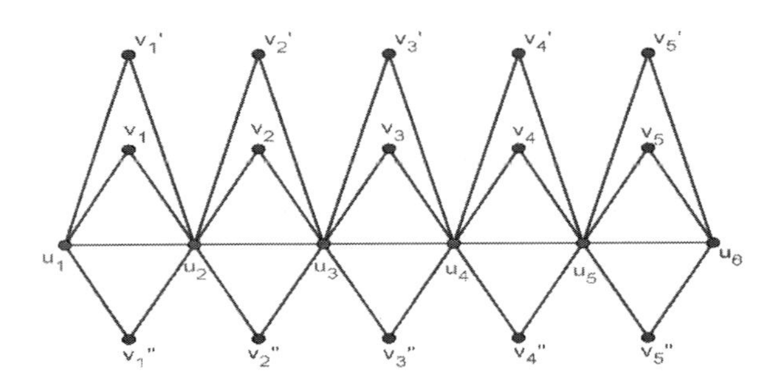

Figure 27.3: $T(T_n)$

Theorem 27.2

If $T(T_n)$ is a triple triangular snake graph then $rvc(T(T_n)) = n-2, n \geq 2$. ■

Proof. Let P_n be the path $u_1, u_2, \ldots, u_n$. Then it is clear that $diam(T(T_n)) = n-1$. We can obtain a triple triangular snake graph from path $u_1, u_2, \ldots, u_n$ by joining u_i and u_{i+1} to a new vertex $v_i, v_i', v_i''; 1 \leq i \leq n-1$. Hence we get new vertices $v_1, v_2, \ldots, v_{n-1}, v_1', v_2', \ldots, v_{n-1}', v_1'', v_2'' \ldots v_{n-1}''$. Vertex coloring algorithm:
(1) Color the vertices u_1 and u_n by color c_1.
(2) Color the vertices u_i by color $c_{i-1}, 2 \leq i \leq n-1$.
(3) Color the vertices v_i, v_i' and v_i'' by color $c_1, 1 \leq i \leq n-1$.
Consider any path in $T(T_n)$.
Case-1 If any path with end vertices u_i and $u_j, 1 \leq i < j \leq n; i \neq j$ then the shortest path between u_i and $u_j, (P_1 : u_i, u_{i+1}, \ldots u_j)$ is a rainbow path.
Case-2 If any path with end vertices v_i and $v_j, 1 \leq i < j < n$ then the shortest path between v_i and $v_j, (P_2 : v_i, u_{i+1}, \ldots u_j, v_j)$ is a rainbow path.
Case-3 The shortest path between u_i and v_j,

$$P_3 = \begin{cases} u_i, u_{i+1}, \ldots u_j, v_j, & i < j; \\ u_i, u_{i-1}, \ldots, u_{j+1}, v_j, & i > j; \end{cases} \text{ is rainbow path } (1 \leq i \leq n, 1 \leq j < n).$$

Case -4 The shortest path between u_i and v'_j,

$$P_4 = \begin{cases} u_i, u_{i+1}, \ldots u_j, v'_j, & i < j; \\ u_i, u_{i-1}, \ldots, u_{j+1}, v'_j, & i > j; \end{cases} \text{ is rainbow path } (1 \leq i \leq n, 1 \leq j < n).$$

Case-5 The shortest path between u_i and v''_j,

$$P_5 = \begin{cases} u_i, u_{i+1}, \ldots u_j, v''_j, & i < j; \\ u_i, u_{i-1}, \ldots, u_{j+1}, v''_j, & i > j; \end{cases} \text{ is rainbow path } (1 \leq i \leq n, 1 \leq j < n).$$

Case-6 The shortest path between v_i and v'_j,

$$P_6 = \begin{cases} v_i, u_{i+1}, \ldots u_j, v'_j, & i < j; \\ v_i, u_{i+1}, v'_j, & i = j; \\ v_i, u_i, u_{i-1}, \ldots, u_{j+1}, v'_j, & i > j; \end{cases} \text{ is rainbow path } (1 \leq i, j < n).$$

Case-7 The shortest path between v_i and v''_j,

$$P_7 = \begin{cases} v_i, u_{i+1}, \ldots u_j, v''_j, & i < j; \\ v_i, u_i, v''_j, & i = j = 1; \\ v_i, u_i, u_{i-1}, \ldots, u_{j+1}, v''_j, & i > j; \\ v_i, u_{i+1}, v''_j, & i = j \neq 1; \end{cases}$$

Case-8 The shortest path between v'_i and $v'_j, 1 \leq i < j < n, i \neq j$ then the shortest path between v'_i and $v'_j, (P_8 : v'_i, u_{i+1}, \ldots u_j, v'_j)$ is a rainbow path.
Case-9 The shortest path between v'_i and v''_j,

$$P_9 = \begin{cases} v'_i, u_{i+1}, \ldots u_j, v''_j, & i < j; \\ v'_i, u_{i+1}, v''_j, & i = j; \\ v'_i, u_i, u_{i-1}, \ldots, u_{j+1}, v''_j, & i > j; \end{cases} \text{ is rainbow path } (1 \leq i, j < n).$$

Case-10 The shortest path between v''_i and $v''_j, 1 \leq i < j < n, i \neq j$ then the shortest path between v''_i and $v''_j, (P_{10} : v''_i, u_{i+1}, \ldots u_j, v''_j)$ is a rainbow path. Thus, if we consider any path connecting the vertices there exists at least one path in $T(T_n)$ that forms a rainbow path. Hence $rvc(T(T_n)) = n - 2$. □

Illustration 27.2.2. Triple triangular snake graph $T(T_6)$ and its rainbow vertex coloring are shown in Figure 27.4.

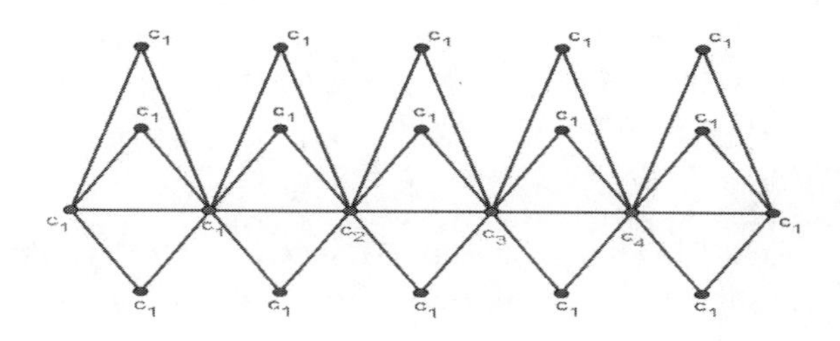

Figure 27.4: $T(T_6)$

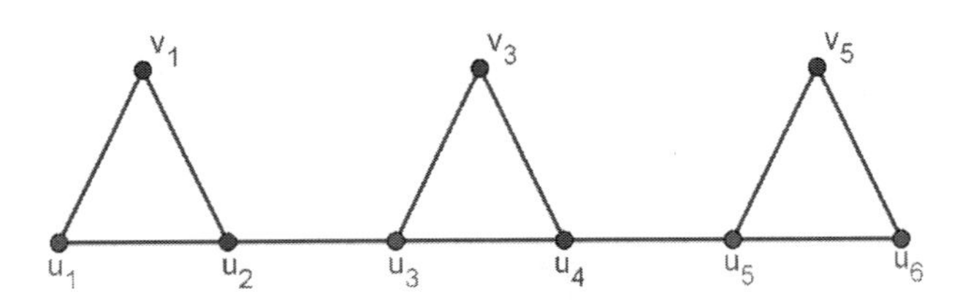

Figure 27.5: $A(T_n)$

Definition. An alternate triangular snake graph $A(T_n)$ is obtained from a path $u_1, u_2, \ldots, u_n$ by joining u_i and u_{i+1} alternatively $(i = 1,3,5,\ldots)$ to a new vertex v_i. That is every alternate edge of a path is replaced by C_3. See Figure 27.5.

Theorem 27.3

If $A(T_n)$ is an alternate triangular snake graph then $rvc(A(T_n)) = n-2, n \geq 2$.
■

Proof. Let P_n be the path $u_1, u_2, \ldots, u_n$. Then it is clear that $diam(A(T_n)) = n-1$. We can obtain an alternate triangular snake graph from path $u_1, u_2, \ldots, u_n$ by joining u_i and u_{i+1} alternatively $(i = 1,3,5,\ldots)$ to a new vertex $v_i, 1 \leq i \leq n-1$. Hence we get new vertices $v_1, v_3, \ldots, v_{\frac{n}{2}}$ (n is even) or $v_1, v_3, \ldots, v_{\frac{n-1}{2}}$ (n is odd).

Vertex coloring algorithm:

(1) Color the vertices u_1 and u_n by color c_1.

(2) Color the vertices u_i by color $c_{i-1}, 2 \leq i \leq n-1$.

(3) Color the vertices v_i by color $c_1, 1 \leq i \leq n-1$.

Consider any path in $A(T_n)$.

Case-1 If any path with end vertices u_i and $u_j, 1 \leq i < j \leq n; i \neq j$ then the shortest path between u_i and $u_j, (P_1 : u_i, u_{i+1}, \ldots, u_j)$ is a rainbow path.

Case-2 If any path with end vertices v_i and $v_j, 1 \leq i < j < n$ then the shortest path between v_i and $v_j, (P_2 : v_i, u_{i+1}, \ldots, u_j, v_j)$ is a rainbow path.

Case-3 The shortest path between u_i and v_j,

$$P_3 = \begin{cases} u_i, u_{i+1}, \ldots u_j, v_j, & i < j; \\ u_i, u_{i-1}, \ldots, u_{j+1}, v_j, & i > j; \end{cases} \text{ is rainbow path } (1 \leq i \leq n, 1 \leq j < n).$$

Thus, if we consider any path connecting the vertices there exists at least one path in $A(T_n)$ that forms a rainbow path. Hence $rvc(A(T_n)) = n-2$. □

Illustration 27.2.3. Alternate triangular snake graph $A(T_7)$ and its rainbow vertex coloring are shown in Figure 27.6.

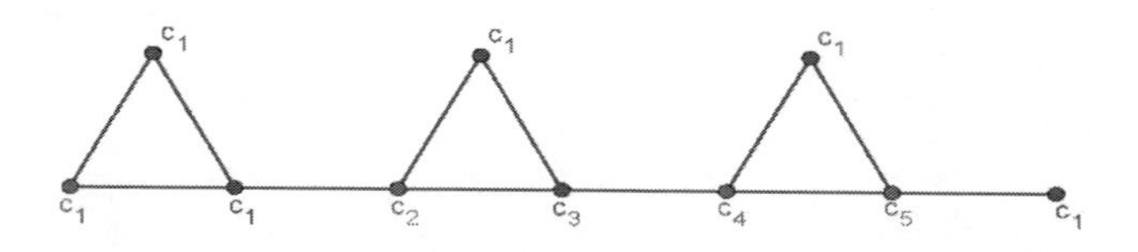

Figure 27.6: $A(T_7)$

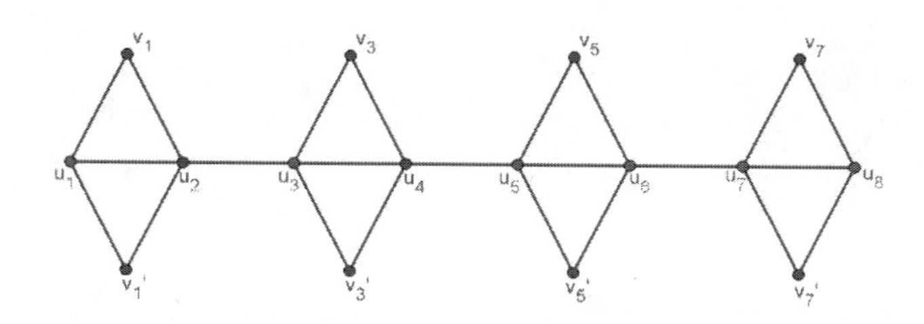

Figure 27.7: $A(D(T_n))$

Definition. An alternate double triangular snake graph $A(D(T_n))$ consists of two alternate triangular snakes that have a common path. See Figure 27.7.

Theorem 27.4

The rainbow vertex connection number for double alternate triangular snake graph $D(A(T_n)) = n-2, n \geq 2$. ■

Proof. Let P_n be the path $u_1, u_2, \ldots, u_n$. Then it is clear that *diam* $(D(A(T_n))) = n-1$.
We can obtain an alternate double triangular snake graph from path $u_1, u_2, \ldots, u_n$ by joining u_i and u_{i+1} alternatively $(i = 1, 3, 5, \ldots)$ to a new vertex $v_1, v_3, \ldots, v_{\frac{n}{2}}$ and $v'_1, v'_3, \ldots, v'_{\frac{n}{2}}$ (n is even) or $v_1, v_3, \ldots, v_{\frac{n-1}{2}}$ and $v'_1, v'_3, \ldots, v'_{\frac{n-1}{2}}$ (n is odd).
Vertex coloring algorithm:
(1) Color the vertices u_1 and u_n by color c_1.
(2) Color the vertices u_i by color $c_{i-1}, 2 \leq i \leq n-1$.
(3) Color the vertices v_i by color $c_1, 1 \leq i \leq n-1$.
(4) Color the vertices v'_i by color $c_1, 1 \leq i \leq n-1$. Consider any path in $D(A(T_n))$.

Case-1 If any path with end vertices u_i and $u_j, 1 \leq i < j \leq n; i \neq j$ then the shortest path between u_i and $u_j, (P_1 : u_i, u_{i+1}, \ldots u_j)$ is a rainbow path.
Case-2 If any path with end vertices v_i and $v_j, 1 \leq i < j < n$ then the shortest path between v_i and $v_j, (P_2 : v_i, u_{i+1}, \ldots, u_j, v_j)$ is a rainbow path.
Case-3 If any path with end vertices u_i and v_j,

$$P_3 = \begin{cases} u_i, u_{i+1}, \ldots, u_j, v_j, & i < j; \\ u_i, u_{i-1}, \ldots, u_{j+1}, v_j, & i > j; \end{cases} \text{ is rainbow path } (1 \leq i \leq n, 1 \leq j < n).$$

Case-4 If any path with end vertices u_i and $v'_j, 1 \leq i < j < n$ then the shortest path between u_i and $v'_j, (P_4 : u_i, u_{i+1}, \ldots, u_j, v'_j)$ is a rainbow path.
Case-5 If any path with end vertices v_i and $v'_j, 1 \leq i < j < n$ then the shortest path between v_i and $v'_j, (P_5 : v_i, u_{i+1}, \ldots, u_j, v'_j)$ is a rainbow path.
Case-6 If any path with end vertices v'_i and $v'_j, 1 \leq i < j < n$ then the shortest path between v'_i and $v'_j, (P_6 : v'_i, u_{i+1}, \ldots, u_j, v'_j)$ is a rainbow path.

Thus, if we consider any path connecting the vertices there exists at least one path in $D(A(T_n))$ that forms a rainbow path. Hence $rvc(D(A(T_n))) = n - 2$. □

Illustration 27.2.4. Double alternate triangular snake graph $D(A(T_9))$ and its rainbow vertex coloring are shown in Figure 27.8.

Definition. A quadrilateral snake graph Q_n is obtained from a path $u_1, u_2, \ldots, u_n$ by joining u_i and u_{i+1} to new vertex v_i and v'_i for $1 \leq i \leq n-1$ in such a way that u_i, u_{i+1}, v_i and v'_i form a quadrilateral. That is, every edge of a path is replaced by a quadrilateral. See Figure 27.9.

Theorem 27.5

If Q_n is quadrilateral snake graph then $rvc(Q_n) = n, n \geq 2$. ■

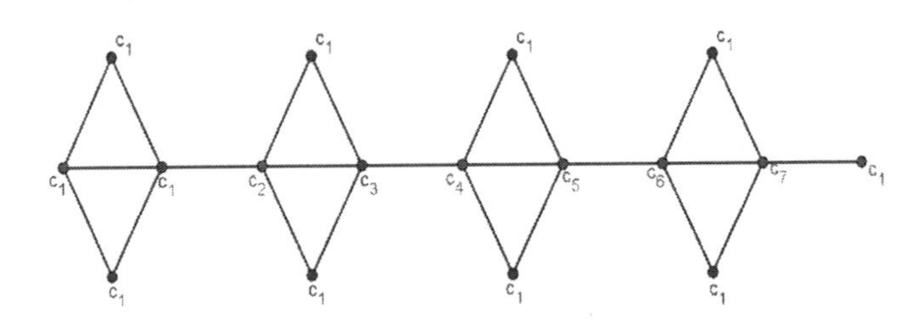

Figure 27.8: $A(D(T_9))$

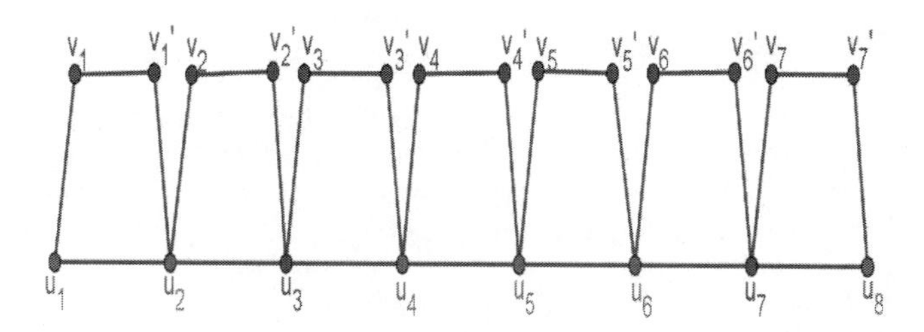

Figure 27.9: Q_n

Proof. Let P_n be the path $u_1, u_2, \ldots, u_n$.
Then it is clear that $diam(Q_n) = n+1$.
We can obtain a quadrilateral snake graph from path $u_1, u_2, \ldots, u_n$ by joining u_i and u_{i+1} to two new vertices v_i and $v_i', 1 \leq i < n$ by considering the edge between u_i to v_i, u_{i+1} to v_i' and v_i to v_i'; i.e every edge of a path is replaced by a quadrilateral.
Hence we get new vertices $v_1, v_2, \ldots, v_{n-1}, v_1', v_2', \ldots, v_{n-1}'$.
Vertex coloring algorithm:
(1) Color the vertices u_i by color $c_i, 1 \leq i \leq n$.
(2) Color the vertices v_i and v_i' by color $c_1, 1 \leq i \leq n-1$.
Consider any path in Q_n. Case-1 If any path with end vertices u_i and $u_j, 1 \leq i < j \leq n$ then the shortest path between u_i and $u_j, (P_1 : u_i, u_{i+1}, \ldots u_j)$ is a rainbow path.
Case-2 If any path with end vertices v_i and $v_j, 1 \leq i < j < n$ then the shortest path between v_i and $v_j, (P_2 : v_i, v_i', u_{i+1} \ldots, u_j, v_j)$ is a rainbow path.
Case-3 If any path with end vertices v_i' and $v_j', 1 \leq i < j < n$ then the shortest path between v_i' and $v_j', (P_3 : v_i', u_{i+1}, \ldots, u_{j+1}, v_j')$ is a rainbow path.
Case-4 If any path with end vertices u_i and $v_j, 1 \leq i < j < n$ then the shortest path between u_i and $v_j, (P_4 : u_i, u_{i+1} \ldots, u_j, v_j)$ is a rainbow path.
Case-5 If any path with end vertices u_i and $v_j', 1 \leq i < j < n$ then the shortest path between u_i and $v_j', (P_5 : u_i, u_{i+1} \ldots u_{j+1}, v_j')$ is a rainbow path.
Case-6 If any path with end vertices v_i and $v_j', 1 \leq i < j < n$ then the shortest path between v_i and $v_j, (P_6 : v_i, v_i', u_{i+1} \ldots u_{j+1}, v_j')$ is a rainbow path.
Thus, if we consider any path connecting the vertices there exists at least one path in Q_n that forms a rainbow path. Hence $rvc(Q_n) = n$. □

Illustration 27.2.5. Quadrilateral snake graph (Q_8)and its rainbow vertex coloring are shown in Figure 27.10.

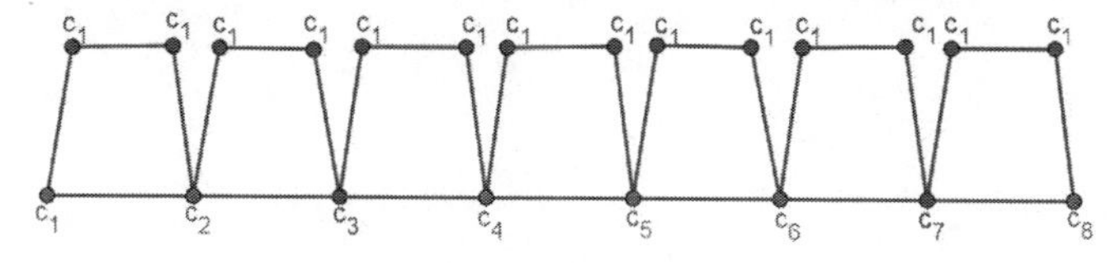

Figure 27.10: Q_8

27.3 CONCLUSIONS

The rainbow vertex coloring of the triangular snake graph, double triangular snake graph, triple triangular snake graph and alternating triangular snake graph has been defined and their rainbow connection numbers have been computed using rainbow edge colorings.

REFERENCES

1. A. Annammal and D. Angel. Rainbow coloring of certain classes of graphs. *International Journal of Advanced Information Science and Technology*, 4(1).
2. A. Annammal and M. Mercy. Rainbow coloring of shadow graph. *International Journal of Pure and Applied Mathematics*, 101(6): 873-881, 2015.
3. N. S. Simamora Dian and A. N. M Salman. The rainbow (vertex) connection number of pencil graphs. *Procedia Computer Science*, 74, 138-142. Doi: 10.1016/j.procs.2015.12.089.
4. G. Chartrand, G. L. Johns, K. A. McKeon and P. Zhang. Rainbow connection in graphs. *Math. Bohem.*, 133, 85-98, 2008.
5. J. Gross and J. Yellen. *Graph Theory and its Applications.* CRC Press, 2005.
6. Li. Hengzhe and Ma. Yingbin. Rainbow connection number and graph operations. *Discrete Applied Mathematics*, 2017.
7. M. Krivelevich and R. Yuster. Rainbow connection of a graph is reciprocal to its minimum degree three. *J.Graph Theory*, 63(3): 185-191, 2009.
8. S. Chakraborty, E. Fischer, A. Matsliah and R. Yuster. Hardness and algorithms for rainbow connectivity. *J.Comb.Optim.*, 21, 330-347, 2011.
9. S. S. Sandhya, E. Merly and S. Kavitha. Stolarsky-3 mean labeling on triangular snake graphs. *International Journal of Mathematics Trends and Technology (IJMTT)*, 53(2), 2018.
10. X. Li. and Y. Sun. *Rainbow Connections of Graphs.* Springer Briefs in Math., Springer, New York, 2012.

28 Hamiltonian Chromatic Number of Trees

Devsi Bantva
Department of Mathematics
Lukhdhirji Engineering College, Morvi
Gujarat (INDIA)
E-mail: devsi.bantva@gmail.com

S. K. Vaidya
Department of Mathematics
Saurashtra University,
Rajkot, Gujarat (INDIA)
E-mail: samirkvaidya@yahoo.co.in

Let G be a simple finite connected graph of order n. The detour distance between two distinct vertices u and v denoted by $D(u,v)$ is the length of a longest uv-path in G. A hamiltonian coloring h of a graph G of order n is a mapping $h : V(G) \to \{0,1,2,\ldots\}$ such that $D(u,v) + |h(u) - h(v)| \geq n-1$, for every two distinct vertices u and v of G. The span of h, denoted by span(h), is $\max\{|h(u) - h(v)| : u,v \in V(G)\}$. The *hamiltonian chromatic number* of G is defined as $\text{hc}(G) := \min\{\text{span}(h)\}$ with minimum taken over all hamiltonian coloring h of G. In this chapter, we give an improved lower bound for the hamiltonian chromatic number of trees and give a necessary and sufficient condition to achieve the improved lower bound. Using this result, we determine the hamiltonian chromatic number of two families of trees.

28.1 INTRODUCTION

Let G be a simple finite connected graph with vertex set $V(G)$ and edge set $E(G)$. The *order* of a graph G written as $|G|$ is the number of vertices in G. For a vertex $v \in V(G)$, the *neighborhood* of v denoted by $N(v)$ is the set of vertices adjacent to v. The *distance* $d(u,v)$ between two vertices u and v is the length of a shortest path joining u and v. The *detour distance* between two vertices u and v denoted by $D(u,v)$ is the length of a longest path joining u and v (refer to [6] for more details on it). The *diameter* of a graph G denoted by diam(G) or simply d is $\max\{d(u,v) : u,v \in V(G)\}$. The *eccentricity* $\epsilon(v)$ of a vertex $v \in V(G)$ is the distance from v to a vertex farthest from v. The *center* $C(G)$ of graph G is the subgraph of G induced by the vertex/vertices of G whose eccentricity is minimum. Moreover, for standard graph theoretic terminology and notation we follow [10].

A *hamiltonian coloring* h of a graph G, introduced by Chartrand *et al.* in [4], is a mapping $h : V(G) \to \{0,1,2,\ldots\}$ such that for every pair of distinct vertices u,v of G,

$$D(u,v)+|h(u)-h(v)| \geq n-1. \tag{28.1}$$

The *span of* h, denoted by span(h), is defined as $\max\{|h(u)-h(v)| : u,v \in V(G)\}$. The *hamiltonian chromatic number* hc(G) of G is

$$n\text{hc}(G) := \min\{\text{span}(h)\}$$

with minimum taken over all hamiltonian coloring h of G. A hamiltonian coloring h of G is called *optimal* if $\text{span}(h) = \text{hc}(G)$.

It is clear from the definition that if G contains a hamiltonian uv-path between two distinct vertices u and v then the same color can be assigned to both u and v. Hence, a graph G is hamiltonian-connected if and only if G can be hamiltonian colored by a single color. In [4], Chartrand *et al.* proved that for any two integers j and n with $2 \leq j \leq (n+1)/2$ and $n \geq 6$, there is a hamiltonian graph of order n with hamiltonian chromatic number $n-j$. Thus the hamiltonian chromatic number of a connected graph G measures how close G is to being hamiltonian-connected. Without loss of generality we allow 0 as a color in the definition of the hamiltonian coloring, then the *span* of any hamiltonian coloring h is the maximum integer used for coloring while in [4, 5, 9] only positive integers are used as colors. Therefore, the hamiltonian chromatic number defined in this article is one less than that defined in [4, 5, 9] and hence we will make necessary adjustments when we present the results of [4, 5, 9] in this article.

The hamiltonian chromatic number was introduced by Chartrand *et al.* in [4] as a variation of *radio antipodal coloring* of graphs but it is less explored compare to it. A very few results have been presented by researchers for hamiltonian chromatic number of graphs. The hamiltonian chromatic numbers of some well-known graph families are determined by Chartrand *et al.* in [4] which are as follows: $\text{hc}(K_n) = 0$ for $n \geq 1$, $\text{hc}(C_n) = n-3$ for $n \geq 3$, $\text{hc}(K_{1,n-1}) = (n-2)^2$ for $n \geq 3$, $\text{hc}(K_{r,r}) = r-1$ for $r \geq 1$, $\text{hc}(K_{r,s}) = (s-1)^2-(r-1)^2-1$ for $2 \leq r < s$ and they gave an upper bound for $\text{hc}(P_n)$. However, it is noted that $hc(P_n)$ is the same as the radio antipodal number $ac(P_n)$ which is determined in [7]. They proved that if T is a spanning tree of a connected graph G then $\text{hc}(G) \leq \text{hc}(T)$ and for any tree T of order $n \geq 2$, $\text{hc}(T) \leq (n-2)^2$. They also proved that for any two integers j and n with $2 \leq j \leq (n+1)/2$ and $n \geq 6$, there is a hamiltonian graph of order n with hamiltonian chromatic number $n-j$. In [5], the same group of authors gave a lower bound for the circumference of G (the circumference of a graph G is the length of a longest cycle in G) which is given in terms of the number of vertices that receive colors between two specified colors in a hamiltonian coloring of G. The authors also proved that for a connected graph G of order $n \geq 3$, if $(n+2)/2$ vertices receive the same color in a hamiltonian coloring then G is hamiltonian. In [9], Shen *et al.* determined the hamiltonian chromatic number of graph G with $\max\{D(u,v) : u,v \in V(G), u \neq v\} \leq n/2$ and illustrated the result with a special class of caterpillars and double stars.

This chapter is organized as follows. In Section 28.2, we define all necessary terms, notations and terminologies for present work. In Section 28.3, we give an improved lower bound for the hamiltonian chromatic number of trees and present a necessary and sufficient condition to achieve the improved lower bound. We determine the hamiltonian chromatic number of two families of trees in Section 28.4. In this section, we show that for a special type of broom trees (see Section 28.4.2 for the definition and detail on it) the lower bound given in the present work is better than one given in [3, Theorem 4]. We explain the strength of our results in the concluding remarks section.

28.2 PRELIMINARIES

A tree T is a connected graph that contains no cycle. In [8] and later used in [2], the *weight of* T from $v \in V(T)$ is defined as $w_T(v) = \sum_{u \in V(T)} d(u,v)$ and the *weight of* T as $w(T) = \min\{w_T(v) : v \in V(T)\}$. A vertex $v \in V(T)$ is a *weight center* of T if $w_T(v) = w(T)$. Denote the set of weight centers of T by $W(T)$. It was proved in [8] that every tree T has either one or two weight centers, and T has two weight centers, say $W(T) = \{w, w'\}$, if and only if w and w' are adjacent and $T - \{ww'\}$ consists of two equal-sized components. We set $W(T) = \{w\}$ as a root if T has only one weight center w and $W(T) = \{w, w'\}$ as a root if T has two adjacent weight centers w and w'. In either case, if a vertex u is on the path joining weight center and a vertex $v \in V(T)$ then u is called the *ancestor* of v and v is called the *descendent* of u. If u is ancestor of v which is adjacent to v then u is called the *parent* of v and v is called a *child* of u. The subtree induced by a child u of a weight center and all descendents of u is called *branch* at u. Two branches are called *different* if they are at two vertices adjacent to the same weight center, and *opposite* if they are at two vertices adjacent to different weight centers. Note that the concept of opposite branches occurs only when T has two weight centers.

Define

$$\mathcal{L}(u) := \min\{D(u,w) : w \in W(T)\}, u \in V(T)$$

to indicate the *detour level* of u in T. Define the *total detour level of* T as

$$\mathcal{L}_W(T) := \sum_{u \in V(T)} \mathcal{L}(u).$$

For any $u, v \in V(T)$, define

$$\phi(u,v) := \max\{\mathcal{L}(t) : t \text{ is a common ancestor of } u \text{ and } v\},$$

$$\delta(u,v) := \begin{cases} 1, & \text{If } W(T) = \{w, w'\} \text{ and } uv\text{-path contains the edge } ww', \\ 0, & \text{otherwise.} \end{cases}$$

Lemma 28.1

Let T be a tree with diameter $d \geq 2$. Then for any $u,v \in V(T)$, the following hold:

1. $\phi(u,v) \geq 0$;
2. $\phi(u,v) = 0$ if and only if u and v are in different or opposite branches;
3. $\delta(u,v) = 1$ if and only if T has two weight centers and u and v are in opposite branches.
4. the detour distance $D(u,v)$ in T between u and v can be expressed as

$$D(u,v) = \mathcal{L}(u) + \mathcal{L}(v) - 2\phi(u,v) + \delta(u,v). \tag{28.2}$$

■

Note that the detour distance $D(u,v)$ is the same as the ordinary distance $d(u,v)$ between any two vertices u and v in a tree T but we continue to use this terminology to remain consistent with the concept of hamiltonian coloring in general.

Define

$$\varsigma(T) := \begin{cases} 0, & \text{If } W(T) = \{w\}, \\ 1, & \text{If } W(T) = \{w,w'\}, \end{cases}$$

and

$$\varsigma'(T) := 1 - \varsigma(T).$$

28.3 A LOWER BOUND FOR HC(T)

In this section, we continue to use terms, notations and terminologies defined in the previous section.

A hamiltonian coloring h of a tree T is injective if T has at least one vertex of degree 3, that is, T is not a path as in this case no two vertices of trees contain a hamiltonian path. So we assume all trees with at least one vertex of degree 3 throughout this discussion unless otherwise specified. Observe that a hamiltonian coloring h on $V(T)$, induces an ordering of $V(T)$, which is a line-up of the vertices with increasing images. We denote this ordering by $V(G) = \{x_0, x_1, x_2, \ldots, x_{n-1}\}$ with

$$0 = h(x_0) < h(x_1) < \ldots < h(x_{n-1}) = \text{span}(h).$$

The next result gives a lower bound for the hamiltonian chromatic number of trees.

Theorem 28.1

Let T be a tree of order $n \geq 4$ and $\Delta(T) \geq 3$. Then

$$\text{hc}(T) \geq (n-1)(n-1-\zeta(T)) + \zeta'(T) - 2\mathcal{L}_W(T). \tag{28.3}$$

■

The proof is similar to that of [3, Theorem 4] except for only one change by which is $\mathcal{L}(T)$ is replaced by $\mathcal{L}_W(T)$ but it should be noted that technically this lower bound is more useful than the one given in [3, Theorem 4]. We support this fact by giving an example in Section 28.4.

Theorem 28.2

Let T be a tree of order $n \geq 4$ and $\Delta(T) \geq 3$. Then

$$\text{hc}(T) \geq (n-1)(n-1-\zeta(T)) + \zeta'(T) - 2\mathcal{L}_W(T) \tag{28.4}$$

holds if and only if there exists an ordering $\{x_0, x_1, \ldots, x_{n-1}\}$ of the vertices of T, with $\mathcal{L}(x_0) = 0$ and $\mathcal{L}(x_{n-1}) = 1$ when $W(T) = \{w\}$ and $\mathcal{L}(x_0) = \mathcal{L}(x_{n-1}) = 0$ when $W(T) = \{w, w'\}$, such that for all $0 \leq i < j \leq n-1$,

$$D(x_i, x_j) \geq \sum_{t=i}^{j-1} (\mathcal{L}(x_t) + \mathcal{L}(x_{t+1})) - (j-i)(n-1-\zeta(T)) + (n-1). \tag{28.5}$$

Moreover, under this condition the mapping h is defined by

$$h(x_0) = 0 \tag{28.6}$$

$$h(x_{i+1}) = h(x_i) + n - 1 - \zeta(T) - \mathcal{L}(x_i) - \mathcal{L}(x_{i+1}). \tag{28.7}$$

■

Proof. Suppose that (28.4) holds. Let h be an optimal hamiltonian coloring of T; then h induces an ordering of vertices of T, say $0 = h(x_0) < h(x_1) < \ldots < h(x_{n-1})$. The span of h is $\text{span}(h) = \text{hc}(T) = (n-1)(n-1-\zeta(T)) + \zeta'(T) - 2\mathcal{L}_W(T)$. Note that this is possible if equality holds in (28.1) together with (a) $\mathcal{L}(x_0) = 0$, $\mathcal{L}(x_{n-1}) = 1$ and $\phi(x_i, x_{i+1}) = \delta(x_i, x_{i+1}) = 0$ when T has only one weight center, (b) $\mathcal{L}(x_0) = \mathcal{L}(x_{n-1}) = 0$ and $\phi(x_i, x_{i+1}) = 0$ and $\delta(x_i, x_{i+1}) = 1$ when T has two adjacent weight centers. Note that this turns into the definition of hamiltonian coloring $h(x_{i+1}) - h(x_i) = n - 1 - D(x_i, x_{i+1})$ to $h(x_0) = 0$ and $h(x_{i+1}) = h(x_i) + n - 1 - \zeta(T) - \mathcal{L}(x_i) - \mathcal{L}(x_{i+1})$ for $0 \leq i \leq$

$n-2$. Moreover, for any two vertices x_i and x_j (without loss of generality, assume $j > i$), summing the latter equality for index i to j, we have

$$nh(x_j) - h(x_i) = \sum_{t=i}^{j-1}[n-1-\zeta(T)-\mathcal{L}(x_t)-\mathcal{L}(x_{t+1})].$$

Now h is a hamiltonian coloring so that $h(x_j)-h(x_i) \geq n-1-D(x_i,x_j)$ which turns the above equation into the following form.

$$nD(x_i,x_j) \geq \sum_{t=i}^{j-1}[\mathcal{L}(x_t)+\mathcal{L}(x_{t+1})]-(j-i)(n-1-\zeta(T))+(n-1).$$

Sufficiency: Suppose that an ordering $\{x_0, x_1, \ldots, x_{n-1}\}$ of vertices of T satisfies (28.5), and h is defined by (28.6) and (28.7) together with $\mathcal{L}(x_0) = 0$, $\mathcal{L}(x_{n-1}) = 1$ when $W(T) = \{w\}$ and $\mathcal{L}(x_0) = \mathcal{L}(x_{n-1}) = 0$ when $W(T) = \{w, w'\}$. Note that it is enough to prove that h is a hamiltonian coloring with span equal to the right-hand side of (28.4). Let x_i and x_j $(j > i)$ be two arbitrary vertices; then by (28.7), we have

$$nh(x_j) - h(x_i) = (j-i)(n-1-\zeta(T)) - \sum_{t=i}^{j-1}[\mathcal{L}(x_t)+\mathcal{L}(x_{t+1})]$$

Now an ordering satisfies (28.5) which turns the above equation into the following form which shows that h is a hamiltonian coloring.

$$nh(x_j) - h(x_i) \geq n-1-D(x_i,x_j).$$

The span of h is given by

$$\begin{aligned} n\text{span}(h) &= h(x_{n-1}) - h(x_0) \\ &= \sum_{i=0}^{n-2}(h(x_{i+1})-h(x_i)) \\ &= (n-1)(n-1-\zeta(T)) - \sum_{i=0}^{n-2}(\mathcal{L}(x_i)+\mathcal{L}(x_{i+1})) \\ &= (n-1)(n-1-\zeta(T)) - 2\mathcal{L}_W(T) + \mathcal{L}(x_0) + \mathcal{L}(x_{n-1}) \\ &= (n-1)(n-1-\zeta(T)) + \zeta'(T) - 2\mathcal{L}_W(T) \end{aligned}$$

which completes the proof. □

Note that the following results give sufficient conditions with optimal hamiltonian coloring for the equality in (28.3). Note that these results are identical with [3, Theorem 5] and [3, Corollary 1] when $W(T) = C(T)$ for a tree T. But it should be noted that these results are more efficient than [3, Theorem 5] and [3, Corollary 1] (see the case of broom trees in Section 28.4.2).

Theorem 28.3

Let T be a tree of order $n \geq 4$ and $\Delta(T) \geq 3$. Then

$$\mathrm{hc}(T) = (n-1)(n-1-\varsigma(T)) + \varsigma'(T) - 2\mathcal{L}_W(T) \tag{28.8}$$

holds if there exists an ordering $\{x_0, x_1, \ldots, x_{n-1}\}$ of the vertices of T such that for all $0 \leq i \leq n-2$,

1. $\mathcal{L}(x_0) + \mathcal{L}(x_{n-1}) = 1$ when $W(T) = \{w\}$ and $\mathcal{L}(x_0) + \mathcal{L}(x_{n-1}) = 0$ when $W(T) = \{w, w'\}$,
2. x_i and x_{i+1} are in different branches when $W(T) = \{w\}$ and in opposite branches when $W(T) = \{w, w'\}$,
3. $D(x_i, x_{i+1}) \leq n/2$.

Moreover, under these conditions the mapping h defined by (28.6) and (28.7) is an optimal hamiltonian coloring of T. ■

Proof. Suppose there exists an ordering $\{x_0, x_1, \ldots, x_{n-1}\}$ of vertices of T such that (a), (b) and (c) holds and h is defined by (28.6) and (28.7). By Theorem 28.2, it is enough to prove that an ordering $\{x_0, x_1, \ldots, x_{n-1}\}$ satisfies (28.5). Let x_i and x_j be two arbitrary vertices, where $0 \leq i < j \leq n-1$. Without loss of generality, we assume that $j - i \geq 2$ and for simplicity let the right-hand side of (28.5) be $x_{i,j}$. Then, we obtain

$$\begin{aligned}
n x_{i,j} &= \sum_{t=i}^{j-1} [\mathcal{L}(x_t) + \mathcal{L}(x_{t+1})] - (j-i)(n-1-\varsigma(T)) + (n-1) \\
&\leq (j-i)(n/2 - \varsigma(T)) - (j-i)(n-1-\varsigma(T)) + (n-1) \\
&= (j-i)(1-n/2) + (n-1) \\
&\leq 2(1-n/2) + (n-1) \\
&= 1 \leq D(x_i, x_j)
\end{aligned}$$

which completes the proof. □

For a tree T of order n, if $\max\{d(u,v) : u, v \in V(T), u \neq v\} \leq n/2$ then such a tree T is called a tree with *maximum distance bound* $n/2$ or $DB(n/2)$ tree.

Corollary 28.3.1. *Let T be a $DB(n/2)$ tree of order $n \geq 4$ and $\Delta(T) \geq 3$. Then*

$$\mathrm{hc}(T) = (n-1)(n-1-\varsigma(T)) + \varsigma'(T) - 2\mathcal{L}_W(T) \tag{28.9}$$

holds if and only if there exists an ordering $\{x_0, x_1, \ldots, x_{n-1}\}$ of the vertices of T such that for all $0 \leq i \leq n-2$,

1. *$\mathcal{L}(x_0) + \mathcal{L}(x_{n-1}) = 1$ when $W(T) = \{w\}$ and $\mathcal{L}(x_0) + \mathcal{L}(x_{n-1}) = 0$ when $W(T) = \{w, w'\}$,*

2. *x_i and x_{i+1} are in different branches when $W(T) = \{w\}$ and in opposite branches when $W(T) = \{w, w'\}$,*

Moreover, under these conditions the mapping h defined by (28.6) and (28.7) is an optimal hamiltonian coloring of T.

28.4 THEOREM 1 VS. [3, THEOREM 4]

In this section, we determine the hamiltonian chromatic number of two families of trees namely A_d and B_d, where $d = 2k, 2k+1$ and $k \geq 1$ are any integer. We show that in the case of $W(T) = C(T)$, the lower bound for hc(T) given in Theorem 28.3 and [3, Theorem 4] are identical but in the case of $W(T) \neq C(T)$ then the lower bound for hc(T) given in Theorem 28.3 is better than [3, Theorem 4]. Note that in the case of A_d, $W(A_d) = C(A_d)$ but in the case of B_d, $W(B_d) \neq C(B_d)$, where $d = 2k, 2k+1$.

28.4.1 SPECIAL TREES A_{2K} AND A_{2K+1}

In [1], a special class of trees, namely A_{2k} and A_{2k+1}, were introduced. The trees A_{2k} with diameter $2k-1$ are defined as follows. (a) A_2 is an edge K_2, (b) $A_{2k}(k \geq 2)$ is obtained from A_{2k-2} in the following manner: Add three pendent vertices (two vertical and one horizontal) to the left most and the right most pendent vertices of A_{2k-2}; next, add a pendent vertex (vertically) to every other pendent vertex of A_{2k-2}. In a manner analogous to the above, define the class of trees $A_{2k+1}(k \geq 1)$ with diameter $2k$ starting with A_3 which is $K_{1,4}$. The trees A_{2k} and A_{2k+1} have $2k^2$ and $2k(k+1)+1$ vertices respectively. Note that $|W(A_{2k})| = 2$ and $|W(A_{2k+1})| = 1$. Denote $W(A_{2k}) = \{w, w'\}$ and $W(A_{2k+1}) = \{w\}$. Moreover, it is clear from the definition that A_{2k} and A_{2k+1} are $DB(n/2)$ trees.

Theorem 28.4

Let $k \geq 1$ be any integer. Then

$$\text{hc}(A_d) := \begin{cases} \frac{2}{3}(k-1)(6k^3+2k^2-4k-3), & \text{If } d = 2k, \\ \frac{1}{3}k(k+1)(3k^2+k-1)+1, & \text{If } d = 2k+1. \end{cases} \tag{28.10}$$

■

Proof. The order n and the total detour level $\mathcal{L}_W(A_d)$ of A_d $(d = 2k, 2k+1)$ are given by

$$n := \begin{cases} 2k^2, & \text{If } d = 2k, \\ 2k(k+1)+1, & \text{If } d = 2k+1. \end{cases} \tag{28.11}$$

$$\mathcal{L}_W(A_d) := \begin{cases} \frac{1}{3}k(k-1)(4k+1), & \text{If } d = 2k, \\ \frac{2}{3}k(k+1)(2k+1), & \text{If } d = 2k+1. \end{cases} \tag{28.12}$$

Substituting (28.11) and (28.12) into (28.3) we obtain that the right-hand side of (28.10) is a lower bound for $\text{hc}(A_d)$ $(d = 2k, 2k+1)$. Now we prove that this lower bound is tight and for this purpose it suffices to give a hamiltonian coloring whose span is equal to the right-hand side of (28.10). Since A_d $(d = 2k, 2k+1)$ are $DB(n/2)$ tree, by Corollary 28.3.1 it is enough to give an ordering $\{x_0, x_1, \ldots, x_{n-1}\}$ of $V(A_d)$ $(d = 2k, 2k+1)$ satisfying conditions (a) and (b) of Corollary 28.3.1 and the hamiltonian coloring h defined by (28.6)-(28.7) is an optimal hamiltonian coloring. For this we consider the following two cases for A_d.

Case-1: $d = 2k$. In this case, note that $|W(A_{2k})| = 2$ and A_{2k} has six branches. Denote the branches of A_{2k} attached to w by T_2, T_4, T_6 and w' by T_1, T_3, T_5 such that $|T_2| > |T_4| = |T_6|$ and $|T_1| > |T_3| = |T_5|$. Define an ordering $\{x_0, x_1, \ldots, x_{n-1}\}$ of $V(A_{2k})$ as follows: Let $x_0 = w$ and $x_{n-1} = w'$. For $1 \le i \le n-4k-1$, let

$$x_i := \begin{cases} v \in V(T_1), & \text{If } i \text{ is odd}, \\ v \in V(T_2), & \text{If } i \text{ is even}. \end{cases}$$

For $n-4k \le i \le n-2k-1$, let

$$x_i := \begin{cases} v \in V(T_3), & \text{If } i \text{ is odd}, \\ v \in V(T_4), & \text{If } i \text{ is even}. \end{cases}$$

For $n-2k \le i \le n-2$, let

$$x_i := \begin{cases} v \in V(T_5), & \text{If } i \text{ is odd}, \\ v \in V(T_6), & \text{If } i \text{ is even}. \end{cases}$$

In the above ordering $v \in V(T_i)$, where $i = 1, 2, \ldots, 6$, we mean any v from $V(T_i)$ without repetition. Note that $\mathcal{L}(x_0) + \mathcal{L}(x_{n-1}) = 0$ and x_i, x_{i+1} $(0 \le i \le n-1)$ are in opposite branches of A_{2k}.

Case-2: $d = 2k+1$. In this case, note that $|W(A_{2k+1})| = 1$ and A_{2k+1} has four branches. Denote the branches of A_{2k+1} attached to w by T_1, T_2, T_3, T_4 such that $|T_1| = |T_2| > |T_3| = |T_4|$. Define an ordering $\{x_0, x_1, \ldots, x_{n-1}\}$ of $V(A_{2k+1})$ as follows: Let $x_0 = w$ and $x_{n-1} = v \in V(T_4)$ such that v is adjacent to w. For $1 \le i \le n-2k-1$, let

$$x_i := \begin{cases} v \in V(T_1), & \text{If } i \text{ is odd}, \\ v \in V(T_2), & \text{If } i \text{ is even}. \end{cases}$$

For $n-2k \le i \le n-2$, let

$$x_i := \begin{cases} v \in V(T_3), & \text{If } i \text{ is odd,} \\ v \in V(T_4), & \text{If } i \text{ is even.} \end{cases}$$

Again in the above ordering $v \in V(T_i)$, where $i = 1,2,3,4$, we mean any v from $V(T_i)$ without repetition. Note that $\mathcal{L}(x_0) + \mathcal{L}(x_{n-1}) = 1$ and x_i, x_{i+1} $(0 \le i \le n-1)$ are in different branches of A_{2k+1}. □

Remark. Note that in the case of A_d $(d = 2k, 2k+1)$, $W(A_d) = C(A_d)$ and hence both the lower bounds given in Theorem 28.1 and [3, Theorem 4] are identical. Moreover, $\Delta(A_2) = 1$ but it is easy to verify that $\mathrm{hc}(A_2) = 0$ and the case of A_2 is also consistent with the formula in (28.10).

Illustration 28.4.1. An optimal hamiltonian coloring of A_2, A_4, A_6 and A_3, A_5, A_7 is shown in Figure 28.1.

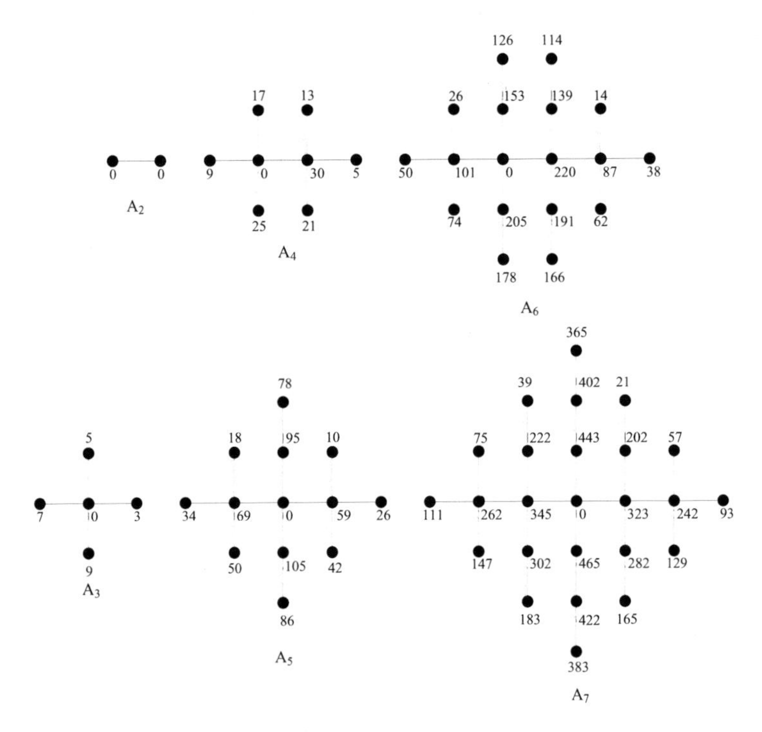

Figure 28.1: An optimal Hamiltonian coloring of A_2, A_4, A_6 and A_3, A_5, A_7

28.4.2 BROOM TREES B_{2K} AND B_{2K+1}

The broom tree $B_{n,d}$ consists of a path P_d, together with $(n-d)$ end vertices all adjacent to the same end vertex of P_d. Let $k \ge 1$ be any integer. Define the broom trees B_{2k} and B_{2k+1} are broom tree $B_{k(2k+1),2k}$ and $B_{(k+1)(2k+1),2k+1}$, respectively. It is clear that $|W(B_{2k})| = |W(B_{2k+1})| = 1$. Denote $W(B_{2k}) = W(B_{2k+1}) = \{w\}$. Moreover, it is clear from definition that B_{2k} and B_{2k+1} are $DB(n/2)$ trees.

Theorem 28.5

Let $k \geq 1$ be any integer. Then

$$\text{hc}(B_d) := \begin{cases} 2k(2k^3 + 2k^2 - \frac{11}{2}k + 1) + 2, & \text{If } d = 2k, \\ (2k+1)(2k^3 + 5k^2 - 2k - 1) + 2, & \text{If } d = 2k+1. \end{cases} \tag{28.13}$$

■

Proof. The order n and the total detour level $\mathcal{L}_W(B_d)$ of B_d $(d = 2k, 2k+1)$ are given by

$$n := \begin{cases} k(2k+1), & \text{If } d = 2k, \\ (k+1)(2k+1), & \text{If } d = 2k+1. \end{cases} \tag{28.14}$$

$$\mathcal{L}_W(B_d) := \begin{cases} 2k(2k-1), & \text{If } d = 2k, \\ 2k(2k+1), & \text{If } d = 2k+1. \end{cases} \tag{28.15}$$

Substituting (28.14) and (28.15) into (28.3) we obtain that the right-hand side of (28.13) is a lower bound for $\text{hc}(B_d)$ $(d = 2k, 2k+1)$. Now we prove that this lower bound is tight and for this purpose; it is suffices to give a hamiltonian coloring whose span is equal to the right-hand side of (28.13). Since B_d $(d = 2k, 2k+1)$ are $DB(n/2)$ tree, by Corollary 28.3.1 it is enough to give an ordering $\{x_0, x_1, \ldots, x_{n-1}\}$ of $V(B_d)$ $(d = 2k, 2k+1)$ satisfying conditions (a) and (b) of Corollary 28.3.1 and the hamiltonian coloring h defined by (28.6)-(28.7) is an optimal hamiltonian coloring. For this we consider the following two cases for B_d.

Case-1: $d = 2k$. In this case, note that B_{2k} has $k(2k-1)+1$ branches. Denote the branches of B_{2k} by T_i, $i = 1, 2, \ldots, k(2k-1)+1$ such that $|T_1| > |T_2| = \ldots = |T_{k(2k-1)+1}|$. Let $S = \{v : v \in V(T_1)\}$ and $S' = \{v : v \in V(T_i), i = 2, 3, \ldots, k(2k-1)+1\}$. Define an ordering $\{x_0, x_1, \ldots, x_n\}$ of $V(B_{2k})$ as follows: Let $x_0 = w$ and for $1 \leq i \leq 4k-4$, let

$$x_i := \begin{cases} v \in S, & \text{If } i \text{ is odd}, \\ v \in S', & \text{If } i \text{ is even}. \end{cases}$$

For $4k-3 \leq i \leq n-1$, let

$$x_i := v \in S'.$$

In the above ordering $v \in X$, where $X = S, S'$ we mean any v from X without repetition. Note that $\mathcal{L}(x_0) + \mathcal{L}(x_{n-1}) = 1$ and x_i, x_{i+1} $(0 \leq i \leq n-1)$ are in different branches of B_{2k}.

Case-2: $d = 2k+1$. In this case, note that B_{2k+1} has $k(2k+1)+1$ branches. Denote the branches of B_{2k+1} by T_i, $i = 1, 2, \ldots, k(2k+1)+1$ such that $|T_1| > |T_2| = \ldots = |T_{k(2k+1)+1}|$. Let $S = \{v : v \in V(T_1)\}$ and $S' = \{v : v \in V(T_i), i = 2, 3, \ldots, k(2k+1)+1\}$. Define an ordering $\{x_0, x_1, \ldots, x_n\}$ of $V(B_{2k+1})$ as follows: Let $x_0 = w$ and for $1 \leq i \leq 4k$, let

$$x_i := \begin{cases} v \in S, & \text{If } i \text{ is odd,} \\ v \in S', & \text{If } i \text{ is even.} \end{cases}$$

For $4k+1 \leq i \leq n-1$, let

$$x_i := v \in S'.$$

Again in the above ordering $v \in X$, where $X = S, S'$ we mean any v from X without repetition. Note that $\mathcal{L}(x_0) + \mathcal{L}(x_{n-1}) = 1$ and x_i, x_{i+1} $(0 \leq i \leq n-1)$ are in different branches of B_{2k+1} which completes the proof. □

Illustration 28.4.2. An optimal hamiltonian coloring of B_3, B_5 and B_4, B_6 is shown in Figure 28.2.

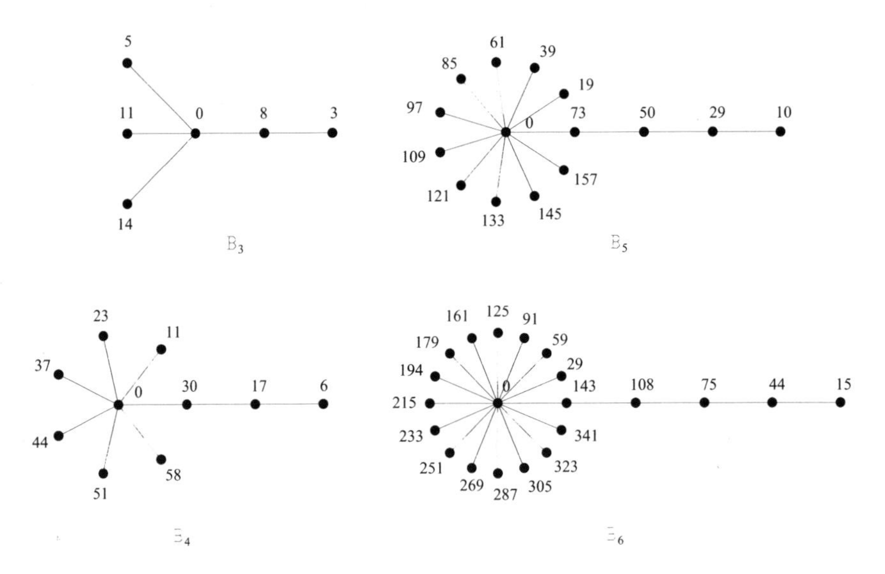

Figure 28.2: An optimal hamiltonian coloring of B_3, B_5 and B_4, B_6

We claim that the lower bound given in Theorem 28.1 is better than the lower bound given in [3, Theorem 4] for broom trees B_d $(d = 2k, 2k+1)$ except

B_2. For broom tree B_2, both lower bounds are identical. Denote the right-hand side of (28.3) and [3, Equation-(2)] by $lb_W(T)$ and $lb(T)$ respectively. That is, $lb_W(T) = (n-1)(n-1-\zeta(T)) + \zeta'(T) - 2\mathcal{L}_W(T)$ and $lb(T) = (n-1)(n-1-\varepsilon(T)) + \varepsilon'(T) - 2\mathcal{L}(T)$. Since $hc(B_d) = lb_W(B_d)$ $(d = 2k, 2k+1 \geq 3)$, it is enough to prove that $lb_W(B_d) - lb(B_d) > 0$ to justify our claim.

Theorem 28.6

Let $k \geq 1$ be any integer. Then $lb_W(B_d) - lb(B_d) > 0$, where $d = 2k, 2k+1 \geq 3$. ■

Proof. We consider the following two cases.

Case-1: $d = 2k$. In this case, $lb_W(B_d) - lb(B_d) = 4k(k-1)^2 > 0$.

Case-2: $d = 2k+1$. In this case, $lb_W(B_d) - lb(B_d) = 4k^3 - 2k^2 - k + 1 = k[(2k-1)^2 + 2(k-1)] + 1 > 0$ which completes the proof. □

28.5 CONCLUDING REMARKS

In [4], Chartrand *et al.* proved that for $n \geq 1$,

$$\mathrm{hc}(K_{1,n-1}) = (n-2)^2. \tag{28.16}$$

This result can be proved using Corollary 28.3.1 as follows. The total detour level of $K_{1,n-1}$ is $\mathcal{L}(K_{1,n-1}) = n-1$ and $|W(K_{1,n-1})| = 1$. Substituting $\mathcal{L}(K_{1,n-1}) = n-1$ in (28.3), we obtain that the right-hand side of (28.16) is a lower bound for $\mathrm{hc}(K_{1,n-1})$ and it is easy to find a hamiltonian coloring whose span is equal to this lower bound (refer to [4]).

A tree is said to be a caterpillar C if it consists of a path $v_1, v_2, \ldots, v_m (m \geq 3)$, called the spine of C, with some hanging edges known as legs, which are incident to the inner vertices $v_2, v_3, \ldots, v_{m-1}$. If $d(v_i) = d$ for $i = 2, 3, \ldots, m-1$, then denote the caterpillar by $C(m,d)$. In [9], Shen *et al.* proved that for any positive integers $m \geq 3$ and $d \geq 3$,

$$\mathrm{hc}(C(m,d)) := \begin{cases} \frac{2d-3}{2d-2}(n-2)^2 + \frac{d-1}{2}, & \text{If } m \text{ is odd,} \\ \frac{2d-3}{2d-2}(n-2)^2, & \text{If } m \text{ is even.} \end{cases} \tag{28.17}$$

This result can also be proved using Corollary 28.3.1 as follows. The order n and total detour level $\mathcal{L}(C(m,d))$ are given by

$$n := \begin{cases} (2k-1)(d-1) + 2, & \text{If } m = 2k+1, \\ 2k(d-1) - 2(d-2), & \text{If } m = 2k. \end{cases} \tag{28.18}$$

$$\mathcal{L}(C(m,d)) := \begin{cases} (k(k+1)-1)(d-1)+1, & \text{If } m = 2k+1, \\ k(k-1)(d-1), & \text{If } m = 2k. \end{cases} \tag{28.19}$$

Substituting (28.18) and (28.19) into (28.3) we obtain that the right-hand side of (28.17) is a lower bound for $\text{hc}(C(m,d))$ and it is easy to find a hamiltonian coloring whose span is equal to this lower bound (refer to [9]).

REFERENCES

1. R. Balakrishnan, K. Viswanathan Iyer and K. Raghavendra. Wiener index of two special trees. *MATCH Commun. Math. Comput. Chem.*, 57, 385-392, 2007.
2. D. Bantva, S. Vaidya and S. Zhou. Radio number of trees. *Discrete Applied Math.*, 217, 110-122, 2017.
3. D. Bantva. On hamiltonian colorings of trees. In: S. Govindrajan and A. Maheshwari (eds.). *Algorithms and Discrete Applied Math.*, CALDAM 2016, LNCS, Springer, Heidelberg, 9602, 49-60, 2016.
4. G. Chartrand, L. Nebeský and P. Zhang. Hamiltonian coloring of graphs. *Discrete Applied Math.*, 146, 257-272, 2005.
5. G. Chartrand, L. Nebeský and P. Zhang. On hamiltonian colorings of graphs. *Discrete Math.*, 290, 133-143, 2005.
6. G. Chartrand and P. Zhanf. Distance in graphs - taking the long view. *AKCE J. Graphs. Combin.*, 1(1): 1-13, 2004.
7. R. Khennoufa and O. Togni. A note on radio antipodal colourings of paths. *Mathematica Bohemica*, 130(3): 277-282, 2005.
8. D. Liu. Radio number of trees. *Discrete Math.*, 308, 1153-1164, 2008.
9. Y. Shen, W. He, X. Li, D. He and X. Yang. On hamiltonian colorings for some graphs. *Discrete Applied Math.*, 156, 3028-3034, 2008.
10. D. West. *Introduction to Graph Theory.* Prentice-Hall of India, 2001.

29 Some Results on Degree Sum Energy of a Graph

Mitesh J. Patel
Tolani College of Arts and Science,
Adipur- Kachchh, Gujarat (INDIA)
E-mail: miteshmaths1984@gmail.com

G. V. Ghodasara
H. & H. B. Kotak Institute of Science,
Rajkot, Gujarat (INDIA)
E-mail: gaurang_enjoy@yahoo.co.in

Let $G = (V, E)$ be a graph with order n and size m. The ordinary energy of the graph is defined as the sum of the absolute values of the eigenvalues of its adjacency matrix. In chemistry, the energy of a graph is used to approximate the total $\pi-$electron energy of a molecule. In this chapter we evaluate the degree sum energy of complete graph K_n, wheel graph W_n, star graph $K_{1,n}$ and Petersen graph P(5,2).

29.1 INTRODUCTION

Let G be a simple graph with vertex set $V(G) = \{v_1, v_2, \ldots, v_n\}$ and edge set $E(G)$. If v_i and v_j are adjacent vertices of G, then the edge connecting them is denoted by v_iv_j. The number of edges incident with vertex v_i is called degree of v_i; we denote d_i as a degree of the vertex $v_i \in V(G)$. The adjacency matrix $A(G)$ of the graph G is a square matrix of order n whose (i, j) - entry is equal to 1 if the vertices v_i and v_j are adjacent, and is equal to zero otherwise. Assume that the eigenvalues of graph G (i.e eigenvalues of adjacency matrix $A(G)$) are $\lambda_1, \lambda_2, \ldots, \lambda_n$ in decreasing order; then the energy of graph G, denoted as $\varepsilon(G)$, is the sum of absolute values of its eigenvalues. i.e.

$$\varepsilon(G) = \sum_{i=1}^{n} |\lambda_i|.$$

The energy of graph G was first introduced by Ivan Gutman[3] in 1978 and then he briefly outlined the connection between the energy of a graph and the total $\pi-$electron energy of organic molecules. He also presented some fundamental results on energy, the relation between energy $\varepsilon(G)$ of the graph G and the characteristic polynomial of G.
H. S. Ramane, D. S. Revankar and J. B. Patil[6] defined degree sum energy of a graph as follows.

Definition. Let G be a simple graph with n vertices $v_1, v_2, \ldots, v_n$ and let d_i be the degree of $v_i, i = 1, 2, \ldots, n$. Then the degree sum adjacency matrix of a graph G is $DS(G) = [d_{ij}]$, where

$$d_{ij} = \begin{cases} d_i + d_j; & v_i v_j \in E(G). \\ 0; & otherwise \end{cases}$$

The characteristic polynomial of degree sum matrix of G is $|\lambda I_n - DS(G)|$, where I_n is an identity matrix of order n. The roots of the characteristic equation $|\lambda I_n - DS(G)| = 0$ are called the degree sum eigenvalues of G. If $\lambda_1, \lambda_2, \ldots, \lambda_n$ are the degree sum eigenvalues of G then the degree sum energy of a graph G is defined as

$$\varepsilon_{DS}(G) = \sum_{i=1}^{n} |\lambda_i|.$$

From the motivation of this belief, we study degree sum energy of some special graphs.

29.2 MAIN RESULTS

Theorem 29.1

For the complete graph K_n, $\varepsilon_{DS}(K_n) = 4(n-1)^2$. ■

Proof. Let K_n be the complete graph with $V(K_n) = \{v_1, v_2, \ldots, v_n\}$ and $E(K_n) = \{v_i v_j | 1 \leq i, j \leq n, \ i \neq j\}$.
Here, $|V(K_n)| = n$ and $|E(K_n)| = \frac{n(n-1)}{2}$.
The degree sum adjacency matrix of complete graph K_n is

$$DS(K_n) = \begin{bmatrix} 0 & 2(n-1) & 2(n-1) & 2(n-1) & \ldots & 2(n-1) \\ 2(n-1) & 0 & 2(n-1) & 2(n-1) & \ldots & 2(n-1) \\ 2(n-1) & 2(n-1) & 0 & 2(n-1) & \ldots & 2(n-1) \\ \vdots & \vdots & \vdots & \ddots & \vdots & \\ 2(n-1) & 2(n-1) & 2(n-1) & 2(n-1) & \ldots & 0 \end{bmatrix}$$

So, the characteristic polynomial of $DS(K_n)$ is

$$|\lambda I_n - DS(K_n)| = \begin{vmatrix} \lambda & -2(n-1) & -2(n-1) & -2(n-1) & \ldots & -2(n-1) \\ -2(n-1) & \lambda & -2(n-1) & -2(n-1) & \ldots & -2(n-1) \\ -2(n-1) & -2(n-1) & \lambda & -2(n-1) & \ldots & -2(n-1) \\ \vdots & \vdots & \vdots & \ddots & \vdots & \\ -2(n-1) & -2(n-1) & -2(n-1) & -2(n-1) & \ldots & \lambda \end{vmatrix}$$

$$= (\lambda + 2(n-1))^{n-1}(\lambda - 2(n-1)^2).$$

The characteristic equation is $(\lambda+2(n-1))^{n-1}(\lambda-2(n-1)^2)=0$.
So, the spectrum of complete graph K_n is

$$\begin{pmatrix} -2(n-1) & 2(n-1)^2 \\ n-1 & 1 \end{pmatrix}.$$

Hence,

$$\varepsilon_{DS}(K_n)=4(n-1)^2.$$

□

Illustration 29.2.1. The degree sum adjacency matrix of complete graph K_5 is

$$DS(K_5)=\begin{bmatrix} 0 & 8 & 8 & 8 & 8 \\ 8 & 0 & 8 & 8 & 8 \\ 8 & 8 & 0 & 8 & 8 \\ 8 & 8 & 8 & 0 & 8 \\ 8 & 8 & 8 & 8 & 0 \end{bmatrix}.$$

So, the characteristic polynomial of $DS(K_5)$ is

$$\begin{vmatrix} \lambda & -8 & -8 & -8 & -8 \\ -8 & \lambda & -8 & -8 & -8 \\ -8 & -8 & \lambda & -8 & -8 \\ -8 & -8 & -8 & \lambda & -8 \\ -8 & -8 & -8 & -8 & \lambda \end{vmatrix}=(\lambda+8)^4(\lambda-32).$$

The characteristic equation is $(\lambda+8)^4(\lambda-32)=0$.
So, the spectrum of complete graph K_5 is

$$\begin{pmatrix} -8 & 32 \\ 4 & 1 \end{pmatrix}.$$

Hence, $\varepsilon_{DS}(K_5)=64$.

Remark. A graph G of order n is said to be degree sum hyperenergetic if $\varepsilon_{DS}(G)>\varepsilon_{DS}(K_n)$.

Theorem 29.2

For the wheel graph W_n, $\varepsilon_{DS}(W_n)=6(n-1)+2\sqrt{9(n-1)^2+n(n+3)^2}$. ■

Proof. Let W_n be a wheel graph with $V(W_n)=\{v_0,v_1,v_2,\ldots,v_n\}$ and $E(W_n)=\{v_0v_i|1\leq i\leq n\}\cup\{v_iv_{i+1}|1\leq i\leq n-1\}\cup\{v_1v_n\}$, where v_0 is the apex and $v_1,v_2,\ldots,v_n$ are rim vertices.

Here, $|V(W_n)| = n+1$ and $|E(W_n)| = 2n$.
The degree sum adjacency matrix of W_n is

$$DS(W_n) = \begin{bmatrix} 0 & n+3 & n+3 & n+3 & \dots & n+3 \\ n+3 & 0 & 6 & 6 & \dots & 6 \\ n+3 & 6 & 0 & 6 & \dots & 6 \\ \vdots & \vdots & \vdots & \ddots & \vdots & \\ n+3 & 6 & 6 & 6 & \dots & 0 \end{bmatrix}$$

So, the characteristic polynomial of $DS(W_n)$ is

$$|\lambda I_n - DS(W_n)| = \begin{vmatrix} \lambda & -(n+3) & -(n+3) & -(n+3) & \dots & -(n+3) \\ -(n+3) & \lambda & -6 & -6 & \dots & -6 \\ -(n+3) & -6 & \lambda & -6 & \dots & -6 \\ \vdots & \vdots & \vdots & \ddots & \vdots & \\ -(n+3) & -6 & -6 & -6 & \dots & \lambda \end{vmatrix}$$

$= (\lambda+6)^{n-1}[\lambda^2 - 6(n-1)\lambda - n(n+3)^2]$.
The characteristic equation is $(\lambda+6)^{n-1}[\lambda^2 - 6(n-1)\lambda - n(n+3)^2] = 0$.
So, the spectrum of W_n is

$$\begin{pmatrix} -6 & 6(n-1)+\sqrt{9(n-1)^2+n(n+3)^2} & 6(n-1)-\sqrt{9(n-1)^2+n(n+3)^2} \\ n-1 & 1 & 1 \end{pmatrix}$$

For $n \geq 3$, it is easy to check that $6(n-1) < \sqrt{9(n-1)^2+n(n+3)^2}$.
Hence, $\varepsilon_{DS}(W_n) = 6(n-1)+2\sqrt{9(n-1)^2+n(n+3)^2}$. □

Illustration 29.2.2. The degree sum adjacency matrix of wheel graph W_5 is

$$DS(W_5) = \begin{bmatrix} 0 & 8 & 8 & 8 & 8 & 8 \\ 8 & 0 & 6 & 6 & 6 & 6 \\ 8 & 6 & 0 & 6 & 6 & 6 \\ 8 & 6 & 6 & 0 & 6 & 6 \\ 8 & 6 & 6 & 6 & 0 & 6 \\ 8 & 6 & 6 & 6 & 6 & 0 \end{bmatrix}$$

So, the characteristic polynomial of $DS(W_5)$ is

$$\begin{vmatrix} \lambda & -8 & -8 & -8 & -8 & -8 \\ -8 & \lambda & -6 & -6 & -6 & -6 \\ -8 & -6 & \lambda & -6 & -6 & -6 \\ -8 & -6 & -6 & \lambda & -6 & -6 \\ -8 & -6 & -6 & -6 & \lambda & -6 \\ -8 & -6 & -6 & -6 & -6 & \lambda \end{vmatrix} = (\lambda+6)^4(\lambda^2 - 24\lambda - 320).$$

The characteristic equation is $(\lambda+6)^4(\lambda^2-24\lambda-320)=0$.
So, the spectrum of W_5 is

$$\begin{pmatrix} -6 & -9.5407 & 33.5407 \\ 4 & 1 & 1 \end{pmatrix}$$

Hence, $\varepsilon_{DS}(W_5)=67.0814$.

Theorem 29.3

For the star graph $K_{1,n}$, $\varepsilon_{DS}(K_{1,n})=2(n-1)+2\sqrt{n^3+3n^2-n+1}$. ■

Proof. Let $K_{1,n}$ be a star graph with $V(K_{1,n})=\{v_0,v_1,v_2,\ldots,v_n\}$ and $E(K_{1,n})=\{v_0v_i|1\leq i\leq n\}\}$, where v_0 is the vertex of degree n and $v_1,v_2,\ldots,v_n$ are pendent vertices.
Here, $|V(K_{1,n})|=n+1$ and $|E(K_{1,n})|=n$.
The degree sum adjacency matrix of star graph $K_{1,n}$ is

$$DS(K_{1,n})=\begin{bmatrix} 0 & (n+1) & (n+1) & (n+1) & \ldots & (n+1) \\ (n+1) & 0 & 2 & 2 & \ldots & 2 \\ (n+1) & 2 & 0 & 2 & \ldots & 2 \\ \vdots & \vdots & \vdots & \ddots & \vdots & \\ (n+1) & 2 & 2 & 2 & \ldots & 0 \end{bmatrix}$$

So, the characteristic polynomial of $DS(K_{1,n})$ is

$$\begin{vmatrix} \lambda & -(n+1) & -(n+1) & -(n+1) & \ldots & -(n+1) \\ -(n+1) & \lambda & -2 & -2 & \ldots & -2 \\ -(n+1) & -2 & \lambda & -2 & \ldots & -2 \\ \vdots & \vdots & \vdots & \ddots & \vdots & \\ -(n+1) & -2 & -2 & -2 & \ldots & \lambda \end{vmatrix}$$

$=(\lambda+2)^{n-1}(\lambda^2-2(n-1)\lambda-n(n+1)^2)$.
The characteristic equation is $(\lambda+2)^{n-1}(\lambda^2-2(n-1)\lambda-n(n+1)^2)=0$.
So, the spectrum of star graph $K_{1,n}$ is

$$\begin{pmatrix} -2 & (n-1)+\sqrt{n^3+3n^2-n+1} & (n-1)-\sqrt{n^3+3n^2-n+1} \\ n-1 & 1 & 1 \end{pmatrix}.$$

For $n\geq 1$, it is easy to check that $(n-1)<\sqrt{n^3+3n^2-n+1}$.
Hence, $\varepsilon_{DS}(K_{1,n})=2(n-1)+2\sqrt{n^3+3n^2-n+1}$. □

Illustration 29.2.3. The degree sum adjacency matrix of star graph $K_{1,5}$ is

$$DS(K_{1,5}) = \begin{bmatrix} 0 & 6 & 6 & 6 & 6 & 6 \\ 6 & 0 & 2 & 2 & 2 & 2 \\ 6 & 2 & 0 & 2 & 2 & 2 \\ 6 & 2 & 2 & 0 & 2 & 2 \\ 6 & 2 & 2 & 2 & 0 & 2 \\ 6 & 2 & 2 & 2 & 2 & 0 \end{bmatrix}$$

So, the characteristic polynomial of $DS(K_{1,5})$ is

$$\begin{vmatrix} \lambda & -6 & -6 & -6 & -6 & -6 \\ -6 & \lambda & -2 & -2 & -2 & -2 \\ -6 & -2 & \lambda & -2 & -2 & -2 \\ -6 & -2 & -2 & \lambda & -2 & -2 \\ -6 & -2 & -2 & -2 & \lambda & -2 \\ -6 & -2 & -2 & 2 & -2 & \lambda \end{vmatrix} = (\lambda+2)^4(\lambda^2-16\lambda-180).$$

So, the characteristic equation is $(\lambda+2)^4(\lambda^2-16\lambda-180)=0$.
So, the spectrum of $K_{1,5}$ is

$$\begin{pmatrix} -2 & -10 & 18 \\ 4 & 1 & 1 \end{pmatrix}.$$

Hence, $\varepsilon_{DS}(K_{1,5})=36$.

Theorem 29.4

For Petersen graph $P(5,2)$, $\varepsilon_{DS}(P(5,2))=96$. ■

Proof. A generalized Petersen graph, denoted by $P(n,k)(n \geq 5,\ 1 \leq k \leq n)$ is a graph with vertex set $\{v_0, v_1, \ldots v_{n-1}, u_0, u_1, \ldots u_{n-1}\}$ and edge set $\{v_i v_{i+1} \setminus i=0,1,\ldots n-1\} \cup \{v_i u_i \setminus i=0,1,\ldots n-1\} \cup \{u_i u_{i+k} \setminus i=0,1,\ldots n-1\}$, where all subscripts are taken over modulo n. The standard Petersen graph is $P(5,2)$ with $|V(P(5,2))|=10$ and $|E(P(5,2))|=15$.
The degree sum adjacency matrix of $P(5,2)$ is

$$DS(P(5,2)) = \begin{bmatrix} 0 & 6 & 0 & 0 & 6 & 6 & 0 & 0 & 0 & 0 \\ 6 & 0 & 6 & 0 & 0 & 0 & 6 & 0 & 0 & 0 \\ 0 & 6 & 0 & 6 & 0 & 0 & 0 & 6 & 0 & 0 \\ 0 & 0 & 6 & 0 & 6 & 0 & 0 & 0 & 6 & 0 \\ 6 & 0 & 0 & 6 & 0 & 0 & 0 & 0 & 0 & 6 \\ 6 & 0 & 0 & 0 & 0 & 0 & 0 & 6 & 6 & 0 \\ 0 & 6 & 0 & 0 & 0 & 0 & 0 & 0 & 6 & 6 \\ 0 & 0 & 6 & 0 & 0 & 6 & 0 & 0 & 0 & 6 \\ 0 & 0 & 0 & 6 & 0 & 6 & 6 & 0 & 0 & 0 \\ 0 & 0 & 0 & 0 & 6 & 0 & 6 & 6 & 0 & 0 \end{bmatrix}$$

So, the characteristic polynomial of $DS(P(5,2))$ is

$$\begin{vmatrix} \lambda & -6 & 0 & 0 & -6 & -6 & 0 & 0 & 0 & 0 \\ -6 & \lambda & -6 & 0 & 0 & 0 & -6 & 0 & 0 & 0 \\ 0 & -6 & \lambda & -6 & 0 & 0 & 0 & -6 & 0 & 0 \\ 0 & 0 & -6 & \lambda & -6 & 0 & 0 & 0 & -6 & 0 \\ -6 & 0 & 0 & -6 & \lambda & 0 & 0 & 0 & 0 & -6 \\ -6 & 0 & 0 & 0 & 0 & \lambda & 0 & -6 & -6 & 0 \\ 0 & -6 & 0 & 0 & 0 & 0 & \lambda & 0 & -6 & -6 \\ 0 & 0 & -6 & 0 & 0 & -6 & 0 & \lambda & 0 & -6 \\ 0 & 0 & 0 & -6 & 0 & -6 & -6 & 0 & \lambda & 0 \\ 0 & 0 & 0 & 0 & -6 & 0 & -6 & -6 & 0 & \lambda \end{vmatrix}$$

$= (\lambda+12)^4(\lambda-6)^5(\lambda-18)$.
The characteristic equation is $(\lambda+12)^4(\lambda-6)^5(\lambda-18) = 0$.
So, the spectrum of Petersen graph $P(5,2)$ is

$$\begin{pmatrix} -12 & 6 & 18 \\ 4 & 5 & 1 \end{pmatrix}.$$

Hence, $\varepsilon_{DS}(P(5,2)) = 96$. □

29.3 CONCLUSION

In this chapter we study the degree sum energy of complete graph, star graph, Petersen graph P(5,2). To evaluate the similar results for a generalised Petersen graph and for other graph families is an open area of research.

REFERENCES

1. C. Adiga and M. Smitha. On maximum degree energy of a graph. *International Journal of Contemporary Mathematical Sciences*, 4(8): 385-396, 2009.
2. J. A. Bondy and U. S. Murty. *Graph Theory with Applications.* Elsevier Science Publication, 1982.
3. I. Gutman. The energy of a graph. *Ber. Math-Statist. Sekt. Forschungszentrum Graz*, 103, 1-22, 1978.
4. S. M. Hosamani and H. S. Ramane. On degree sum energy of a graph. *European Journal of Pure and Applied Mathematics*, 9(3): 340-345, 2016.
5. K. I. Ramchandran, G. Deeta and K. Namboori. *Computational Chemistry and Molecular Modeling Principals and Applications.* Springer, 2008.
6. H. S. Ramane, D. S. Revankar and J. B. Patil. Bounds for the degree sum eigenvalues and degree sum energy of a graph. *International Journal of Pure and Applied Mathematical Sciences*, 6(2): 161-167, 2013.

30 Randić Energy of Some Graphs

S. K. Vaidya
Saurashtra University,
Rajkot-360005,Gujarat(India).
E-mail: samirkvaidya@yahoo.co.in

G. K. Rathod
V.V.P. Engineering College,
Rajkot-360005,Gujarat(India).
Email: gopalrathod852@gmail.com

Let d_i be the degrees of vertices $v_i, i = 1, 2, \ldots, n$ of graph G; then the Randić matrix $R(G)$ of graph G is an $n \times n$ matrix whose entries are given by $r_{ij} = 1/\sqrt{d_i d_j}$, if the vertices v_i and v_j are adjacent in G and 0 otherwise. The Randić energy of the graph G is defined to be the sum of absolute values of eigenvalues of $R(G)$. Here,we explore this concept for extended m-shadow, extended m-splitting, double cover and extended double cover of graph G.

30.1 INTRODUCTION

Let G be a simple connected undirected graph with $v_1, v_2, \ldots, v_n$ as its vertices; then the adjacency matrix of graph G is a symmetric matrix of order n which is denoted and defined as $A(G) = [a_{ij}]$, where

$$a_{ij} = \begin{cases} 1 & ;\text{if } v_i \text{ and } v_j \text{ are adjacent} \\ 0 & ;\text{otherwise} \end{cases}$$

If $\lambda_1, \lambda_2, \ldots, \lambda_n$ are eigenvalues of $A(G)$ then they are also known as eigenvalues of graph G. Since $A(G)$ is a symmetric matrix so all the eigenvalues of $A(G)$ are real number with their sum zero. The concept energy of graphs was introduced by I. Gutman[8] in 1978. The energy of graph is defined as the sum of absolute values of eigenvalues of graph G.

$$\mathcal{E}(G) = \sum_{i=1}^{n} |\lambda_i|.$$

A brief account of graph energy can be found in Balakrishnan [2], Li *et al.* [2] and Cvetković *et al.* [6].

In 1975, Milan Randić[13] defined the concept named Randić index which is defined as

$$R = \sum_{i \sim j} \frac{1}{\sqrt{d_i d_j}},$$

where summation is taken over all pairs of adjacent vertices v_i and v_j. The Randić index is certainly the most widely applied in chemistry and pharmacology, in particular for designing quantitative structure property and structure activity relations. The brief account of these applications can be found in [14, 15, 16]

In 2010, Bozkurt *et al.* [3, 4] defined the concepts of Randić matrix and Randić energy. The Randić matrix $R(G) = [r_{ij}]$ of a graph G is a symmetric matrix of order n which is defined as,

$$r_{ij} = \begin{cases} \frac{1}{\sqrt{d_i d_j}} & ;\text{if } v_i \text{ and } v_j \text{ are adjacent} \\ 0 & ;\text{otherwise} \end{cases}$$

The connection between the Randić index and Randić matrix is obvious as the sum of all elements of $R(G)$ is equal to $2R$.
If $\rho_1, \rho_2, \ldots, \rho_n$ are eigenvalues of matrix $R(G)$ then the Randić energy [3, 4] is defined as the sum of absolute values of eigenvalues of graph G which is denoted as $RE(G)$. That is,

$$RE(G) = \sum_{i=1}^{n} |\rho_i|.$$

Many results on Randić matrix and Randić energy can be found in [1, 3, 4, 9, 17].

We state some definitions and propositions which are useful for the next discussion.

Definition. [10] *The Cartesian product of $G \times H$ of two graphs $G = (V(G), E(G))$ and $H = (V(H), E(H))$, is the graph with vertex set $V(G \times H) = \{(u,v)/\ u \in V(G) \text{ and } v \in V(H)\}$ and edge set $E(G \times H) = \{(u,v)(u',v')/u = u', (v,v') \in E(H) \text{ or } (u,u') \in E(G), v = v'\}$.*

Definition. [11] *The Kronecker product of $A = [a_{ij}]_{m \times n}$, $B = [b_{ij}]_{p \times q}$ is defined as the matrix*

$$A \otimes B = \begin{bmatrix} a_{11}B & \cdots & a_{1n}B \\ \vdots & \ddots & \vdots \\ a_{m1}B & \cdots & a_{mn}B \end{bmatrix}$$

Proposition 30.1.1. *[11] If λ is an eigenvalue of matrix $A = [a_{ij}]_{m \times m}$ with corresponding eigenvector x, and μ is an eigenvalue of matrix $B = [b_{ij}]_{n \times n}$ with corresponding eigenvector y, then $\lambda\mu$ is an eigenvalue of $A \otimes B$ with corresponding eigenvector $x \otimes y$.*

Proposition 30.1.2. *[7] Let*

$$A = \begin{bmatrix} A_0 & A_1 \\ A_1 & A_0 \end{bmatrix}$$

be a symmetric block matrix. Then the spectrum of A is the union of spectra of $A_0 + A_1$ and $A_0 - A_1$.

For any undefined term in graph theory, we follow Balakrishnan *et al.* [2] while for matrix algebra we rely upon Horn [11]. In this chapter we obtain the Randić energy of an extended m-shadow graph, extended m-splitting graph, double cover of graph and extended double cover of given graph G.

30.2 RANDIĆ ENERGY OF EXTENDED M-SHADOW GRAPH

Definition. The extended $m-$*Shadow* graph $D_m^*(G)$ of a connected graph G is constructed by taking m copies of G, say $G_1, G_2, \ldots, G_m$, then joining each vertex u in G_i to the neighbors of the corresponding vertex v and with v in $G_j, 1 \leqslant i,j \leqslant m$.

In Figure 30.1. the extended 2-shadow graph of C_4 is shown.

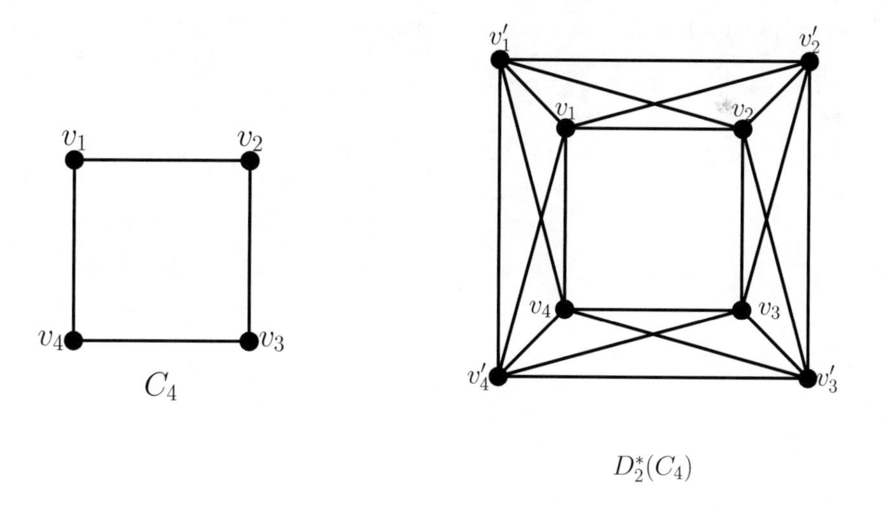

Figure 30.1: cycle C_4 and its 2-shadow graph

Now, we find the Randić energy for an extended $m-$shadow of graph G.

Theorem 30.1

Let G be an r-regular graph with eigenvalues $\lambda_1, \lambda_2, \ldots, \lambda_n$ with $|\lambda_i| \geqslant \frac{m-1}{m}$, for all $1 \leqslant i \leqslant n$ then

$$RE(D_m^*(G)) = \frac{m}{mr+(m-1)}\left[\mathcal{E}(G) + \frac{m-1}{m}\theta\right] + \frac{mn-m}{mr+(m-1)},$$

where θ is the difference between the number of positive and negative eigenvalues of graph G. ■

Proof. Let G be a graph of order n; then the Randić matrix of G can be written as

$$R(G) = \begin{bmatrix} 0 & r_{12} & r_{13} & \cdots & r_{1n} \\ r_{21} & 0 & r_{23} & \cdots & r_{2n} \\ r_{31} & r_{32} & 0 & \cdots & r_{3n} \\ \vdots & \vdots & \vdots & \ddots & \vdots \\ r_{n1} & r_{n2} & r_{n3} & \cdots & 0 \end{bmatrix}$$

Now, consider m-copies $G_1, G_2, \ldots, G_m$ of graph G and then join each vertex of u of graph G_i to the neighbors of the corresponding vertex v and also with v in graph G_j, $1 \leqslant i, j \leqslant m$ to obtain extended $m-$Shadow $D_m^*(G)$. Then the Randić matrix of graph $D_m^*(G)$ can be written as
$R(D_m^*(G))$

$$= \begin{bmatrix} \frac{1}{mr+(m-1)}A(G) & \frac{1}{mr+(m-1)}[A(G)+I] & \cdots & \frac{1}{mr+(m-1)}[A(G)+I] \\ \frac{1}{mr+(m-1)}[A(G)+I] & \frac{1}{mr+(m-1)}A(G) & \cdots & \frac{1}{mr+(m-1)}[A(G)+I] \\ \vdots & \vdots & \ddots & \vdots \\ \frac{1}{mr+(m-1)}[A(G)+I] & \frac{1}{mr+(m-1)}[A(G)+I] & \cdots & \frac{1}{mr+(m-1)}A(G) \end{bmatrix}$$

so,

$$R(D_m^*(G)) + \frac{1}{mr+(m-1)}I_{mn} = \begin{bmatrix} \frac{1}{mr+(m-1)}[A(G)+I] & \cdots & \frac{1}{mr+(m-1)}[A(G)+I] \\ \frac{1}{mr+(m-1)}[A(G)+I] & \cdots & \frac{1}{mr+(m-1)}[A(G)+I] \\ \vdots & \ddots & \vdots \\ \frac{1}{mr+(m-1)}[A(G)+I] & \cdots & \frac{1}{mr+(m-1)}[A(G)+I] \end{bmatrix}$$

$$= scriptstyle\frac{1}{mr+(m-1)}J_m \otimes [A(G)+I]$$

where J_m is a matrix of order m with all the entries being 1. Since, we know that $spec(J_m) = \begin{pmatrix} m & 0 \\ 1 & m-1 \end{pmatrix}$.

so, $spec\left(\frac{1}{mr+(m-1)}J_m\right) = \begin{pmatrix} \frac{m}{mr+(m-1)} & 0 \\ 1 & m-1 \end{pmatrix}$. Also, we know that eigenvalues of the matrix $A(G)+I$ are λ_i+1, for all $1 \leq i \leq n$.
Hence, from Proposition 30.1.1 we have,

$$spec\left(R(D_m^*(G)) + \frac{1}{mr+(m-1)}I_{mn}\right) = \begin{pmatrix} \frac{m}{mr+(m-1)}(\lambda_i+1) & 0 \\ n & mn-n \end{pmatrix}.$$

$$\Rightarrow spec(R(D_m^*(G))) = \begin{pmatrix} \frac{m}{mr+(m-1)}(\lambda_i+1) - \frac{1}{mr+(m-1)} & \frac{-m}{mr+(m-1)} \\ n & mn-n \end{pmatrix}$$
$$= \begin{pmatrix} \frac{m\lambda_i+(m-1)}{mr+(m-1)} & \frac{-m}{mr+(m-1)} \\ n & mn-n \end{pmatrix}$$

Suppose that $|\lambda_i| \geq \frac{m-1}{m}$ for all $1 \leq i \leq n$ then

$$\left|\lambda_i + \frac{m-1}{m}\right| = |\lambda_i| + \frac{m-1}{m}, \ if\ \lambda_i \geq 0$$
$$= |\lambda_i| - \frac{m-1}{m}, \ if\ \lambda_i < 0$$

Therefore,

$$RE(D_m^*(G)) = \sum_{i=1}^{n}\left|\frac{m\lambda_i+(m-1)}{mr+(m-1)}\right| + \sum_{i=1}^{mn-n}\left|\frac{-1}{mr+(m-1)}\right|$$
$$= \frac{m}{mr+(m-1)}\sum_{i=1}^{n}\left|\lambda_i + \frac{m-1}{m}\right| + \frac{mn-m}{mr+(m-1)}$$
$$= \frac{m}{mr+(m-1)}\left(\sum_{\lambda_i \geq 0}(|\lambda_i| + \tfrac{m-1}{m}) + \sum_{\lambda_i<0}(|\lambda_i| - \tfrac{m-1}{m})\right) + \frac{mn-m}{mr+(m-1)}$$
$$= \frac{m}{mr+(m-1)}\left(\sum_{\lambda_i \geq 0}|\lambda_i| + \sum_{\lambda_i<0}|\lambda_i| + \tfrac{m-1}{m}\left(\sum_{\lambda_i\geq 0}1 + \sum_{\lambda_i<0}1\right)\right)$$
$$= \frac{m}{mr+(m-1)}\left(\mathcal{E}(G) + \tfrac{m-1}{m}\theta\right) + \frac{mn-m}{mr+(m-1)},$$

where θ is the difference between the number of positive and negative eigenvalues of graph G. □

30.3 RANDIĆ ENERGY OF EXTENDED M-SPLITTING GRAPH

Definition. The extended m-splitting graph $Spl_m^*(G)$of a graph G is obtained by adding to each vertex v of G new m vertices, say $v_1, v_2, \ldots, v_m$ such that v_i, $1 \leq i \leq m$ is adjacent to each vertex that is adjacent to v in G and also adjacent to v and v_j with $i \neq j$.

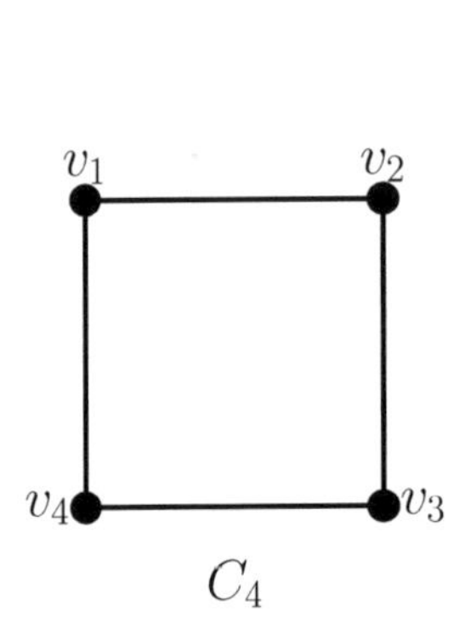

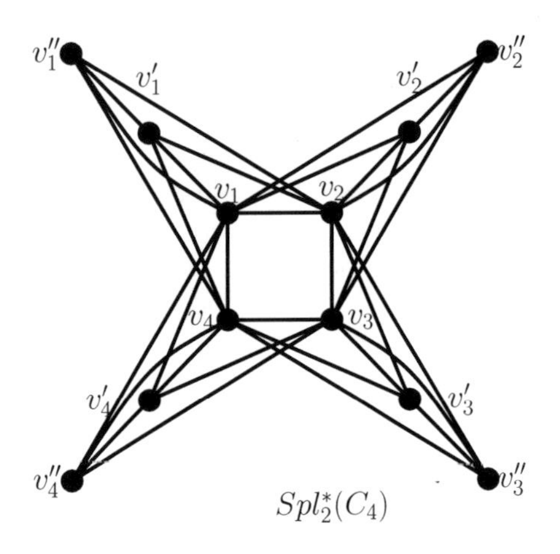

Figure 30.2: cycle C_4 and its 2-splitting graph

In Figure 30.2 the extended 2-splitting graph of C_4 is shown.

Now, we find the Randić energy for extended $m-$splitting of graph G.

Theorem 30.2

Let G be an $r-$regular graph of order n with eigenvalues $\lambda_1, \lambda_2, \ldots, \lambda_n$ with $|\lambda_i| \geqslant 1$,for all $1 \leqslant i \leqslant n$ then

$$RE(Spl_m^*(G)) = \frac{\sqrt{1 + \frac{4m}{m+r}(mr+m+r)}}{mr+m+r}[\mathcal{E}(G)+\theta] + \frac{n}{mr+m+r},$$

where θ is the difference between the number of positive and negative eigenvalues of graph G. ■

Proof. Let G be a graph with $v_1, v_2, \ldots, v_n$ being its vertices; then the Randić matrix of G can be written as given in the last theorem. Now, consider $m-$copies of vertex v_i for $1 \leq i \leq n$, say $v_i', v_i'', \ldots, v_i^{m'}$ and then join each vertex $v_i^{k'}$, for $1 \leq k \leq m$ to neighbors of vertex v_i and also join to v_i and $v_i^{j'}$ with $k \neq j$ to obtain an extended $m-$splitting of given graph G.
Then the Randić matrix of an extended $m-$splitting of graph G can be written as,

$$R(Spl_m^*(G)) = \begin{bmatrix} \frac{A(G)}{mr+(m+r)} & \frac{[A(G)+I]}{\sqrt{(m+r)(mr+m+r)}} & \cdots & \frac{[A(G)+I]}{\sqrt{(m+r)(mr+m+r)}} \\ \frac{[A(G)+I]}{\sqrt{(m+r)(mr+m+r)}} & 0 & \cdots & 0 \\ \vdots & \vdots & \ddots & \vdots \\ \frac{[A(G)+I]}{\sqrt{(m+r)(mr+m+r)}} & 0 & \cdots & 0 \end{bmatrix}$$

Therefore,
$R(Spl_m^*(G)) + \frac{1}{mr+m+r}B =$

$$\begin{bmatrix} \frac{[A(G)+I]}{mr+(m+r)} & \frac{[A(G)+I]}{\sqrt{(m+r)(mr+m+r)}} & \cdots & \frac{[A(G)+I]}{\sqrt{(m+r)(mr+m+r)}} \\ \frac{[A(G)+I]}{\sqrt{(m+r)(mr+m+r)}} & 0 & \cdots & 0 \\ \vdots & \vdots & \ddots & \vdots \\ \frac{[A(G)+I]}{\sqrt{(m+r)(mr+m+r)}} & 0 & \cdots & 0 \end{bmatrix}$$

where $A(G)$ is the adjacency matrix of graph G and B is the block matrix of order $mn \times mn$ of the form $B = \begin{bmatrix} I & 0 & \cdots & 0 \\ 0 & 0 & \cdots & 0 \\ \vdots & \vdots & \ddots & \vdots \\ 0 & 0 & \cdots & 0 \end{bmatrix}$. So, the above matrix can be written as, $R(Spl_m^*(G)) + \frac{1}{mr+m+r}B = A \otimes [A(G)+I]$, where

$$A = \begin{bmatrix} \frac{1}{mr+(m+r)} & \frac{1}{\sqrt{(m+r)(mr+m+r)}} & \cdots & \frac{1}{\sqrt{(m+r)(mr+m+r)}} \\ \frac{1}{\sqrt{(m+r)(mr+m+r)}} & 0 & \cdots & 0 \\ \vdots & \vdots & \ddots & \vdots \\ \frac{1}{\sqrt{(m+r)(mr+m+r)}} & 0 & \cdots & 0 \end{bmatrix}$$

Since A is a matrix of rank 2, it has only two non-zero eigenvalues, say μ_1, μ_2.Thus,

$$\mu_1 + \mu_2 = tr(A) = \frac{1}{mr+m+r}.$$

Also, we have

$$\mu_1^2 + \mu_2^2 = tr(A^2) = \frac{1}{(mr+m+r)^2} + \frac{2m}{(m+r)(m+r+mr)}.$$

By solving the above two equations we get, $\mu_1 = \frac{1+\sqrt{1+\frac{4m}{m+r}(mr+m+r)}}{2(mr+m+r)}$ and $\mu_2 = \frac{1-\sqrt{1+\frac{4m}{m+r}(mr+m+r)}}{2(mr+m+r)}$.

$$\text{Therefore, } spec(A) = \begin{pmatrix} 0 & \mu_1 & \mu_2 \\ m-1 & 1 & 1 \end{pmatrix}$$

So, by Proposition 30.1.1

$$spec(A \otimes [A(G)+I]) = \begin{pmatrix} 0 & \mu_1(\lambda_i+1) & \mu_2(\lambda_i+1) \\ n(m-1) & n & n \end{pmatrix}.$$

Hence,

$$spec(R(Spl_m^*(G))) = \begin{pmatrix} \frac{-1}{mr+m+r} & 0 & \mu_1(\lambda_i+1) & \mu_2(\lambda_i+1) \\ n & n(m-2) & n & n \end{pmatrix}.$$

Also, suppose that $|\lambda_i| \geq 1$ for all $1 \leq i \leq n$ then

$$\begin{aligned} |\lambda_i+1| &= |\lambda_i|+1, \ if\ \lambda_i \geq 0 \\ &= |\lambda_i|-1, \ if\ \lambda_i < 0 \end{aligned}$$

Therefore,

$$\begin{aligned}
&RE(Spl_m^*(G)) \\
&= \sum_{i=1}^n |\mu_1(\lambda_i+1)| + \sum_{i=1}^n |\mu_2(\lambda_i+1)| + \sum_{i=1}^n \left|\frac{-1}{mr+m+r}\right| \\
&= \sum_{i=1}^n \left| \frac{1 \pm \sqrt{1+\frac{4m}{m+r}(mr+m+r)}}{2(mr+m+r)}(\lambda_i+1) \right| + \frac{n}{mr+m+r} \\
&= \frac{\sqrt{1+\frac{4m}{m+r}(mr+m+r)}}{(mr+m+r)} \sum_{i=1}^n |\lambda_i+1| + \frac{n}{mr+m+r} \\
&= \frac{\sqrt{1+\frac{4m}{m+r}(mr+m+r)}}{(mr+m+r)} \left[\sum_{\lambda_i>0} |\lambda_i+1| + \sum_{\lambda_i<0} |\lambda_i+1| \right] + \frac{n}{mr+m+r} \\
&= \frac{\sqrt{1+\frac{4m}{m+r}(mr+m+r)}}{(mr+m+r)} \left[\sum_{\lambda_i>0} (|\lambda_i|+1) + \sum_{\lambda_i<0} (|\lambda_i|-1) \right] + \frac{n}{mr+m+r} \\
&= \frac{\sqrt{1+\frac{4m}{m+r}(mr+m+r)}}{(mr+m+r)} \left[\left(\sum_{\lambda_i>0} |\lambda_i| + \sum_{\lambda_i<0} |\lambda_i| \right) + \left(\sum_{\lambda_i>0} 1 - \sum_{\lambda_i<0} 1 \right) \right] \\
&+ \frac{n}{mr+m+r} \\
&= \frac{\sqrt{1+\frac{4m}{m+r}(mr+m+r)}}{(mr+m+r)} [\mathcal{E}(G)+\theta] + \frac{n}{mr+m+r},
\end{aligned}$$

where θ is the difference between the number of positive and negative eigenvalues of graph G. □

30.4 RANDIĆ ENERGY OF DOUBLE COVER OF GRAPH

Definition. [5] The double cover of graph G with vertex set $V(G) = \{v_1, v_2, \dots, v_n\}$ is a bipartite graph G' with bipartition $(X, Y), X = \{x_1, x_2, \dots, x_n\}$ and $Y = \{y_1, y_2, \dots, y_n\}$, where two vertices x_i and y_j are adjacent if and only if v_i is adjacent to v_j in G.

In Figure 30.3 the double cover graph of C_4 is shown.

Theorem 30.3

Let G be an r-regular graph with $\lambda_1, \lambda_2, \dots, \lambda_n$ eigenvalues then,

$$RE(G') = \frac{2}{r}\left[\mathcal{E}(G)\right].$$

■

Proof. Let G be an r-regular graph on n vertices; then construct a double cover graph G' of graph G by considering bipartition of the vertex set (X, Y), where $X = \{x_1, x_2, \dots, x_n\}$ and $Y = \{y_1, y_2, \dots, y_n\}$ with two vertices x_i and y_j are adjacent if and only if v_i is adjacent to v_j in G.
Then the Randić matrix of double cover G' of graph G can be written as

$$R(G') = \begin{bmatrix} 0 & \frac{1}{r}A(G) \\ \frac{1}{r}A(G) & 0 \end{bmatrix},$$

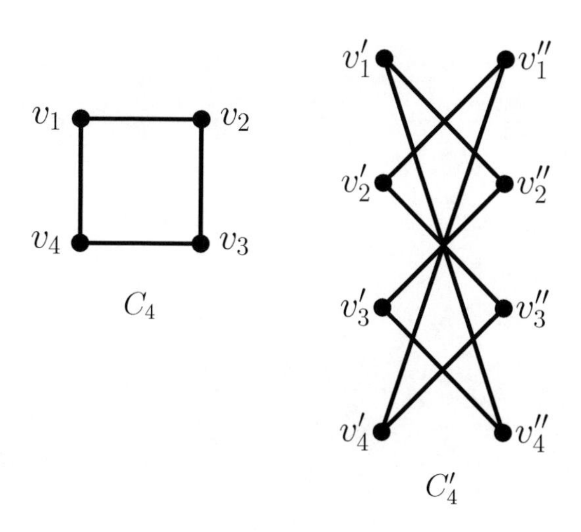

Figure 30.3: Cycle C_4 and its double cover graph

where $A(G)$ is the adjacency matrix of graph G.
So, from Proposition 30.1.1 the spectra of $R(G')$ can be given by

$$spec(R(G')) = spec\left(\frac{1}{r}A(G)\right) \cup spec\left(\frac{-1}{r}A(G)\right)$$

Therefore,

$$spec(R(G')) = \begin{pmatrix} \frac{1}{r}\lambda_1 & \frac{1}{r}\lambda_2 & \cdots & \frac{1}{r}\lambda_n & \frac{-1}{r}\lambda_1 & \frac{-1}{r}\lambda_2 & \cdots & \frac{-1}{r}\lambda_n \\ 1 & 1 & \cdots & 1 & 1 & 1 & \cdots & 1 \end{pmatrix}$$

Hence,

$$\begin{aligned} RE(G') &= \sum_{i=1}^{n}\left|\frac{1}{r}\lambda_i\right| + \sum_{i=1}^{n}\left|\frac{-1}{r}\lambda_i\right| \\ &= \frac{1}{r}\sum_{i=1}^{n}|\lambda_i| + \frac{1}{r}\sum_{i=1}^{n}|\lambda_i| \\ &= \frac{2}{r}\sum_{i=1}^{n}|\lambda_i| \\ &= \frac{2}{r}\mathcal{E}(G). \end{aligned}$$

□

Definition. Let G' be the double cover of graph G; then define $G'' = (G')'$, and in general $G^{k'} = G^{(k-1')'}$, for $k \geq 1$ called the the k-th iterated double cover of graph G.

Theorem 30.4

Let G be an r-regular graph with $\lambda_1, \lambda_2, \ldots, \lambda_n$ eigenvalues then,

$$RE(G'') = \frac{4}{r}\left[\mathcal{E}(G)\right].$$

■

Proof. Let G be an r-regular graph on n vertices; then construct a second iterated double cover graph G'' of graph G by constructing a double cover of graph G'.
Then the Randić matrix of the second iterated double cover graph G'' of graph G can be written as

$$R(G'') = \begin{bmatrix} 0 & R(G') \\ R(G') & 0 \end{bmatrix},$$

where $R(G')$ is the Randić matrix of double cover G' of graph G.

So, from Proposition 30.1.1 the spectra of $R(G'')$ can be given by

$$spec(R(G'')) = spec\left(\frac{2}{r}A(G)\right) \cup spec\left(\frac{-2}{r}A(G)\right)$$

Therefore,

$$spec(R(G')) = \begin{pmatrix} \frac{2}{r}\lambda_1 & \frac{2}{r}\lambda_2 & \cdots & \frac{2}{r}\lambda_n & \frac{-2}{r}\lambda_1 & \frac{-2}{r}\lambda_2 & \cdots & \frac{-2}{r}\lambda_n \\ 1 & 1 & \cdots & 1 & 1 & 1 & \cdots & 1 \end{pmatrix}$$

Hence,

$$\begin{aligned} RE(G'') &= \sum_{i=1}^{n}\left|\frac{2}{r}\lambda_i\right| + \sum_{i=1}^{n}\left|\frac{-2}{r}\lambda_i\right| \\ &= \frac{2}{r}\sum_{i=1}^{n}|\lambda_i| + \frac{2}{r}\sum_{i=1}^{n}|\lambda_i| \\ &= \frac{4}{r}\sum_{i=1}^{n}|\lambda_i| \\ &= \frac{4}{r}\mathcal{E}(G). \end{aligned}$$

□

Theorem 30.5

Let G be an r-regular graph with $\lambda_1, \lambda_2, \ldots, \lambda_n$ eigenvalues then,

$$RE(G^{k'}) = \frac{2k}{r}\left[\mathcal{E}(G)\right].$$

■

Proof. Let G be an r-regular graph on n vertices; then in general construct a kth iterated double cover graph $G^{k'}$ of graph G by constructing a double cover of graph $G^{(k-1')'}$.

Then the Randić matrix of kth iterated double cover graph $G^{k'}$ of graph G can be written as

$$R(G^{k'}) = \begin{bmatrix} 0 & R(G^{(k-1')'}) \\ R(G^{(k-1')'}) & 0 \end{bmatrix},$$

where $R(G^{(k-1')'})$ is the Randić matrix of $(k-1)^{th}$ iterated double cover $G^{(k-1')'}$ of graph G.

So, from Proposition 30.1.1 the spectra of $R(G^{k'})$ can be given by

$$spec(R(G^{k'})) = spec\left(R(G^{(k-1')'})\right) \cup spec\left(-R(G^{(k-1')'})\right)$$

Therefore,

$$spec(R(G^{k'})) = spec\left(\frac{k}{r}A(G)\right) \cup spec\left(\frac{-k}{r}A(G)\right)$$

Therefore,

$$spec(R(G^{k'})) = \begin{pmatrix} \frac{k}{r}\lambda_1 & \frac{k}{r}\lambda_2 & \cdots & \frac{k}{r}\lambda_n & \frac{-k}{r}\lambda_1 & \frac{-k}{r}\lambda_2 & \cdots & \frac{-k}{r}\lambda_n \\ 1 & 1 & \cdots & 1 & 1 & 1 & \cdots & 1 \end{pmatrix}$$

Hence,

$$\begin{aligned} RE(G^{k'}) &= \sum_{i=1}^{n}\left|\frac{k}{r}\lambda_i\right| + \sum_{i=1}^{n}\left|\frac{-k}{r}\lambda_i\right| \\ &= \frac{k}{r}\sum_{i=1}^{n}|\lambda_i| + \frac{k}{r}\sum_{i=1}^{n}|\lambda_i| \\ &= \frac{2k}{r}\sum_{i=1}^{n}|\lambda_i| \\ &= \frac{2k}{r}\mathcal{E}(G). \end{aligned}$$

□

30.5 RANDIĆ ENERGY OF EXTENDED DOUBLE COVER OF GRAPH

Definition. [5] The extended double cover of graph G with vertex set $V(G) = \{v_1, v_2, \ldots, v_n\}$ is a bipartite graph G^* with bipartition $(X,Y), X = \{x_1, x_2, \ldots, x_n\}$ and $Y = \{y_1, y_2, \ldots, y_n\}$, where two vertices x_i and y_j are adjacent if and only if $i = j$ or v_i is adjacent to v_j in G.

In Figure 30.4 the extended double cover graph of C_4 is shown.

Theorem 30.6

Let G be an r-regular graph and $\lambda_1, \lambda_2, \ldots, \lambda_n$ are eigenvalues of G with $|\lambda_i| \geqslant 1$,for all $1 \leqslant i \leqslant n$ then,

$$RE(G^*) = \frac{2}{r+1}\left[\mathcal{E}(G) + \theta\right].$$

■

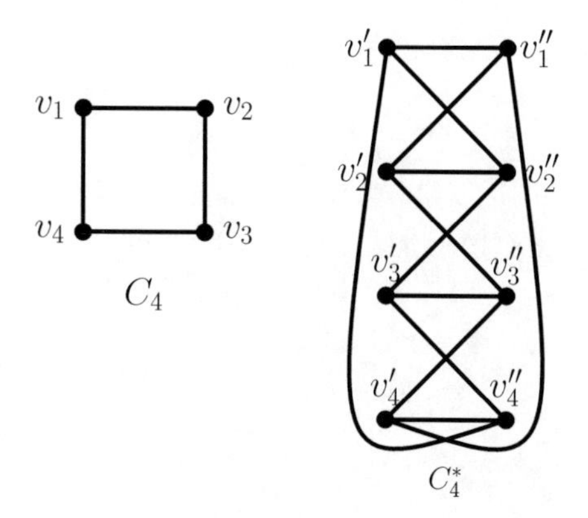

Figure 30.4: Cycle C_4 and its extended double cover graph

Proof. Let G be an r-regular graph on n vertices; then construct an extended double cover graph G^* of graph G by considering bipartition of vertex set (X,Y), where $X=\{x_1,x_2,\ldots,x_n\}$ and $Y=\{y_1,y_2,\ldots,y_n\}$ with two vertices x_i and y_j adjacent if and only if $i=j$ or v_i is adjacent to v_j in G.
Then the Randić matrix of extended double cover G^* of graph G can be written as

$$R(G^*)=\begin{bmatrix} 0 & \frac{1}{r+1}[A(G)+I]) \\ \frac{1}{r+1}[A(G)+I] & 0 \end{bmatrix},$$

where $A(G)$ is the adjacency matrix of graph G.
So, from Proposition 30.1.1 the spectra of $R(G^*)$ can be given by

$$spec(R(G^*))=spec\left(\frac{1}{r+1}[A(G)+I]\right)\cup spec\left(\frac{-1}{r+1}[A(G)+I]\right)$$

$$\text{Therefore, } spec(R(G^*))=\begin{pmatrix} \frac{1}{r+1}(\lambda_1+1) & \cdots & \frac{1}{r+1}(\lambda_n+1) \\ 1 & \cdots & 1 \\ \\ \frac{-1}{r+1}(\lambda_1+1) & \cdots & \frac{-1}{r+1}(\lambda_n+1) \\ \\ 1 & \cdots & 1 \end{pmatrix}$$

Hence,

$$\begin{aligned} RE(G^*) &= \sum_{i=1}^{n}\left|\frac{1}{r+1}(\lambda_i+1)\right|+\sum_{i=1}^{n}\left|\frac{-1}{r+1}(\lambda_i+1)\right| \\ &= \frac{1}{r+1}\sum_{i=1}^{n}|\lambda_i+1|+\frac{1}{r+1}\sum_{i=1}^{n}|\lambda_i+1| \end{aligned}$$

$$= \frac{2}{r+1}\sum_{i=1}^{n}|\lambda_i+1|$$

Also, suppose that $|\lambda_i| \geq 1$ for all $1 \leq i \leq n$ then

$$\begin{aligned}|\lambda_i+1| &= |\lambda_i|+1, \ if\ \lambda_i \geq 0\\ &= |\lambda_i|-1, \ if\ \lambda_i < 0\end{aligned}$$

$$\begin{aligned}\text{Therefore, } RE(G^*) &= \frac{2}{r+1}\left[\sum_{\lambda_i>0}|\lambda_i+1| + \sum_{\lambda_i<0}|\lambda_i+1|\right]\\ &= \frac{2}{r+1}\left[\sum_{\lambda_i>0}(|\lambda_i|+1) + \sum_{\lambda_i<0}(|\lambda_i|-1)\right]\\ &= \frac{2}{r+1}\left[\left(\sum_{\lambda_i>0}|\lambda_i| + \sum_{\lambda_i<0}|\lambda_i|\right) + \left(\sum_{\lambda_i>0}1 - \sum_{\lambda_i<0}1\right)\right]\\ &= \frac{2}{r+1}[\mathcal{E}(G)+\theta],\end{aligned}$$

where θ is the difference between the number of positive and negative eigenvalues of graph G. □

Theorem 30.7

Let G be an r-regular graph and $\lambda_1, \lambda_2, \ldots, \lambda_n$ be eigenvalues of G with $|\lambda_i| \geqslant 1$,for all $1 \leqslant i \leqslant n$ then, $RE(G \times P_2) = \frac{2}{r+1}\mathcal{E}(G)$. ■

Proof. Let G be an $r-$regular graph of order n and take the Cartesian product of graph G with path on 2 vertices P_2; then the Randić matrix of graph $G \times P_2$ can be written as

$$R(G \times P_2) = \begin{bmatrix} \frac{1}{r+1}A(G) & \frac{1}{r+1}I_n \\ \frac{1}{r+1}I_n & \frac{1}{r+1}A(G) \end{bmatrix},$$

where $A(G)$ is the adjacency matrix of graph G.
So, from Proposition 30.1.1 the spectra of $R(G \times P_2)$ can be given by

$$spec(R(G \times P_2)) = spec\left(\frac{1}{r+1}[A(G)+I_n]\right) \cup spec\left(\frac{1}{r+1}[A(G)-I_n]\right)$$

$$\text{Therefore, } spec(R(G\times P_2)) = \begin{pmatrix} \frac{1}{r+1}(\lambda_1+1) & \cdots & \frac{1}{r+1}(\lambda_n+1) \\ 1 & \cdots & 1 \\ \frac{1}{r+1}(\lambda_1-1) & \cdots & \frac{1}{r+1}(\lambda_n-1) \\ 1 & \cdots & 1 \end{pmatrix}$$

Also, suppose that $|\lambda_i| \geq 1$ for all $1 \leq i \leq n$ then

$$\begin{aligned} |\lambda_i+1| &= |\lambda_i|+1, \ if\ \lambda_i > 0 \\ &= |\lambda_i|-1, \ if\ \lambda_i < 0 \end{aligned}$$

and

$$\begin{aligned} |\lambda_i-1| &= |\lambda_i|-1, \ if\ \lambda_i > 0 \\ &= |\lambda_i|+1, \ if\ \lambda_i < 0 \end{aligned}$$

Hence,

$$\begin{aligned} RE(G\times P_2) &= \sum_{i=1}^{n}\left|\tfrac{1}{r+1}(\lambda_i+1)\right| + \sum_{i=1}^{n}\left|\tfrac{1}{r+1}(\lambda_i-1)\right| \\ &= \tfrac{1}{r+1}\left[\sum_{i=1}^{n}|\lambda_i+1| + \sum_{i=1}^{n}|\lambda_i-1|\right] \\ &= \tfrac{1}{r+1}\left[\sum_{\lambda_i>0}|\lambda_i+1| + \sum_{\lambda_i<0}|\lambda_i+1| + \sum_{\lambda_i>0}|\lambda_i-1| + \sum_{\lambda_i<0}|\lambda_i-1|\right] \\ &= \tfrac{1}{r+1}\left[\sum_{\lambda_i>0}(|\lambda_i|+1) + \sum_{\lambda_i<0}(|\lambda_i|-1) + \sum_{\lambda_i>0}(|\lambda_i|-1) + \sum_{\lambda_i<0}(|\lambda_i|+1)\right] \\ &= \tfrac{1}{r+1}\left[2\sum_{\lambda_i>0}|\lambda_i| + 2\sum_{\lambda_i<0}|\lambda_i| + \sum_{\lambda_i>0}(1-1) + \sum_{\lambda_i<0}(1-1)\right] \\ &= \tfrac{2}{r+1}\left[\sum_{\lambda_i>0}|\lambda_i| + \sum_{\lambda_i<0}|\lambda_i|\right] \\ &= \tfrac{2}{r+1}\mathcal{E}(G). \end{aligned}$$

□

Corollary 30.5.1. *For an r-regular graph G, $RE(G\times P_2) = RE(G^*)$ if and only if G is a bipartite graph.*

Proof. Let G be an $r-$regular graph and suppose that $RE(G\times P_2) = RE(G^*)$, then we have

$$\frac{2}{r+1}\mathcal{E}(G) = \frac{2}{r+1}[\mathcal{E}(G)+\theta] \Rightarrow \theta = 0$$

Also, it is well known that graph G is a bipartite graph if and only if its spectrum is symmetric to 0.That is, the difference between number of positive and negative is 0. Hence, graph G is bipartite.

Conversely, suppose that G is an r-regular bipartite graph; then all their eigenvalues are symmetric about 0. So, the difference between number of positive and negative eigenvalues is 0. Therefore, $\theta = 0$.
Hence, $RE(G\times P_2) = RE(G^*)$ □

Definition. [5] Let G^* be the extended double cover of the graph G, then define $G^{**} = (G^*)^*$, called the 2-nd extended iterated double cover of graph G.

Theorem 30.8

Let G be an r-regular graph and $\lambda_1, \lambda_2, \ldots, \lambda_n$ be eigenvalues of G with $|\lambda_i| \geqslant 2$, for all $1 \leqslant i \leqslant n$ then,

$$RE(G^{**}) - \frac{4}{r+2}\left[\mathcal{E}(G) + \theta\right].$$

■

Proof. Let G be an $r-$regular graph on n vertices and construct second extended iterated double cover of graph G by taking extended double cover of graph G^*.
Then the Randić matrix of graph G^{**} can be written as,

$$R(G^{**}) = \begin{bmatrix} 0 & \frac{1}{r+2}[A(G^*) + I_{2n}]) \\ \frac{1}{r+2}[A(G^*) + I_{2n}] & 0 \end{bmatrix},$$

where $A(G^*)$ is the adjacency matrix of graph G^*.
So, from Proposition 30.1.1 the spectra of $R(G^{**})$ can be given by

$$spec(R(G^{**})) = spec\left(\frac{1}{r+2}[A(G^*) + I_{2n}]\right) \cup spec\left(\frac{-1}{r+2}[A(G^*) + I_{2n}]\right)$$

Therefore, $spec(R(G^{**})) = \begin{pmatrix} \frac{1}{r+2}[(\lambda_1+1)+1] & \cdots & \frac{1}{r+2}[(\lambda_n+1)+1] \\ 1 & \cdots & 1 \\ \frac{-1}{r+2}[-(\lambda_1+1)+1] & \cdots & \frac{-1}{r+2}[-(\lambda_n+1)+1] \\ 1 & \cdots & 1 \\ \frac{-1}{r+2}[(\lambda_1+1)+1] & \cdots & \frac{-1}{r+2}[(\lambda_n+1)+1] \\ 1 & \cdots & 1 \\ \frac{1}{r+2}[-(\lambda_1+1)+1] & \cdots & \frac{1}{r+2}[-(\lambda_n+1)+1] \\ 1 & \cdots & 1 \end{pmatrix}$

Therefore,

$$spec(R(G^{**})) = \begin{pmatrix} \frac{1}{r+2}[\lambda_1+2] & \cdots & \frac{1}{r+2}[\lambda_n+2] & \frac{1}{r+2}[\lambda_1] & \cdots & \frac{1}{r+2}[\lambda_n] \\ 1 & \cdots & 1 & 1 & \cdots & 1 \\ \frac{-1}{r+2}[\lambda_1+2] & \cdots & \frac{-1}{r+2}[\lambda_n+2] & \frac{1}{r+2}[-\lambda_1] & \cdots & \frac{1}{r+2}[-\lambda_n] \\ 1 & \cdots & 1 & 1 & \cdots & 1 \end{pmatrix}$$

Also, suppose that $|\lambda_i| \geq 2$ for all $1 \leq i \leq n$ then

$$\begin{aligned} |\lambda_i + 1| &= |\lambda_i| + 2, \ if\ \lambda_i > 0 \\ &= |\lambda_i| - 2, \ if\ \lambda_i < 0 \end{aligned}$$

Hence,

$$\begin{aligned} RE(G^{**}) &= \sum_{i=1}^{n} \left|\frac{1}{r+2}[\lambda_i+2]\right| + \sum_{i=1}^{n} \left|\frac{1}{r+2}\lambda_i\right| + \sum_{i=1}^{n} \left|\frac{-1}{r+2}[\lambda_n+2]\right| + \sum_{i=1}^{n} \left|\frac{-1}{r+2}[-\lambda_i]\right| \\ &= \frac{1}{r+2}\left[\sum_{i=1}^{n}|\lambda_i+2| + \sum_{i=1}^{n}|\lambda_i| + \sum_{i=1}^{n}|\lambda_i+2| + \sum_{i=1}^{n}|\lambda_i|\right] \\ &= \frac{2}{r+2}\left[\sum_{i=1}^{n}|\lambda_i+2| + \sum_{i=1}^{n}|\lambda_i|\right] \end{aligned}$$

Therefore,

$$\begin{aligned} RE(G^{**}) &= \frac{2}{r+2}\left[\sum_{\lambda_i>0}|\lambda_i+2| + \sum_{\lambda_i<0}|\lambda_i+2| + \sum_{i=1}^{n}|\lambda_i|\right] \\ &= \frac{2}{r+2}\left[\sum_{\lambda_i>0}(|\lambda_i|+2) + \sum_{\lambda_i<0}(|\lambda_i|-2) + \sum_{i=1}^{n}|\lambda_i|\right] \\ &= \frac{2}{r+2}\left[\left(\sum_{\lambda_i>0}|\lambda_i| + \sum_{\lambda_i<0}|\lambda_i|\right) + 2\left(\sum_{\lambda_i>0}1 - \sum_{\lambda_i<0}1\right) + \sum_{i=1}^{n}|\lambda_i|\right] \\ &= \frac{2}{r+2}[2E(G)+2\theta] \\ &= \frac{4}{r+2}[E(G)+\theta], \end{aligned}$$

where θ is the difference between the number of positive and negative eigenvalues of graph G. □

Theorem 30.9

Let G be an r-regular graph and $\lambda_1, \lambda_2, \ldots, \lambda_n$ be eigenvalues of G with $|\lambda_i| \geqslant 1$,for all $1 \leqslant i \leqslant n$ then,

$$RE((G \times P_2)^*) = \frac{4}{r+2} E(G). \qquad \blacksquare$$

Proof. Let G be an $r-$regular graph of order n; then construct a graph $(G \times P_2)^*$ by taking first, the Cartesian product of graphs G and P_2 and then taking the extended double cover of graph $G \times P_2$. So, the Randić matrix of graph $(G \times P_2)^*$ can be written as

$$R((G \times P_2)^*) = \begin{bmatrix} 0 & \frac{1}{r+2}[A(G \times P_2) + I]) \\ \frac{1}{r+2}[A(G \times P_2) + I] & 0 \end{bmatrix},$$

where $A(G \times P_2)$ is the adjacency matrix of graph $G \times P_2$.
So, from Proposition 30.1.1 the spectra of $R((G \times P_2)^*)$ can be given by

$$spec(R((G \times P_2)^*)) = spec\left(\frac{1}{r+2}[A(G \times P_2) + I]\right) \cup spec\left(\frac{-1}{r+2}[A(G \times P_2) + I]\right)$$

$$\text{Therefore, } spec(R((G \times P_2)^*)) = \begin{pmatrix} \frac{1}{r+2}(\lambda_1 + 1) & \cdots & \frac{1}{r+2}(\lambda_n + 1) \\ 1 & \cdots & 1 \\ \frac{1}{r+2}(\lambda_1 - 1) & \cdots & \frac{1}{r+2}(\lambda_n - 1) \\ 1 & \cdots & 1 \\ \frac{-1}{r+2}(\lambda_1 + 1) & \cdots & \frac{-1}{r+2}(\lambda_n + 1) \\ 1 & \cdots & 1 \\ \frac{-1}{r+2}(\lambda_1 - 1) & \cdots & \frac{-1}{r+2}(\lambda_n - 1) \\ 1 & \cdots & 1 \end{pmatrix}$$

Also, suppose that $|\lambda_i| \geq 1$ for all $1 \leq i \leq n$ then

$$\begin{aligned} |\lambda_i + 1| &= |\lambda_i| + 1, \ if\ \lambda_i \geq 0 \\ &= |\lambda_i| - 1, \ if\ \lambda_i < 0 \end{aligned}$$

and

$$\begin{aligned} |\lambda_i - 1| &= |\lambda_i| - 1, \ if\ \lambda_i > 0 \\ &= |\lambda_i| + 1, \ if\ \lambda_i < 0 \end{aligned}$$

Hence,

$$
\begin{aligned}
&RE((G\times P_2)^*)\\
&=\sum_{i=1}^{n}\left|\tfrac{1}{r+2}(\lambda_i+1)\right|+\sum_{i=1}^{n}\left|\tfrac{1}{r+2}(\lambda_i-1)\right|+\sum_{i=1}^{n}\left|\tfrac{-1}{r+2}(\lambda_i+1)\right|\\
&\quad+\sum_{i=1}^{n}\left|\tfrac{-1}{r+2}(\lambda_i-1)\right|\\
&=\tfrac{2}{r+2}\sum_{i=1}^{n}|\lambda_i+1|+\tfrac{2}{r+2}\sum_{i=1}^{n}|\lambda_i-1|\\
&=\tfrac{2}{r+2}\left[\sum_{i=1}^{n}|\lambda_i+1|+\sum_{i=1}^{n}|\lambda_i-1|\right]\\
&=\tfrac{2}{r+2}\left[\sum_{\lambda_i>0}|\lambda_i+1|+\sum_{\lambda_i<0}|\lambda_i+1|+\sum_{\lambda_i>0}|\lambda_i-1|+\sum_{\lambda_i<0}|\lambda_i-1|\right]\\
&=\tfrac{2}{r+2}\left[\sum_{\lambda_i>0}(|\lambda_i|+1)+\sum_{\lambda_i<0}(|\lambda_i|-1)+\sum_{\lambda_i>0}(|\lambda_i|-1)+\sum_{\lambda_i<0}(|\lambda_i|+1)\right]\\
&=\tfrac{2}{r+2}\left[2\sum_{\lambda_i>0}|\lambda_i|+2\sum_{\lambda_i<0}|\lambda_i|+\sum_{\lambda_i>0}(1-1)+\sum_{\lambda_i<0}(1-1)\right]\\
&=\tfrac{4}{r+2}\left[\sum_{\lambda_i>0}|\lambda_i|+\sum_{\lambda_i<0}|\lambda_i|\right]\\
&=\tfrac{4}{r+2}E(G).
\end{aligned}
$$

□

Corollary 30.5.2. *For an r-regular graph G, $RE((G\times P_2)^*)=RE(G^{**})$ if and only if G is a bipartite graph.*

Proof. Let G be an $r-$regular graph and suppose that $RE((G\times P_2)^*)=RE(G^{**})$, then we have

$$\frac{4}{r+2}E(G)=\frac{4}{r+2}[E(G)+\theta]\Rightarrow\theta=0$$

Also, it is well known that graph G is a bipartite graph if and only if its spectrum is symmetric to 0.That is, the difference between number of positive and negative is 0. Hence, graph G is bipartite.

Conversely, suppose that G is an r-regular bipartite graph; then all their eigenvalues are symmetric about 0, so the difference between number of positive and negative eigenvalues is 0. Therefore, $\theta=0$.
Hence, $RE((G\times P_2)^*)=RE(G^{**})$ □

30.6 CONCLUDING REMARKS

The concept of Randić energy is one of the concepts which is a frontier between graph theory and chemistry. We have investigated the Randić energy of larger graphs obtained from standard graphs by means of various graph operations.

REFERENCES

1. Saeid Alikhani and Nima Ghanbari. Randić energy of specific graphs. 10 Nov., 2014, arXiv:1411.2544v1 [math.CO].

2. R. Balakrishnan. The energy of graph. *Linear Algebra Appl.*, 387, 287-295, 2004.
3. Ş. B. Bozkurt, A. D. G̈ungör, I. Gutman and A. S. Çevik, Randić matrix and Randić energy. *MATCH Math. Comput. Chem.*, 64, 239-250, 2010.
4. Ş. B. Bozkurt, A. D. G̈ungör, I. Gutman. A. S. Çevik. Randić spectral radius and Randić energy. *MATCH Math. Comput. Chem.*, 64, 321-334, 2010.
5. Z. Chen. Spectra of extended double cover graphs. *Czechoslovak Math. J.*, 54(4): 1077-1082, 2004.
6. D. Cvetković, P. Rowlinson and S. Simić. *An Introduction to the Theory of Graph Spectra.* 1st ed., Cambridge University Press, 2010.
7. P. J. Davis, *Circulant Matrices*, Wiley, New York, 1978.
8. I. Gutman. The energy of graphs. *Ber. Math. Statit. Sekt. Forshugsz. Graz*, 103, 1-22, 1978.
9. I. Gutman, B. Furtula and Ş. B. Bozkurt. On Randić energy. *Lin. Algebra Appl.*, 442, 50-57, 2014.
10. F. Harary. *Graph Theory.* Addison Wesley, Massachusetts, 1972.
11. R. A. Horn and C. R. Johnson. *Topics In Matrix Analysis.* Cambridge University Press, Cambridge, 1991.
12. X. Li, Y. Shi and I. Gutman. *Graph Energy.* 1st ed., Springer, New York, 2010.
13. M. Randić. On characterization of molecular branching. *J. Am. Chem. Soc.*, 97, 6609-6615, 1975.
14. M. Randić. The connectivity index 25 years after. *J. Mol. Graph. Model.*, 20, 19-35, 2001.
15. M. Randić. On history of Randić index and emerging hostility toward chemical graph theory. *MATCH Commun. Math. Comput. Chem.*, 59, 5-124, 2008.
16. M. Randić, M. Novič and D. Plavšić. *Solved and Unsolved Problems in Structural Chemistry.* CRC Press, Boca Raton, 2016.
17. O. Rojo and L. Medina. Construction of bipartite graphs having the same Randić energy. *MATCH Commun. Math. Comput. Chem.*, 68, 805-814, 2012.

31 L-Spectra of Graphs Obtained by Duplicating Graphs Elements

S. K. Vaidya
Saurashtra University,
Rajkot, Gujarat (INDIA)
Email: samirkvaidya@yahoo.co.in

K. M. Popat
Atmiya University,
Rajkot, Gujarat (INDIA)
Email: kalpeshmpopat@gmail.com

The concept of L-energy is a concept that stems from energy of graphs. The energy of a graph G is defined with the help of adjacency matrix $A(G)$. The Laplacian matrix of G is defined as $L(G) = D(G) - A(G)$, where $D(G)$ is the diagonal matrix with $(ii)^{th}$ entry as a degree of vertex v_i. The collection of eigenvalues of $L(G)$ together with their multiplicities is known as L-spectra of G. We have investigated L-spectra for the graphs obtained by means of duplication of graph elements.

31.1 INTRODUCTION

We begin with simple, finite and undirected graphs. For standard terminology and notations related to graph theory we follow Balakrishnan and Ranganathan [5] while for algebra we follow Lang [6]. The adjacency matrix $A(G)$ of a graph G with vertices $v_1, v_2, \cdots, v_n$ is an $n \times n$ matrix $A(G) = [a_{ij}]$ such that,

$$\begin{aligned} a_{ij} &= 1, \text{ if } v_i \text{ is adjacent with } v_j \\ &= 0, \text{ otherwise} \end{aligned}$$

The spectra of $A(G)$ is called the spectra of graph G. If $\lambda_1, \lambda_2, \cdots, \lambda_n$ are eigenvalues of graph G then the *energy of graph* G is $E(G) = \sum_{i=1}^{n} |\lambda_i|$. The concept of graph energy was introduced by Gutman [1] in 1978. A brief account of graph energy can be found in Cvetkovič [2] and Li [9].

Consider $D(G)$ to be the diagonal matrix of vertex degrees of G. The matrix $L(G) = D(G) - A(G)$ is called the Laplacian matrix of G [7, 8]. The characteristic polynomial of $L(G)$ is denoted by $\phi(L(G), x)$. Roots of $\phi(L(G), x)$

are known as Laplacian eigenvalues of G. Clearly $L(G)$ is a symmetric matrix and therefore all of its eigenvalues are real and non-negative as well as the sum of entries in every row is zero and hence zero is always an eigenvalue of $L(G)$. Laplacian eigenvalues are denoted by $\mu_1, \mu_2, \cdots, \mu_n = 0$. The Laplacian eigenvalue together with their multiplicity forms the Laplacian spectrum(L-spectra) of graph G. For convenience we represent the L-spectra of graph by writing each Laplacian eigenvalue together with multiplicity as its power.

Fiedler [4] proved that $\mu_n = 0$ with multiplicity equal to the number of connected components of G. It is easy to see that

$$tr(L(G)) = \sum_{i=1}^{n} \mu_i = 2m \qquad tr(L^+(G)) = \sum_{i=1}^{n} \mu_i^+ = 2m$$

where tr is the trace of the matrix.

Let G be a graph with n vertices, e edges, and no self loops. The incidence matrix $B(G)$ of graph G is an $n \times e$ matrix $[b_{ij}]$, such that $b_{ij} - 1$, if j^{th} edge e_j is incident on i^{th} vertex v_i, and 0 otherwise. It is well known that if $B(G)$ is the incidence matrix of graph G then $B(G)B(G)^T = D(G)$.

Let $A \in R^{m \times n}$, $B \in R^{p \times q}$ then the *Kronecker product* (or tensor product) of A and B is defined as the matrix

$$A \otimes B = \begin{bmatrix} a_{11}B & \cdots & a_{1n}B \\ \vdots & \ddots & \vdots \\ a_{m1}B & \cdots & a_{mn}B \end{bmatrix}$$

Proposition 31.1.1. *[12, 13] Let M,N,P,Q be matrices, and let M be invertible. Let*

$$S = \begin{bmatrix} M & N \\ P & Q \end{bmatrix}$$

Then, $det\ S = det\ Q\ .\ det\ \left[M - NQ^{-1}P\right]$

31.2 L-SPECTRA OF GRAPH OBTAINED BY DUPLICATING EACH VERTEX BY AN EDGE

Definition. Duplication of a vertex v_k by a new edge $e = v'v''$ in a graph G produces a new graph G_1 such that $N(v') = \{v_k, v''\}$ and $N(v'') = \{v_k, v'\}$.

Theorem 31.1

Let G be a k-regular graph with eigenvalues $\lambda_1, \lambda_2, \cdots, \lambda_n$ and G_1 be the graph obtained from G by duplicating each vertex by a new edge then the

L-spectra of G_1 is

$$3^n, \left(\frac{(k-\lambda_i+3)+\sqrt{k^2+\lambda_i^2-2k\lambda_i+2k-2\lambda_i+9}}{2}\right)^1 \text{ and}$$

$$\left(\frac{(k-\lambda_i+3)-\sqrt{k^2+\lambda_i^2-2k\lambda_i+2k-2\lambda_i+9}}{2}\right)^1,$$

for $i = 1, 2, 3\cdots, n$ ■

Proof. Let $v_1, v_2, \cdots, v_n$ be the vertices of a graph G; then the adjacency matrix $A(G)$ is given by

$$A(G) = \begin{array}{c|ccccc} & v_1 & v_2 & v_3 & \cdots & v_n \\ \hline v_1 & 0 & a_{12} & a_{13} & \cdots & a_{1n} \\ v_2 & a_{21} & 0 & a_{23} & \cdots & a_{2n} \\ v_3 & a_{31} & a_{32} & 0 & \cdots & a_{3n} \\ \vdots & \vdots & \vdots & \vdots & \ddots & \vdots \\ v_n & a_{n1} & a_{n2} & a_{n3} & \cdots & 0 \end{array}$$

We duplicate vertices $v_1, v_2, \cdots, v_n$ by edges $e_1, e_2, \cdots, e_n$ respectively such that, $e_1 = v_1'v_1''$, $e_2 = v_2'v_2''$, $\cdots$, $e_n = v_n'v_n''$ to obtain graph G_1. $L(G_1)$ is given in terms of the block matrix as follow

$$L(G_1) = \begin{array}{c|cccc|ccccccc} & v_1 & v_2 & \cdots & v_n & v_1' & v_1'' & v_2' & v_2'' & \cdots & v_n' & v_n'' \\ \hline v_1 & k+2 & -a_{12} & \cdots & -a_{1n} & -1 & -1 & 0 & 0 & \cdots & 0 & 0 \\ v_2 & -a_{21} & k+2 & \cdots & -a_{2n} & 0 & 0 & -1 & -1 & \cdots & 0 & 0 \\ \vdots & \vdots & \vdots & \ddots & \vdots & \vdots & \vdots & \vdots & \vdots & \vdots & \vdots & \vdots \\ v_n & -a_{n1} & -a_{n2} & \cdots & k+2 & 0 & 0 & 0 & 0 & \cdots & -1 & -1 \\ \hline v_1' & -1 & 0 & \cdots & 0 & 2 & -1 & 0 & 0 & \cdots & 0 & 0 \\ v_1'' & -1 & 0 & \cdots & 0 & -1 & 2 & 0 & 0 & \cdots & 0 & 0 \\ v_2' & 0 & -1 & \cdots & 0 & 0 & 0 & 2 & -1 & \cdots & 0 & 0 \\ v_2'' & 0 & -1 & \cdots & 0 & 0 & 0 & -1 & 2 & \cdots & 0 & 0 \\ \vdots & \vdots & \vdots & \ddots & \vdots & \vdots & \vdots & \vdots & \vdots & \vdots & \vdots & \vdots \\ v_n' & 0 & 0 & \cdots & -1 & 0 & 0 & 0 & 0 & \cdots & 2 & -1 \\ v_n'' & 0 & 0 & \cdots & -1 & 0 & 0 & 0 & 0 & \cdots & -1 & 2 \end{array}$$

$$\text{Let } B = \begin{bmatrix} -1 & -1 & 0 & 0 & \cdots & 0 & 0 \\ 0 & 0 & -1 & -1 & \cdots & 0 & 0 \\ \vdots & \vdots & \vdots & \vdots & \vdots & \vdots & \vdots \\ 0 & 0 & 0 & 0 & \cdots & -1 & -1 \end{bmatrix}$$

Then,

$$L(G_1) = \begin{bmatrix} -A(G)+(k+2)I & B \\ B^T & I_n \otimes (2I - A(K_2)) \end{bmatrix}$$

The characteristic polynomial of the above matrix is given by

$$\begin{aligned}
\phi(L(G_1):x) &= |xI - L(G_1)| \\
&= \begin{vmatrix} (x-(k+2))I + A(G) & -B \\ -B^T & I_n \otimes ((x-2)I + A(K_2)) \end{vmatrix} \\
&= |I_n \otimes ((x-2)I + A(K_2))| \\
&\quad |(x-(k+2)I + A(G) - B(I_n \otimes ((x-2)I + A(K_2)))^{-1} B^T| \\
&= ((x-2)^2 - 1)^n \\
&\quad \left|(x-(k+2))I + A(G) - B\left[I_n \otimes \left(\frac{1}{((x-2)^2-1)}((x-2)I - A(K_2))\right)\right] B^T\right| \\
&= (x^2 - 4x + 3)^n \\
&\quad \left|(x-(k+2))I + A(G) - B\left[I_n \otimes \left(\frac{1}{((x^2-4x+3)}((x-2)I - A(K_2))\right)\right] B^T\right| \\
&= |(x^2-4x+3)((x-(k+2)I + A(G)) - B(I_n \otimes ((x-2)I - A(K_2)))B^T|
\end{aligned}$$

Now, $B(I_n \otimes ((x-2)I - A(K_2))B^T$

$$= \begin{bmatrix} -1 & -1 & 0 & 0 & \cdots & 0 & 0 \\ 0 & 0 & -1 & -1 & \cdots & 0 & 0 \\ \vdots & \vdots & \vdots & \vdots & \cdots & \vdots & \vdots \\ 0 & 0 & 0 & 0 & \cdots & -1 & -1 \end{bmatrix}$$

$$\begin{bmatrix} x-2 & -1 & 0 & 0 & \cdots & 0 & 0 \\ -1 & x-2 & 0 & 0 & \cdots & 0 & 0 \\ 0 & 0 & x-2 & -1 & \cdots & 0 & 0 \\ 0 & 0 & -1 & x-2 & \cdots & 0 & 0 \\ \vdots & \vdots & \vdots & \vdots & \cdots & \vdots & \vdots \\ 0 & 0 & 0 & 0 & \cdots & x-2 & -1 \\ 0 & 0 & 0 & 0 & \cdots & -1 & x-2 \end{bmatrix} \begin{bmatrix} -1 & 0 & \cdots & 0 \\ -1 & 0 & \cdots & 0 \\ 0 & -1 & \cdots & 0 \\ 0 & -1 & \cdots & 0 \\ \vdots & \vdots & \vdots & \vdots \\ 0 & 0 & \cdots & -1 \\ 0 & 0 & \cdots & -1 \end{bmatrix}$$

$$= \begin{bmatrix} 3-x & 3-x & 0 & 0\cdots & 0 & 0 \\ 0 & 0 & 3-x & 3-x\cdots & 0 & 0 \\ \vdots & \vdots & \vdots & \vdots\cdots & \vdots & \vdots \\ 0 & 0 & 0 & 0\cdots & 3-x & 3-x \end{bmatrix} \begin{bmatrix} -1 & 0 & \cdots & 0 \\ -1 & 0 & \cdots & 0 \\ 0 & -1 & \cdots & 0 \\ 0 & -1 & \cdots & 0 \\ \vdots & \vdots & \vdots & \vdots \\ 0 & 0 & \cdots & -1 \\ 0 & 0 & \cdots & -1 \end{bmatrix}$$

$$= \begin{bmatrix} 2x-6 & 0 & 0 & \cdots & 0 \\ 0 & 2x-6 & 0 & \cdots & 0 \\ 0 & 0 & 2x-6 & \cdots & 0 \\ \cdots & \cdots & \ddots & \cdots & 0 \\ 0 & 0 & 0 & \cdots & 2x-6 \end{bmatrix}$$

$$= (2x-6)I_n$$

Continuing the proof of the theorem

$$\begin{aligned}\phi(L(G_1):x) &= |(x^2-4x+3)((x-(k+2)I+A(G))-B(I_n\otimes((x-2)I-A(K_2)))B^T| \\ &= |(x^2-4x+3)((x-(k+2)I+A(G))-(2x-6)I_n)|\end{aligned}$$

It follows that if $\lambda_1, \lambda_2, \cdots, \lambda_n$ are eigenvalues of A then

$$\begin{aligned}\phi(L(G_1):x) &= \prod_{i=1}^{n}(x^2-4x+3)(x-k-2+\lambda_i)-(2x-6) \\ &= \prod_{i=1}^{n}(x-1)(x-3)(x-k-2+\lambda_i)-2(x-3) \\ &= (x-3)^n\prod_{i=1}^{n}x^2-x(k-\lambda_i+3)+(k-\lambda_i)\end{aligned}$$

The roots of the above characteristic polynomial are

$$x = 3(n\,\text{times}),\ x = \frac{(k-\lambda_i+3)\pm\sqrt{k^2+\lambda_i^2-2k\lambda_i+2k-2\lambda_i+9}}{2},$$

for $i = 1, 2, 3\cdots, n$
Hence, the L-spectra of G_1 is,

$$3^n, \left(\frac{(k-\lambda_i+3)+\sqrt{k^2+\lambda_i^2-2k\lambda_i+2k-2\lambda_i+9}}{2}\right)^1 \text{ and}$$

$$\left(\frac{(k-\lambda_i+3)-\sqrt{k^2+\lambda_i^2-2k\lambda_i+2k-2\lambda_i+9}}{2}\right)^1,$$

for $i = 1, 2, 3\cdots, n$ □

Illustration 31.2.1. Consider cycle C_3 and a graph (say G_1) obtained from C_3 by duplicating each vartex by an edge which is given in the following Figure 31.1. It is obvious that the $L-$ spectra of C_3 is

$$2^1, (-1)^2$$

$$L(G_1) = \begin{array}{c|ccccccccc} & v_1 & v_2 & v_3 & v_1' & v_1'' & v_2' & v_2'' & v_3' & v_3'' \\ \hline v_1 & 4 & -1 & -1 & -1 & -1 & 1 & 0 & 0 & 0 \\ v_2 & -1 & 4 & -1 & 0 & 0 & -1 & -1 & 0 & 0 \\ v_3 & -1 & -1 & 4 & 0 & 0 & 0 & 0 & -1 & -1 \\ v_1' & -1 & 0 & 0 & 2 & -1 & 0 & 0 & 0 & 0 \\ v_1'' & -1 & 0 & 0 & -1 & 2 & 0 & 0 & 0 & 0 \\ v_2' & 0 & -1 & 0 & 0 & 0 & 2 & -1 & 0 & 0 \\ v_2'' & 0 & -1 & 0 & 0 & 0 & -1 & 2 & 0 & 0 \\ v_3' & 0 & 0 & -1 & 0 & 0 & 0 & 0 & 2 & -1 \\ v_3'' & 0 & 0 & -1 & 0 & 0 & 0 & 0 & -1 & 2 \end{array}$$

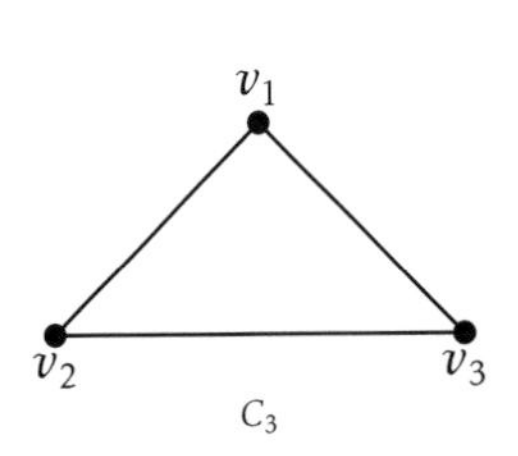

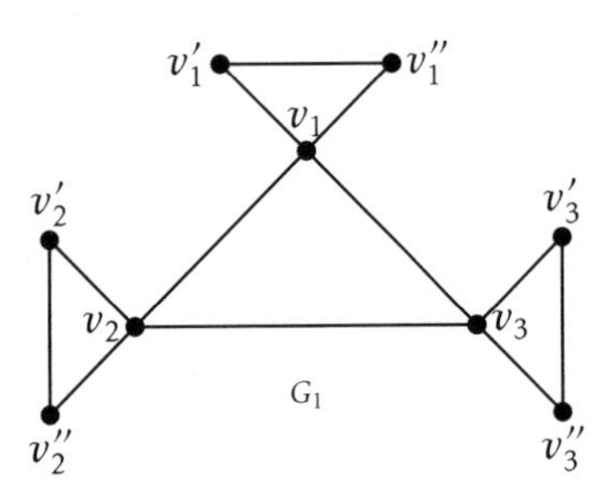

Figure 31.1: C_3 and graph obtained from C_3 by duplicating each vertex by an edge

Therefore, the L-spectra of G_1 is,

$$0^1, 3^4, (3-\sqrt{6})^2, (3+\sqrt{6})^2$$

The following table compares the spectrum of C_3 and $L-$ spectra of G_1

L-spectra of C_3	**L-spectra of G_1** $= \frac{(k-\lambda_i+3)\pm\sqrt{k^2+\lambda_i^2-2k\lambda_i+2k-2\lambda_i+9}}{2}$
$\lambda_1 = 2$	$0, 3$
$\lambda_2 = -1$	$3 \pm \sqrt{6}$

31.3 L-SPECTRA OF GRAPH OBTAINED BY DUPLICATING EACH EDGE BY A VERTEX

Definition. Duplication of an edge $e = v_i v_{i+1}$ by a vertex v' in a graph G produces a new graph G_1 such that $N(v') = \{v_i, v_{i+1}\}$.

Theorem 31.2

Let G be a k-regular graph with m edges and eigenvalues $\lambda_1, \lambda_2, \cdots, \lambda_n$. Let G_1 be the graph obtained from G by duplicating each edge by a new vertex; then the L-spectra of G_1 is

$$2^{m-n}, \left(\frac{(2k-\lambda_i+2)+\sqrt{4k^2+\lambda_i^2-4k\lambda_i+8\lambda_i-4k+4}}{2}\right)^1 \text{ and}$$

,

$$\left(\frac{(2k-\lambda_i+2)-\sqrt{4k^2+\lambda_i^2-4k\lambda_i+8\lambda_i-4k+4}}{2}\right)^1$$

■

Proof. Let $v_1, v_2, \cdots, v_n$ be the vertices and $e_1, e_2, \cdots, e_m$ be the edges of k-regular graph G; then the adjacency matrix $A(G)$ and incidence matrix $X(G)$ are given by

$$A(G) = \begin{array}{c} \\ v_1 \\ v_2 \\ v_3 \\ \vdots \\ v_n \end{array} \begin{array}{c} \begin{array}{ccccc} v_1 & v_2 & v_3 & \cdots & v_n \end{array} \\ \begin{bmatrix} 0 & a_{12} & a_{13} & \cdots & a_{1n} \\ a_{21} & 0 & a_{23} & \cdots & a_{2n} \\ a_{31} & a_{32} & 0 & \cdots & a_{3n} \\ \vdots & \vdots & \vdots & \ddots & \vdots \\ a_{n1} & a_{n2} & a_{n3} & \cdots & 0 \end{bmatrix} \end{array}$$

$$B(G) = \begin{array}{c} \\ v_1 \\ v_2 \\ v_3 \\ \vdots \\ v_n \end{array} \begin{array}{c} \begin{array}{ccccc} e_1 & e_2 & e_3 & \cdots & e_m \end{array} \\ \begin{bmatrix} b_{11} & b_{12} & b_{13} & \cdots & b_{1m} \\ b_{21} & b_{22} & b_{23} & \cdots & b_{2m} \\ b_{31} & b_{32} & b_{33} & \cdots & a_{3m} \\ \vdots & \vdots & \vdots & \ddots & \vdots \\ b_{n1} & b_{n2} & b_{n3} & \cdots & b_{nm} \end{bmatrix} \end{array}$$

We duplicate edges $e_1, e_2, \cdots, e_m$ by vertices $e'_1, e'_2, \cdots, e'_m$ respectively to obtain a graph G_1. $L(G_1)$ is given in terms of the block matrix as follows

$$L(G_1) = \begin{array}{c} \\ v_1 \\ v_2 \\ \vdots \\ v_n \\ e'_1 \\ e'_2 \\ e'_3 \\ \vdots \\ e'_m \end{array} \begin{array}{c} \begin{array}{cccccccccc} v_1 & v_2 & \cdots & v_n & e'_1 & e'_2 & e'_3 & \cdots & e'_m \end{array} \\ \left[\begin{array}{cccc|ccccc} 2k & -a_{12} & \cdots & -a_{1n} & -b_{11} & -b_{12} & -b_{13} & \cdots & -b_{1m} \\ -a_{21} & 2k & \cdots & -a_{2n} & -b_{21} & -b_{22} & -b_{23} & \cdots & -b_{2m} \\ \vdots & \vdots & \ddots & \vdots & \vdots & \vdots & \vdots & \vdots & \vdots \\ -a_{n1} & -a_{n2} & \cdots & 2k & -b_{n1} & -b_{n2} & -b_{n3} & \cdots & -b_{nm} \\ \hline -b_{11} & -b_{21} & \cdots & -b_{n1} & 2 & 0 & 0 & \cdots & 0 \\ -b_{12} & -b_{22} & \cdots & -b_{n2} & 0 & 2 & 0 & \cdots & 0 \\ -b_{13} & -b_{23} & \cdots & -b_{n3} & 0 & 0 & 2 & \cdots & 0 \\ \vdots & \vdots & \vdots & \vdots & \vdots & \vdots & \vdots & \ddots & \vdots \\ -b_{1m} & -b_{2m} & \cdots & -b_{nm} & 0 & 0 & 0 & \cdots & 2 \end{array}\right] \end{array}$$

That is,

$$L(G_1) = \begin{bmatrix} 2kI - A(G) & -B(G) \\ -B(G)^T & 2I \end{bmatrix}$$

The characteristic polynomial of the above matrix is given by

$$\begin{aligned} \phi(L(G_1) : x) &= |xI - A(G_1)| \\ &= \begin{vmatrix} (x-2k)I_n + A(G) & B(G) \\ B(G)^T & (x-2)I_m \end{vmatrix} \\ &= |(x-2)I_m||(x-2k)I + A(G) - B(G)((x-2)I_m)^{-1}B(G)^T| \\ &= (x-2)^m|(x-2k)I_n + A(G) - \frac{1}{x-2}B(G)B(G)^T| \\ &= (x-2)^{m-n}|(x-2)(x-2k)I_n + (x-2)A(G) - (A(G) + kI_n)| \end{aligned}$$

It follows that if $\lambda_1, \lambda_2, \cdots, \lambda_n$ are eigenvalues of A then

$$\phi(G_1 : x) = (x-2)^{m-n} \prod_{i=1}^{n} x^2 - x(2k - \lambda_i + 2) + 3(k - \lambda_i)$$

The roots of the above characteristic polynomial are

$$x = 2(m-n\,\text{times}),\, x = \frac{(2k - \lambda_i + 2) \pm \sqrt{4k^2 + \lambda_i^2 - 4k\lambda_i + 8\lambda_i - 4k + 4}}{2}$$

For each $i = 1, 2, \cdots, n$
Hence, the L-spectra of G_1 is ,

$$2^{m-n}, \left(\frac{(2k - \lambda_i + 2) + \sqrt{4k^2 + \lambda_i^2 - 4k\lambda_i + 8\lambda_i - 4k + 4}}{2}\right)^1 \text{ and}$$

,

$$\left(\frac{(2k - \lambda_i + 2) - \sqrt{4k^2 + \lambda_i^2 - 4k\lambda_i + 8\lambda_i - 4k + 4}}{2}\right)^1$$

□

Illustration 31.3.1. Consider cycle C_3 and a graph (say G_1) obtained from C_3 by duplicating each edge by the vertex which is given in Figure 31.2. As discussed in Illustration 31.2.1 that L-spectra of C_3 is

$$2^1, (-1)^2$$

$$L(G_1) = \begin{array}{c} \\ v_1 \\ v_2 \\ v_3 \\ e_1' \\ e_2' \\ e_3' \end{array} \begin{array}{c} \begin{array}{cccccc} v_1 & v_2 & v_3 & e_1' & e_2' & e_3' \end{array} \\ \begin{bmatrix} 4 & -1 & -1 & -1 & 0 & -1 \\ -1 & 4 & -1 & -1 & -1 & 0 \\ -1 & -1 & 4 & 0 & -1 & -1 \\ -1 & -1 & 0 & 2 & 0 & 0 \\ 0 & -1 & -1 & 0 & 2 & 0 \\ -1 & 0 & -1 & 0 & 0 & 2 \end{bmatrix} \end{array}$$

Therefore,the L-spectra of G_1 is ,

$$0^1, 4^1, \left(\frac{7+\sqrt{13}}{2}\right)^1, \left(\frac{7-\sqrt{13}}{2}\right)^1$$

The following table compares the spectrum of C_4 and G_1

spectrum of C_3	**spectrum of** $G_1 = \frac{(2k-\lambda_i+2)\pm\sqrt{4k^2+\lambda_i^2-4k\lambda_i+8\lambda_i-4k+4}}{2}$
$\lambda_1 = 2$	4, 0
$\lambda_2 = -1$	$\frac{7\pm\sqrt{13}}{2}$

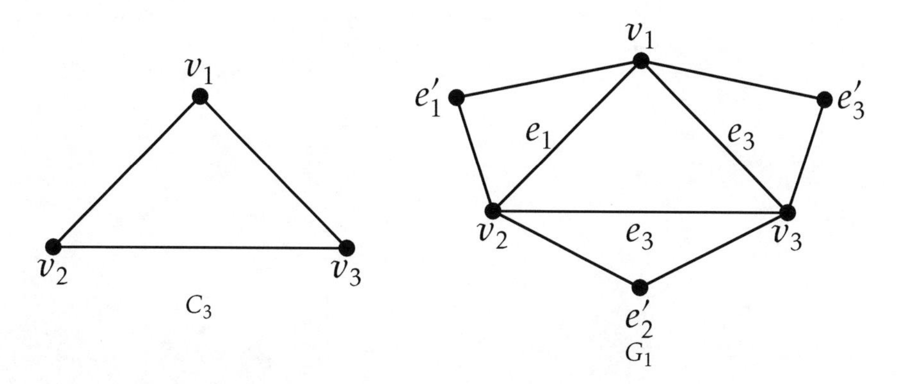

Figure 31.2: C_3 and graph obtained from C_3 by duplicating each edge by a vertex

31.4 CONCLUSION

We have investigated L-spectra for the larger graphs obtained from the given graphs by means of duplication of graph elements for any k-regular graph.

REFERENCES

1. I. Gutman. The energy of a graph. *Ber. Math. Statist. Sekt. Forschungszentram Graz.*, 103, 1-22, 1978.
2. X. Li, Y. Shi and I. Gutman. *Graph Energy.* Springer, New York, 2012.
3. D. Cvetkoviȇ, P. Rowlison and S. Simiȇ. *An Introduction to the Theory of Graph Spectra.* Cambridge University Press, 2010.
4. M. Fiedler. Algebraic connectivity of graphs. *Czechoslovak Math. J.*, 23(2): 298-305, 1973.
5. R. Balakrishnan and K. Ranganathan. *A Textbook of Graph Theory.* Springer, New York, 2000.
6. S. Lang. *Algebra*, Springer, New York, 2002.
7. R. Merris. Laplacian matrices of graphs: a survey. *Linear Algebra Appl.*, 143-176, 1994.
8. R. Merris. A survey of graph Laplacians. *Linear and Multilinear Algebra*, 39, 19-31, 1995.
9. D. M. Cvetkoviȇ, M. Doob and H. Sachs. *Spectra of Graphs, Theory and Application.* Academic Press, NewYork, 1980.
10. B. Mohar. The Laplacian spectrum of graphs, in graph theory, combinatorics, and applications. *In Proceedings of the Sixth Quadrennial International Conference on the Theory and Applications of Graphs, Western Michigan University,* Kalamazoo, (Edited by Y. Alavi, G. Chartrand, O.R. Oellermann and A.J. Schwenk), Wiley, New York, 871-898, 1991.
11. R. Grone and R. Merris. The Laplacian spectrum of a graph II*. *SIAM J. Discrete Math.*, 7(2): 221-229, 1994.

12. R. A. Horn and C. R. Johnson. *Topics In Matrix Analysis.* Cambridge University Press, Cambridge, 1991.
13. F. Z. Zhang. *The Schur Complement and its Applications.* Springer, 2005.

Index

Printed in the United States
By Bookmasters